READINGS IN ECOLOGY

To G. E. Hutchinson, who gave us so much ecology

READINGS IN ECOLOGY

Selected by
Stanley I. Dodson
Timothy F. H. Allen
Stephen R. Carpenter
Kandis Elliot
Anthony R. Ives
Robert L. Jeanne
James F. Kitchell
Nancy E. Langston
Monica G. Turner

University of Wisconsin–Madison

New York Oxford
OXFORD UNIVERSITY PRESS
1999

Oxford University Press

Oxford • New York
Athens • Auckland • Bangkok • Bogota • Bombay • Buenos Aires
Calcutta • Cape Town • Dar es Salaam • Delhi • Florence • Hong Kong
Istanbul • Karachi • Kuala Lumpur • Madras • Madrid • Melbourne
Mexico City • Nairobi • Paris • Singapore • Taipei • Tokyo • Toronto

and associated companies in
Berlin • Ibadan

Published by Oxford University Press, Inc.,
198 Madison Avenue, New York, New York, 10016

Library of Congress Cataloging-in-Publication Data
Readings In Ecology / selected by Stanley I. Dodson ... [et al.].
p. cm.
Includes bibliographical references.
ISBN 0-19-513309-9 (pbk.)
1. Ecology. I. Dodson, Stanley I.
QH541.R33 1999 99-17689
577—dc21 CIP

9 8 7 6 5 4 3 2 1

Printed in the United States of America

Contents

Preface

This book was created as a companion reader for *Ecology,* edited by Stanley I. Dodson and written by eight distinguished ecologists currently working at the University of Wisconsin. The papers herein were selected and are introduced by the *Ecology* authors as examples of excellent and insightful research that greatly contribute to our grasp and exploration of ecological questions. These studies span the range of ecological perspective and are significant reading for every student in the life sciences. The authors of these papers inspired those who followed after, often changing the very way investigators thought about living things and their interactions with the world—including human impact on the resources we now dominate and change.

Besides being an *Ecology* accompaniment, *Readings in Ecology* is also a useful and exciting supplement for any college-level ecology, biology, environmental science or other natural history course. For students and professionals alike, it serves as a primer collection of highlights in ecological investigation.

Editors' notes:

Readings in Ecology follows the design and format of its parent volume, but every attempt was made to preserve each original paper's identity. British or American usage and spelling were maintained, as were contractions, mathematical notation and other specifications used by the original journal. Art was rerendered for the sake of legibility and format consistency, but otherwise figures replicate the originals. Color images were rendered in grays as clearly as possible, but readers interested in more rigorous analyses are directed to view the original color images. For ease of use, references were put into a standardized format, or as much so as feasible. Footnotes were placed at the end of the articles. We corrected the rare, obvious, typesetter's error.

We did not include out-of-date authors' addresses, or summaries and abstracts in foreign languages. Readers are cautioned about such statements as "in press" and "see (another paper) in this journal," which of course are dated and, in this compilation, not relevant.

READINGS IN ECOLOGY

Readings for

WHAT IS ECOLOGY?

Looking at Nature from Different Perspectives

Stanley I. Dodson

- **Aldo Leopold**
  Thinking Like A Mountain
  Escudilla
  Prairie Birthday
- **John Henry Fabre**
  The Pine Processionary
  More Inquiries Into Mason-Bees

Introduction to
Aldo Leopold
Thinking Like A Mountain
Escudilla
Prairie Birthday

by Stanley I. Dodson

Aldo Leopold wrote these essays based on his observations and experiences both as a young man in the Southwest, and as a professor engaged in restoring an abandoned farm in central Wisconsin. Often photographed in an outdoor setting with a dog, gun, and smoking a pipe, Leopold was a conservationist and hunter who explored new dimensions in the ecology of hunting. During his life, he experienced a shift from bird watching with a gun to bird watching with binoculars; a shift from relentless exploitation of nature to a time of concern about management and restoration. In 1948 he was a successful ecologist and conservationist at the University of Wisconsin, laying the foundations of wildlife and conservation ecology, when he died fighting a grass fire on a neighbor's farm.

Leopold's essays provide vivid images of how communities work. Known by many North American ecologists, these stories provide a common starting place for discussions of community and population ecology, as well as conservation, wildlife management, and restoration. Leopold's essays are also romantic in that they evoke emotions: sadness, regret for what is lost, even, perhaps, despair.

Leopold was impressed with how much of nature had been lost to "mechanized man." He foresaw the possibility of redirecting this trend, and although he did not live to see his ideas implemented, we are now reaping benefits from Leopold's insights. He wrote in 1940 that wolves had been extirpated from much of their range; we are currently reintroducing wolves to their previous habitat, thanks in no small part to thousands of ecologists who have read and thought about the ideas in the essays featured here.

Leopold doesn't explicitly mention scale—a term not in fashion when he wrote the essays—nonetheless he tells his story over a breath-taking range of scales. His prose zooms from the smallest detail (a glint of green in a wolf's eye, a single plant in a graveyard) to vast prairies, mountains, or desert. Leopold uses the range of physical scale to resonate with a range of scale for ecological concepts when he generalizes a single, often simple and fleeting, observation into a principle of ecology or management.

When Leopold wrote these essays, ecology was divided into plant and animal, or autecology and synecology. Today, there are many additional ecological perspectives; nonetheless all are based on observations of organisms in nature, and this connection remains however much ecologists focus on chemical cycles or computer models. Leopold's essays are filled with excellent examples of careful observation, whether of the deserts of the Southwest or Wisconsin backyards. Leopold makes himself comfortable outside, uses all his senses to observe nature, and practices a great deal of patience.

• Thinking Like A Mountain • Escudilla • Prairie Birthday

Aldo Leopold

THINKING LIKE A MOUNTAIN

A deep chesty bawl echoes from rimrock to rimrock, rolls down the mountain, and fades into the far blackness of the night. It is an outburst of wild defiant sorrow, and of contempt for all the adversities of the world.

Every living thing (and perhaps many a dead one as well) pays heed to that call. To the deer it is a reminder of the way of all flesh, to the pine a forecast of midnight scuffles and of blood upon the snow, to the coyote a promise of gleanings to come, to the cowman a threat of red ink at the bank, to the hunter a challenge of fang against bullet. Yet behind these obvious and immediate hopes and fears there lies a deeper meaning, known only to the mountain itself. Only the mountain has lived long enough to listen objectively to the howl of a wolf.

Those unable to decipher the hidden meaning know nevertheless that it is there, for it is felt in all wolf country, and distinguishes that country from all other land. It tingles in the spine of all who hear wolves by night, or who scan their tracks by day. Even without sight or sound of wolf, it is implicit in a hundred small events: the midnight whinny of a pack horse, the rattle of rolling rocks, the bound of a fleeing deer, the way shadows lie under the spruces. Only the ineducable tyro can fail to sense the presence or absence of wolves, or the fact that mountains have a secret opinion about them.

My own conviction on this score dates from the day I saw a wolf die. We were eating lunch on a high rimrock, at the foot of which a turbulent river elbowed its way. We saw what we thought was a doe fording the torrent, her breast awash in white water. When she climbed the bank toward us and shook out her tail, we realized our

error: it was a wolf. A half-dozen others, evidently grown pups, sprang from the willows and all joined in a welcoming melee of wagging tails and playful maulings. What was literally a pile of wolves writhed and tumbled in the center of an open flat at the foot of our rimrock.

In those days we had never heard of passing up a chance to kill a wolf. In a second we were pumping lead into the pack, but with more excitement than accuracy: how to aim a steep downhill shot is always confusing. When our rifles were empty, the old wolf was down, and a pup was dragging a leg into impassable slide-rocks.

We reached the old wolf in time to watch a fierce green fire dying in her eyes. I realized then, and have known ever since, that there was something new to me in those eyes—something known only to her and to the mountain. I was young then, and full of trigger-itch; I thought that because fewer wolves meant more deer, that no wolves would mean hunters' paradise. But after seeing the green fire die, I sensed that neither the wolf nor the mountain agreed with such a view.

Since then I have lived to see state after state extirpate its wolves. I have watched the face of many a newly wolfless mountain, and seen the south-facing slopes wrinkle with a maze of new deer trails. I have seen every edible bush and seedling browsed, first to anaemic desuetude, and then to death. I have seen every edible tree defoliated to the height of a saddlehorn. Such a mountain looks as if someone had given God a new pruning shears, and forbidden Him all other exercise. In the end the starved bones of the hoped-for deer herd, dead of its own too-much, bleach with the bones of the dead sage, or molder under the high-lined junipers.

I now suspect that just as a deer herd lives in mortal fear of its wolves, so does a mountain live in mortal fear of its deer. And perhaps with better cause, for while a buck pulled down by wolves can be replaced in two or three years, a range pulled down by too many deer may fail of replacement in as many decades.

So also with cows. The cowman who cleans his range of wolves does not realize that he is taking over the wolf's job of trimming the herd to fit the range. He has not learned to think like a mountain. Hence we have dustbowls, and rivers washing the future into the sea.

We all strive for safety, prosperity, comfort, long life, and dullness. The deer strives with his supple legs, the cowman with trap and poison, the statesman with pen, the most of us with machines, votes, and dollars, but it all comes to the same thing: peace in our time. A measure of success in this is all well enough, and perhaps is a requisite to objective thinking, but too much safety seems to yield only danger in the long run. Perhaps this is behind Thoreau's dictum: wildness is the salvation of the world. Perhaps this is the hidden meaning in the howl of the wolf, long known among mountains, but seldom perceived among men.

ESCUDILLA

Life in Arizona was bounded under foot by grama grass, overhead by sky, and on the horizon by Escudilla.

To the north of the mountain you rode on honey-colored plains. Look up anywhere, any time, and you saw Escudilla.

To the east you rode over a confusion of wooded mesas. Each hollow seemed its own small world, soaked in sun, fragrant with juniper, and cozy with the chatter of piñon jays. But top out on a ridge and you at once became a speck in an immensity. On its edge hung Escudilla.

To the south lay the tangled canyons of Blue River, full of whitetails, wild turkeys, and wilder cattle. When you missed a saucy buck waving his goodbye over the skyline, and looked down your sights to wonder why, you looked at a far blue mountain: Escudilla.

To the west billowed the outliers of the Apache National Forest. We cruised timber there, converting the tall pines, forty by forty, into notebook figures representing hypothetical lumber piles. Panting up a canyon, the cruiser felt a curious incongruity between the remoteness of his notebook symbols and the immediacy of sweaty fingers, locust thorns, deer-fly bites, and scolding squirrels. But on the next ridge a cold wind, roaring across a green sea of pines, blew his doubts away. One the far shore hung Escudilla.

The mountain bounded not only our work and our play, but even our attempts to get a good dinner. On winter evenings we often tried to ambush a mallard on the river flats. The wary flocks circled the rosy west, the steel-blue north, and then disappeared into the inky black of Escudilla. If they reappeared on set wings, we had a fat drake for the Dutch oven. If they failed to reappear, it was bacon and beans again.

There was, in fact, only one place from which you did not see Escudilla on the skyline: that was the top of Escudilla itself. Up there you could not see the mountain, but you could feel it. The reason was the big bear.

Old Bigfoot was a robber-baron, and Escudilla was his castle. Each spring, when the warm winds had softened the shadows on the snow, the old grizzly crawled out of his hibernation den in the rock slides and, descending the mountain, bashed in the head of a cow. Eating his fill, he climbed back to his crags, and there summered peaceably on marmots, conies, berries, and roots.

I once saw on of his kills. The cow's skull and neck were pulp, as if she had collided head-on with a fast freight.

No one ever saw the old bear, but in the muddy springs about the base of the cliffs you saw his incredible tracks. Seeing them made the most hard-bitten cowboys aware of bear. Wherever they rode they saw the mountain, and when they saw the mountain they thought of bear. Campfire conversation ran to beef, *bailes*, and bear. Bigfoot claimed for his own only a cow a year, and a few square miles of useless rocks, but his personality pervaded the county.

Those were the days when progress first came to the cow country. Progress had various emissaries.

One was the first transcontinental automobilist. The cowboys understood this breaker of roads; he talked the same breezy bravado as any breaker of broncos.

They did not understand, but they listened to and looked at, the pretty lady in black velvet who came to enlighten them, in a Boston accent, about woman suffrage.

They marveled, too, at the telephone engineer who strung wires on the junipers and brought instantaneous messages from town. An old man asked whether the wire could bring him a side of bacon.

One spring, progress sent still another emissary, a government trapper, a sort of St. George in overalls, seeking dragons to slay at government expense. Were there, he asked, any destructive animals in need of slaying? Yes, there was the big bear.

The trapper packed his mule and headed for Escudilla.

In a month he was back, his mule staggering under a heavy hide. There was only one barn in town big enough to dry it on. He had tried traps, poison, and all his usual wiles to no avail. Then he had erected a set-gun in a defile through which only the bear could pass, and waited. The last grizzly walked into the string and shot himself.

It was June. The pelt was foul, patchy, and worthless. It seemed to us rather an insult to deny the last grizzly the chance to leave a good pelt as a memorial to his race. All he left was a skull in the National Museum, and a quarrel among scientists over the Latin name of the skull.

It was only after we pondered on these things that we began to wonder who wrote the rules for progress.

Since the beginning, time had gnawed at the basaltic hulk of Escudilla, wasting, waiting, and building. Time built three things on the old mountain, a venerable aspect, a community of minor animals and plants, and a grizzly.

The government trapper who took the grizzly knew he had made Escudilla safe for cows. He did not know he had toppled the spire off an edifice a-building since the morning stars sang together.

The bureau chief who sent the trapper was a biologist versed in the architecture of evolution, but he did not know that spires might be as important as cows. He did not foresee that within two decades the cow country would become tourist country, and as such have a greater need of bears than of beefsteaks.

The Congressmen who voted money to clear the ranges of bears were the sons of pioneers. They acclaimed the superior virtues of the frontiersman, but they strove with might and main to make an end of the frontier.

We forest officers, who acquiesced in the extinguishment of the bear, knew a local rancher who had plowed up a dagger engraved with the name of one of Coronado's captains. We spoke harshly of the Spaniards who, in their zeal for gold and converts, had needlessly extinguished the native Indians. It did not occur to us that we, too, were the captains of an invasion too sure of its own righteousness.

Escudilla still hangs on the horizon, but when you see it you no longer think of bear. It's only a mountain now.

PRAIRIE BIRTHDAY

During every week from April to September there are, on the average, ten wild plants coming into first bloom. In June as many as a dozen species may burst their buds on a single day. No man can heed all of these anniversaries; no man can ignore all of them. He who steps unseeing on May dandelions may be hauled up short by August ragweed pollen; he who ignores the ruddy haze of April elms may skid his car on the fallen corollas of June catalpas. Tell me of what plant-birthday a man takes notice, and I shall tell you a good deal about his vocation, his hobbies, his hay fever, and the general level of his ecological education.

Every July I watch eagerly a certain country graveyard that I pass in driving to and from my farm. It is time for a prairie birthday, and in one corner of this graveyard lives a surviving celebrant of that once important event.

It is an ordinary graveyard, bordered by the usual spruces, and studded with the usual pink granite or white marble headstones, each with the usual Sunday bouquet of red or pink geraniums. It is extraordinary only in being triangular instead of square, and in harboring, within the sharp angle of its fence, a pin-point remnant of the native prairie on which the graveyard was established in the 1840's. Heretofore unreachable by scythe or mower, this yard-square relic of original Wisconsin gives birth, each July, to a man-high stalk of compass plant or cutleaf Silphium, spangled with saucer-sized yellow blooms resembling sunflowers. It is the sole remnant of this plant along this highway, and perhaps the sole remnant in the western half of our county. What a thousand acres of Silphiums looked like when they tickled the bellies of the buffalo is a question never again to be answered, and perhaps not even asked.

This year I found the Silphium in first bloom on 24 July, a week later than usual; during the last six years the average date was 15 July.

When I passed the graveyard again on 3 August, the fence had been removed by a road crew, and the Silphium cut. It is easy now to predict the future; for a few years my Silphium will try in vain to rise above the mowing machine, and then it will die. With it will die the prairie epoch.

The Highway Department says that 100,000 cars pass yearly over this route during the three summer months when the Silphium is in bloom. In them must ride at least 100,000 people who have 'taken' what is called history, and perhaps 25,000 who have 'taken' what is called botany. Yet I doubt whether a dozen have seen the Silphium, and of these hardly one will notice its demise. If I were to tell a preacher of the adjoining church that the road crew has been burning history books in his cemetery, under the guise of mowing weeds, he would be amazed and uncomprehending. How could a weed be a book?

This is one little episode in the funeral of the native flora, which in turn is one episode in the funeral of the floras of the world. Mechanized man, oblivious of floras, is proud of his progress in cleaning up the landscape on which, willy-nilly, he must live out his days. It might be wise to prohibit at once all teaching of real botany and real history, lest some future citizen suffer qualms about the floristic price of his good life.

Thus it comes to pass that farm neighborhoods are good in proportion to the poverty of their floras. My own farm was selected for its lack of goodness and its lack of highway; indeed my whole neighborhood lies in a backwash of the River Progress. My road is the original wagon track of the pioneers, innocent of grades or gravel, brushings or bulldozers. My neighbors bring a sigh to the County Agent. Their fencerows go unshaven for years on end. Their marshes are neither dyked nor drained. As between going fishing and going forward, they are prone to prefer fishing. Thus on week ends my floristic standard of living is that of the backwoods, while on week days I subsist as best I can on the flora of the university farms, the university campus, and the adjoining suburbs. For a decade I have kept, for pastime, a record of the wild plant species in first bloom on these two diverse areas:

Species First Blooming in	Suburb and Campus	Backward Farm
April	14	26
May	29	59
June	43	70
July	25	56
August	9	14
September	0	1
Total visual diet	120	226

It is apparent that the backward farmer's eye is nearly twice as well fed as the eye of the university student or businessman. Of course neither sees his flora as yet, so we are confronted by the two alternatives already mentioned: either insure the continued blindness of the populace, or examine the question whether we cannot have both progress and plants.

The shrinkage in the flora is due to a combination of clean-farming, woodlot grazing, and good roads. Each of these necessary changes of course requires a larger reduction in the acreage available for wild plants, but none of them requires, or benefits by, the erasure of species from whole farms, townships, or counties. There are idle spots on every farm, and every highway is bordered by an idle strip as long as it is; keep cow, plow, and mower out of these idle spots, and the full native flora, plus dozens of interesting stowaways from foreign parts, could be part of the normal environment of every citizen.

The outstanding conservator of the prairie flora, ironically enough, knows little and cares less about such frivolities: it is the railroad with its fenced right-of-way. Many of these railroad fences were erected before the prairie had been plowed. Within these linear reservations, oblivious of cinders, soot, and annual clean-up fires, the prairie flora still splashes its calendar of colors, from pink shooting-star in May to blue aster in October. I have long wished to confront some hard-boiled railway president with the physical evidence of his soft-heartedness. I have not done so because I haven't met one.

The railroads of course use flame-throwers and chemical sprays to clear the track of weeds, but the cost of such necessary clearance is still too high to extend it much beyond the actual rails. Perhaps further improvements are in the offing.

The erasure of a human subspecies is largely painless—to us—if we know little enough about it. A dead Chinaman is of little import to us whose awareness of things Chinese is bounded by an occasional dish of chow mein. We grieve only for what we know. The erasure of Silphium from western Dane County is no cause for grief if one knows it only as a name in a botany book.

Silphium first became a personality to me when I tried to dig one up to move to my farm. It was like digging an oak sapling. After half an hour of hot grimy labor the root was still enlarging, like a great vertical sweet-potato. As far as I know, that Silphium root went clear through to bedrock. I got no Silphium, but I learned by what elaborate underground stratagems it contrives to weather the prairie drouths.

I next planted Silphium seeds, which are large, meaty, and taste like sunflower seeds. They came up promptly, but after five years of waiting the seedlings are still juvenile, and have not yet borne a flower-stalk. Perhaps it takes a decade for a Silphium to reach flowering age; how old, then, was my pet plant in the cemetery? It may have been older than the oldest tombstone, which is dated 1850. Perhaps it watched the fugitive Black Hawk retreat from the Madison lakes to the Wisconsin River; it stood on the route of that famous march. Certainly it saw the successive funerals of the local pioneers as they retired, one by one, to their repose beneath the bluestem.

I once saw a power shovel, while digging a roadside ditch, sever the 'sweet-potato' root of a Silphium plant. The root soon sprouted new leaves, and eventually it again produced a flower stalk. This explains why this plant, which never invades new ground, is nevertheless sometimes seen on recently graded roadsides. Once established, it apparently withstands almost any kind of mutilation except continued grazing, mowing, or plowing.

Why does Silphium disappear from grazed areas? I once saw a farmer turn his cows into a virgin prairie meadow previously used only sporadically for mowing wild hay. The cows cropped the Silphium to the ground before any other plant was visibly eaten at all. One can imagine that the buffalo once had the same preference for Silphium, but he brooked no fences to confine his nibblings all summer long to one meadow. In short, the buffalo's pasturing was discontinuous, and therefore tolerable to Silphium.

It is a kind providence that has withheld a sense of history from the thousands of species of plants and animals that have exterminated each other to build the present world. The same kind providence now withholds it from us. Few grieved when the last buffalo left Wisconsin, and few will grieve when the last Silphium follows him to the lush prairies of the never-never land.

Introduction to
Jean Henri Fabre
The Pine Processionary
More Inquiries Into Mason-Bees

by Kandis Elliot

Jean Henri Fabre lived from 1823 to 1915, an era that saw biological sciences wrestle furiously to escape religious dogma and begin interpretation of the world as a consequence of natural, rather than supernatural, phenomena. Intolerance by clergy, apathy of the masses, and wrenching poverty followed Fabre throughout his youth and middle age. He was often ridiculed by passers-by as he sat in the hot sun, studying animal behavior, especially that of insects; he outraged his superiors with his advanced ideas about instinctive behavior and about teaching it (going on field trips). In 1870, when he admitted girls to his science classes at the Lycée, the clergy denounced him from the pulpit. Dismissed from his low-paying job, Fabre, soon penniless, was evicted from his house with his wife and five small children.

The world was to gain from this misfortune. With money loaned from his English friend John Stuart Mill, Fabre rented a house near weedfields on the edge of Orange, where he spent nine years supporting his family through writing popular science books for young people. Though a miserly living, it enabled Fabre to roam the fields and hone his pen. When he was 55 he at last managed to buy a bit of wasteland at the edge of the village of Sérignan in the south of France. There, over the next twenty years, he would produce some of the late nineteenth and early twentieth centuries' most important—and most beautifully written—studies on insect biology and behavior, the *Souvenirs Entomologiques*—a work that galvanized a generation of young biologists.

Fabre was a quintessential observer, his best instruments "Time and Patience." He challenged long-held beliefs about the ecology of insects, and sought to record accurate data about their behavior. He painstakingly tested his own questions and hypotheses with repeated experiments, some of which, like the grueling trek of the Pine Processionary caterpillars along the circular lip of a vase, comprise our most well-known accounts. He used insects as models to test then-current dogma about instinct and reasoning. Moreover, his writing, translated by Alexander Teixiera de Mattos, is a delightful respite from contemporary "objective" prose that is, all too often, eye-glazing and uninspired.

Like his contemporary and admiring friend Charles Darwin, Fabre challenged concepts of the day with facts and reason. Nonetheless, working in self-imposed and often bitter isolation, Fabre was not unaffected by his own cultural lenses and philosophy. A lifetime of studying insects developed his firm belief in the immutable nature of instinct. He could not accept Darwin's theory of evolution. Fabre's pioneering greatness, however, towers above these foibles. With crude equipment, an endless tide of questions, and unresting inventiveness, he struggled to solve mysteries of the commonplace, and in so doing crated the blueprints for studies of the nature of instinct, insect behavior, comparative psychology and experimental biology. Students of ecology would do well to avail themselves of the following classic lessons in investigation, both scientific and philosophic, from the Poet of Science.

• The Pine Processionary
• More Inquiries Into Mason-Bees

J. Henri Fabre

It is not the animal that we are now consulting upon the nature of its aptitudes, upon the primary motives of its activity, but our own opinions, which always yield a reply in favour of our cherished notions. As I have already repeatedly shown, observation in itself is often a snare: we interpret its data according to the exigencies of our theories.

THE PINE PROCESSIONARY

Original publication in *The Life Of The Caterpillar,* Chapter III, Dodd, Mead & Co. 1914.

[This essay is probably Fabre's best known, having been reprinted a number of times in the popular literature. In addition to describing observations on animal behavior, it imparts something to the reader that is often lacking in the cogent, scholarly style of modern scientific papers: an intriguing glimpse of the writer's own sense of wonder.]

Drover Dingdong's Sheep followed the Ram which Panurge had maliciously thrown overboard and leapt nimbly into the sea, one after the other, "for you know," says Rabelais, "it is the nature of the sheep always to follow the first, wheresoever it goes; which makes Aristotle mark them for the most silly and foolish animals in the world."

The Pine Caterpillar is even more sheeplike, not from foolishness, but from necessity: where the first goes all the others go, in a regular string, with not an empty space between them.

They proceed in single file, in a continuous row, each touching with its head the rear of the one in front of it. The complex twists and turns described in his vagaries by the caterpillar leading the van are scrupulously described by all the others. No Greek theoria winding its way to the Eleusinian festivals was ever more orderly. Hence the name of Processionary given to the gnawer of the pine.

His character is complete when we add that he is a rope-dancer all his life long: he walks only on the tight-rope, a silken rail placed in position as he advances. The caterpillar who chances to be at the head of the procession dribbles his thread without ceasing and fixes it on the path which his fickle preferences cause him to take. The thread is so tiny that the eye, though armed with a magnifying glass, suspects it rather than sees it.

But a second caterpillar steps on the slender footboard and doubles it with his thread; a third trebles it; and all the others, however many there be, add the sticky spray from their spinnerets, so much so that, when the procession has marched by, there remains, as a record of its passing, a narrow white ribbon whose dazzling whiteness shimmers in the sun. Very much more sumptuous than ours, their system of road-making consists in upholstering with silk instead of macadamizing. We sprinkle our roads with broken stones and level them by the pressure of a heavy steamroller; they lay over their paths a soft satin rail, a work of general interest to which each contributes his thread.

What is the use of all this luxury? Could they not, like other caterpillars, walk about without these costly preparations? I see two reasons for their mode of progression. It is night when the Processionaries sally forth to browse upon the pine-leaves. They leave their nest, situated at the top of a bough, in profound darkness; they go down the denuded pole till they come to the nearest branch that has not yet been gnawed, a branch which becomes lower and lower by degrees as the consumers finish stripping the upper storeys; they climb up this untouched branch and spread over the green needles.

When they have had their suppers and begin to feel the keen night air, the next thing is to return to the shelter of the house. Measured in a straight line, the distance is not great, hardly an arm's length; but it cannot be covered in this way on foot. The caterpillars have to climb down from one crossing to the next, from the needle to the twig, from the twig to the branch, from the branch to the bough and from the bough, by a no less angular path, to go back home. It is useless to rely upon sight as a guide on this long and erratic journey. The Processionary, it is true, has five ocular specks on either side of his head, but they are so infinitesimal, so difficult to make out through the magnifying glass, that we cannot attribute to them any great power of vision. Besides, what good would those short-sighted lenses be in the absence of light, in black darkness?

It is equally useless to think of the sense of smell. Has the Processional any olfactory powers or has he not? I do not know. Without giving a positive answer to the question, I can at least declare that his sense of smell is exceedingly dull and in no way suited to help him find his way. This is proved, in my experiments, by a number of hungry caterpillars that, after a long fast, pass close beside a pine-branch without betraying any eagerness or showing a sign of stopping. It is the sense of touch that tells them where they are. So long as their lips do not chance to light upon the pasture-land, not one of them settles there, though he be ravenous. They do not hasten to food which they have scented from afar; they stop at a branch which they encounter on their way. Apart from sight and smell, what remains to guide them in returning to the nest? The ribbon spun on the road. In the Cretan labyrinth, Theseus would have been lost but for the clue of thread with which Ariadne supplied him. The spreading maze of the pine-needles is, especially at night, as inextricable a labyrinth as that constructed for Minos. The Processionary finds his way through it, without the possibility of a mistake, by the aid of his bit of silk. At the time for going home, each easily recovers either his own thread or one or other of the neighbouring threads, spread fanwise by the diverging herd, one by one the scattered tribe line up on the common ribbon, which started from the nest; and the sated caravan finds its way back to the manor with absolute certainty.

Longer expeditions are made in the daytime, even in winter, if the weather be fine. Our caterpillars then come down from the tree, venture on the ground, march in procession for a distance of thirty yards or so. The object of these sallies is not to look for food, for the native pine-tree is far from being exhausted: the shorn branches hardly count amid the vast leafage. Moreover, the caterpillars observe complete abstinence till nightfall. The trippers have no other object than a constitutional, a pilgrimage to the outskirts to see what these are like, possibly an inspection of the locality where, later on, they mean to bury themselves in the sand for their metamorphosis.

It goes without saying that, in these greater evolutions, the guiding cord is not neglected. It is now more necessary than ever. All contribute to it from the produce of their spinnerets, as is the invariable rule whenever there is a progression. Not one takes a step forward without fixing to the path the thread hanging from his lip.

If the series forming the procession be at all long, the ribbon is dilated sufficiently to make it easy to find; nevertheless, on the homeward journey, it is not picked up without some hesitation. For observe that the caterpillars when on the march never turn completely; to wheel round on their tight-rope is a method utterly unknown to them. In order therefore to regain the road already covered, they have to describe a zig-zag whose windings and extent are determined by the leader's fancy. Hence come gropings and roamings which are sometimes prolonged to the point of causing the herd to spend the night out of doors. It is not a serious matter. They collect into a motionless cluster. To-morrow the search will start afresh and will sooner or later be successful. Oftener still the winding curve meets the guide-thread at the first attempt. As soon as the first caterpillar has the rail between his legs, all hesitation ceases; and the band makes for the nest with hurried steps.

The use of this silk-tapestried roadway is evident from a second point of view. To protect himself against the severity of the winter which he has to face when working, the Pine Caterpillar weaves himself a shelter in which he spends his bad hours, his days of enforced idleness. Alone, with none but the meagre resources of his silk-glands, he would find difficulty in protecting himself on the top of a branch buffeted by the winds. A substantial dwelling, proof against snow, gales and icy fogs, requires the cooperation of a large number. Out of the individual's piled-up atoms, the community obtains a spacious and durable establishment.

The enterprise takes a long time to complete. Every evening, when the weather permits, the building has to be strengthened and enlarged. It is indispensable, therefore, that the corporation of workers should not be dissolved while the stormy season continues and the insects are still in the caterpillar stage. But, without special arrangements, each nocturnal expedition at grazing-time would be a cause of separation. At that moment of appetite for food there is a return to individualism. The caterpillars become more or less scattered, settling singly on the branches around; each browses his pine-needle separately. How are they to find one another afterwards and become a community again?

The several threads left on the road make this easy. With that guide, every caterpillar, however far he may be, comes back to his companions without ever missing the way. They come hurrying from a host of twigs, from here, from there, from above, from below; and soon the scattered legion reforms into a group. The silk thread is something more than a road-making expedient: it is the social bond, the system that keeps the members of the community indissolubly united.

At the head of every procession, long or short, goes a first caterpillar whom I will call the leader of the march or file, though the word leader, which I use for want of a better, is a little out of place here. Nothing, in fact, distinguishes this caterpillar from the others: it just depends upon the order in which they happen to line up; and mere chance brings him to the front. Among the Processionaries, every captain is an officer of fortune. The actual leader leads; presently he will be a subaltern, if the file should break up in consequence of some accident and be formed anew in a different order.

His temporary functions give him an attitude of his own. While the others follow passively in a close file, he, the captain, tosses himself about and with an abrupt movement flings the front of his body hither and thither. As he marches ahead he seems to be seeking his way. Does he in point of fact explore the country? Does he choose the most practicable places? Or are his hesitations merely the result of the absence of a guiding thread on ground that has not yet been covered? His subordinates follow very placidly, reassured by the cord which they hold between their legs; he, deprived of that support, is uneasy.

Why cannot I read what passes under his black, shiny skull, so like a drop of tar? To judge by actions, there is here a small dose of discernment which is able, after experimenting, to recognize excessive roughnesses, over-slippery surfaces, dusty places that offer no resistance and, above all, the threads left by other excursionists. This is all or nearly all that my long acquaintance with the Processionaries has taught me as to their mentality. Poor brains, indeed; poor creatures, whose commonwealth has its safety hanging upon a thread!

The processions vary greatly in length. The finest that I have seen maneuvering on the ground measured twelve or thirteen yards and numbered about three hundred caterpillars, drawn up with absolute precision in a wavy line. But, if there were only two in a row, the order would still be perfect: the second touches and follows the first.

By February I have processions of all lengths in the greenhouse. What tricks can I play upon them? I see only two: to do away with the leader; and to cut the thread.

The suppression of the leader of the file produces nothing striking. If the thing is done without creating a disturbance, the procession does not alter its ways at all. The second caterpillar, promoted to captain, knows the duties of his rank off-hand: he selects and leads, or rather he hesitates and gropes.

The breaking of the silk ribbon is not very important either. I remove a caterpillar from the middle of the file. With my scissors, so as not to cause a commotion in the ranks, I cut the piece of ribbon on which he stood and clear away every thread of it. As a result of this breach, the procession acquires two marching leaders, each independent of the other. It may be that the one in the rear joins the file ahead of him, from which he is separated by but a slender interval; in that case, things return to their original condition. More frequently, the two parts do not become reunited. In that case, we have two distinct processions, each of which wanders where it pleases and diverges from the other. Nevertheless, both will be able to return to the nest by discovering sooner or later, in the course of their peregrinations, the ribbon on the other side of the break.

These two experiments are only moderately interesting. I have thought out another, one more fertile in possibilities. I propose to make the caterpillars describe a closed circuit, after the ribbons running from it and liable to bring about a change of direction have been destroyed. The locomotive engine pursues its invariable course so long as it is not shunted on to a branch-line. If the Processionaries find the silken rail always clear in front of them, with no switches anywhere, will they continue on the same track, will they persist in following a road that never comes to an end? What we have to do is to produce this circuit, which is unknown under ordinary conditions, by artificial means.

The first idea that suggests itself is to seize with the forceps the silk ribbon at the back of the train, to bend it without shaking it and to bring the end of it ahead of the file. If the caterpillar marching in the van steps upon it, the thing is done: the others will follow him faithfully. The operation is very simple in theory but very difficult in practice and produces no useful results. The ribbon, which is extremely slight, breaks under the weight of the grains of sand that stick to it and are lifted with it. If it does not break, the caterpillars at the back, however delicately we may go to work, feel a disturbance which makes them curl up or even let go.

There is a yet greater difficulty: the leader refuses the ribbon laid before him; the cut end makes him distrustful. Failing to see the regular, uninterrupted road, he slants off to the right or left, he escapes at a tangent. If I try to interfere and to bring him back to the path of my choosing, he persists in his refusal, shrivels up, does not budge; and soon the whole procession is in confusion. We will not insist: the method is a poor one, very wasteful of effort for at best a problematical success.

We ought to interfere as little as possible and obtain a natural closed circuit. Can it be done? Yes. It lies in our power, without the least meddling, to see a procession march along a perfect circular track. I owe this result, which is eminently deserving of our attention, to pure chance.

On the shelf with the layer of sand in which the nests are planted stand some big palm-vases measuring nearly a yard and a half in circumference at the top. The caterpillars often scale the sides and climb up to the moulding which forms a cornice around the opening. This place suits them for their processions, perhaps because of the absolute firmness of the surface, where there is no fear of landslides, as on the loose, sandy soil below; and also, perhaps, because of the horizontal position, which is favorable to repose after the fatigue of the ascent. It provides me with a circular track all ready-made. I have nothing to do but wait for an occasion propitious to my plans. This occasion is not long in coming.

On the 30th of January, 1896, a little before twelve o'clock in the day, I discover a numerous troop making their way up and gradually reaching the popular cornice. Slowly, in single file, the caterpillars climb the great vase, mount the ledge and advance in regular procession, while others are constantly arriving and continuing the series. I wait for the string to close up, that is to say, for the leader, who keeps following the circular moulding, to return to the point from which he started. My object is achieved in a quarter of an hour. The closed circuit is realized magnificently, in something very nearly approaching a circle. The next thing is to get rid of the rest of the ascending column, which would disturb the fine order of the procession by an excess of newcomers; it is also important that we should do away with all the silken paths, both new and old, that can put the cornice into communication with the ground. With a thick hair-pencil I sweep away the surplus climbers; with a big brush, one that leaves no smell behind it—for this might afterwards prove confusing—I carefully rub down the vase and get rid of every thread which the caterpillars have laid on the march. When these preparations are finished, a curious sight awaits us.

In the uninterrupted circular procession there is no longer a leader. Each caterpillar is preceded by another on whose heels he follows, guided by the silk track, the work of the whole party; he again has a companion close behind him, following him in the same orderly way. And this is repeated without variation throughout the length of the chain. None commands, or rather none modifies the trail according to his fancy; all obey, trusting in the guide who ought normally to lead the march and who in reality has been abolished by my trickery.

From the first circuit of the edge of the tub the rail of silk has been laid in position and is soon turned into a narrow ribbon by the procession, which never ceases dribbling its thread as it goes. The rail is simply doubled and has no branches anywhere, for my brush has destroyed them all. What will the caterpillars do on this deceptive, closed path? Will they walk endlessly round and round until their strength gives out entirely?

The old schoolmen were fond of quoting Buridan's Ass, that famous Donkey who, when placed between two bundles of hay, starved to death because he was unable to decide in favour of either by breaking the equilibrium between two equal but opposite attractions. They slandered the worthy animal. The Ass, who is no more foolish than any one else, would reply to the logical snare by feasting off both bundles. Will my caterpillars show a little of his mother wit? Will they, after many attempts, be able to break the equilibrium of their closed circuit, which keeps them on a road without a turning? Will they make up their minds to swerve to this side or that, which is the only method of reaching their bundle of hay, the green branch yonder, quite near, not two feet off?

I thought that they would and I was wrong. I said to myself: "The procession will go on turning for some time, for an hour, two hours perhaps; then the caterpillars will perceive their mistake. They will abandon the deceptive road and make their descent somewhere or other."

That they should remain up there, hard pressed by hunger and the lack of cover, when nothing prevented them from going away, seemed to me inconceivable imbecility. Facts, however, forced me to accept the incredible. Let us describe them in detail.

The circular procession begins, as I have said, on the 30th of January, about midday, in splendid weather. The caterpillars march at an even pace, each touching the stern of the one in front of him. The unbroken chain eliminates the leader with his changes of direction, and all follow mechanically, as faithful to their circle as are the hands of a watch. The headless file has no liberty left, no will; it has become mere clock-work. And this continues for hours and hours. My success goes far beyond my wildest suspicions. I stand amazed at it, or rather I am stupefied.

Meanwhile, the multiplied circuits change the original rail into a superb ribbon a twelfth of an inch broad. I can easily see it glittering on the red ground of the pot. The day is drawing to a close and no alteration has yet taken place in the position of the trail. A striking proof confirms this.

The trajectory is not a plane curve, but one which, at a certain point, deviates and goes down a little way to the lower surface of the cornice, returning to the top some eight inches farther. I marked these two points of deviation in pencil on the vase at the outset. Well, all that afternoon and, more conclusive still, on the following days, right to the end of this mad dance, I see the string of caterpillars dip under the ledge at the first point and come to the top again at the second. Once the first thread is laid, the road to be pursued is permanently established. If the road does not vary, the speed does. I measure nine centimetres a minute as the average distance covered. But there are more or less lengthy halts; the pace slackens at times, especially when the temperature falls. At ten o'clock in the evening the walk is little more than a lazy swaying of the body. I foresee an early halt, in consequence of the cold, of fatigue and doubtless also of hunger.

Grazing-time has arrived. The caterpillars have come crowding from all the nests in the greenhouse to browse upon the pine-branches planted by myself beside the silken purses. Those in the garden do the same, for the temperature is mild. The others, lined up along the earthenware cornice, would gladly take part in the feast; they are bound to have an appetite after a ten hours' walk. The branch stands green and tempting not a hand's breadth away. To reach it they need but go down; and the poor wretches, foolish slaves of their ribbon that they are, cannot make up their minds to do so. I leave the famished ones at half-past ten, persuaded that they will take counsel with their pillow and that on the morrow things will have resumed their ordinary course.

I was wrong. I was expecting too much of them when I accorded them that faint gleam of intelligence which the tribulations of a distressful stomach ought, one would think, to have aroused. I visit them at dawn. They are lined up as on the day before, but motionless. When the air grows a little warmer, they shake off their torpor, revive and start walking again. The circular procession begins anew, like that which I have already seen. There is nothing more and nothing less to be noted in their machine-like obstinacy.

This time it is a bitter night. A cold snap has supervened, was indeed foretold in the evening by the garden caterpillars, who refused to come out despite appearances which to my duller senses seemed to promise a continuation of the fine weather. At daybreak the rosemary-walks are all asparkle with rime and for the second time this year there is a sharp frost. The large pond in the garden is frozen over. What can the caterpillars in the conservatory be doing? Let us go and see.

All are ensconced in their nests, except the stubborn processionists on the edge of the vase, who, deprived of shelter as they are, seem to have spent a very bad night. I find them clustered in two heaps, without any attempt at order. They have suffered less from the cold, thus huddled together.

'Tis an ill wind that blows nobody any good. The severity of the night has caused the ring to break into two segments which will, perhaps, afford a chance of safety. Each group, as it revives and resumes its walk, will presently be headed by a leader who, not being obliged to follow a caterpillar in front of him, will possess some liberty of movement and perhaps be able to make the procession swerve to one side. Remember that, in the ordinary processions, the caterpillar walking ahead acts as a scout. While the others, if nothing occurs to create excitement, keep to their ranks, he attends to his duties as a leader and is continually turning his head to this side and that, investigating, seeking, groping, making his choice. And things happen as he decides: the band follows him faithfully. Remember also that, even on a road which has already been travelled and beribboned, the guiding caterpillar continues to explore.

There is reason to believe that the Processionaries who have lost their way on the ledge will find a chance of safety here. Let us watch them. On recovering from their torpor, the two groups line up by degrees into two distinct files. There are therefore two leaders, free to go where they please, independent of each other. Will they succeed in leaving the enchanted circle? At the sight of their large black heads swaying anxiously from side to side, I am inclined to think so for a moment. But I am soon undeceived. As the ranks fill out, the two sections of the chain meet and the circle is reconstituted. The momentary leaders once more become simple subordinates; and again the caterpillars march round and round all day.

For the second time in succession, the night, which is very calm and magnificently starry, brings a hard frost. In the morning the Processionaries on the tub, the only ones who have camped out unsheltered, are gathered into a heap which largely overflows both sides of the fatal ribbon. I am present at the awakening of the numbed ones. The first to take the road is, as luck will have it, outside the track. Hesitatingly he ventures into unknown ground. He reaches the top of the rim and descends upon the other side on the earth in the vase. He is followed by six others, no more. Perhaps the rest of the troop, who have not fully recovered from their nocturnal torpor, are too lazy to bestir themselves.

The result of this brief delay is a return to the old track. The caterpillars embark on the silken trail and the circular march is resumed, this time in the form of a ring with a gap in it. There is no attempt, however, to strike a new course on the part of the guide whom this gap has placed at the head. A chance of stepping outside the magic circle has presented itself at last; and he does not know how to avail himself of it.

As for the caterpillars who have made their way to the inside of the vase, their lot is hardly improved. They climb to the top of the palm, starving and seeking for food. Finding nothing to eat that suits them, they retrace their steps by following the thread which they have left on the way, climb the ledge of the pot, strike the procession again and, without further anxiety, slip back into the ranks. Once more the ring is complete, once more the circle turns and turns.

Then when will the deliverance come? There is a legend that tells of poor souls dragged along in an endless round until the hellish charm is broken by a drop of holy water. What drop will good fortune sprinkle on my Processionaries to dissolve their circle and bring them back to the nest? I see only two means of conjuring the spell and obtaining a release from the circuit. These two means are two painful ordeals. A strange linking of cause and effect: from sorrow and wretchedness good is to come.

And, first, shrivelling as the result of cold. The caterpillars gather together without any order, heap themselves some on the path, some, more numerous these, outside it. Among the latter there may be, sooner or later, some revolutionary who, scorning the beaten track, will trace out a new road and lead the troop back home. We have just seen an instance of it. Seven penetrated to the interior of the vase and climbed the palm. True, it was an attempt with no result, but still an attempt. For complete success, all that need be done would have been to take the opposite slope. An even chance is a great thing. Another time we shall be more successful.

In the second place, the exhaustion due to fatigue and hunger. A lame one stops, unable to go farther. In front of the defaulter the procession still continues to wend its way for a short time. The ranks close up and an empty space appears. On coming to himself and resuming the march, the caterpillar who has caused the breach becomes a leader, having nothing before him. The least desire for emancipation is all that he wants to make him launch the band into a new path which perhaps will be the saving path.

In short, when the Processionaries' train is in difficulties, what it needs, unlike ours, is to run off the rails. The side-tracking is left to the caprice of a leader who alone is capable of turning to the right or left; and this leader is absolutely non-existent so long as the ring remains unbroken. Lastly, the breaking of the circle, the one stroke of luck, is the result of a chaotic halt, caused principally by excess of fatigue or cold.

The liberating accident, especially that of fatigue, occurs fairly often. In the course of the same day, the moving circumference is cut up several times into two or three sections; but continuity soon returns and no change takes place. Things go on just the same. The bold innovator who is to save the situation has not yet had his inspiration.

There is nothing new on the fourth day, after an icy night like the previous one; nothing to tell except the following detail. Yesterday I did not remove the trace left by the few caterpillars who made their way to the inside of the vase. This trace, together with a junction connecting it with the circular road, is discovered in the course of the morning. Half the troop takes advantage of it to visit the earth in the pot and climb the palm; the other half remains on the ledge and continues to walk along the old rail. In the afternoon the band of emigrants rejoins the others, the circuit is completed and things return to their original condition.

We come to the fifth day. The night frost becomes more intense, without however as yet reaching the greenhouse. It is followed by bright sunshine in a calm and limpid sky. As soon as the sun's rays have warmed the panes a little, the caterpillars, lying in heaps, wake up and resume their evolutions on the ledge of the vase. This time the fine order of the beginning is disturbed and a certain disorder becomes manifest, apparently an omen of deliverance near at hand. The scouting-path inside the vase, which was upholstered in silk yesterday and the day before, is to-day followed to its origin on the rim by a part of the band and is then abandoned after a short loop. The other caterpillars follow the usual ribbon. The result of this bifurcation is two almost equal files, walking along the ledge in the same direction, at a short distance from each other, sometimes meeting, separating farther on, in every case with some lack of order.

Weariness increases the confusion. The crippled, who refuse to go on, are many. Breaches increase; files are split up into sections each of which has its leader, who pokes the front of his body this way and that to explore the ground. Everything seems to point to the disintegration which will bring safety. My hopes are once more disappointed. Before the night the single file is reconstituted and the invincible gyration resumed.

Heat comes, just as suddenly as the cold did. To-day, the 4th of February, is a beautiful, mild day. The greenhouse is full of life. Numerous festoons of caterpillars, issuing from the nests, meander along the sand on the shelf. Above them, at every moment, the ring on the ledge of the vase breaks up and comes together again. For the first time I see daring leaders who, drunk with heat, standing only on their hinder prolegs at the extreme edge of the earthenware rim, fling themselves forward into space, twisting about, sounding the depths. The endeavour is frequently repeated, while the whole troop stops. The caterpillars' heads give sudden jerks; their bodies wriggle.

One of the pioneers decides to take the plunge. He slips under the ledge. Four follow him. The others, still confiding in the perfidious silken path, dare not copy him and continue to go along the old road.

The short string detached from the general chain gropes about a great deal, hesitates long on the side of the vase; it goes half-way down, then climbs up again slantwise, rejoins and takes its place in the procession. This time the attempt has failed, though at the foot of the vase, not nine inches away, there lay a bunch of pine-needles which I had placed there with the object of enticing the hungry ones. Smell and sight told them nothing. Near as they were to the goal, they went up again.

No matter, the endeavour has its uses. Threads were laid on the way and will serve as a lure to further enterprise. The road of deliverance has its first landmarks. And two days later, on the eighth day of the experiment, the caterpillars—now singly, anon in small groups, then again in strings of some length—come down from the ledge by following the staked-out path. At sunset the last of the laggards is back in the nest.

Now for a little arithmetic. For seven times twenty-four hours the caterpillars have remained on the ledge of the vase. To make an ample allowance for stops due to the weariness of this one or that and above all for the rest taken during the colder hours of the night, we will deduct one-half of the time. This leaves eighty-four hours' walking. The average pace is nine centimetres a minute.

The aggregate distance covered, therefore, is 453 metres, a good deal more than a quarter of a mile, which is a great walk for these little crawlers. The circumference of the vase, the perimeter of the track, is exactly 1 m. 35. Therefore the circle covered, always in the same direction and always without result, was described three hundred and thirty-five times.

These figures surprise me, though I am already familiar with the abysmal stupidity of insects as a class whenever the least accident occurs. I feel inclined to ask myself whether the Processionaries were not kept up there so long by the difficulties and dangers of the descent rather than by the lack of any gleam of intelligence in their benighted minds. The facts, however, reply that the descent is as easy as the ascent.

The caterpillar has a very supple back, well adapted for twisting round projections or slipping underneath. He can walk with the same ease vertically or horizontally, with his back down or up. Besides, he never moves forward until he has fixed his thread to the ground. With this support to his feet, he has no falls to fear, no matter what his position.

I had a proof of this before my eyes during a whole week. As I have already said, the track, instead of keeping on one level, bends twice, dips at a certain point under the ledge of the vase and reappears at the top a little farther on. At one part of the circuit, therefore, the procession walks an the lower surface of the rim; and this inverted position implies so little discomfort or danger that it is renewed at each turn for all the caterpillars from first to last.

It is out of the question then to suggest the dread of a false step on the edge of the rim which is so nimbly turned at each point of inflexion. The caterpillars in distress, starved, shelterless, chilled with cold at night, cling obstinately to the silk ribbon covered hundreds of times, because they lack the rudimentary glimmers of reason which would advise them to abandon it.

Experience and reflection are not in their province. The ordeal of a five hundred yards' march and three to four hundred turns teach them nothing; and it takes casual circumstances to bring them back to the nest. They would perish on their insidious ribbon if the disorder of the nocturnal encampments and the halts due to fatigue did not cast a few threads outside the circular path. Some three or four move along these trails, laid without an object, stray a little way and, thanks to their wanderings, prepare the descent, which is at last accomplished in short strings favoured by chance.

The school most highly honoured to-day is very anxious to find the origin of reason in the dregs of the animal kingdom. Let me call its attention to the Pine Processionary.

MORE INQUIRIES INTO MASON-BEES

Original publication in *The Mason-bees*, Chapter IV, Dodd, Mead & Co. 1916.

[Fabre had no end of tribulations in conducting his experiments, in part because of his lack of funding, and in part because he was so ahead of his time. This passage describes his efforts at finding the homing mechanism of mason bees, which make one-celled clay nests provisioned with honey. His efforts at applying Charles Darwin's suggestions for experiments were commendable, but suffered from equipment failure. Scientists often have to invent workable ways and means before they can answer a question via experimentation. Effects of magnetic currents on bees were eventually documented, but had to await the invention of cyanoacrylic ("super") glue and microtransistors. Notice Fabre's underlying, and unquestioned, assumption that insects have similar, if not identical, senses, and thus perceptions, as humans. Charles Darwin died on April 19, 1882.]

This chapter and the next were to have taken the form of a letter addressed to Charles Darwin, the illustrious naturalist, who now lies buried beside Newton in Westminster Abbey. It was my task to report to him the result of some experiments which he had suggested to me in the course of our correspondence: a very pleasant task, for, though facts, as I see them, disincline me to accept his theories, I have none the less the deepest veneration for his noble character and his scientific honesty. I was drafting my letter when the sad news reached me: Darwin was dead; after searching the mighty question of origins, he was now grappling with the last and darkest problem of the hereafter. I therefore abandon the epistolary form, which would be unwarranted in view of that grave at Westminster. A free and impersonal statement shall set forth what I intended to relate in a more academic manner

One thing, above all, had struck the English scientist on reading the first volume of my Souvenirs Entomologiques, namely, the Mason-bees' faculty of knowing the way back to their nests after being carried to great distances from home. What sort of compass do they employ on their return journeys? What sense guides them? The profound observer thereupon spoke of an experiment which he had always longed to make with pigeons and which he had always neglected making, absorbed as he was by other interests. This experiment, he thought, I might attempt with my Bees. Substitute the insect for the bird, and the problem remained the same. I quote from his letter the passage referring to the trial which he wished made:

"Allow me to make a suggestion in relation to your wonderful account of insects finding their way home. I formerly wished to try it with pigeons; namely, to carry the insects in their paper cornets about a hundred paces in the opposite direction to that which you intended ultimately to carry them, before turning round to return, to put the insects in a circular box with an axle which could be made to revolve very rapidly first in one direction and then in another, so as to destroy for a time all sense of direction in the insects. I have sometimes imagined that animals may feel in which direction they were at the first start carried."

This method of experimenting seemed to me very ingeniously conceived. Before going west, I walk eastwards. In the darkness of their paper-bags, the mere fact that I am moving them gives my prisoners a sense of the direction in which I am taking them. If nothing happened to disturb this first impression, the insect would be guided by it in returning. This would explain the homing of my Mason-bees carried to a distance of two or three miles amid strange surroundings. But, when the insects have been sufficiently impressed by their conveyance to the east, there comes the rapid twirl, first this way round, then that. Bewildered by all these revolutions first in one direction and then in another, the insect does not know that I have turned round and remains under its original impression. I am now taking it to the west, when it believes itself to be still travelling towards the east. Under the influence of this impression, the insect is bound to lose its bearings. When set free, it will fly in the opposite direction to its home, which it will never find again.

This result seemed to me the more probable inasmuch as the statements of the country-folk around me were all of a nature to confirm my hopes. Favier [Fabre's assistant], the very man for this sort of information, was the first to put me on the track. He told me that, when people want to move a Cat from one farm to another at some distance, they place the animal in a bag, which they twirl rapidly at the moment of starting, thus preventing the animal from returning to the house which it has quitted. Many others, besides Favier, described the same practice to me. According to them, this twirling round in a bag was an infallible expedient: the bewildered Cat never returned. I communicated what I had learnt to England, I wrote to the sage of Down and told him how the peasant had anticipated the researches of science. Charles Darwin was amazed; so was I; and we both of us almost reckoned on a success.

These preliminaries took place in the winter; I had plenty of time to prepare for the experiment which was to be made in the following May... [Fabre describes at length establishing a new colony of mason bees on his property, to eliminate the possibility of the bees' possessing a possible ancestral memory of their home site.]

The question of the rotary machine remains. Darwin advised me to use a circular box with an axle and a handle. I have nothing of the kind in the house. It will be simpler and quite as effective to employ the method of the countryman who tries to lose his Cat by swinging him in a bag. My insects, each one placed by itself in a paper cornet, or screw, shall be placed in a tin box; the screws of paper shall be wedged in so as to avoid collisions during the rotation; lastly, the box shall be tied to a cord and I will whirl the whole thing round like a sling. With this contrivance, it will be quite easy to obtain any rate of speed that I wish, any variety of inverse movements that I consider likely to make my captives lose their bearings. I can whirl my sling first in one direction and then in another, turn and turn about; I can slacken or increase the pace; if I like, I can make it describe figures of eight, combined with circles; if I spin on my heels at the same time, I am able to make the process still more complicated by compelling my sling to trace every known curve. That is what I shall do.

On the 2d of May, 1880, I make a white mark on the thorax of ten Mason-bees busied with various tasks: some are exploring the slabs of clay in order to select a site; others are bricklaying; others are laying in stores. When the mark is dry, I catch them and pack them as I have described. I first carry them a quarter of a mile in the opposite direction to the one which I intend to take. A path skirting my house favours this preliminary manaeuvre; I have every hope of being alone when the time comes to make play with my sling. There is a wayside cross at the end; I stop at the foot of the cross. Here I swing my Bees in every direction. Now, while I am making the box describe inverse circles and loops, while I am pirouetting on my heels to achieve the various curves, up comes a woman from the village and stares at me. Oh, how she stares at me, what a look she gives me! At the foot of the cross! Acting in such silly way! People talked about it. It was sheer witchcraft. Had I not dug up a dead body, only a few days before? Yes, I had been to a prehistoric burial-place, I had take from it a pair of venerable, well-developed tibias, a set of funerary vessels and a few shoulders of horse, placed there as a viaticum for the great journey. I had done this thing and people knew it. And now, to crown all, the man of evil reputation is found at the foot of a cross indulging in unhallowed antics.

No matter—and it shows no small courage on my part—the gyrations are duly accomplished in the presence of this unexpected witness. Then I retrace my steps and walk west ward of Serignan. I take the least-frequented paths, I cut across country so as, if possible, to avoid a second meeting. It would be the last straw if I were seen opening my paper bags and letting loose my insects! When halfway, to make my experiment more decisive still, I repeat the rotation, in as complicated a fashion as before. I repeat it for the third time at the spot chosen for the release.

I am at the end of a flint-strewn plain, with here and there a scanty curtain of almond-trees and holm-oaks. Walking at a good pace, I have taken thirty minutes to cover the ground in a straight line. The distance, therefore, is, roughly, two miles. It is a fine day, under a clear sky, with a very light breeze blowing from the north. I sit down on the ground, facing the south, so that the insects may be free to take either the direction of their nest or the opposite one. I let them loose at a quarter past two. When the bags are opened, the Bees, for the most part, circle several times around me and then dart off impetuously in the direction of Serignan, as far as I can judge. It is not easy to watch them, because they fly off suddenly, after going two or three times round my body, a suspicious-looking object which they wish, apparently, to reconnoitre before starting. A quarter of an hour later, my eldest daughter, Antonia, who is on the look-out beside the nests, sees the first traveller arrive. On my return, in the course of the evening, two others come back. Total: three of my Masons home on the same day, out of ten scattered abroad.

I resume the experiment next day. I mark ten Mason-bees with red, which will enable me to distinguish them from those who returned on the day before and from those who may still return with the white spot uneffaced. The same precautions, the same rotations, the same localities as on the first occasion; only, I make no rotation on the way, confining myself to swinging my box round on leaving and on arriving. The insects are released at a quarter past eleven. I prefer the morning, as this was the busiest time at the works. One Bee was seen by Antonia to be back at the nest by twenty minutes past eleven. Supposing her to be the first let loose, it took her just five minutes to cover the distance. But there is nothing to tell me that it is not another, in which case she needed less. It is the fastest speed that I have succeeded in noting. I myself am back at twelve and, within a short time, catch three others. I see no more during the rest of the evening. Total: four home, out of ten.

The 4th of May is a very bright, calm, warm day, weather highly propitious for my experiments. I take fifty Chalicodomae marked with blue. The distance to be travelled remains the same. I make the first rotation after carrying my insects a few hundred steps in the direction opposite to that which I finally take; in addition, three rotations on the road; a fifth rotation at the place where they are set free. If they do not lose their bearings this time, it will not be for lack of twisting and turning. I begin to open my screws of paper at twenty minutes past nine. It is rather early, for which reason my Bees, on recovering their liberty, remain for a moment undecided and lazy; but, after a short sun-bath on a stone where I place them, they take wing. I am sitting on the ground, facing the south, with Serignan on my left and Piolenc on my right. When the flight is not too swift to allow me to perceive the direction taken, I see my released captives disappear to my left. A few, but only a few, go south; two or three go west, or to right of me. I do not speak of the north, against which I act as a screen. All told, the great majority take the left, that is to say, the direction of the nest. The last is released at twenty minutes to ten. One of the fifty travellers has lost her mark in the paper bag. I deduct her from the total, leaving forty-nine.

According to Antonia, who watches the home-coming, the earliest arrivals appeared at twenty-five minutes to ten, say fifteen minutes after the first was set free. By twelve o'clock midday, there are eleven back; and, by four o'clock in the evening, seventeen. That ends the census. Total: seventeen, out of forty-nine.

I resolved upon a fourth experiment, on the 4th of May. The weather is glorious, with a light northerly breeze. I take twenty Mason-bees marked in pink, at eight o'clock in the morning. Rotations at the start, after a preliminary backing in a direction opposite to that which I intend to take; two rotations on the road; a fourth on arriving. All those whose flight I am able to follow with my eyes turn to my left, that is to say, towards Serignan. Yet I had taken care to leave the choice free between the two opposite directions: in particular, I had sent away my Dog, who was on my right. To-day, the Bees do not circle round me: some fly away at once; the others, the greater number, feeling giddy perhaps after the pitching of the journey and the rolling of the sling, alight on the ground a few yards away, seem to wait until they are somewhat recovered and then fly off to the left. I perceived this to be the general flight, whenever I was able to observe at all. I was back at a quarter to ten. Two Bees with pink marks were there before me, of whom one was engaged in building, with her pellet of mortar in her mandibles. By one o'clock in the afternoon there were seven arrivals; I saw no more during the rest of the day. Total: seven, out of twenty.

Let us be satisfied with this: the experiment has been repeated often enough, but it does not conclude as Darwin hoped, as I myself hoped, especially after what I have been told about the Cat. In vain, adopting the advice given, do I carry my insects, first in the opposite direction to the place at which I intend to release them; in vain, when about to retrace my steps, do I whirl my sling with every complication in the way of whirls and twists that I am able to imagine; in vain, thinking to increase the difficulties, do I repeat the rotation as often as five times over: at the start, on the road, on arriving; it makes no difference: the Mason-bees return and the proportion of returns on the same day fluctuates between thirty and forty per cent. It goes to my heart to abandon an idea suggested by so famous a man of science and cherished all the more readily inasmuch as I thought it likely to provide a final solution. The facts are there, more eloquent than any number of ingenious views; and the problem remains as mysterious as ever.

In the following year, 1881, I began experimenting again, but in a different way. Hitherto, I had worked on the level. To return to the nest, my lost Bees had only to cross slight obstacles, the hedges and spinneys of the tilled fields. To-day, I propose to add to the difficulties of distance those of the ground to be traversed. Discontinuing all my backing- and whirling-tactics, things which I recognize as useless, I think of releasing my Chalicodomae in the thick of the Serignan Woods. How will they escape from that labyrinth, where, in the early days, I needed a compass to find my way? Moreover, I shall have an assistant with me, a pair of eyes younger than mine and better-fitted to follow my insects' first flight. That immediate start in the direction of the nest has already been repeated very often and is beginning to interest me more than the return itself. A pharmaceutical student, spending a few days with his parents, shall be my eyewitness. With him, I feel at ease; science and he are no strangers.

The trip to the woods takes place on the 16th of May. The weather is hot and hints at a coming storm. There is a perceptible breeze from the south, but not enough to upset my travellers. Forty Mason-bees are caught. To shorten the preparations, because of the distance, I do not mark them while they are on the nests; I shall mark them at the starting-point, as I release them. It is the old method, prolific of stings; but I prefer it to-day, in order to save time. It takes me an hour to reach the place. The distance, therefore, allowing for windings, is about three miles.

The site selected must permit me to recognize the direction of the insects' first flight. I choose a clearing in the middle of the copses. All around is a great expanse of dense woods, shutting out the horizon on every side; on the south, in the direction of the nests, a curtain of hills rises to a height of some three hundred feet above the spot at which I stand. The wind is not strong, but it is blowing in the opposite direction to that which my insects will have to take in order to reach their home. I turn my back on Serignan, so that, when leaving my fingers, the Bees, to return to the nest, will be obliged to fly sideways, to right and left of me; I mark the insects and release them one by one. I begin operations at twenty minutes past ten.

One-half of the Bees seem rather indolent, flutter about for a while, drop to the ground, appear to recover their spirits and then start off. The other half show greater decision. Although the insects have to fight against the soft wind that is blowing from the south, they make straight for the nest. All go south, after describing a few circles, a few loops around us. There is no exception in the case of any of those whose departure we are able to follow. The fact is noted by myself and my colleague beyond dispute or doubt. My Mason-bees head for the south as though some compass told them which way the wind was blowing.

I am back at twelve o'clock. None of the strays is at the nest; but, a few minutes later, I catch two. At two o'clock, the number has increased to nine. But now the sky clouds over, the wind freshens and the storm is approaching. We can no longer rely on any further arrivals. Total: nine, out of forty, or twenty-two per cent.

The proportion is smaller than in the former cases, when it varied between thirty and forty per cent. Must we attribute this result to the difficulties to be overcome ? Can the Mason-bees have lost their way in the maze of the forest? It is safer not to give an opinion: other causes intervened which may have decreased the number of those who returned. I marked the insects at the starting place; I handled them; and I am not prepared to say that they were all in the best of condition on leaving my stung and smarting fingers. Besides, the sky has become overcast, a storm is imminent. In the month of May, so variable, so fickle, in my part of the world, we can hardly ever count on a whole day of fine weather. A splendid morning is swiftly followed by a fitful afternoon; and my experiments with Mason-bees have often suffered by these variations. All things considered, I am inclined to think that the homeward journey across the forest and the mountain is effected just as readily as across the corn-fields and the plain.

I have one last resource left whereby to try and put my Bees out of their latitude. I will first take them to a great distance; then, describing a wide curve, I will return by another road and release my captives when I am near enough to the village, say, about two miles. A conveyance is necessary, this time. My collaborator of the day in the woods offers me the use of his gig. The two of us set off, with fifteen Mason-bees, along the road to Orange, until we come to the viaduct. Here, on the right, is the straight ribbon of the old Roman road, the Via Domitia. We take it, driving north towards the Uchaux Mountains, the classic home of superb Turonian fossils. We next turn back towards Serignan, by the Piolenc Road. A halt is made by the stretch of country known as Font-Claire, the distance from which to the village is about one mile and five furlongs. The reader can easily follow my route on the ordnance-survey map; and he will see that the loop described measures not far short of five miles and a half.

At the same time, Favier came and joined me at Font-Claire, by the direct road, the one that runs through Piolenc. He brought with him fifteen Mason-bees, intended for purposes of comparison with mine. I am, therefore, in possession of two sets of insects. Fifteen marked in pink, have taken the five-mile bend; fifteen, marked in blue, have come by the straight road, the shortest road for returning to the nest. The weather is warm, exceedingly bright and very calm; I could not hope for a better day for my experiment. The insects are given their freedom at mid-day.

At five o'clock, the arrivals number seven of the pink Mason-bees, whom I thought that I had bewildered by a long and circuitous drive, and six of the blue Mason-bees, who came to Font-Claire by the direct route. The two proportions, forty-six and forty per cent, are almost equal; and the slight excess in favour of the insects that went the roundabout way is evidently an accidental result which we need not take into consideration. The bend described cannot have helped them to find their way home; but it has also certainly not hampered them.

There is no need of further proof. The intricate movements of a rotation such as I have described; the obstacle of hills and woods; the pitfalls of a road which moves on, moves back and returns after making a wide circuit: none of these is able to disconcert the Chalicodomae or prevent them from going back to the nest.

I had written to Charles Darwin telling him of my first, negative results, those obtained by swinging the Bees round in a box. He expected a success and was much surprised at the failure. Had he had time to experiment with his pigeons, they would have behaved just like my Bees; the preliminary twirling would not have affected them. The problem called for another method; and what he proposed was this:

"To place the insect within an induction coil, so as to disturb any magnetic or diamagnetic sensibility which it seems just possible that they may possess."

To treat an insect as you would a magnetic needle and to subject it to the current from an induction-coil in order to disturb its magnetism or diamagnetism appeared to me, I must confess, a curious notion, worthy of an imagination in the last ditch. I have but little confidence in our physics, when they pretend to explain life; nevertheless, my respect for the great man would have made me resort to the induction-coils, if I had possessed the necessary apparatus. But my village boasts no scientific resources: if I want an electric spark, I am reduced to rubbing a sheet of paper on my knees. My physics cupboard contains a magnet; and that is about all. When this penury was realized; another method was suggested, simpler than the first and more certain in its results, as Darwin himself considered:

"To make a very thin needle into a magnet; then breaking it into very short pieces, which would still be magnetic, and fastening one of these pieces with some cement on the thorax of the insects to be experimented on. I believe that such a little magnet, from its close proximity to the nervous system of the insect, would affect it more than would the terrestrial currents."

There is still the same idea of turning the insect into a sort of bar magnet. The terrestrial currents guide it when returning to the nest. It becomes a living compass which, withdrawn from the action of the earth by the proximity of a loadstone, loses its sense of direction. With a tiny magnet fastened on its thorax, parallel with the nervous system and more powerful than the terrestrial magnetism by reason of its comparative nearness, the insect will lose its bearings. Naturally, in setting down these lines, I take shelter behind the mighty reputation of the learned begetter of the idea. It would not be accepted as serious coming from a humble person like myself. Obscurity cannot afford these audacious theories.

The experiment seems easy; it is not beyond the means at my disposal. Let us attempt it. I magnetize a very fine needle by rubbing it with my bar magnet; I retain only the slenderest part, the point, some five or six millimetres long.This broken piece is a perfect magnet: it attracts and repels another magnetized needle hanging from a thread. I am a little puzzled as to the best way to fasten it on the insect's thorax. My assistant of the moment, the pharmaceutical student, requisitions all the adhesives in his laboratory. The best is a sort of cerecloth which he prepares specially with a very fine material. It possesses the advantage that it can be softened at the bowl of one's pipe when the time comes to operate out of doors.

I cut out of this cerecloth a small square the size of the Bee's thorax; and I insert the magnetized point through a few threads of the material. All that we now have to do is to soften the gum a little and then dab the thing at once on the Mason-bee's back, so that the broken needle runs parallel with the spine. Other engines of the same kind are prepared and due note taken of their poles, so as to enable me to point the south pole at the insect's head in some cases and at the opposite end in others.

My assistant and I began by rehearsing the performance; we must have a little practice before trying the experiment away from home. Besides, I want to see how the insect will behave in its magnetic harness. I take a Mason-bee at work in her cell, which I mark. I carry her to my study, at the other end of the house. The magnetized outfit is fastened on the thorax; and the insect is let go The moment she is free, the Bee drops to the ground and rolls about, like a mad thing, on the floor of the room. She resumes her flight, flops down again, turns over on her side, on her back, knocks against the things in her way, buzzes noisily, flings herself about desperately and ends by daring through the open window in headlong flight.

What does it all mean? The magnet appears to have a curious effect on my patient's system! What a fuss she makes! How terrified she is! The Bee seemed utterly distraught at losing her bearings under the influence of my knavish tricks. Let us go to the nests and see what happens. We have not long to wait: my insect returns, but rid of its magnetic tackle. I recognize it by the traces of gum that still cling to the hair of the thorax. It goes back to its cell and resumes its labours.

Always on my guard when searching the unknown, unwilling to draw conclusions before weighing the arguments for and against, I feel doubt creeping in upon me with regard to what I have seen. Was it really the magnetic influence that disturbed my Bee so strangely? When she struggled and kicked on the floor, fighting wildly with both legs and wings, when she fled in terror, was she under the sway of the magnet fastened on her back? Can my appliance have thwarted the guiding influence of the terrestrial currents on her nervous system? Or was her distress merely the result of an unwonted harness? This is what remains to be seen and that without delay.

I construct a new apparatus, but provide it with a short straw in place of the magnet. The insect carrying it on its back rolls on the ground, kicks and flings herself about like the first, until the irksome contrivance is removed, taking with it a part of the fur on the thorax. The straw produces the same effects as the magnet, in other words, magnetism had nothing to do with what happened. My invention, in both cases alike, is a cumbrous tackle of which the Bee tries to rid herself at once by every possible means. To look to her for normal actions so long as she carries an apparatus, magnetized or not, upon her back is the same as expecting to study the natural habits of a Dog after tying a kettle to his tail.

The experiment with the magnet is impracticable. What would it tell us if the insect consented to it? In my opinion, it would tell us nothing. In the matter of the homing instinct, a magnet would have no more influence than a bit of straw.

Jean Henri Fabre

Readings for

PEOPLE AND NATURE

Understanding the Changing Interactions Between People and Ecological Systems

Nancy E. Langston

- **Norman L. Christensen**
  Landscape History and Ecological Change
- **Alfred W. Crosby**
  Ecological Imperialism
- **Michael Pollan**
  The Idea of a Garden
- **Donald Worster**
  The Ecology of Order and Chaos

Introduction to

Norman L. Christensen

Landscape History and Ecological Change

by Nancy E. Langston

Scientists as well as historians have been calling for a renewed attention to history. One example of this trend is the article by the forest ecologist Norman Christensen (1989), in which he summarizes some of the ways that forest succession theory can be enriched by a deeper understanding of history. One of the critical insights of modern ecology is that we cannot understand environmental dynamics without paying attention to human influences, for human disturbances are now among the most important factors shaping ecosystem change. Ecologists long assumed that they could study the laws of nature apart from human history. Human-influenced systems seemed to be aberrations because human history seemed too brief in evolutionary time to worry about. Most ecologists believed that over time, natural processes would eventually erase the effects of different initial stages. Any change would occur in a predictable manner and would lead to a constant endpoint.

This was a deterministic model: ecologists believed that the direction of change was determined by laws that people could understand (and manipulate, if they were applied forest ecologists). But Christensen's article shows that past environmental conditions play a continuing role in most ecological systems: you cannot erase or ignore history, and people are one of many sources of historical disturbances that shape environments.

Landscape History and Ecological Change

Norman L. Christensen

Journal of Forest History 33:116–124, reprinted with permission from the Forest History Society, Durham, NC.

North American ecologists have a long tradition of studying "natural" or undisturbed ecosystems. In addition to the innate affection for wilderness that likely initially enticed most ecologists to their discipline, practical reasons have prompted this focus. Ecologists seek to identify and understand the factors regulating the distribution and abundance of organisms. Interpreting the complex relationships between organisms and their environments is complicated by the variation in ecological communities caused by disturbances and past human interventions. Ecologists can control the effects of these external variables on their research by studying undisturbed ecosystems.

Yet ecologists have for some time recognized that their concentration on natural ecosystems is, in many cases, rather naive or even misleading. During the past several years, the emphasis of their research has shifted to the consequences of past historical events for the current structure and function of ecosystems. Several factors account for this shift.

First, it has become obvious that the research emanating from the focus on natural ecosystems suffers from serious limitations. Although statistically significant relationships between the distribution of organisms and existing environmental conditions have allowed us to construct predictive models, such models often account for only a small part of the observed variation among ecosystems.[1] We are forced to conclude that the unexplained variation is due either to unmeasured variables or to past events.

Second, we are increasingly aware of the longevity of historical impacts. For many years, ecologists assumed that the importance of the historical impacts of human interventions would diminish to insignificance given sufficient time for succession to "heal the wounds." In recent years, however, this assumption has been seriously questioned. Many plants, especially forest trees, live a long time. Because of this longevity, past environmental conditions play a significant and continuing role in the structure of most forest ecosystems.

Third, we are beginning to understand that the processes regulating the structure and function of ecosystems function on a spatial scale much larger than the conventional ecological study unit—the population or stand. As ecologists turn their attention to landscape-scale patterns and processes, they cannot avoid accounting for influences stemming from human uses of the land. Even segments of a landscape that have not been directly disturbed are influenced by the human factors affecting surrounding areas. A small plot of undisturbed woods preserved in the midst of an otherwise urban landscape differs markedly from a similar plot surrounded by undeveloped forest. Thus past land use patterns may influence the structure of wilderness remnants that appear not to have been directly altered by human activities.

Finally, the influence of human interventions in the environment will continue to grow in the future. Thus the most compelling reason to study the effects of past history on the current structure of ecosystems is our desire to make informed predictions about future changes in the ecosystem.

This essay presents an overview of this new area of ecological research. I shall discuss the use of the term "history" among ecologists and assert that our views of the effects of historical events on ecosystems are colored by our understanding of the nature and mechanisms of ecosystem change. Having established that past history does affect current ecosystems, I shall discuss several approaches to deciphering the historical record from the structure of the ecosystem. In addition, I shall describe examples that demonstrate how purportedly simple historical effects are often brought about by complex influences. I shall also demonstrate the importance of historical studies to our understanding and management of current ecological problems. I shall argue, in conclusion, that the history of land use and the succession of ecosystems are directly linked: future land use depends in large part on the nature of ecosystem change, which, in turn, is often initiated and affected by patterns of land use.

The Ecologist's View of History

Ecologists often use the terms "history" and "historical effect" in their most generic senses. In the ecological literature, any property of an ecosystem that is a consequence of some past event, whether or not caused by humans, is often said to be "due to history."[2] Indeed, variations in ecosystem characteristics not attributable to current environmental conditions are often simply attributed to "historical effects." I do not wish to defend ecologists' lax use of these terms. Rather, I feel it illustrates a very basic philosophical difference between the central interests of the disciplines of history and ecology.

The historian most often focuses on humans and the human condition; the ecologist on the causes of ecological variation. This is not to say that ecologists are not interested in humans, but rather that ecologists view humans simply as one factor altering the environment. When, for example, an ecologist studies the effects of a forest fire on the forest ecosystems, the event has the same ecological consequences regardless of its origin. Whether the forest fire was started by a lightning strike or by human activity is thus of little consequence to the ecologist. Furthermore, the ecologist may not be equipped to answer the question of exact causes. Many "natural" and human-caused events have similar ecological consequences, so ecological data alone are often insufficient to identify the specific agents of past disturbance.

The Role of History in Changing Landscapes

Since Henry Chandler Cowles's description of the patterns of vegetation change on sand dunes surrounding Lake Michigan,[3] American plant ecologists have been preoccupied with studies of succession, or ecosystem change following disturbance. Frederick E. Clements, influenced by Cowles's work and his own observations of prairie ecosystems, developed a comprehensive theory describing the mechanisms and patterns of successional change that was the mainstay of plant community ecology for nearly half a century.[4] According to this theory, the primary forces driving successional change are biotic reaction and competition. Clements coined the phrase "biotic reaction" to describe the influence of organisms on their environment. An example of a biotic reaction is the mechanical and chemical weathering of soil caused by plant roots. "Competition" refers to the effects of one organism on another that uses a common shared resource in short supply.

Figure 1 is a diagram of the pattern of succession as described by Clements's theory. The process is initiated by a disturbance resulting in "nudation." Clements realized that the considerable variation among disturbances in post-disturbance environments would greatly affect the rate and trajectory of this process. Disturbances that leave habitats devoid of soil (e.g., volcanic eruptions, glaciation, etc.) initiate so-called primary succession and typically produce very slow rates of invasion and successional change. Successional change is much more rapid following disturbances such as fire and land clearing in which soil and even some surviving plants are left (secondary succession). Nevertheless, Clements thought that the general mechanisms of successional change were the same in both cases. Early invading organisms (pioneers) establish and alter the environment in such a way as to favor the establishment and growth of other potential migrants over themselves. In the same manner, these migrants change their environment so that they too eventually are replaced. This process of replacement continues until a community of plants (and animals) is established that alters the environment to one that favors their own reproduction. This is the climax.

One of Clements's central assertions was that regardless of the types or varieties of disturbances that initiated successional change on a landscape, succession inexorably leads to a single climax community whose composition is determined by the characteristics of the region's climate.[5] Although the pioneer vegetation surrounding a kettle lake and that growing on an outwash plain on a recently deglaciated landscape initially have little in common, biotic reactions result in increased environmental similarity, and this results in increasing similarity among later successional

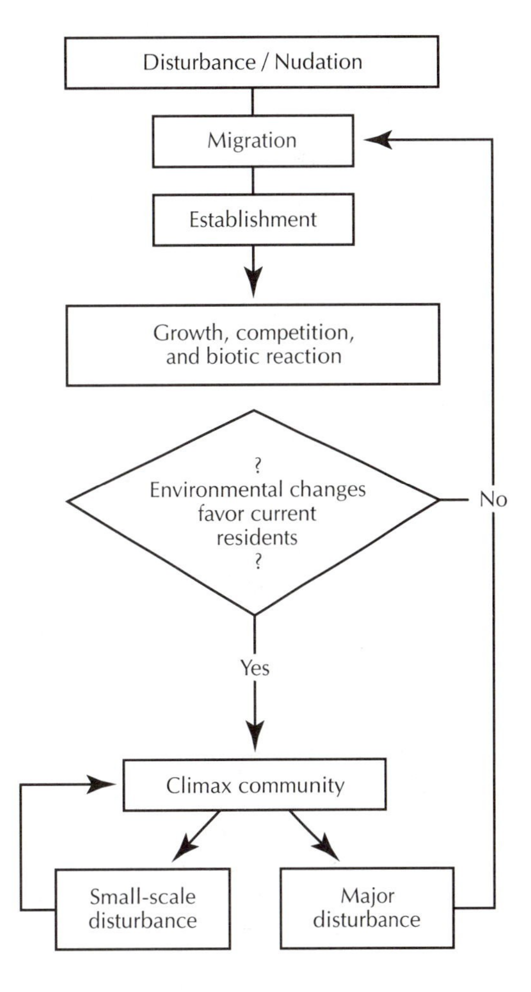

Figure 1. Diagram of processes in ecological succession as proposed by Frederick E. Clements.

communities. Similarly, the vegetation of a field recently abandoned from row crops and another abandoned from pasture may be quite different, but successive communities in these two fields will become increasingly similar.

Clements's notion of ecological convergence has important consequences for both the historian and the ecologist. If true, it suggests that historical effects are eventually erased by succession and that the structure and composition of relatively late successional ecosystems may contain little information about their past history. To the ecologist, it also indicates that if one simply waits long enough, historical events and patterns of past disturbance will become relatively unimportant.

Certain of Clements's assertions, such as the notion of a single ultimate climax or "monoclimax," were questioned rather quickly.[6] Ecologists recognized that local soil and hydrological conditions greatly influenced the structure and function of ecosystems and that these conditions were not likely to be changed by biotic reaction, even over long time spans. Thus developed the theory of "polyclimax," an accommodation to explain why vegetation was not homogeneous over large regions, but rather varied continuously along environmental gradients.[7] Nevertheless, many of the central tenets of Clements's paradigm were widely accepted by ecologists until the 1960s.[8]

Nearly all successional studies up to the 1960s were based on comparing at a single point in time ecosystems of varying successional age. Such studies assumed that the younger samples would eventually look like the older samples and that the older samples once looked like the younger ones. The mechanisms that drive successional change were, for the most part, taken for granted: very few studies actually rigorously demonstrated the roles of biotic reaction or competition in driving successional change. Gradually, however, some ecologists came to realize that the available data were not sufficient to support the major assumptions of Clements's model.

Some of the critics began to argue that pioneer species usurped available resources and prevented subsequent invasion until they began to decline, instead of preparing the way for subsequent invaders. Thus they argued that succession was driven by dispersal and competitive exclusion; if later successional species had access to a site, they could invade recently disturbed habitats.[9] Other ecologists argued that many successional sequences were a simple consequence of differential longevity. Most species that eventually invaded a site arrived early, and those that lived longest (e.g., broad-leafed trees) formed the eventual climax community.[10]

Suffice it here to say that both of these patterns have been shown to be important in one or another ecosystem. The notion of convergence has not been entirely abandoned, but patterns of convergence, if and when they occur, have been shown to be far more complex than originally envisioned by Clements. Consequently, the last twenty years have witnessed the nearly complete demise of Clements's theory of succession.[11]

The alternative successional models postulate an important difference in the role past history plays in ecosystem structure. For example, if most eventual climax species do indeed arrive early and simply live longer, disturbances may have long-range ecological consequences—lasting for centuries and longer.

One of the most significant changes to occur in ecologists' views of successional change is the realization that, even on the scale of decades and certainly on the scale of centuries, environments are not static. Palynologists have demonstrated that the vegetation of North America has been in a continuous state of flux for the past forty thousand years. Most ecologists agree that such variation has been the norm since the Pleistocene.[12] It is now generally recognized that natural disturbance cycles, involving such agents as fire, wind, or pathogens, are a normal part of most landscapes and that few ecosystems ever achieve a steady-state climax [13] Furthermore, abundant data convince us that humankind has influenced the composition and structure of nearly all North American ecosystems for nearly ten thousand years. Because the properties of environment (i.e., the factors affecting the relative success and performance of organisms) are not static, ecologists have come to recognize that it is generally hopeless to try to understand environmental dynamics in the absence of human influences.

The Historical Record in the Ecosystem

Ecologists generally agree that past events influence the properties of current ecosystems. They vary, however, in deciding which features of ecosystems ought to be used to reconstruct past history. What follows is a brief review of ecological approaches to identifying the influences of past events. It is important to note that these approaches may tell us a good deal about what has happened in the past, but they only rarely tell us why. For example, forest data may identify a period of high tree mortality fifty years ago, but such data are unlikely to indicate the cause of tree death.

The architecture or physical structure of a forest stand or the individual trees within it may provide information about its past. In southeastern pine forests the diameter-to-height ratio of the dominant trees correlates highly with growth conditions (such as initial stand density) in the abandoned field in which they establish.[14]

Mark McDonnell and Edward Stiles show that variations in the physical structure of forest stands that regenerate from abandoned agricultural fields may determine the spatial patterns of different tree species within such stands.[15] Such significant historical variations include the size of the field and the presence or absence of agricultural debris, fence posts, and other structures affecting the behavior of animals that disperse tree seeds into such fields.

As forests develop and mature, a forest floor of litter and woody debris begins to accumulate. Fires add charcoal to the soil. Trees blown over by wind create tip-up mounds and pits, churning the soil and altering microtopography. In order to reconstruct the disturbance history of a forest stand in the Pisgah Forest of southwestern New Hampshire, J. D. Henry and J. M. A. Swan excavated a tenth-acre site using much the same techniques an archaeologist might use to excavate a site of suspected past human habitation. They took a census of the living vegetation, measured woody debris, and used various techniques to infer the year of death for downed and decayed logs. They excavated the soil to determine the location of long-decayed logs and deposits of charcoal. Then they analyzed their data to reconstruct the history of disturbance in this stand back to 1665.[16]

Carbon dating of charcoal from soils beneath forests in the Amazon Basin has been used to document patterns of human land use and burning over the past several centuries [17] Thomas Bonnicksen and Edward Stone use a similar approach to reconstruct past stand structure and history in the giant sequoia forests of the central Sierra Nevada.[18] By linking their reconstructions to features of the forest that can be discerned from aerial photographs, they then extend their analysis to a landscape encompassing hundreds of square kilometers. These researchers suggest that this method might be used by agencies such as the National Park Service to determine the precolonial structure of ecosystems when designating wilderness in areas that have been altered by human activities.

Although such studies are of some use in documenting past episodes of disturbance, they suffer from several shortcomings. Of these, perhaps the most important is their limited scope. They typically focus on relatively small areas (usually a few hundred square meters), and thus they seldom provide a statistically meaningful history of change for entire forest stands, much less landscapes or regions.

Another problem is the accuracy of the data collected. Several researchers have questioned the precision of dating disturbances on the basis of dead debris on and in forest soils.[19] Because disturbances are often spatially heterogeneous, they will not leave behind uniform accumulated debris or charcoal throughout a forest. Furthermore, the technical problems in making such estimates are numerous and are complicated by the varying rate of decay of such debris among locations and tree species. From a practical standpoint, the labor-intensive and destructive nature of such methods is another major drawback.

The species that compose a forest community can also provide information about its past. The presence of shade intolerant trees in the forest canopy is usually a sign of past disturbance. This is true today in the southeastern United States, where shade-intolerant trees such as pine or tulip tree in the forest canopy provide a reliable measure of the nature and extent of past disturbance.[20] In the western United States, however, this generalization does not hold. Western land management policies that excluded natural wildfire eventually allowed shade-tolerant trees to invade the understory of relatively open mixed-conifer forests.[21]

In forest ecosystems in which species that dominate late in succession actually establish relatively early, the distribution of such species may provide evidence of the nature of disturbance or the immediate postdisturbance environment. In southern New England, for example, secondary forests dominated by birch likely became established in fields that had been row cropped before they were abandoned. On the other hand, red cedar predominates in forests established in old pastures.[22]

Even some herbaceous plants indicate past forest history. For example, in forests of the southeastern U.S. piedmont, the common garden periwinkle *(Vinca minor)* is a faithful indicator of the location of old homesites and graveyards. The distribution of specific herbs in the woods of Great Britain is an important tool used to recognize "ancient woods" (i.e., forest stands that have never been totally cleared and cultivated).[23]

Our primary source of information about changes in ecosystems over hundreds or thousands of years is fossil pollen. The often ornately sculpted walls of pollen grains are unique to specific plant taxa and are very resistant to decay in anaerobic sediments such as those found in bogs and lake bottoms. The pollen grains accumulating in the sediment of such sites record changes in the vegetation of such an environment through time. Scientists compare the patterns of such change among different sites to reconstruct vegetation change on whole continents and over very long time spans.[24] In some cases lake sediments are "varved" or deposited in such a way that scientists can differentiate individual years. Such records have been used to reconstruct the history of fires on boreal forest landscapes in Canada.[25]

During the past decade a number of researchers have used palynology to chronicle changes in forests and forested landscapes caused by human land use. For example, Emily Russell has coupled palynological and archival studies to reconstruct patterns of vegetation change in northern New Jersey from precolonial times to the present. James Clark has recently published a similar study for New York.[27]

The studies by Grace Brush and Frank Davis of sediment cores taken from multiple locations in Chesapeake Bay provide an excellent example of the power of these techniques.[28] Brush and Davis discern changes in the extent of deforestation surrounding the bay by examining fossil pollens and comparing the relative amounts of oak and ragweed pollen. They also use the pollen record to trace major forest events such as the decline of the American chestnut. They analyze fossil-diatom populations to document variation in salinity and relate these to changes in rainwater runoff caused by changing land uses. They use the sediment record to trace changing agricultural practices as evidenced by widespread eutrophication and rapid siltation.

Several factors affect the value of palynological research. First, it may be limited by the availability of appropriate depositional environments. Second, it may fail to isolate short time spans adequately. Third, it may fail to isolate the extent of the landscape contributing materials to the sediment core. For example, evidence of pollen of a certain species may sometimes reflect highly localized activities, and sometimes reflect landscape-scale patterns of land use. Despite the difficulties, however, this approach has proven to be very useful, especially when coupled with archival data documenting land use changes.

Dendrochronology, the study of annual growth rings in trees, is a very useful tool for studying forest history. Tree reproduction in many forest ecosystems occurs in episodes associated with major disturbances. Thus the distribution of ages in a population (trees of one species within a stand) is often a sensitive indicator of the history of disturbance in the stand.[29] In even-aged forests that regenerate following disturbance (fire, land clearance, windthrow), the pattern and distribution of stands of different ages reveal the history of disturbance.[30]

Properly calibrated, the relative width of tree rings provides a measure of changes in tree growth. In habitats where tree growth is tightly correlated with climatic variation (usually extreme environments), dendrochronology has been used to reconstruct patterns of climatic change over millennia.[31] In more hospitable habitats, tree growth is often regulated by competition from other trees. Thus the changing widths of tree rings in such forests may indicate an event such as cutting or disturbance that reduces competition among surviving trees.[32]

The history of fire on forested landscapes can often be discerned by studying fire scars. The date of a fire can be determined by the location in the tree-ring record of scars left by a fire-caused wound. Successive fires tend to rescar a tree in the same location, providing a continuous record of fire near that tree. Collected and mapped for many forest stands and large spatial scales, such data provide a history fire frequency and behavior over entire landscapes.[33]

Problems in Reconstructing the Past From the Present

One of the most active areas of research on forest succession is the development of computer simulation to depict and predict patterns of forest change following disturbance.[34] Such models couple the physiological and life history characteristics of forest species with data on climate, site environment, and disturbance to simulate change in forest structure. They allow us to predict with some degree of confidence the future status of a particular forest stand or forested landscape from appropriate data on the current status of that stand or landscape. It should be noted that our confidence in such models declines with the specificity of the prediction required and the length of time over which we try to predict.

Some researchers have suggested running the prediction models backward to reconstruct a description of a site at some past time.[35] Unfortunately, this technique is unlikely to be effective, primarily because successional change results in convergence of forest properties only in a manner far more complex than was envisioned by Frederick Clements.[36] Thus on most landscapes, the range of past situations that might have led to an existing forest structure is considerably greater than the range of possible future states. In predicting the future we assume the absence of historical "accidents" such as fires, wind storms, or human actions; we can make no such assumption for the past. Nevertheless, reconstructive computer simulation may be useful in defining the range of possible past forest states or disturbances or in exploring the possible consequences of past historical accidents.

Causation and Correlation

Although the current structure of a forest may allow us to infer specific properties of that forest at various times in the past, we can do little more than speculate about the specific causes of those properties. We may be able to attribute some characteristic of a forest or forest landscape to an event in its past history and yet be able to say little about that event. Thus we may infer from an even-aged population of shade-intolerant trees that a stand was initiated by a disturbance as old as the trees themselves. But lacking any additional information, we have no way of knowing the exact nature of this initial disturbance (e.g., "natural" or man-caused).

Variations in the relative abundance of shortleaf pine *(Pinus echinata)* and loblolly pine *(Pinus tueda)* in forest stands regenerating on abandoned agricultural land on the eastern piedmont of North Carolina illustrate the complexity of sorting out the causes behind even relatively simple observations. Both of these pines invade recently abandoned fields and eventually form even-aged stands. Because they are unable to reproduce in their own shade, they are replaced in 75 to 150 years by broadleafed deciduous hardwoods (Figure 2).

A census of stands in the piedmont region reveals rather marked differences in the relative abundance of these two species during the last two centuries. These differences seem to depend on the age of the pine stand, i.e., the time when it was abandoned (see table). There is no doubt that shortleaf pine was a more successful invader in fields abandoned during the

Figure 2. Successional change following agricultural land abandonment in the North Carolina piedmont. All photographs provided by author. **A.** Immediate post-abandonment field; historical factors such as agricultural practices and season of abandonment significantly influence patterns of plant invasion. **B.** Five-year-old field; patterns of field disturbance, year-to-year variations in climate, and activities adjacent to the field influence establishment of tree species. **C.** A forty-year-old pine stand in which agricultural furrows are still obvious. **D.** A mature hardwood forest, species composition and population structure bear witness to past historical events.

Variations in the Distribution and Abundance of Loblolly and Shortleaf Pine

	Stands established before 1900		Stands established after 1900	
	Frequency (%)	Biomass (%)	Frequency (%)	Biomass (%)
Shortleaf small seeds, uplands, adapted to "poor" sites	92	45	17	10
Loblolly larger seeds, bottomlands, grows best on fertile sites	37	50	100	90

nineteenth century than in those abandoned after 1930. But at least five distinct types of historical change might account for this observation: in landscape pattern, in silviculture and forest demography, in site conditions, in patterns of land use and abandonment, and in the general environment.

Fields abandoned during the nineteenth century stood in landscapes very different from the landscapes surrounding fields abandoned during the twentieth century. Prior to 1900 a newly abandoned field was something of an island in a "sea" of other fields (abandoned and cultivated), so there was unlikely to be a nearby seed source of forest trees. By this century, reforestation of the landscape was well underway, and newly abandoned fields were likely to be close to a source of tree seeds.[37]

The "natural" or precolonial habitats of the shortleaf and loblolly pines differ. Shortleaf pine occurs naturally on rocky, shallow upland soils; loblolly in bottomland areas. Seeds of shortleaf pine are considerably smaller and more widely dispersed than those of loblolly. The shortleafs preferred upland location and wider dispersal range may have given it an advantage as an invader in the isolated fields of the nineteenth-century landscape. But after dispersal became less important (such as on the reforested landscape of the mid-twentieth century), loblolly pine may have been the better competitor.

On all but the poorest sites, loblolly pine grows more quickly than shortleaf pine. For this reason loblolly is generally favored for silvicultural plantations. Thus rapid growth plus silvicultural selection may have greatly increased the "seed rain" of loblolly pine compared to shortleaf pine.

The environmental characteristics of fields abandoned a century ago may have been quite different from those of fields abandoned more recently. For example, nitrogen and phosphate fertilizers and lime were rarely used on cropland prior to about 1920. Thus fields abandoned during the nineteenth century were probably considerably more sterile than their twentieth-century counterparts.[38] Shortleaf pine is known to compete more effectively with loblolly pine on sterile soils than on nutrient-rich soils.

As a consequence of variations in the chemical makeup of the parent rock underlying the eastern piedmont, soil conditions vary considerably over spatial scales of tens and hundreds of meters. The distribution of sites favoring one species over another was further complicated by economic and other factors, which produced regional patterns of land clearing and development that were sometimes very different from patterns of subsequent abandonment. The poorest sites (least fertile, driest, etc.) were most likely to be abandoned first; somewhat more fertile sites were probably not abandoned until later. Given shortleaf pine's increased competitive ability on poor sites, its abundance in old stands may simply reflect the bias in the pattern of abandonment.

Finally, we cannot rule out the possibility that general environmental conditions during the latter half of the nineteenth century might have favored shortleaf, whereas those of 1930–89 might have favored loblolly pine. There is, for example, ample indication that climatic conditions have changed between these two periods. Then too, the relative success of these two species over the past century might have been influenced by changes in competition from other tree species or changes in the importance of fungal or insect pathogens.

Forest Decline: An Example of the Importance of Understanding History

Forest decline (regionwide decreased production or increased mortality of trees) is one of the most daunting problems currently facing forest ecologists. Where pathological symptoms are clear and impacts obvious, forest ecologists are occasionally successful in identifying the specific causes of such decline—even if they cannot remedy the problem. Such is certainly the case with respect to the impact of atmospheric pollutants and acidic precipitation on forested areas near industrial centers in central Europe.

In other areas, evidence of decline is more subtle and potential causes more complex.[39] It is clear, however, that historical patterns of land use are often a significant factor explaining forest decline.

The alleged decline of successional pine forests on the piedmont of the southeastern United States illustrates the potential contribution of land use history to explaining forest declines. Analysis of regional forest census data (gathered over the past thirty years) in the piedmont region suggests that the growth increment of several important southern pines has declined by as much as 25 percent in some areas.[40] Recall that most of these forests have regenerated on land abandoned by humans from nonsilvicultural land uses. Given the current dependence of the southeastern timber industry on these species, the financial consequences of a decline of this magnitude are substantial.

When forests have declined on other landscapes, scientists have frequently been able to help diagnose the cause. This is not the case in the southeastern piedmont: the southern pines have displayed no cause-specific symptoms. But assuming that the data are reliable, there are several possible explanations for the decline: competition from invading broad-leafed hardwoods, pests, widespread drought, increased atmospheric deposition (acid rain and ozone), forest aging, and land use.[41] Although atmospheric deposition has received the most attention, historical explanations may be much more likely.

Change in the successional forests of the southeastern piedmont follows a very predictable pattern for secondary forests in general (Figure 2). In this pattern, tree populations are initiated during an establishment phase (the old field). As trees grow and their canopies begin to intersect, competition begins to regulate growth and result in mortality of smaller, slower-growing individuals. This initiates a thinning phase during which overall stand growth is rapid, but mortality continues. Eventually such stands thin to the point where additional mortality begins to create openings in the forest canopy. Growth of the surviving trees is not sufficient to replace the biomass lost through mortality of large stems, and stand biomass begins to decline. Growth in height of surviving trees may also decline at this time. During this transition phase, if there is no intervention, deciduous hardwoods replace pines to form a mixed deciduous forest. In short, decline is a normal expectation in forest stands in which the successional process is not arrested by cutting.

Foresters are very familiar with the trajectory of change described above, and they ordinarily harvest a stand before it reaches the transition stage. However, most of the forested land in the Southeast is not being managed by foresters.

Furthermore, abandonment of land from agriculture or other nonforest uses has not been continuous during the past 150 years. Rather, reforestation has been initiated in pulses associated with historically important events such as Reconstruction and the Great Depression. Thus the various stages of succession represented in Figure 2 are not distributed naturally over the landscape; rather, their abundance reflects the history of land use and abandonment. What appears to be widespread decline may simply reflect the synchronous onset of the transition stage of stand development in many forest stands affected by the same large-scale historical events.

Whether this is the correct explanation for the phenomenon has yet to be decided. Only a comprehensive study of the history of land use and abandonment in this region can resolve the issue. The example does, however, demonstrate the potential contribution of historical studies to our understanding of contemporary environmental problems.

Historical Ecology as a Predictive Endeavor

Much of our preoccupation with the past is related to our interest in predicting the future. By studying past ecosystem change, ecologists have developed elegant models that can, with reasonable accuracy, predict the properties of future forests. Similarly, studies of patterns of past human behavior on landscapes provide us with likely scenarios for patterns of future land use. The necessity of understanding the interrelationships between ecological and historical change in order to create realistic predictive models is becoming increasingly clear. Again, the example of the relationship between land use and old-field succession in the southeastern piedmont illustrates this point.

Old-field succession is a creation of human land use and agricultural development. Although old fields may in some ways mimic natural disturbances that might have occurred on this landscape prior to human intervention, ecosystems such as broom-sedge fields and even-aged pine forests are unique to abandoned agricultural fields and are a consequence of past human history. Reciprocally, successional processes have altered the course of human history on the landscape. Rapid invasion of herbs in abandoned fields provided the basis for the fallow-field system of crop rotation used in the early nineteenth century.[42] The permanent abandonment of land later in that century resulted in the establishment of the extensive pine stands that now sustain the modern timber industry in this region.

For the foreseeable future, extensive abandonment of agricultural land such as occurred during the first four decades of this century is unlikely. Furthermore, unless considerable postharvest site preparation is done, cut pine stands will be replaced with a mixture of hardwood species and some pines. In addition, forest stands are now developing in a matrix of increasing urban development. Just as surely as our use of the landscape will affect the course of ecosystem change, that change will alter the values of the landscape for future human use. Can there be any doubt of the value and need for the collaboration between historians and ecologists?

Notes/References

The author is thankful to the participants in the "Landscape History and Ecological Succession" workshop held at Duke University in January 1988. They stimulated and clarified many of the points presented here. He is also grateful to the National Science Foundation for support of that workshop. Many of the ideas presented here grew out of conversations with Judson Edeburn, Rachel Frankel, Frank B. Golley, Robert K. Peet, and John Richards.

1. Christensen, N. L and Robert K. Peet. 1984. Convergence During Secondary Forest Succession. Journal of Ecology 72:25–36.
2. See for example William H. Drury and lan C. T. Nisbet, Succession, Journal of the Arnold Arboretum 54 (July 1973):331–68, or several of the papers in Darrel C. West, Herman H. Shugart, Jr. and Daniel B. Botkin, eds., *Forest Succession: Concepts and Application.* Springer-Verlag, New York. 1981.
3. Cowles, H. C. 1899. The Ecological Relations of the Vegetation on the Sand Dunes of Lake Michigan. Botanical Gazette 27:95–391.
4. Clements, F. C. 1935. Plant Succession: An Analysis of the Development of Vegetation, Carnegie Institute of Washington Publication 242 (Washington, D.C., Carnegie Institute of Washington). For a lucid history of the Clementsian school of ecology see Ronald C. Tobey, *Salving the Prairies: The Life Cycle of the Founding School of American Plant Ecology, 1895–1955.* University of California Press, Berkeley. 1981.
5. Clements, F. C. 1936. Nature and Structure of the Climax. Journal of Ecology 24:252–84.
6. Gleason, H. A. 1926. The Individualistic Concept of the Plant Association. Bulletin of the Torrey Botanical Club 53:7–26.
7. For discussions of polyclimax see Henry I. Oosting, *The Study of Plant Communities,* 2nd ed., Freeman, San Francisco, California, 1956, pp. 252–64; and Rexford E. Danbenmire, *Plant Communities,* Harper and Row, New York, 1968, pp. 240.
8. Odum, E. R. 1969. The Strategy of Ecosystem Development. Science 164:262–70.
9. Connell. J. H., and Ralph.O. Slatyer. 1977. Mechanisms of Succession in Natural Communities and Their Role in Stability and Organization. American Naturalist 111:1119–44.
10. Egler, F. G. 1954. Vegetation Science Concepts: 1. Initial Floristic Composition—A Factor in Old Field Vegetation Development. Vegetatio 4:412–17; Egler, F. G. 1976. Nature of Vegetation: Its Management and Mismanagement. Pp. 165–66.
11. Unlike the scientific revolutions described by Thomas Kuhn in *The Structure of Scientific Revolutions* (University of Chicago Press, Chicago, Illinois. 1962).
12. Delcourt, P. A., and Hazel R. Delcourt. (1987). *Long-Term Forest Dynamics of the Temperate Zone.* Springer-Verlag, New York. Pp. 5–16.
13. This recognition has led ecologists to view landscapes as mosaics of patches recovering from disturbance. See Peter S. White, Pattern, Process, and Natural Disturbance in Vegetation, Botanical Review 45 July–September 1979):229–99; and several chapters in Steward T. A. Pickett and Peter S. White, eds., *The Ecology of Natural Disturbance and Patch Dynamics.* Academic Press, New York. 1985.
14. Peet, R. K., and Norman L. Christensen. 1987. Competition and Tree Death. BioScience 37:586–95.
15. McDonnell, M. I., and Edward J. Stiles. 1983. The Structural Complexity of Old-Field Vegetation and the Recruitment of Bird-Dispersed Species. Oecologia 56:109–16.
16. Henry, J. D., and J. M. A. Swan. 1974. Reconstructing Forest History from Live and Dead Plant Material: An Approach to the Study of Forest Succession in Southwest New Hampshire. Ecology 55:772–83.
17. Sanford, Robert L., Juan Soldarriaga, Kathleen E. Clark, Christopher Uhl, and Raphael Herrera. 1985. Amazon Rainforest Fires. Science 227:53– 55.
18. Bonnicksen, Thomas M., and Edward C. Stone. 1982. Reconstruction of a Presettlement Giant Sequoia-Mixed Conifer Forest Community Using the Aggregation Approach. Ecology 63:1134–48.
19. This approach is reviewed by Nathan R. Stephenson, 1987, Use of Tree Aggregations in Forest Ecology and Management, Environmental Management 11:1–5.
20. Greller, Andrew M. 1988. Deciduous Forest. In: *North American Terrestrial Vegetation,* Michael G. Barbour and William D. Billings, eds. Cambridge University Press. Cambridge, England. Pp. 288–316. Runkle, James R. Disturbance Regimes in Temperate Forests. In: *The Ecology of Natural Disturbance and Patch Dynamics.* Pickett and White, eds.
21. Leopold, A. Starker, Stanley A. Cain, C. M. Cottam, J. N. Gabrielson, and T. L. Kimball 1963. Wildlife Management in the National Parks. American Forests 69:32–35 and 61–63.
22. Russell, Emily W. B. 1979. Vegetation Change in Northern New Jersey since 1500 A.D.: A Palynological, Vegetational, and Historical Synthesis. Ph.D. thesis, Rutgers University.
23. For examples see Peterken, George E. 1976. Long-Term Changes in Woodlands of Rockingham Forest and Other Areas. Journal of Ecology 64:123–46; Rackham, Oliver. 1980. *Ancient Woodland: Its History, Vegetation and Uses in England.* E. Arnold, London, England.
24. The use of these techniques for temperate forests of North America has been recently reviewed by Paul A. Delcourt and Hazel R. Delcourt, *Long-Term Forest Dynamics of the Temperate Zone.*
25. Cwynar, Les C. 1977. The Recent Fire History of Barron

Township, Algonquin Park. Canadian Journal of Botany 55:1524–38. Cwynar, Les C. 1978. Recent History of Fire and Vegetation from Laminated Sediment of Greenleaf Lake, Algonquin Park, Ontario. Canadian Journal of Botany 56:10–21.
26. Russell, Emily W. B. 1980. Vegetation Change from Precolonization to the Present: A Palynological Interpretation. Bulletin of the Torrey Botanical Club 107:432–46.
27. Clark, James S. 1986. Coastal Forest Tree Populations in a Changing Environment, Southeastern Long Island, New York. Ecological Monographs 56:259–77.
28. Brush, Grace S., and Frank W. Davis. 1984. Stratigraphic Evidence of Human Disturbance in an Estuary. Quaternary Research 22:91–108.
29. Lorimer, Craig G. 1980. Age Structure and Disturbance History of a Southern Appalachian Virgin Forest. Ecology 61:1169-1184. This paper is an excellent example of the use of these techniques to decipher the past history of a single forest.
30. The application of such methods to reconstructing the fire history of a large portion of the Yellowstone National Park is described by Romme, William H. 1982. Fire and Landscape Diversity in Subalpine Forests of Yellowstone National Park. Ecological Monographs 52:199–221.
31. Harold C. Fritts, *Tree Rings and Climate.* Academic Press, New York.
32. See, for example, Stephen H. Spurr and Burton V. Barnes, 1980, *Forest Ecology,* 3d ed. John Wiley, New York. P. 381.
33. Romme, 1973. Fire and Landscape Diversity in Subalpine Forests; Heinselman, Myron L. 1973. Fire in the Virgin Forests of the Boundary Waters Canoe Area, Minnesota. Quaternary Research 3:329–82; Kilgore, Bruce M., and Dean Taylor. 1979. Fire History of a Sequoia-Mixed Conifer Forest. Ecology 60:129–42.
34. Shugart, Herman H. Jr., 1984. *A Theory of Forest Dynamics: The Ecological Implications of Forest Succession Models.* Springer-Verlag, New York.
35. This approach has been suggested by Bonnicksen and Stone, *Reconstruction of a Presettlement Giant Sequoia-Mixed Conifer Forest Community,* as a means of determining past forest structure. They propose that such a method would allow the National Park Service to know the presettlement forest structure of ecosystems that have been altered by humans.
36. For example, see Norman L. Christensen and Robert K. Peet, "Convergence during Secondary Forest Succession."
37. Oosting, Henry J. 1942. An Ecological Analysis of the Plant Communities of Piedmont, North Carolina. The American Midland Naturalist 28:1–126.
38. Trimble, Stanley W. 1973. Man-induced Soil Erosion on the Southern Piedmont 1700–1970. Ph.D. thesis, University of Wisconsin, Milwaukee. 180 pp.
39. Pietelka, Louis E., and Dudley J. Raynal. 1989. Forest Decline and Acidic Deposition. Ecology 70:2–10.
40. Sheffield, R. M., Noel D. Cost, William A. Becktold, and Joseph R McClure. 1985. Pine Growth Reductions in the Southeast. USDA Forest Service Resource Bulletin SE-83. P. 90.
41. *Ibid.,* Alan A. Luaer 1986. Summary and Interpretation of USDA Forest Service Report on Pine Growth Reductions in the Southeast. Technical Bulletin of the National Council of the Paper Industry for Air and Stream Improvement 508. P. 12.
42. Cowdrey, Albert. 1983. T*his Land, This South: An Environmental History.* University of Kentucky Press, Lexington. 1983.

Introduction to
Alfred W. Crosby
Ecological Imperialism: The Overseas Migration of Western Europeans as a Biological Phenomenon

by Nancy E. Langston

Alfred Crosby's "Ecological Imperialism" (1978) illustrates early work by ecological historians that helped define the new field. In the following paper, Alfred Crosby asks a question that many people have long pondered: Why did Europeans managed to conquer and colonize much (but certainly not all) of the world? Why are whites now over 90% of the population of places as diverse and geographically distant as Europe, North America, Argentina, Australia, and New Zealand? Why did so few of the natives in those places survive conquest by whites? And even more puzzling, why are there now so many fewer whites in other places that were also conquered by whites, such as Africa and Asia? For generations, historians tried to answer these questions from a purely human perspective. Some historians argued that Europeans' greater technology led them to conquer native peoples around the world; others argued that the causes lay with military might, politics, or the will of God.

Crosby turns this debate on its head by placing human invasions in their ecological context. He argues that associated ecological invasions were the key to the success of human invasions. Europeans did so well in the northern and southern temperate zones (but ultimately failed in the tropics) because their livestock, weeds, and pathogens did so well. For example, Europeans brought with them to the New World sheep, cattle, horses, rats, rabbits, honeybees, and most importantly, pathogens. Well before numbers of Europeans in the New World substantially increased, populations of their fellow travelers exploded, paving the way for the human invaders. For example, smallpox and other epidemics probably decimated over 90% of the native human populations in North America, making conquest by Europeans far easier. Yet few North American animals or pathogens made the reverse journey to Europe. In such places as the tropics, where European livestock and pathogens did not thrive, guns and political institutions alone could not ultimately ensure the success of Europeans.

In this essay, Crosby argues for the importance of weeds, livestock, and pathogens to human history. In later works, he developed a controversial set of epidemiological arguments to explain the trends he identified. Although scholars and scientists are still arguing over his theories, the fact remains that Crosby was one of the first to persuade historians that we could not understand human culture without paying close attention to the network of ecological relationships within which human history occurs. In asking historical questions from an ecological perspective, he revolutionized the ways historians think about big questions in human history.

Ecological Imperialism: The Overseas Migration of Western Europeans as a Biological Phenomenon

ALFRED W. CROSBY

Original publication in *Texas Quarterly* (1978), 21:10–22. Reprinted with permission.

Europeans in North America, especially those with an interest in gardening and botany, are often stricken with fits of homesickness at the sight of certain plants which, like themselves, have somehow strayed thousands of miles eastward across the Atlantic. Vladimir Nabokov, the Russian exile, had such an experience on the mountain slopes of Oregon:

> Do you recognize that clover?
> Dandelions, *l'or du pauvre*?
> (Europe, nonetheless, is over.)

A century earlier the success of European weeds in America inspired Charles Darwin to goad the American botanist, Asa Gray: "Does it not hurt your Yankee pride that we thrash you so confoundly? I am sure Mrs. Gray will stick up for your own weeds. Ask her whether they are not more honest, downright good sort of weeds."

The common dandelion, *l'or du pauvre*, despite its ubiquity and its bright yellow flower, is not at all the most visible of the Old World immigrants in North America. Vladimir Nabokov was a prime example of the most visible kind: the *Homo sapiens* of European origin. Europeans and their descendants, who comprise the majority of human beings in North America and in a number of other lands outside of Europe, are the most spectacularly successful overseas migrants of all time. How strange it is to find Englishmen, Germans, Frenchmen, Italians, and Spaniards comfortably ensconced in places with names like Wollongong (Australia), Rotorua (New Zealand), and Saskatoon (Canada), where obviously other peoples should dominate, as they must have at one time.

None of the major genetic groupings of humankind is as oddly distributed about the world as European, especially Western European, whites. Almost all the peoples we call Mongoloids live in the single contiguous land mass of Asia. Black Africans are divided between three continents—their homeland and North and South America—but most of them are concentrated in their original latitudes, the tropics, facing each other across one ocean. European whites were all recently concentrated in Europe, but in the last few centuries have burst out, as energetically as if from a burning building, and have created vast settlements of their kind in the South Temperate Zone and North Temperate Zone (excepting Asia, a continent already thoroughly and irreversibly tenanted). In Canada and the United States together they amount to nearly ninety percent of the population; in Argentina and Uruguay together to over ninety-five percent; in Australia to ninety-eight percent; and in New Zealand to ninety percent. The only nations in the Temperate Zones outside of Asia which do not have enormous majorities of European whites are Chile, with a population of two-thirds mixed Spanish and Indian stock, and South Africa, where blacks outnumber whites six to one. How odd that these two, so many thousands of miles from Europe, should be exceptions in not being predominantly pure European.

Europeans have conquered Canada, the United States, Argentina, Uruguay, Australia, and New Zealand not just militarily and economically and technologically—as they did India, Nigeria, Mexico, Peru, and other tropical lands, whose native peoples have long since expelled or interbred with and even absorbed the invaders. In the Temperate Zone lands listed above, Europeans conquered and triumphed demographically. These, for the sake of convenience, we will call the Lands of the Demographic Takeover.

There is a long tradition of emphasizing the contrasts between Europeans and Americans—a tradition honored by such names as Henry James and Frederick Jackson Turner—but the vital question is really why Americans are so European. And why the Argentinians, the Uruguayans, the Australians, and the New Zealanders are so European in the obvious genetic sense?

The reasons for the relative failure of the European demographic takeover in the tropics are clear. In tropical Africa, until recently, Europeans died in droves of the fevers; in tropical America they died almost as fast of the same diseases, plus a few native American additions. Furthermore, in neither region did European agricultural techniques, crops, and animals prosper. Europeans did try to found colonies for settlement, rather than merely exploitation, but they failed or achieved only partial success in the hot lands. The Scots left their bones as monument to their short-lived colony at Darien at the turn of the eighteenth century. The English Puritans who skipped Massachusetts Bay Colony to go to Providence Island in the Caribbean Sea did not even achieve a permanent settlement, much less a Commonwealth of God. The Portuguese who went to northeastern Brazil created viable settlements, but only by perching themselves on top of first a population of native Indian laborers and then, when these faded away, a population of laborers imported from Africa. They did achieve a demographic takeover, but only by interbreeding with their servants. The Portuguese in Angola, who helped supply those servants, never had a breath of a chance to achieve a demographic takeover. There was much to repel and little to attract the mass of Europeans to the tropics, and so they stayed home or went to the lands where life was healthier, labor more rewarding, and where white immigrants, by their very number, encouraged more immigration.

In the cooler lands, the colonies of the Demographic Takeover, Europeans achieved very rapid population growth by means of immigration, by increased life span, and by maintaining very high birth rates. Rarely has population expanded more rapidly than it did in the eighteenth and nineteenth centuries in these lands. It is these lands, especially the United States, that enabled Europeans and their overseas offspring to expand from something like eighteen percent of the human species in 1650 to well over thirty percent in 1900. Today 670 million Europeans live in Europe, and 250 million or so other Europeans—genetically as European as any left behind in the Old World—live in the Lands of the Demographic Takeover, an ocean or so from home. What the Europeans have done with unprecedented success in the past few centuries can accurately be described by a term from apiculture: they have swarmed.

They swarmed to lands which were populated at the time of European arrival by peoples as physically capable of rapid increase as the Europeans, and yet who are now small minorities in their homelands and sometimes no more than relict populations. These population explosions among colonial Europeans of the past few centuries coincided with population crashes among the aborigines. If overseas Europeans have historically been less fatalistic and grim than their relatives in Europe, it is because they have viewed the histories of their nations very selectively. Charles Darwin, as a biologist rather than a historian, when he returned from his world voyage on the Beagle in the 1830s, wrote, "Wherever the European has trod, death seems to pursue the aboriginal."

Any respectable theory which attempts to explain the Europeans' demographic triumphs has to provide explanations for at least two phenomena. The first is the decimation and demoralization of the aboriginal populations of Canada, the United States, Argentina, and others. The obliterating defeat of these populations was not simply due to European technological superiority. The Europeans who settled in temperate South Africa seemingly had the same advantages as those who settled in Virginia and New South Wales, and yet how different was their fate. The Bantu-speaking peoples, who now overwhelmingly outnumber the whites in South Africa, were superior to their American, Australian, and New Zealand counterparts in that they possessed iron weapons, but how much more inferior to a musket or a rifle is a stone-pointed spear than an iron-pointed spear? The Bantu have prospered demographically not because of their numbers at the time of first contact with whites, which were probably not greater per square mile than those of the Indians east of the Mississippi River. Rather, the Bantu have prospered because they survived military conquest, avoided the conquerors, or became their indispensable servants—and in the long run because they reproduced faster than the whites. In contrast, why did so few of the natives of the Lands of the Demographic Takeover survive?

Second, we must explain the stunning, even awesome success of European agriculture, that is, the European way of manipulating the environment in the Lands of the Demographic Takeover. The difficult progress of the European frontier in the Siberian *taiga* or the Brazilian *sertão* or the South African *veldt* contrasts sharply with its easy, almost fluid advance in North America. Of course, the pioneers of North America would never have characterized their progress as easy: their lives were filled with danger, deprivation, and unremitting labor; but as a group they always succeeded in taming whatever portion of North America they wanted within a few decades and usually a good deal less time. Many individuals among them failed—they were driven mad by blizzards and dust storms, lost their crops to locusts and their flocks to cougars and wolves, or lost their scalps to understandably inhospitable Indians—but as a group they always succeeded—and in terms of human generations, very quickly.

In attempting to explain these two phenomena, let us examine four categories of organisms deeply involved in European expansion: (1) human beings; (2) animals closely associated with human beings—both the desirable animals like horses and cattle and undesirable varmints like rats and mice; (3) pathogens or microorganisms that cause disease in humans; and (4) weeds. Is there a pattern in the histories of these groups which suggests an overall explanation for the phenomenon of the Demographic Takeover or which at least suggests fresh paths of inquiry?

Europe has exported something in excess of sixty million people in the past few hundred years. Great Britain alone exported over twenty million. The great mass of these white emigrants went to the United States, Argentina, Canada, Australia, Uruguay, and New Zealand. (Other areas to absorb comparable quantities of Europeans were Brazil and Russia east of the Urals. These would qualify as Lands of the Demographic Takeover except that very large fractions of their populations are non-European.)

In stark contrast, very few aborigines of the Americas, Australia, or New Zealand ever went to Europe. Those who did often died not long after arrival. The fact that the flow of human migration was almost entirely from Europe to her colonies and not visa versa is not startling—or very enlightening. Europeans controlled overseas migration, and Europe needed to export, not import, labor. But this pattern of one-way migration is significant in that it reappears in other connections.

The vast expanses of forests, savannahs, and steppes in the Lands of the Demographic Takeover were inundated by animals from the Old World, chiefly from Europe. Horses, cattle, sheep, goats, and pigs have for hundreds of years been among the most numerous of the quadrupeds of these lands, which were completely lacking in these species at the time of first contact with the Europeans. By 1600 enormous feral herds of horses and cattle surged over the pampas of the Rio de la Plata (today's Argentina and Uruguay) and over the plains of northern Mexico. By the beginning of the seventeenth century, packs of Old World dogs gone wild were among the predators of these herds.

In the forested country of British North America population explosions among imported animals were also spectacular, but only by European standards, not by those of Spanish America. In 1700 in Virginia feral hogs, said one witness, "swarm like vermaine upon the Earth," and young gentlemen were entertaining themselves by hunting wild horses of the inland counties. In Carolina the herds of cattle were "incredible, being from one to two thousand head in one Man's Possession." In the eighteenth and early nineteenth centuries the advancing European frontier from New England to the Gulf of Mexico was preceded into Indian territory by an *avant-garde* of semi-wild herds of hogs and cattle tended, now and again, by semi-wild herdsmen, white and black.

The first English settlers landed in Botany Bay, Australia, in January of 1788 with livestock, most of it from the Cape of Good Hope. The pigs and poultry thrived; the cattle did well enough; the sheep, the future source of the colony's good fortune, died fast. Within a few months two bulls and four cows strayed away. By 1804 the wild herds they founded numbered from three to five thousand head and were in possession of much of the best land between the settlements and the Blue Mountains. If they had ever found their way through the mountains to the grasslands beyond, the history of Australia in the first decades of the nineteenth century might have been one dominated by cattle rather than sheep. As it is, the colonial government wanted the land the wild bulls so ferociously defended, and considered the growing practice of convicts running away to live off the herds as a threat to the whole colony; so the adult cattle were shot and salted down and the calves captured and tamed. The English settlers imported woolly sheep from Europe and sought out the interior pastures for them. The animals multiplied rapidly, and when Darwin made his visit to New South Wales in 1836, there were about a million sheep there for him to see.

The arrival of Old World livestock probably affected New Zealand more radically than any other of the Lands of the Demographic Takeover. Cattle, horses, goats, pigs and—in this land of few or no large predators—even the usually timid sheep went wild. In New Zealand herds of feral farm animals were practicing the ways of their remote ancestors as late as the 1940s and no doubt still run free. Most of the sheep, though, stayed under human control, and within a decade of Great Britain's annexation of New Zealand in 1840, her new acquisition was home to a quarter million sheep. In 1974 New Zealand had over fifty-five million sheep, about twenty times more sheep than people.

In the Lands of the Demographic Takeover the European pioneers were accompanied and often preceded by their domesticated animals, walking sources of food, leather, fiber, power, and wealth, and these animals often adapted more rapidly to the new surroundings and reproduced much more rapidly than their masters. To a certain extent, the success of Europeans as colonists was automatic as soon as they put their tough, fast, fertile, and intelligent animals ashore. The latter were sources of capital that sought out their own sustenance, improvised their own protection against the weather, fought their own battles against predators and, if their masters were smart enough to allow calves, colts, and lambs to accumulate, could and often did show the world the amazing possibilities of compound interest.

The honey bee is the one insect of world-wide importance which human beings have domesticated, if we may use the word in a broad sense. Many species of bees and other insects produce honey, but the one which does so in greatest quantity and which is easiest to control is a native of the Mediterranean area and the Middle East, the honey bee *(Apis mellifera)*. The European has probably taken this sweet and short-tempered servant to every colony he ever established, from Arctic to Antarctic Circle, and the honey bee has always been one of the first immigrants to set off on its own. Sometimes the advance of the bee frontier could be very rapid: the first hive in Tasmania swarmed sixteen times in the summer of 1832.

Thomas Jefferson tells us that the Indians of North America called the honey bees "English flies," and St. John de Crevecoeur, his contemporary, wrote that "the Indians look upon them with an evil eye, and consider their progress into the interior of the continent as an omen of the white man's approach: thus, as they discover the bees, the news of the event, passing from mouth to mouth, spreads sadness and consternation on all sides."

Domesticated creatures that traveled from the Lands of the Demographic Takeover to Europe are few. Australian aborigines and New Zealand Maoris had a few tame dogs, unimpressive by Old World standards and unwanted by the whites. Europe happily accepted the American Indians' turkeys and guinea pigs, but had no need for their dogs, llamas, and alpacas. Again the explanation is simple: Europeans, who controlled the passage of large animals across the oceans, had no need to reverse the process.

It is interesting and perhaps significant, though, that the exchange was just as one sided for varmints, the small mammals whose migrations Europeans often tried to stop. None of the American or Australian or New Zealand equivalents of rats have become established in Europe, but Old World varmints, especially rats, have colonized right alongside the Europeans in the Temperate Zones. Rats of assorted sizes, some of them almost surely European immigrants, were tormenting Spanish Americans by at least the end of the sixteenth century. European rats established a beachhead in Jamestown, Virginia, as early as 1609, when they almost starved out the colonists by eating their food stores. In Buenos Aires the increase in rats kept pace with that of cattle, according to an early nineteenth-century witness. European rats proved as aggressive as the Europeans in New Zealand, where they completely replaced the local rats in the North Islands as early as the 1840s. Those poor creatures are probably completely extinct today or exist only in tiny relict populations.

The European rabbits are not usually thought of as varmints, but where there are neither diseases nor predators to hold down their numbers they can become the worst of pests. In 1859 a few members of the species *Orytolagus cuniculus* (the scientific name for the protagonists of all the Peter Rabbits of literature) were released in southeast Australia. Despite massive efforts to stop them, they reproduced—true to their reputation—and spread rapidly all the way across Australia's southern half to the Indian Ocean. In 1950 the rabbit population of Australia was estimated at 500 million, and they were outcompeting the nation's most important domesticated animals, sheep, for the grasses and herbs. They have been brought under control, but only by means of artificially fomenting an epidemic of myxomatosis, a lethal American rabbit disease. The story of rabbits and myxomatosis in New Zealand is similar.

Europe, in return for her varmints, has received muskrats and gray squirrels and little else from America, and nothing at all of significance from Australia or New Zealand, and we might well wonder if muskrats and squirrels really qualify as varmints. As with other classes of organisms, the exchange has been a one-way street.

None of Europe's emigrants were as immediately and colossally successful as its pathogens, the microorganisms that make human beings ill, cripple them, and kill them. Whenever and wherever Europeans crossed the oceans and settled, the pathogens they carried created prodigious epidemics of smallpox, measles, tuberculosis, influenza, and a number of other diseases. It was this factor, more than any other, that Darwin had in mind as he wrote of the Europeans' deadly tread.

The pathogens transmitted by the Europeans, unlike the Europeans themselves or most of their domesticated animals, did at least as well in the tropics as in the temperate Lands of the Demographic Takeover. Epidemics devastated Mexico, Peru, Brazil, Hawaii, and Tahiti soon after the Europeans made the first contact with aboriginal populations. Some of these populations were able to escape demographic defeat because their initial numbers were so large that a small fraction was still sufficient to maintain occupation of, if not title to, the land, and also because the mass of Europeans were never attracted to the tropical lands, not even if they were partially vacated. In the Lands of the Demographic Takeover the aboriginal populations were too sparse to rebound from the onslaught of disease or were inundated by European immigrants before they could recover.

The First Strike Force of the white immigrants to the Lands of the Demographic Takeover were epidemics. A few examples from scores of possible examples follow. Smallpox first arrived in the Rio de la Plata region in 1558 or 1560 and killed, according to one chronicler possibly more interested in effect than accuracy, "more than a hundred thousand Indians" of the heavy Averine population there. An epidemic of plague or typhus decimated the Indians of the New England coast immediately before the founding of Plymouth. Smallpox or something similar struck the aborigines of Australia's Botany Bay in 1789, killed half, and rolled on into the interior. Some unidentified disease or diseases spread through the Maori tribes of the North Island of New Zealand in the 1790s, killing so many in a number of villages that the survivors were not able to bury the dead. After a series of such lethal and rapidly moving epidemics, then came the slow, unspectacular but thorough cripplers and killers like venereal disease and tuberculosis. In conjunction with the large numbers of white settlers these diseases were enough to smother aboriginal chances of recovery. First the blitzkrieg, then the mopping up.

The greatest of the killers in these lands was probably smallpox. The exception is New Zealand, the last of these lands to attract permanent European settlers. They came to New Zealand after the spread of vaccination in Europe, and so were poor carriers. As of the 1850s smallpox still had not come ashore, and by that time two-thirds of the Maori had been vaccinated. The tardy arrival of smallpox in these islands may have much to do with the fact that the Maori today comprise a larger percentage (nine percent) of their country's population than that of any other aboriginal people in any European colony or former European colony in either Temperate Zone, save only South Africa.

American Indians bore the full brunt of smallpox, and its mark is on their history and folklore. The Kiowa of the southern plains of the United States have a legend in which a Kiowa man meets Smallpox on the plain, riding a horse. The man asks, "Where do you come from and what do you do and why are you here?" Smallpox answers, "I am one with the white men—they are my people as the Kiowas are yours. Sometimes I travel ahead of them and sometimes behind. But I am always their companion and you will find me in their camps and their houses." "What can you do?" the Kiowa asks. "I bring death," Smallpox replies. "My breath causes children to wither like young plants in spring snow. I bring destruction. No matter how beautiful a woman is, once she has looked at me she becomes as ugly as death. And to men I bring not death alone, but the destruction of their children and the blighting of their wives. The strongest of warriors go down before me. No people who have looked on me will ever be the same."

In return for the barrage of diseases that Europeans directed overseas, they received little in return. Australia and New Zealand provided no new strains of pathogens to Europe—or none that attracted attention. And of America's native diseases none had any real influence on the Old World—with the likely exception of venereal syphilis, which almost certainly existed in the New World before 1492 and probably did not occur in its present form in the Old World.

Weeds are rarely history makers, for they are not as spectacular in their effects as pathogens. But they, too, influence our lives and migrate over the world despite human wishes. As such, like varmints and germs, they are better indicators of certain realities than human beings or domesticated animals.

The term *weed* in modern botanical usage refers to any type of plant which—because of especially large numbers of seeds produced per plant, or especially effective means of distributing those seeds, or especially tough roots and rhizomes from which new plants can grow, or especially tough seeds that survive the alimentary canals of animals to be planted with their droppings—spread rapidly and outcompete others on disturbed, bare soil. Weeds are plants that tempt the botanist to use such anthropomorphic words as aggressive and opportunistic.

Many of the most successful weeds in the well-watered regions of the Lands of the Demographic Takeover are of European or Eurasian origin. French and Dutch and English farmers brought with them to North America their worst enemies, weeds, "to exhaust the land, hinder and damnify the Crop." By the last third of the seventeenth century at least twenty different types were widespread enough in New England to attract the attention of the English visitor, John Josselyn, who identified couch grass, dandelion, nettles, mallowes, knot grass, shepherd's purse, sow thistle, and clot burr and others. One of the most aggressive was plantain, which the Indians called "English-Man's Foot."

European weeds rolled west with the pioneers, in some cases spreading almost explosively. As of 1823 corn chamomile and maywood had spread up to but not across the Muskingum River in Ohio. Eight years later they were over the river. The most prodigiously imperialistic of the weeds in the eastern half of the United States and Canada were probably Kentucky bluegrass and white clover. They spread so fast after the entrance of Europeans into a given area that there is some suspicion that they may have been present in pre-Columbian America, although the earliest European accounts do not mention them. Probably brought to the Appalachian area by the French, these two kinds of weeds preceded the English settlers there and kept up with the movement westward until reaching the plains across the Mississippi.

Old World plants set up business on their own on the Pacific coast of North America just as soon as the Spaniards and Russians did. The climate of coastal southern California is much the same as that of the Mediterranean, and the Spaniards who came to California in the eighteenth century brought their own Mediterranean weeds with them via Mexico: wild oats, fennel, wild radishes. These plants, plus those brought in later by the Forty-niners, muscled their way to dominance in the coastal grasslands. These immigrant weeds followed Old World horses, cattle, and sheep into California's interior prairies and took over there as well.

They did not push so swiftly into the coastal Northwest because the Spanish, their reluctant patrons, were slow to do so, and because those shores are cool and damp. Most of the present-day weeds in that region had to come with the Russians or Anglo-Americans from similar areas on other coasts. The Northwest has a semi-arid interior, however, into which some European plants like redstem filaree spread quite early, presumably from the prairies of California.

The region of Argentina and Uruguay was almost as radically altered in its flora as in its fauna by the coming of the Europeans. The ancient Indian practice, taken up immediately by the whites, of burning off the old grass of the pampa every year, as well as the trampling and cropping to the ground of indigenous grasses and fortes by the thousands of imported quadrupeds who also changed the nature of the soil with their droppings, opened the whole countryside to European plants. In the 1780s Félix de Azara observed that the pampa, already radically altered, was changing as he watched. European weeds sprang up around every cabin, grew up along roads, and pressed into the open steppe. Today only a quarter of the plants growing wild in the pampa are native, and in the well-watered eastern portions, the "natural" ground cover consists almost entirely of Old World grasses and clovers.

The invaders were not, of course, always desirable. When Darwin visited Uruguay in 1832, he found large expanses, perhaps as much as hundreds of square miles, monopolized by the immigrant wild artichoke and transformed into a prickly wilderness fit neither for man nor his animals.

The onslaught of foreign and specifically European plants on Australia began abruptly in 1778 because the first expedition that sailed from Britain to Botany Bay carried some livestock and considerable quantities of seed. By May of 1803 over two hundred foreign plants, most of them European, had been purposely introduced and planted in New South Wales, undoubtedly along with a number of weeds. Even today so-called clean seed characteristically contains some weed seeds, and this was much more so two hundred years ago. By and large, Australia's north has been too tropical and her interior too hot and dry for European weeds and grasses, but much of her southern coasts and Tasmania have been hospitable indeed to Europe's willful flora.

Thus, many—often a majority—of the most aggressive plants in the temperate humid regions of North America, South America, Australia, and New Zealand are of European origin. It may be true that in every broad expanse of the world today where there are dense populations, with whites in the majority, there are also dense populations of European weeds. Thirty-five of eighty-nine weeds listed in 1953 as common in the state of New York are European. Approximately sixty percent of Canada's worst weeds are introductions from Europe. Most of New Zealand's weeds are from the same source, as are many, perhaps most, of the weeds of Australia's well-watered coasts. Most of the European plants that Josselyn listed as naturalized in New England in the seventeenth century are growing wild today in Argentina and Uruguay, and are among the most widespread and troublesome of all weeds in those countries.

In return for this largesse of pestiferous plants, the Lands of the Demographic Takeover have provided Europe with only a few equivalents. The Canadian water weed jammed Britain's nineteenth century waterways, and North Americas horseweed and burnweed have spread in Europe's empty lots, and South Americas flowered galinsoga has thrived in her gardens. But the migratory flow of a whole group of organisms between Europe and the Lands of the Demographic Takeover has been almost entirely in one direction. Englishman's foot still marches in seven league jackboots across every European colony of settlement, but very few American or Australian or New Zealand invaders stride the waste lands and unkempt backyards of Europe.

European and Old World human beings, domesticated animals, varmints, pathogens, and weeds all accomplished demographic takeovers of their own in the temperate, well-watered regions of North and South America, Australia, and New Zealand. They crossed oceans and Europeanized vast territories, often in informal cooperation with each other—the farmer and his animals destroying native plant cover, making way for imported grasses and fortes, many of which proved more nourishing to domesticated animals than the native equivalents; Old World pathogens, sometimes carried by Old World varmints, wiping out vast numbers of aborigines, opening the way for the advance of the

European frontier, exposing more and more native peoples to more and more pathogens. The classic example of symbiosis between European colonists, their animals, and plants comes from New Zealand. Red clover, a good forage for sheep, could not seed itself and did not spread without being annually sown until the Europeans imported the bumblebee. Then the plant and insect spread widely, the first providing the second with food, the second carrying pollen from blossom to blossom for the first, and the sheep eating the clover and compensating the human beings for their effort with mutton and wool.

There have been few such stories of the success in Europe of organisms from the Lands of the Demographic Takeover, despite the obvious fact that for every ship that went from Europe to those lands, another traveled in the opposite direction.

The demographic triumph of Europeans in the temperate colonies is one part of a biological and ecological takeover which could not have been accomplished by human beings alone, gunpowder notwithstanding. We must at least try to analyze the impact and success of all the immigrant organisms together—the European portmanteau of often mutually supportive plants, animals, and microlife which in its entirety can be accurately described as aggressive and opportunistic, an ecosystem simplified by ocean crossings and honed by thousands of years of competition in the unique environment created by the Old World Neolithic Revolution.

The human invaders and their descendants have consulted their egos, rather than ecologists, for explanations of their triumphs. But the human victims, the aborigines of the Lands of the Demographic Takeover, knew better, knew they were only one of many species being displaced and replaced; knew they were victims of something more irresistible and awesome than the spread of capitalism or Christianity. One Maori, at the nadir of the history of his race, knew these things when he said, "As the clover killed off the fern, and the European dog the Maori dog—as the Maori rat was destroyed by the Pakeha [European] rat—so our people, also, will be gradually supplanted and exterminated by the Europeans." The future was not quite so grim as he prophesied, but we must admire his grasp of the complexity and magnitude of the threat looming over his people and over the ecosystem of which they were part.

Introduction to
Michael Pollan
The Idea of a Garden

by Nancy Langston

In his lively essay "The Idea of a Garden," the journalist Michael Pollan (1992) shows the ways that changing ideas about succession and the balance of nature can affect how we treat the earth.

He starts with a local controversy: a hurricane has devastated a wonderful stand of old white pine in his home town, and the citizens have to decide what to do with the mess. Should they log and sell the fallen timber before insects destroy its financial value? Should they do something to prevent fires burning in the piles of dead wood? Should they clear out the mess and replant with young trees? Or should they just stand back and let nature take its course? Pollan shows how all the sides in this debate reflect a long, complicated history of ideas about the balance of nature and the human place in that balance.

Most lay people in the town feel that if people only left nature alone, it would tend toward a healthy and abiding state of equilibrium. If humans could avoid disturbing it, the forest would return to a balanced, natural state. But Pollan shows the ways that this view of nature largely ignores the role of history and chance. His essay attempts to derive an environmental ethic out of insights offered by disturbance theory and nonequilibrial ecology. His ethic is one that sees a role for human history in nature. Pollan finally argues that the idea of a garden may help us far more in our relationship with nature than the idea of nature's balance. While the details of his ethic are debatable, this entertaining and provocative essay illuminates the ways that scientific and cultural ideas intertwine in modern environmental controversies.

The Idea of a Garden

Michael Pollan

The biggest news to come out of my town in many years was the tornado, or tornadoes, that careened through here on July 10, 1989, a Monday. Shooting down the Housatonic River Valley from the Berkshires, it veered east over Coltsfoot Mountain and then, after smudging the sky a weird gray green, proceeded to pinball madly from hillside to hillside for about fifteen minutes before wheeling back up into the sky. This was part of the same storm that ripped open the bark of my ash tree. But the damage was much, much worse on the other side of town. Like a gigantic, skidding pencil eraser, the twister neatly erased whole patches of woods and roughly smeared many other ones, where it wiped out just the tops of the trees. Overnight, large parts of town were rendered unrecognizable.

One place where the eraser came down squarely was in the Cathedral Pines, a famous forest of old-growth white pine trees close to the center of town. A kind of local shrine, this forty-two-acre forest was one of the oldest stands of white pine in New England, the trees untouched since about 1800. To see it was to have some idea how the New World forest must have looked to the first settlers, and in 1985 the federal government designated it a "national natural landmark." To enter Cathedral Pines on a hot summer day was like stepping out of the sun into a dim cathedral, the sunlight cooled and sweetened by the trillions of pine needles as it worked its way down to soft, sprung ground that had been unacquainted with blue sky for the better part of two centuries. The storm came through at about five in the evening, and it took only a few minutes of wind before pines more than one hundred fifty feet tall and as wide around as missiles lay jackstrawed on the ground like a fistful of pencils dropped from a great height. The wind was so thunderous that people in houses at the forest's edge did not know trees had fallen until they ventured outside after the storm had passed. The following morning, the sky now clear, was the first in more than a century to bring sunlight crashing down onto this particular patch of earth.

"It is a terrible mess," the first selectman told the newspapers; "a tragedy," said another Cornwall resident, voicing the deep sense of loss shared by many in town. But in the days that followed, the selectman and the rest of us learned that our responses, though understandable, were shortsighted, unscientific, and, worst of all, anthropocentric. "It may be a calamity to us," a state environmental official told a reporter from the *Hartford Courant,* but "to biology it is not a travesty. It is just a natural occurrence." The Nature Conservancy, which owns Cathedral Pines, issued a press release explaining that "Monday's storm was just another link in the continuous chain of events that is responsible for shaping and changing this forest."

It wasn't long before the rub of these two perspectives set off a controversy heated enough to find its way into the pages of *The New York Times*. The Nature Conservancy, in keeping with its mandate to maintain its lands in a "state of nature," indicated that it would leave Cathedral Pines alone, allowing the forest to take its "natural course," whatever that might be. To town officials and neighbors of the forest this was completely unacceptable. The downed trees, besides constituting an eyesore right at the edge of town, also posed a fire hazard. A few summers of drought, and the timber might go up in a blaze that would threaten several nearby homes and possibly even the town itself. Many people in Cornwall wanted Cathedral Pines cleared and replanted, so that at least the next generation might live to see some semblance of the old forest. A few others had the poor taste to point out the waste of more than a million board-feet of valuable timber, stupendous lengths of unblemished, knot-free pine.

The newspapers depicted it as a classic environmental battle, pitting the interests of man against nature, and in a way it was that. On one side were the environmental purists, who felt that any intervention by man in the disposition of this forest would be unnatural. "If you're going to clean it up," one purist declared in the local press, "you might as well put up condos." On the other side stood the putative interests of man, variously expressed in the vocabulary of safety (the fire hazard), economics (the wasted lumber), and aesthetics (the "terrible mess").

Everybody enjoys a good local fight, but I have to say I soon found the whole thing depressing. This was indeed a classic environmental battle, in that it seemed to exemplify just about everything that's wrong with the way we approach problems of this kind these days. Both sides began to caricature each other's positions: the selectman's "terrible mess" line earned him ridicule for his anthropocentrism in the letters page of *The New York Times;* he in turn charged a Yale scientist who argued for non-interference with "living in an ivory tower."

But as far apart as the two sides seemed to stand, they actually shared more common ground than they realized. Both started from the premise that man and nature were irreconcilably opposed, and that the victory of one necessarily entailed the loss of the other. Both sides, in other words, accepted the premises of what we might call the "wilderness ethic," which is based on the assumption that the relationship of man and nature resembles a zero-sum game. This idea, widely held and yet largely unexamined, has set the terms of most environmental battles in this country since the very first important one: the fight over the building of the Hetch Hetchy Dam in 1907, which pitted John Muir against Gifford Pinchot, whom Muir used to call a "temple destroyer." Watching my little local debate unfold over the course of the summer, and grow progressively more shrill and sterile, I began to wonder if perhaps the wilderness ethic itself, for all that it has accomplished in this country over the past century, had now become part of the problem. I also began to wonder if it might be possible to formulate a different ethic to guide us in our dealings with nature, at least in some places some of the time, an ethic that would be based not on the idea of wilderness but on the idea of wilderness but on the idea of a garden.

Foresters who have examined sections of fallen trees in Cathedral Pines think that the oldest trees in the forest date from 1780 or so, which suggests that the site was probably logged by the first generation of settlers. The Cathedral Pines are not, then, "virgin growth." The rings of felled trees also reveal a significant growth spurt in 1840, which probably indicates that loggers removed hardwood trees in that year, leaving the pines to grow without competition. In 1883, the Calhouns, an old Cornwall family whose property borders the forest, bought the land to protect the trees from the threat of logging; in 1967 they deeded it to the Nature Conservancy, stipulating that it be maintained in its natural state. Since then, and up until the tornado made its paths impassable, the forest has been a popular place for hiking and Sunday outings. Over the years, more than a few Cornwall residents have come to the forest to be married.

Cathedral Pines is not in any meaningful sense a wilderness. The natural history of the forest intersects at many points with the social history of Cornwall. It is the product of early logging practices, which clear-cut the land once and then cut it again, this time selectively, a hundred years later. Other human factors almost certainly played a part in the forest's history; we can safely assume that any fires in the area were extinguished before they reached Cathedral Pines. (Though we don't ordinarily think of it in these terms, fire suppression is one of the more significant effects that the European has had on the American landscape.) Cathedral Pines, then, is in some part a manmade landscape, and it could reasonably be argued that to exclude man at this point in its history would constitute a break with its past.

But both parties to the dispute chose to disregard the actual history of Cathedral Pines, and instead to think of the forest as a wilderness in the commonly accepted sense of that term: a pristine place untouched by white men. Since the romantics, we've prized such places as refuges from the messiness of the human estate, vantages from which we might transcend the vagaries of that world and fix on what Thoreau called "higher laws." Certainly an afternoon in Cathedral Pines fostered such feelings, and its very name reflects the pantheism that lies behind them. Long before science coined the term ecosystem to describe it, we've had the sense that nature undisturbed displays a miraculous order and balance, something the human world can only dream about. When man leaves it alone, nature will tend toward a healthy and abiding state of equilibrium. Wilderness, the purest expression of this natural law, stands out beyond history.

These are powerful and in many ways wonderful ideas. The notion of wilderness is a kind of taboo in our culture, in many cases acting as a check on our inclination to dominate and spoil nature. It has inspired us to set aside such spectacular places as Yellowstone and Yosemite. But wilderness is also a profoundly alienating idea, for it drives a large wedge between man and nature. Set against the foil of nature's timeless cycles, human history appears linear and unpredictable, buffeted by time and chance as it drives blindly into the future. Natural history, by comparison, obeys fixed and legible laws, ones that make the "laws" of human history seem puny, second-rate things scarcely deserving of the label. We have little idea what the future holds for the town of Cornwall, but surely nature has a plan for Cathedral Pines; leave the forest alone and that plan—which science knows by the name of "forest succession"—will unfold inexorably, in strict accordance with natural law. A new climax forest will emerge as nature works to restore her equilibrium—or at least that's the idea.

The notion that nature has a plan for Cathedral Pines is a comforting one, and certainly it supplies a powerful argument for leaving the forest alone. Naturally I was curious to know what that plan was: what does nature do with an old pine forest blown down by a tornado? I consulted a few field guides and standard works of forest ecology hoping to find out.

According to the classical theory of forest succession, set out in the nineteenth century by, among others, Henry Thoreau, a pine forest that has been abruptly destroyed will usually be succeeded by hardwoods, typically oak. This is because squirrels commonly bury acorns in pine forests and neglect to retrieve many of them. The oaks sprout and, because shade doesn't greatly hinder young oaks, the seedlings frequently manage to survive beneath the dark canopy of a mature pine forest. Pine seedlings, on the other hand, require more sunlight than a mature pine forest admits; they won't sprout in shade. So by the time the pine forest comes down, the oak saplings will have had a head start in the race to dominate the new forest. Before any new pines have had a chance to sprout, the oaks will be well on their way to cornering the sunlight and inheriting the forest.

This is what I read, anyway, and I decided to ask around to confirm that Cathedral Pines was expected to behave as predicted. I spoke to a forest ecologist and an expert on the staff of the Nature Conservancy. They told me that the classical theory of pine-forest succession probably does describe the underlying tendency at work in Cathedral Pines. But it turns out that a lot can go, if not "wrong" exactly, then at least differently. For what if there are no oaks nearby? Squirrels will travel only so far in search of a hiding place for their acorns. Instead of oaks, there may be hickory nuts stashed all over Cathedral Pines. And then there's the composition of species planted by the forest's human neighbors to consider; one of these, possibly some exotic (that is, non-native), could conceivably race in and take over. "It all depends," is the refrain I kept hearing as I tried to pin down nature's intentions for Cathedral Pines. Forest succession, it seems, is only a theory, a metaphor of our making, and almost as often as not nature makes a fool of it. The number of factors that will go into the determination of Cathedral Pines' future is almost beyond comprehension. Consider just this small sample of the things that could happen to alter irrevocably its future course:

A lightning storm—or a cigarette butt flicked from a passing car—ignites a fire next summer. Say it's a severe fire, hot enough to damage the fertility of the soil, thereby delaying recovery of the forest for decades. Or say it rains that night, making the fire a mild one, just hot enough to kill the oak saplings and allow the relatively fire-resistant pine seedlings to flourish without competition. A new pine forest after all? Perhaps. But what if the population of deer happens to soar the following year? Their browsing would wipe out the young pines and create an opening for spruce, the taste of which deer happen not to like.

Or say there is no fire. Without one, it could take hundreds of years for the downed pine trees to rot and return their nutrients to the soil. Trees grow poorly in the exhausted soil, but the seeds of brambles, which can lie dormant in the ground for fifty years, sprout and proliferate: we end up with a hundred years of brush. Or perhaps a breeze in, say, the summer of 1997 carries in seedpods from the Norway maple standing in a nearby front yard at the precise moment when conditions for their germination are perfect. Norway maple, you'll recall, is a European species, introduced here early in the nineteenth century and widely planted as a street tree. Should this exotic species happen to prevail, Cathedral Pines becomes one very odd-looking and awkwardly named wilderness area.

But the outcome could be much worse. Let's say the rains next spring are unusually heavy, washing all the topsoil away (the forest stood on a steep hillside). Only exotic weed species can survive now, and one of these happens to the Japanese honeysuckle, a nineteenth-century import of such rampant habit that it can choke out the growth of all trees indefinitely. We end up with no forest at all.

Nobody, in other words, can say what will happen in Cathedral Pines. And the reason is not that forest ecology is a young or imperfect science, but because *nature herself doesn't know what's going to happen here.* Nature has no grand design for this place. An incomprehensibly various and complex set of circumstances—some of human origin, but many not—will determine the future of Cathedral Pines. And whatever that future turns out to be, it would not unfold in precisely the same way twice. Nature may possess certain inherent tendencies, ones that theories such as forest succession can describe, but chance events can divert her course into an almost infinite number of different channels.

It's hard to square this fact with our strong sense that some kind of quasi-divine order inheres in nature's workings. But science lately has been finding that contingency plays nearly as big a role in natural history as it does in human history. Forest ecologists today will acknowledge that succession theories are little more than comforting narratives we impose on a surprisingly unpredictable process; even so-called climax forests are sometimes superseded. (In many places in the northern United States today, mature stands of oak are inexplicably being invaded by maples— skunks at the climax garden party.) Many ecologists will now freely admit that even the concept of an ecosystem is only a metaphor, a human construct imposed upon a much more variable and precarious reality. An ecosystem may be a useful concept, but no ecologist has ever succeeded in isolating one in nature. Nor is the process of evolution as logical or inexorable as we have thought. The current thinking in paleontology holds that the evolution of any given species, our own included, is not the necessary product of any natural laws, but rather the outcome of a concatenation of chance events—of "just history" in the words of Stephen Jay Gould. Add or remove any single happenstance—the asteroid fails to wipe out the dinosaurs; a little chordate worm called *Pikaia* succumbs in the Burgess extinction—and humankind never arrives. Across several disciplines, in fact, scientists are coming to the conclusion that more "just history" is at work in nature than had previously been thought. Yet our metaphors still picture nature as logical, stable, and a historical—more like a watch than, say, an organism or a stock exchange, to name two metaphors that may well be more apt. Chance and contingency, it turns out, are everywhere in nature; she has no fixed goals, no unalterable pathways into the future, no inflexible rules that she herself can't bend or break at will. She is more like us (or we are more like her) than we ever imagined.

To learn this, for me at least, changes everything. I take it to be profoundly good news, though I can easily imagine how it might trouble some people. For many of us, nature is a last bastion of certainty; wilderness, as something beyond the reach of history and accident, is one of the last in our fast-dwindling supply of metaphysical absolutes, those comforting transcendental values by which we have traditionally taken our measure and set our sights. To take away predictable, divinely ordered nature is to pull up one of our last remaining anchors. We are liable to float away on the trackless sea of our own subjectivity.

But the discovery that time and chance hold sway even in nature can also be liberating. Because contingency is an invitation to participate in history. Human choice is unnatural only if nature is deterministic; human change is unnatural only if she is changeless in our absence. If the future of Cathedral Pines is up for grabs, if its history will always be the product of myriad chance events, then why shouldn't we also claim our place among all those deciding factors? For aren't we also one of nature's contingencies? And if our cigarette butts and Norway maples and acid rain are going to shape the future of this place, then why not also our hopes and desires?

Nature will condone an almost infinite number of possible futures for Cathedral Pines. Some would be better than others. True, what we would regard as "better" is probably not what the beetles would prefer. But nature herself has no strong preference. That doesn't mean she will countenance any outcome; she's already ruled out many possible futures (tropical rain forest, desert, etc.) and, all things being equal, she'd probably lean toward the oak. But all things aren't equal (her idea) and she is evidently happy to let the free play of numerous big and little contingencies settle the matter. To exclude from these human desire would be, at least in this place at this time, arbitrary, perverse and, yes, unnatural.

Establishing that we should have a vote in the disposition of Cathedral Pines is much easier than figuring out how we should cast it. The discovery of contingency in nature would seem to fling open a Pandora's box. For if there's nothing fixed or inevitable about nature's course, what's to stop us from concluding that anything goes? It's a whole lot easier to assume that nature left to her own devices knows what's best for a place, to let ourselves be guided by the wilderness ethic.

And maybe that's what we should do. Just because the wilderness ethic is based on a picture of nature that is probably more mythical than real doesn't necessarily mean we have to discard it. In the same way that the Declaration of Independence begins with the useful fiction that "all men are created equal," we could simply stipulate that Cathedral Pines is wilderness, and proceed on that assumption. The test of the wilderness ethic is not how truthful it is, but how useful it is in doing what we want to do—in protecting and improving the environment.

So how good a guide is the wilderness ethic in this particular case? Certainly treating Cathedral Pines as a wilderness will keep us from building condos there. When you don't trust yourself to do the right thing, it helps to have an authority as wise and experienced as nature to decide matters for you. But what if nature decides on Japanese honeysuckle—three hundred years of wall-to-wall brush? We would then have a forest not only that we don't like, but that isn't even a wilderness, since it was man who brought Japanese honeysuckle to Cornwall. At this point in history, after humans have left their stamp on virtually every corner of the Earth, doing nothing is frequently a poor recipe for wilderness. In many cases it leads to a gradually deteriorating environment (as seems to be happening in Yellowstone), or to an environment shaped in large part by the acts and mistakes of previous human inhabitants.

If it's real wilderness we want in Cathedral Pines, and not merely an imagined innocence, we will have to restore it. This is the paradox faced by the Nature Conservancy and most other advocates of wilderness: at this point in history, creating a landscape that bears no marks of human intervention will require a certain amount of human intervention. At a minimum it would entail weeding the exotic species from Cathedral Pines, and that is something the Nature Conservancy's strict adherence to the wilderness ethic will not permit.

But what if the Conservancy *was* willing to intervene just enough to erase any evidence of man's presence? It would soon run up against some difficult questions for which its ethic leaves it ill prepared. For what is the "real" state of nature in Cathedral Pines? Is it the way the forest looked before the settlers arrived? We could restore that condition by removing all traces of European man. Yet isn't that a rather Eurocentric (if not racist) notion of wilderness? We now know that the Indians were not the ecological eunuchs we once thought. They too left their mark on the land: fires set by Indians determined the composition of the New England forests and probably created that "wilderness" we call the Great Plains. For true untouched wilderness we have to go a lot further back than 1640 or 1492. And if we want to restore the landscape to its pre-Indian condition, then we're going to need a lot of heavy ice-making equipment (not to mention a few woolly mammoths) to make it look right.

But even that would be arbitrary. In fact there is no single moment in time that we can point to and say, *this* is the state of nature in Cathedral Pines. Just since the last ice age alone, that "state of nature" has undergone a thorough revolution every thousand years or so, as tree species forced south by the glaciers migrated back north (a process that is still going on), as the Indians arrived and set their fires, as the large mammals disappeared, as the climate fluctuated—as all the usual historical contingencies came on and off the stage. For several thousand years after the ice age, this part of Connecticut was a treeless tundra; is *that* the true state of nature in Cathedral Pines? The inescapable fact is that, if we want wilderness here, we will have to choose *which* wilderness we want—an idea that is inimical to the wilderness ethic. For wasn't the attraction of wilderness precisely the fact that it relieved us of having to make choices—wasn't nature going to decide, letting us off the hook of history and anthropocentrism?

No such luck, it seems. "Wilderness" is not nearly as straightforward or dependable a guide as we'd like to believe. If we do nothing, we may end up with an impoverished weed patch of our own (indirect) creation, which would hardly count as a victory for wilderness. And if we want to restore Cathedral Pines to some earlier condition, we're forced into making the kinds of inevitably anthropocentric choices and distinctions we turned to wilderness to escape. (Indeed, doing a decent job of wilderness restoration would take all the technology and scientific know-how humans can muster.) Either way, there appears to be no escape from history, not even in nature.

The reason that the wilderness ethic isn't very helpful in a place like Cathedral Pines is that it's an absolutist ethic: man or nature, it says, pick one. As soon as history or circumstance blurs that line, it gets us into trouble. There are times and places when man or nature is the right and necessary choice; back at Hetch Hetchy in 1907 that may well have been the case. But it seems to me that these days most of the environmental questions we face are more like the ambiguous ones posed by Cathedral Pines, and about these the wilderness ethic has less and less to say that is of much help.

The wilderness ethic doesn't tell us what to do when Yellowstone's ecosystem begins to deteriorate, as a result not of our interference but of our neglect. When a species threatens to overwhelm and ruin a habitat because history happened to kill off the predator that once kept its population in check, the ethic is mute. It is confounded, too, when the only hope for the survival of another species is the manipulation of its natural habitat by man. It has nothing to say in all those places where development is desirable or unavoidable except: Don't do it. When we're forced to choose between a hydroelectric power plant and a nuclear one, it refuses to help. That's because the wilderness ethic can't make distinctions between one kind of intervention in nature and another—between weeding Cathedral Pines and developing a theme park there. "You might as well put up condos" is its classic answer to any plan for human intervention in nature.

"All or nothing," says the wilderness ethic, and in fact we've ended up with a landscape in America that conforms to that injunction remarkably well. Thanks to exactly this kind of either/or thinking, Americans have done an admirable job of drawing lines around certain sacred areas (we did invent the wilderness area) and a terrible job of managing the rest of our land. The reason is not hard to find: the only environmental ethic we have has nothing useful to say about those areas outside the line. Once a landscape is no longer "virgin" it is typically written off as fallen, lost to nature, irredeemable. We hand it over to the jurisdiction of that other sacrosanct American ethic: laissez-faire economics. "You might as well put up condos." And so we do.

Indeed, the wilderness ethic and laissez-faire economics, antithetical as they might at first appear, are really mirror images of one another. Each proposes a quasi-divine force—Nature, the Market—that, left to its own devices, somehow knows what's best for a place. Nature and the market are both self-regulating, guided by an invisible hand. Worshippers of either share a deep, Puritan distrust of man, taking it on faith that human tinkering with the natural or economic order can only pervert it. Neither will acknowledge that their respective divinities can also err: that nature produces the AIDS virus as well as the rose, that the same markets that produce stupendous wealth can also crash. (Actually, worshippers of the market are a bit more realistic than worshippers of nature: they long ago stopped relying on the free market to supply us with such necessities as food and shelter. Though they don't like to talk about it much, they accept the need for society to "garden" the market.)

Essentially, we have divided our country in two, between the kingdom of wilderness, which rules about 8 percent of America's land, and the kingdom of the market, which rules the rest. Perhaps we should be grateful for secure borders. But what do those of us who care about nature do when we're on the market side, which is most of the time? How do we behave? What are our goals? We can't reasonably expect to change the borders, no matter how many power lines and dams Earth First! blows up. No, the wilderness ethic won't be of much help over here. Its politics are bound to be hopelessly romantic (consisting of impractical schemes to redraw the borders) or nihilistic. Faced with hard questions about how to confront global environmental problems such as the greenhouse effect or ozone depletion (problems that respect no borders), adherents of the wilderness ethic are apt to throw up their hands in despair and declare the "end of nature."

The only thing that's really in danger of ending is a romantic, pantheistic idea of nature that we invented in the first place, one whose passing might well turn out to be a blessing in disguise. Useful as it has been in helping us protect the sacred 8 percent, it nevertheless has failed to prevent us from doing a great deal of damage to the remaining 92 percent. This old idea may have taught us how to worship nature, but it didn't tell us how to live with her. It told us more than we needed to know about virginity and rape, and almost nothing about marriage. The metaphor of divine nature can admit only two roles for man: as worshipper (the naturalist's role) or temple destroyer (the developer's). But that drama is all played out now. The temple's been destroyed—if it ever was a temple. Nature is dead, if by nature we mean something that stands apart from man and messy history. And now that it is, perhaps we can begin to write some new parts for ourselves, ones that will show us how to start out from here, not from some imagined state of innocence, and let us get down to the work at hand.

Thoreau and Muir and their descendants went to the wilderness and resumed with the makings of America's first environmental ethic. Today it still stands, though somewhat strained and tattered. What if now, instead of to the wilderness, we were to look to the garden for the makings of a new ethic? One that would not necessarily supplant the earlier one, but might give us something useful to say in those cases when it is silent or unhelpful?

It will take better thinkers than me to flesh out what such an ethic might look like. But even my limited experience in the garden has persuaded me that the materials needed to construct it—the fresh metaphors about nature we need—may be found there. For the garden is a place with long experience of questions having to do with man in nature. Below are some provisional notes, based on my own experiences and the experiences of other gardeners I've met or read, on the kinds of answers the garden is apt to give.

1. An ethic based on the garden would give local answers. Unlike the wilderness idea, it would propose different solutions in different places and times. This strikes me as both a strength and a weakness. It's a weakness because a garden ethic will never speak as clearly or univocally as the wilderness ethic does. In a country as large and geographically various as this, it is probably inevitable that we will favor abstract landscape ideas—grids, lawns, monocultures, wildernesses—which can be applied across the board, even legislated nationally; such ideas have the power to simplify and unite. Yet isn't this power itself part of the problem? The health of a place generally suffers whenever we impose practices on it that are better suited to another place; a lawn in Virginia makes sense in a way that a lawn in Arizona does not.

So a garden ethic would begin with Alexander Pope's famous advice to landscape designers: "Consult the Genius of the Place in all." It's hard to imagine this slogan ever replacing Earth First!'s "No Compromise in Defense of Mother Earth" on American bumper stickers; nor should it, at least not everywhere. For Pope's dictum suggests that there are places whose "genius" will, if hearkened to, counsel "no compromise." Yet what is right for Yosemite is not necessarily right for Cathedral Pines.

2. The gardener starts out from here. By that I mean, he accepts contingency, his own and nature's. He doesn't spend a lot of time worrying about whether he has a god-given right to change nature. It's enough for him to know that, for some historical or biological rea-

son, humankind finds itself living in places (six of the seven continents) where it must substantially alter the environment in order to survive. If we had remained on African savannas things might be different. And if I lived in zone six I could probably grow good tomatoes without the use of plastic. The gardener learns to play the hand he's been dealt.

3. A garden ethic would be frankly anthropocentric. As I began to understand when I planted my roses and my maple tree, we know nature only through the screen of our metaphors; to see her plain is probably impossible. (And not necessarily desirable, as George Eliot once suggested: "If we could hear the squirrel's heartbeat, the sound of the grass growing, we should die of that roar." Without the editing of our perceptions, nature might prove unbearable.) Melville was describing all of nature when he described the whiteness of the whale, its "dumb blankness, full of meaning." Even wilderness, in both its satanic and benevolent incarnations, is an historical, man-made idea. Every one of our various metaphors for nature—"wilderness," "ecosystem," "Gaia," "resource," "wasteland"—is already a kind of garden, an indissoluble mixture of our culture and whatever it is that's really out there. "Garden" may sound like a hopelessly anthropocentric concept, but it's probably one we can't get past.

The gardener doesn't waste much time on metaphysics on figuring out what a "truer" perspective on nature (such as biocentrism or geocentrism) might look like. That's probably because he's noticed that most of the very long or wide perspectives we've recently been asked to adopt (including the one advanced by the Nature Conservancy in Cathedral Pines) are indifferent to our well-being and survival as a species. On this point he agrees with Wendell Berry—that "it is not natural to be disloyal to one's own kind."

4. That said, though, the gardener's conception of his self-interest is broad and enlightened. Anthropocentric as he may be, he recognizes that he is dependent for his health and survival on many other forms of life, so he is careful to take their interests into account in whatever he does. He is in fact a wilderness advocate of a certain kind. It is when he respects and nurtures the wilderness of his soil and his plants that his garden seems to flourish most. Wildness, he has found, resides not only out there, but right here: in his soil, in his plants, even in himself. Over-cultivation tends to repress this quality, which experience tells him is necessary to health in all three realms. But wildness is more a quality than a place, and though humans can't manufacture it, they can nourish and husband it. That is precisely what I'm doing when I make compost and return it to the soil; it is what we could be doing in Cathedral Pines (and not necessarily by leaving the place alone). The gardener cultivates wildness, but he does so carefully and respectfully, in full recognition of its mystery.

5. The gardener tends not to be romantic about nature. What could be more natural than the storms and droughts and plagues that ruin his garden? Cruelty, aggression, suffering—these too are nature's offspring (and not, as Rousseau tried to convince us, culture's). Nature is probably a poor place to look for values. She was indifferent to humankind's arrival, and she is indifferent to our survival.

It's only in the last century or so that we seem to have forgotten this. Our romance of nature is a comparatively recent idea, the product of the industrial age's novel conceit that nature could be conquered, and probably also of the fact that few of us work with nature directly anymore. But should current weather forecasts prove to be accurate (a rapid, permanent warming trend accompanied by severe storms), our current romance will look like a brief historical anomaly, a momentary lapse of judgment. Nature may once again turn dangerous and capricious and unconquerable. When this happens, we will quickly lose our crush on her.

Compared to the naturalist, the gardener never fell head over heels for nature. He's seen her ruin his plans too many times for that. The gardener has learned, perforce, to live with her ambiguities—that she is neither all good nor all bad, that she gives as well as takes away. Nature's apt to pull the rug out from under us at any time, to make a grim joke of our noblest intention. Perhaps this explains why garden writing tends to be comic, rather than lyrical or elegiac in the way that nature writing usually is: the gardener can never quite forget about the rug underfoot, the possibility of the offstage hook.

6. The gardener feels he has a legitimate quarrel with nature—with her weeds and storms and plagues, her rot and death. What's more, that quarrel has produced much of value, not only in his own time here (this garden, these fruits), but over the whole course of Western history. Civilization itself, as Freud and Frazer and many others have observed, is the product of that quarrel. But at the same time, the gardener appreciates that it would probably not be in his interest, or in nature's, to push his side of this argument too hard. Many points of contention that humankind thought it had won—DDT's victory over insects, say, or medicine's conquest of infectious disease—turned out to be Pyrrhic or illusory triumphs. Better to keep the quarrel going, the good gardener reasons, than to reach for outright victory, which is dangerous in the attempt and probably impossible anyway.

7. The gardener doesn't take it for granted that man's impact on nature will always be negative. Perhaps he's observed how his own garden has made this patch of land a better place, even by nature's own standards. His gardening has greatly increased the diversity and abundance of life in this place. Besides the many exotic species of plants he's introduced, the mammal, rodent, and insect populations have burgeoned, and his soil supports a much richer community of microbes than it did before.

Judged strictly by these standards, nature occasionally makes mistakes. The climax forest could certainly be considered one (a place where the number and variety of living things have declined to a crisis point) and evolution teems with others. At the same time, it should be acknowledged that man occasionally creates new ecosystems much richer than the ones they replaced, and not merely on the scale of a garden: think of the tall-grass prairies of the Midwest, England's hedgerow landscape, the countryside of the Ile de France, the patchwork of fields and forests in this part of New England. Most of us would be happy to call such places "nature," but that does not do them (or us) justice; they are really a kind of garden, a second nature.

The gardener doesn't feel that by virtue of the fact that he changes nature he is somehow outside of it. He looks around and sees that human hopes and desires are by now part and parcel of the landscape. The "environment" is not, and has never been, a neutral, fixed backdrop; it is in fact alive, changing all the time in response to innumerable contingencies, one of these being the presence within it of the gardener. And that presence is neither inherently good nor bad.

8. The gardener firmly believes it is possible to make distinctions between kinds and degrees of human intervention in nature. Isn't the difference between the Ile de France and Love Canal, or a pine forest and a condo development, proof enough that the choice isn't really between "all or nothing"? The gardener doesn't doubt that it is possible to discriminate; it is through experience in the garden that he develops this faculty.

Because of his experience, the gardener is not likely to conclude from the fact that some intervention in nature is unavoidable, therefore "anything goes." This is precisely where his skill and interest lie: in determining what does and does not go in a particular place. How much is too much? What suits this land? How can we get what we want here while nature goes about getting what she wants? He has no doubt that good answers to these questions can be found.

9. The good gardener commonly borrows his methods, if not his goals, from nature herself. For though nature doesn't seem to dictate in advance what we can do in a place—we are free, in the same way evolution is, to try something completely new—in the end she will let us know what does and does not work. She is above all a pragmatist, and so is the successful gardener.

By studying nature's ways and means, the gardener can find answers to the questions, What is apt to work? What avails here? This seems to hold true at many levels of specificity. In one particular patch of my vegetable garden—a low, damp area—I failed with every crop I planted until I stopped to consider what nature grew in a similar area nearby: briars. So I planted raspberries, which are of course a cultivated kind of briar, and they have flourished. A trivial case, but it shows how attentiveness to nature can help us to attune our desires with her ways.

The imitation of nature is of course the principle underlying organic gardening. Organic gardeners have learned to mimic nature's own methods of building fertility in the soil, controlling insect populations and disease, recycling nutrients. But the practices we call "organic" are not themselves "natural," any more than the bird call of a hunter is natural. They are more like man-made analogues of natural processes. But they seem to work. And they at least suggest a way to approach other problems—from a town's decision on what to do with a blown-down pine forest, to society's choice among novel new technologies. In each case, there will be some alternatives that align our needs and desires with nature's ways more closely than others.

It does seem that we do best in nature when we imitate her—when we learn to think like running water, or a carrot, an aphid, a pine forest, or a compost pile. That's probably because nature, after almost four billion years of trial-and-error experience, has wide knowledge of what works in life. Surely we're better off learning how to draw on her experience than trying to repeat it, if only because we don't have that kind of time.

10. If nature is one necessary source of instruction for a garden ethic, culture is the other. Civilization may be part of our problem with respect to nature, but there will be no solution without it. As Wendell Berry has pointed out, it is culture, and certainly not nature, that teaches us to observe and remember, to learn from our mistakes, to share our experiences, and perhaps most important of all, to restrain ourselves. Nature does not teach its creatures to control their appetites except by the harshest of

lessons—epidemics, mass death, extinctions. Nothing would be more natural than for humankind to burden the environment to the extent that it was rendered unfit for human life. Nature in that event would not be the loser, nor would it disturb her laws in the least—operating as it has always done, natural selection would unceremoniously do us in. Should this fate be averted, it will only be because our culture—*our* laws and metaphors, our science and technology, our ongoing conversation about nature and man's place in it—pointed us in the direction of a different future. Nature will not do this for us.

The gardener in nature is that most artificial of creatures, a civilized human being: in control of his appetites, solicitous of nature, self-conscious and responsible, mindful of the past and the future, and at ease with the fundamental ambiguity of his predicament—which is that though he lives in nature, he is no longer strictly *of* nature. Further, he knows that neither his success nor his failure in this place is ordained. Nature is apparently indifferent to his fate, and this leaves him free—indeed, obliges him—to make his own way here as best he can.

What would an ethic based on these ideas—based on the idea of the garden—advise us to do in Cathedral Pines? I don't know enough about the ecology of the place to say with certainty, but I think I have some sense of how we might proceed under its dispensation. We would start out, of course, by consulting "the Genius of the Place." This would tell us, among other things, that Cathedral Pines is not a wilderness, and so probably should not be treated as one. It is a cultural as well as a natural landscape, and to exclude the wishes of the townspeople from our plans for the place would be false. To treat it now as wilderness is to impose an abstract and alien idea on it. Consulting the genius of the place also means inquiring as to what nature will allow us to do here, what this "locale permits, and what [it] denies," as Virgil wrote in *The Georgics*. We know right off, for instance, that this plot of land can support a magnificent forest of white pines. Nature would not object if we decided to replant the pine forest. Indeed, this would be a perfectly reasonable, environmentally sound thing to do.

If we chose to go this route, we would be undertaking a fairly simple act of what is called "ecological restoration." This relatively new school of environmentalism has its roots in Aldo Leopold's pioneering efforts to re-create a tall-grass prairie on the grounds of the University of Wisconsin Arboretum in the 1930s. Leopold and his followers (who continue to maintain the restored prairie today) believed that it is not always enough to conserve the land—that sometimes it is desirable, and possible, for man to intervene in nature in order to improve it. Specifically, man should intervene to re-create damaged ecosystems: polluted rivers, clear-cut forests, vanished prairies, dead lakes. The restorationists also believe, and in this they remind me of the green thumb, that the best way to learn about nature's ways is by trying to imitate them. (In fact much of what we know about the role of fire in creating and sustaining prairies comes from their efforts.) But the most important contribution of the restorationists has been to set forth a positive, active role for man in nature—in their conception, as equal parts gardener and healer. It seems to me that the idea of ecological restoration is consistent with a garden ethic, and perhaps with the Hippocratic Oath as well.

From the work of the ecological restorationists, we now know that it is possible to skip and manipulate the stages of forest succession. They would probably advise us to burn the fallen timber—an act that, though not strictly speaking "natural," would serve as an effective analogue of the natural process by which a forest is regenerated. The fires we set would reinvigorate the soil (thereby enhancing *that* wilderness) and at the same time clear out the weed species, hardwood saplings, and brush. By doing all this, we will have imitated the conditions under which a white pine forest is born, and the pines might then return on their own. Or else—it makes little difference—we could plant them. At that point, our work would be done, and the pine forest could take care of itself. It would take many decades, but restoring the Cathedral Pines would strain neither our capabilities nor nature's sufferance. And in doing so, we would also be restoring the congenial relationship between man and nature that prevailed in this place before the storm and the subsequent controversy. That would be no small thing.

Nature would not preclude more novel solutions for Cathedral Pines—other kinds of forest-gardens or even parks could probably flourish on this site. But since the town has traditionally regarded Cathedral Pines as a kind of local institution, one steeped in shared memories and historical significance, I would argue that the genius of the place rules out doing anything unprecedented here. The past is our best guide in this particular case, and not only on questions of ecology.

But replanting the pine forest is not the only good option for Cathedral Pines. There is another forest we might want to restore on this site, one that is also in keeping with its history and its meaning to the town.

Before the storm, we used to come to Cathedral Pines and imagine that this was how the New World forest looked to the first settlers. We now know that the precolonial forest probably looked somewhat different—for one thing, it was not exclusively pine. But it's conceivable that we could restore Cathedral Pines to something closely resembling its actual precolonial condition. By analyzing historical accounts, the rings of fallen trees, and fossilized pollen grains buried in the soil, we could reconstruct the variety and composition of species that flourished here in 1739, the year when the colonists first settled near this place and formed the town of Cornwall. We know that nature, having done so once before, would probably permit us to have such a forest here. And, using some of the more advanced techniques of ecological restoration, it is probably within our competence to re-create a precolonial forest on this site.

We would do this not because we'd decided to be faithful to the "state of-nature" at Cathedral Pines, but very simply because the precolonial forest happens to mean a great deal to us. It is a touchstone in the history of this town, not to mention this nation. A walk in a restored version of the precolonial forest might recall us to our culture's first, fateful impressions of America, to our thoughts on coming upon what Fitzgerald called the "fresh green breast of the new world." In the contemplation of that scene we might be moved to reconsider what happened next—to us, to the Indians who once hunted here, to nature in this corner of America.

This is pretty much what I would have stood up and said if we'd had a town meeting to decide what to do in Cathedral Pines. Certainly a town meeting would have been a fitting way to decide the matter, nicely in keeping with the genius of *this* place, a small town in New England. I can easily imagine the speeches and the arguments. The people from the Nature Conservancy would have made their plea for leaving the place alone, for "letting nature take her course." Richard Dakin, the first selectman, and John Calhoun, the forest's nearest neighbor, would have warned about the dangers of fire. And then we might have heard some other points of view. I would have tried to make a pitch for restoration, talking about some of the ways we might "garden" the site. I can imagine Ian Ingersoll, a gifted cabinetmaker in town, speaking with feeling about the waste of such rare timbers, and the prospect of sitting down to a Thanksgiving dinner at a table in which you could see rings formed at the time of the American Revolution. Maybe somebody else would have talked about how much she missed her Sunday afternoon walks in the forest, and how very sad the place looked now. A scientist from the Yale School of Forestry might have patiently tried to explain, as indeed one Yale scientist did in the press, why "It's just as pretty to me now as it was then."

This is the same fellow who said, "If you're going to clean it up, you might as well put up condos." I can't imagine anyone actually proposing that, or any other kind of development in Cathedral Pines. But if someone did, he would probably get shouted down. Because we have too much respect for this place; and besides, our sympathies and interests are a lot more complicated than the economists or environmentalists always seem to think. Sooner than a developer, we'd be likely to hear from somebody speaking on behalf of the forest's fauna—the species who have lost out in the storm (like the owls), but also the ones for whom doing nothing would be a boon (the beetles). And so the various interests of the animals would be taken into account, too; indeed, I expect that "nature"—all *those* different (and contradictory) points of view—would be well represented at this town meeting. Perhaps it is naive of me to think so, but I'm confident that in the course of a public, democratic conversation about the disposition of Cathedral Pines, we would eventually arrive at a solution that would have at once pleased us and not offended nature.

But unfortunately that's not what happened. The future of Cathedral Pines was decided in a closed-door meeting at the Nature Conservancy in September, after a series of negotiations with the selectmen and the owners of adjacent property. The result was a compromise that seems to have pleased no one. The fallen trees will remain untouched—except for a fifty-foot swath clear-cut around the perimeter of the forest, a firebreak intended to appease the owners of a few nearby houses. The sole human interest taken into account in the decision was the worry about fire.

I drove up there one day in late fall to have a look around, to see what the truce between the Conservancy and the town had wrought. What a sad sight it is. Unwittingly, and in spite of the good intentions on both sides, the Conservancy and the selectmen have conspired to create a landscape that is a perfect symbol of our perverted relation to nature. The firebreak looks like nothing so much as a no-man's-land in a war zone, a forbidding expanse of blistered ground impounding what little remains of the Cathedral Pines. The landscape we've made here is grotesque. And yet it is the logical outcome of a confrontation between, on the one side, an abstract and mistaken concept of nature's interests and, on the other, a pinched and demeaning notion of our own interests. We should probably not be surprised that the result of such a confrontation is not a wilderness, or a garden, but a DMZ.

Introduction to
Donald Worster
The Ecology of Order and Chaos

by Nancy Langston

In "The Ecology of Order and Chaos" (1993), Donald Worster cautions us to be as critical about the cultural context of new ecological theories of instability as we now are about older climax theories of stability. His article suggests that recent shifts in ecological theory may reflect not just a better grasp of natural processes, but also anti-environmental biases in American political culture.

Ecology, like all sciences, reflects a set of views about the world, a set of choices about what is worth observing and measuring. Scientific theories are models of nature: they are ways of making sense of the world by reducing complexity to a subset of measurable factors. People create those theories, and those people live within a particular culture and society.

Many equilibrium ecologists such as the Odums were quite explicit about the ways they wanted their work to be used to preserve rather than control nature. In contrast, current non-equilibrial ecologists have sometimes been far less interested in conservation or preservation. In this essay, Donald Worster argues that part of the cultural appeal of disequilibrium as the norm of the natural world comes from people who want to blunt some of the more radical implications of environmentalism. Some of the ecologists who developed the initial theories about instability in nature were hostile toward environmentalism. They justified their hostility by arguing that, because nature does not have a holistic balance, any human interference we desire is acceptable, indeed even scientifically validated. The implication is that, if there is no stable climax community, then no ecological relationship needs our concern. Worster argues that the image of the natural world as a place of competition, flux and chaos, and instability provides for some people a dangerous excuse for unrestrained human intervention in nature.

Worster's work, along with the work of Pollan, Crosby, Christensen, and other ecological historians, suggests some of the ways that linking ecology and history can illuminate both disciplines, while also helping us make better sense of our troubled, changing relationships with the natural world.

The Ecology of Order and Chaos

DONALD WORSTER

The science of ecology has had a popular impact unlike that of any other academic field of research. Consider the extraordinary ubiquity of the word itself: it has appeared in the most everyday places and the most astonishing, on day-glo T-shirts, in corporate advertising, and on bridge abutments. It has changed the language of politics and philosophy—springing up in a number of countries are political groups that are self-identified as "Ecology Parties." Yet who ever proposed forming a political party named after comparative linguistics or advanced paleontology? On several continents we have a philosophical movement termed "Deep Ecology," but nowhere has anyone announced a movement for "Deep Entomology" or "Deep Polish Literature." Why has this funny little word, ecology, coined by an obscure nineteenth-century German scientist, acquired so powerful a cultural resonance, so widespread a following?

Behind the persistent enthusiasm for ecology, I believe, lies the hope that this science can offer a great deal more than a pile of data. It is supposed to offer a pathway to a kind of moral enlightenment that we can call, for the purposes of simplicity, "conservation." The expectation did not originate with the public but first appeared among eminent scientists within the field. For instance, in his 1935 book *Deserts on the March,* the noted University of Oklahoma, and later Yale, botanist Paul Sears urged Americans to take ecology seriously, promoting it in their universities and making it part of their governing process. "In Great Britain," he pointed out,

> *...The ecologists are being consulted at every step in planning the proper utilization of those parts of the Empire not yet settled, thus...ending the era of haphazard exploitation. There are hopeful, but all too few signs that our own national government realizes the part which ecology must play in a permanent program.*[1]

Sears recommended that the United States hire a few thousand ecologists at the county level to advise citizens on questions of land use and thereby bring an end to environmental degradation; such a brigade, he thought, would put the whole nation on a biologically and economically sustainable basis.

In a 1947 addendum to his text, Sears added that ecologists, acting in the public interest, would instill in the American mind that "body of knowledge," that "point of view, which peculiarly implies all that is meant by conservation."[2] In other words, by the time of the 1930s and 40s, ecology was being hailed as a much-needed guide to a future motivated by an ethic of conservation. And conservation for Sears meant restoring the biological order, maintaining the health of the land and thereby the well-being of the nation, pursuing by both moral and technical means a lasting equilibrium with nature.

While we have not taken to heart all of Sears's suggestions—have not yet put any ecologists on county payrolls, with an office next door to the tax collector and sheriff—we have taken a surprisingly long step in his direction. Every day in some part of the nation, an ecologist is at work writing an environmental impact report or monitoring a human disturbance of the landscape or testifying at a hearing.

Twelve years ago I published a history, going back to the eighteenth century, of this scientific discipline and its ideas about nature.[3] The conclusions in that book still strike me as being, on the whole, sensible and valid: that this science has come to be a major influence on our perception of nature in modern times; that its ideas, on the other hand, have been reflections of ourselves as much as objective apprehensions of nature; that scientific analysis cannot take the place of moral reasoning; that science, including the science of ecology, promotes, at least in some of its manifestations, a few of our darker ambitions toward nature and therefore itself needs to be morally examined and critiqued from time to time. Ecology, I argued, should never be taken as an all-wise, always trustworthy guide. We must be willing to challenge this authority, and indeed challenge the authority of science in general; not be quick to scorn or vilify or behead, but simply, now and then, to question.

During the period since my book was published, there has accumulated a considerable body of new thinking and new research in ecology. I mean to survey some of that recent thinking, contrasting it with its predecessors, and to raise a few of the same questions I did before. Part of my argument will be that Paul Sears would be astonished, and perhaps dismayed, to hear the kind of advice that ecological experts have got to give these days. Less and less do they offer, or even promise to offer, what he would consider to be a program of moral enlightenment: of "conservation" in the sense of a restored equilibrium between humans and nature.

There is a clear reason for that outcome, I will argue, and it has to do with drastic changes in the ideas that ecologists hold about the structure and function of the natural world. In Sears's day, ecology was basically a study of equilibrium, harmony, and order; it had been so from its beginnings. Today, however, in many circles of scientific research, it has become a study of disturbance, disharmony, and chaos, and coincidentally or not, conservation is often not even a remote concern.

At the time *Deserts on the March* appeared in print, and through the time of its second and even third edition, the dominant name in the field of American ecology was that of Frederic L. Clements, who more than any other individual introduced scientific ecology into our national academic life. He called his approach "dynamic ecology," meaning it was concerned with change and evolution in the landscape. At its heart Clements's ecology dealt with the process of vegetational succession—the sequence of plant communities that appear on a piece of soil, newly made or disturbed, beginning with the first pioneer communities that invade and get a foothold.[4] Here is how I have defined the essence of the Clementsian paradigm:

> *Change upon change became the inescapable principle of Clements's science. Yet he also insisted stubbornly and vigorously on the notion that the natural landscape must eventually reach a vaguely final climax stage. Nature's course, he contended, is not an aimless wandering to and fro but a steady flow toward stability that can be exactly plotted by the scientist.*[5]

Most interestingly, Clements referred to that final climax stage as a "superorganism," implying that the assemblage of plants had achieved the close integration of parts, the self-organizing capability, of a single animal or plant. In some unique sense, it had become a live, coherent thing, not a mere collection of atomistic individuals, and exercised some control over the non-living world around it, as organisms do.

Until well after World War II Clements's climax theory dominated ecological thought in this country.[6] Pick up almost any textbook in the field written forty, or even thirty, years ago, and you will likely find mention of the climax. It was this theory that Paul Sears had studied and took to be the core lesson of ecology that his county ecologists should teach their fellow citizens: that nature tends toward a climax state and that, as far as practicable, they should learn to respect and preserve it. Sears wrote that the chief work of the scientist ought to be to show "the unbalance which man has produced on this continent" and to lead people back to some approximation of nature's original health and stability.[7]

But then, beginning in the 1940s, while Clements and his ideas were still in the ascendant, a few scientists began trying to speak a new vocabulary. Words like "energy flow" and "trophic levels" and "ecosystem" appeared in the leading journals, and they indicated a view of nature shaped more by physics than botany. Within another decade or two nature came to be widely seen as a flow of energy and nutrients through a physical or thermodynamic system. The early figures prominent in shaping this new view included C. Juday, Raymond Lindeman, and G. Evelyn Hutchinson. But perhaps its most influential exponent was Eugene P. Odum, hailing from North Carolina and Georgia, discovering in his southern saltwater marshes, tidal estuaries, and abandoned cotton fields the animating, pulsating force of the sun, the global flux of energy. In 1953 Odum published the first edition of his famous textbook, *The Fundamentals of Ecology*.[8] In 1966 he became president of the Ecological Society of America.

By now anyone in the United States who regularly reads a newspaper or magazine has come to know at least a few of Odum's ideas, for they furnish the main themes in our popular understanding of ecology, beginning with the sovereign idea of the ecosystem. Odum defined the ecosystem as "any unit that includes all of the organisms (i.e., the 'community') in a given area interacting with the physical environment so that a flow of energy leads to clearly defined trophic structure, biotic diversity, and material cycles (i.e., exchange of materials between living and nonliving parts) within the system."[9] The whole earth, he argued, is organized into an interlocking series of such "ecosystems," ranging in size from a small pond to so vast an expanse as the Brazilian rainforest.

What all those ecosystems have in common is a "strategy of development," a kind of game plan that gives nature an overall direction. That strategy is, in Odum's words, "directed toward achieving as large and diverse an organic structure as is possible within the limits set by the available energy input and the prevailing physical conditions of existence."[10] Every single ecosystem, he believed, is either moving toward or has already achieved that goal. It is a clear, coherent, and easily observable strategy; and it ends in the happy state of order.

Nature's strategy, Odum added, leads finally to a world of mutualism and cooperation among the organisms inhabiting an area. From an early stage of competing against one another, they evolve toward a more symbiotic relationship. They learn, as it were, to work together to control their surrounding environment, making it more and more suitable as a habitat, until at last they have the power to protect themselves from its stressful cycles of drought and flood, winter and summer, cold and heat. Odum called that point "homeostasis." To achieve it, the living components of an ecosystem must evolve a structure of interrelatedness and cooperation that can, to some extent, manage the physical world—manage it for maximum efficiency and mutual benefit.

I have described this set of ideas as a break from the past, but that is misleading. Odum may have used different terms than Clements, may even have had a radically different vision of nature at times; but he did not repudiate Clements's notion that nature moves toward order and harmony. In the place of the theory of the "climax" stage he put the theory of the "mature ecosystem." His nature may have appeared more as an automated factory than as a Clementsian super- organism, but like its predecessor it tends toward order.

The theory of the ecosystem presented a very clear set of standards as to what constituted order and disorder, which Odum set forth in the form of a "tabular model of ecological succession." When the ecosystem reaches its end point of homeostasis, his table shows, it expends less energy on increasing production and more on furnishing protection from external vicissitudes: that is, the biomass in an area reaches a steady level, neither increasing nor decreasing, and the emphasis in the system is on keeping it that way—on maintaining a kind of no-growth economy. Then the little, aggressive, weedy organisms common at an early stage in development (the r-selected species) give way to larger, steadier creatures (K-selected species), who may have less potential for fast growth and explosive reproduction but also better talents at surviving in dense settlements and keeping the place on an even keel.[11] At that point there is supposed to be more diversity in the community—i.e., a greater array of species. And there is less loss of nutrients to the outside; nitrogen, phosphorous, and calcium all stay in circulation within the ecosystem rather than leaking out. Those are some of the key indicators of ecological order, all of them susceptible to precise measurement. The suggestion was implicit but clear that if one interfered too much with nature's strategy of development, the effects might be costly: a serious loss of nutrients, a decline in species diversity, an end to biomass stability. In short, the ecosystem would be damaged.

The most likely source of that damage was no mystery to Odum: it was human beings trying to force up the production of useful commodities and stupidly risking the destruction of their life support system.

> *Man has generally been preoccupied with obtaining as much "production" from the landscape as possible, by developing and maintaining early successional types of ecosystems, usually monocultures. But, of course, man does not live by food and fiber alone; he also needs a balanced CO_2–O_2 atmosphere, the climatic buffer provided by oceans and masses of vegetation, and clean (that is, unproductive) water for cultural and industrial uses. Many essential life-cycle resources, not to mention recreational and esthetic needs, are best provided man by the less "productive" landscapes. In other words, the landscape is not just a supply depot but is also the oikos—the home—in which we must live.*[12]

Odum's view of nature as a series of balanced ecosystems, achieved or in the making, led him to take a strong stand in favor of preserving the landscape in as nearly natural a condition as possible. He suggested the need for substantial restraint on human activity—for environmental planning "on a rational and scientific basis." For him as for Paul Sears, ecology must be taught to the public and made the foundation of education, economics, and politics; America and other countries must be "ecologized."

Of course not every one who adopted the ecosystem approach to ecology ended up where Odum did. Quite the contrary, many found the ecosystem idea a wonderful instrument for promoting global technocracy. Experts familiar with the ecosystem and skilled in its manipulation, it was hoped in some quarters, could manage the entire planet for improved efficiency. "Governing" all of nature with the aid of rational science was the dream of these ecosystem technocrats.[13] But technocratic management was not the chief lesson, I believe, the public learned in Professor Odum's classroom; most came away devoted, as he was, to preserving large parts of nature in an unmanaged state and sure that they had been given a strong scientific rationale, as well as knowledge base, to do it. We must defend the world's endangered ecosystems, they insisted. We must safeguard the integrity of the Greater Yellowstone ecosystem, the Chesapeake Bay ecosystem, the Serengeti ecosystem. We must protect species diversity, biomass stability, and calcium recycling. We must make the world safe for K-species.[14]

That was the rallying cry of environmentalists and ecologists alike in the 1960s and early 1970s, when it seemed that the great coming struggle would be between what was left of pristine nature, delicately balanced in Odum's beautifully rational ecosystems, and a human race bent on mindless, greedy destruction. A decade or two later the situation has changed considerably. There are still environmental threats around, to be sure, and they are more dangerous than ever. The newspapers inform of us of continuing disasters like the massive 1989 oil spill in Alaska's Prince William Sound, and reporters persist in using words like "ecosystem" and "balance" and "fragility" in describing such disasters. So do many scientists, who continue to acknowledge their theoretical indebtedness to Odum. For instance, in a recent British poll, 447 ecologists out of 645 questioned ranked the "ecosystem" as one of the most important concepts their discipline has contributed to our understanding of the natural world; indeed, "ecosystem" ranked first on their list, drawing more votes than nineteen other leading concepts.[15] But all the same, and despite the persistence of environmental problems, Odum's ecosystem is no longer the main theme in research or teaching in the science. A survey of recent ecology textbooks shows that the concept is not even mentioned in one leading work and has a much diminished place in the others.[16]

Ecology is not the same as it was. A rather drastic change has been going on in this science of late: a radical shifting away from the thinking of Eugene Odum's generation, away from its assumptions of order and predictability, a shifting toward what we might call a new ecology of chaos.

In July 1973 the *Journal of the Arnold Arboretum* published an article by two scientists associated with the Massachusetts Audubon Society, William Drury and Ian Nisbet, and it challenged Odum's ecology fundamentally. The title of the article was simply "Succession," indicating that old subject of observed sequences in plant and animal associations. With both Frederic Clements and Eugene Odum, succession had been taken to be the straight and narrow road to equilibrium. Drury and Nisbet disagreed completely with that assumption. Their observations, drawn particularly from northeastern temperate forests, strongly suggested that the process of ecological succession does not lead anywhere. Change is without any determinable direction and goes on forever, never reaching a point of stability. They found no evidence of any progressive development in nature: no progressive increase over time in biomass stabilization, no progressive diversification of species, no progressive movement toward a greater cohesiveness in plant and animal communities, nor toward a greater success in regulating the environment. Indeed, they found none of the criteria Odum had posited for mature ecosystems. The forest, they insisted, no matter what its age, is nothing but an erratic, shifting mosaic of trees and other plants. In their words, "most of the phenomena of succession should be understood as resulting from the differential growth, differential survival, and perhaps differential dispersal of species adapted to grow at different points on stress gradients."[17] In other words, they could see lots of individual species, each doing its thing, but they could locate no emergent collectivity, nor any strategy to achieve one.

Prominent among their authorities supporting this view was the nearly forgotten name of Henry A. Gleason, a taxonomist who, in 1926, had challenged Frederic Clements and his organismic theory of the climax in an article entitled "The Individualistic Concept of the Plant Association." Gleason had argued that we live in a world of constant flux and impermanence, not one tending toward Clements's climaxes. There is no such thing, he argued, as balance or equilibrium or steady-state. Each and every plant association is nothing but a temporary gathering of strangers, a clustering of species unrelated to one another, here for a brief while today, on their way somewhere else tomorrow. "Each...species of plant is a law unto itself," he wrote.[18] We look for cooperation in nature and we find only competition. We look for organized wholes, and we can discover only loose atoms and fragments. We hope for order and discern only a mishmash of conjoining species, all seeking their own advantage in utter disregard of others.

Thanks in part to Drury and Nisbet, this "individualistic" view was reborn in the mid-1970s and, by the present decade, it had become the core idea of what some scientists hailed as a new, revolutionary paradigm in ecology. To promote it, they attacked the traditional notion of succession; for to reject that notion was to reject the larger idea that organic nature tends toward order. In 1977 two more biologists, Joseph Connell and Ralph Slatyer, continued the attack, denying the old claim that an invading community of pioneering species, the first stage in Clements's sequence, works to prepare the ground for its successors, like a group of Daniel Boones blazing the trail for civilization. The first comers, Connell and Slatyer maintained, manage in most cases to stake out their claims and successfully defend them; they do not give way to a later, superior group of colonists. Only when the pioneers die or are damaged by natural disturbances, thus releasing the resources they have monopolized, can latecomers find a foothold and get established.[19]

As this assault on the old thinking gathered momentum, the word "disturbance" began to appear more frequently in the scientific literature and be taken far more seriously. "Disturbance" was not a common subject in Odum's heyday, and it almost never appeared in combination with the adjective "natural." Now, however, it was as though scientists were out looking strenuously for signs of disturbance in nature—especially signs of disturbance that were not caused by humans—and they were finding them everywhere. By the present decade these new ecologists have succeeded in leaving little tranquility in primitive nature. Fire is one of the most common disturbances they have noted. So is wind, especially in the form of violent hurricanes and tornadoes. So are invading populations of microorganisms and pests and predators. And volcanic eruptions. And invading ice sheets of the Quaternary Period. And devastating droughts like that of the 1930s in the American West. Above all, it is these last sorts of disturbances, caused by the restlessness of climate, that the new generation of ecologists has emphasized. As one of the most influential of them, Professor Margaret Davis of the University of Minnesota, has written: "For the last 50 years or 500 or 1,000—as long as anyone would claim for 'ecological time'—there has never been an interval when temperature was in a steady state with symmetrical fluctuations about a mean... Only on the longest time scale, 100,000 years, is there a tendency toward cyclical variation, and the cycles are asymmetrical, with a mean much different from today."[20]

One of the most provocative and impressive expressions of the new post-Odum ecology is a book of essays edited by S. T. A. Pickett and P. S. White, *The Ecology of Natural Disturbance and Patch Dynamics* (published in 1985). I submit it as symptomatic of much of the thinking going on today in the field. Though the final section of the book does deal with ecosystems, the word has lost much of its former meaning and implications. Two of the authors in fact open their contribution with a complaint that many scientists assume that "homogeneous ecosystems are a reality," when in truth "virtually all naturally occurring and man-disturbed ecosystems are mosaics of environmental conditions." "Historically," they write, "ecologists have been slow to recognize the importance of disturbances and the heterogeneity they generate." The reason for this slowness? "The majority of both theoretical and empirical work has been dominated by an equilibrium perspective."[21] Repudiating that perspective, these authors take us to the tropical forests of South and Central America and to the Everglades of Florida, showing us instability on every hand: a wet, green world of continual disturbance—or as they prefer to say, "of perturbations." Even the grasslands of North America, which inspired Frederic Clements's theory of the climax, appear in this collection as regularly disturbed environments. One paper describes them as a "dynamic, fine-textured mosaic" that is constantly kept in upheaval by the workings of badgers, pocket gophers, and mound-building ants, along with fire, drought, and eroding wind and water.[22] The message in all these papers is consistent: the climax notion is dead, the ecosystem has receded in usefulness, and in their place we have the idea of the lowly "patch." Nature should be regarded as a landscape of patches, big and little, patches of all textures and colors, a patchwork quilt of living things, changing continually through time and space, responding to an unceasing barrage of perturbations. The stitches in that quilt never hold for long.

Now, of course, scientists have known about gophers and winds and the Ice Age and droughts for a considerable time. Yet heretofore they have not let those disruptions spoil their theories of balanced plant and animal associations, and we must ask why that was so. Why did Clements and Odum tend to dismiss such forces as climatic change, at least of the less catastrophic sort, as threats to the order of nature? Why have their successors, on the other hand, tended to put so much emphasis on those same changes, to the point that they often see nothing but instability in the landscape?

One clue comes from the fact that many of these disturbance boosters are not and have never been ecosystem scientists; they received their training in the subfield of population biology and reflect the growing confidence, methodological maturity, and influence of that subfield.[23] When they look at a forest, the population ecologists see only the trees: see them and count them—so many white pines, so many hemlocks, so many maples and birches. They insist that if we know all there is to know about the individual species that constitute a forest, and can measure their lives in precise, quantitative terms, we will know all there is to know about that forest. It has no "emergent" or organismic properties. It is not some whole greater than the sum of its parts, requiring "holistic" understanding. Outfitted with computers that can track the life histories of individual species, chart the rise and fall of populations, they have brought a degree of mathematical precision to ecology that is awesome to contemplate. And what they see when they look at population histories for any patch of land is wildly swinging oscillations. Populations rise and populations fall, like stock market prices, auto sales, and hemlines. We live, they insist, in a non-equilibrium world.[24]

There is another reason for the paradigmatic shift I have been describing, though I suggest it quite tentatively and can offer only sketchy evidence for it. For some scientists, a nature characterized by highly individualistic associations, constant disturbance, and incessant change may be more ideologically satisfying than Odum's ecosystem, with its stress on cooperation, social organization, and environmentalism. A case in point is the very successful popularizer of contemporary ecology, Paul Colinvaux, author of *Why Big Fierce Animals Are Rare* (1978). His chapter on succession begins with these lines: "If the planners really get hold of us so that they can stamp out all individual liberty and do what they like with our land, they might decide that whole counties full of inferior farms should be put back into forest." Clearly, he is not enthusiastic about land-use planning or forest restoration. And he ends that same chapter with these remarkably revealing and self-assured words:

> *We can now...explain all the intriguing, predictable events of plant successions in simple, matter of fact, Darwinian ways. Everything that happens in successions comes about because all the different species go about earning their livings as best they may, each in its own individual manner. What look like community properties are in fact the summed results of all these bits of private enterprise.*[26]

Apparently, if this example is any indication, the Social Darwinists are back on the scene, and at least some of them are ecologists, and at least some of their opposition to Odum's science may have to do with a revulsion toward what they perceive are its political implications, including its attractiveness for environmentalists. Colinvaux is very clear about the need to get some distance between himself and groups like the Sierra Club.

I am not alone in wondering whether there might be a deeper, half-articulated ideological motive generating the new direction in ecology. The Swedish historian of science Thomas Soderqvist, in his recent study of ecology's development in his country, concludes that the present generation of evolutionary ecologists

> *seem to do ecology for fun only, indifferent to practical problems, including the salvation of the nation. They are mathematically and theoretically sophisticated, sitting indoors calculating on computers, rather than traveling out in the wilds. They are individualists, abhorring the idea of largescale ecosystem projects. Indeed, the transition from ecosystem ecology to evolutionary ecology seems to reflect the generational transition from the politically consciousness generation of the 1960s to the "yuppie" generation of the 1980s.*[26]

That may be an exaggerated characterization, and I would not want to apply it to every scientist who has published on patch dynamics or disturbance regimes. But it does draw our attention to an unmistakable attempt by many ecologists to disassociate themselves from reform environmentalism and its criticisms of human impact on nature.

I wish, however, that the emergence of the new post-Odum ecology could be explained so simply in those two ways: as a triumph of reductive population dynamics over holistic consciousness, or as a triumph of Social Darwinist or entrepreneurial ideology over a commitment to environmental preservation. There is, it seems, more going on than that, and it is going on all through the natural sciences—biology, astronomy, physics—perhaps going on through all modern technological societies. It is nothing less than the discovery of chaos. Nature, many have begun to believe, is fundamentally erratic, discontinuous, and unpredictable. It is full of seemingly random events that elude our models of how things are supposed to work. As a result, the unexpected keeps hitting us in the face. Clouds collect and disperse, rain falls or doesn't fall, disregarding our careful weather predictions, and we cannot explain why. Cars suddenly bunch up on the freeway, and the traffic controllers fly into a frenzy. A man's heart beats regularly year after year, then abruptly begins to skip a beat now and then. A ping-pong ball bounces off the table in an unexpected direction. Each little snowflake falling out of the sky turns out to be completely unlike any other. These are ways in which nature seems, by all our previous theories and methods, to be chaotic. If the ultimate test of any body of scientific knowledge is its ability to predict events, then all the sciences and pseudo-sciences—physics, chemistry, climatology, economics, ecology—fail the test regularly. They all have been announcing laws, designing models, predicting what an individual atom or person is supposed to do; and now, increasingly, they are beginning to confess that the world never quite behaves the way it is supposed to do.

Making sense of this situation is the task of an altogether new kind of inquiry calling itself the science of chaos. Some say it portends a revolution in thinking equivalent to quantum mechanics or relativity. Like those other twentieth-century revolutions, the science of chaos rejects tenets going back as far as the days of Sir Isaac Newton. In fact, what is occurring may be not two or three separate revolutions but a single revolution against all the principles, laws, models, and applications of classical science, the science ushered in by the great Scientific Revolution of the seventeenth century.[27] For centuries we have assumed that nature, despite a few appearances to the contrary, is a perfectly predictable system of linear, rational order. Give us an adequate number of facts, scientists have said, and we can describe that order in complete detail—can plot the lines along which everything moves and the speed of that movement and the collisions that will occur. Even Darwin's theory of evolution, which in the last century challenged much of the Newtonian worldview, left intact many people's confidence that order would prevail at last in the evolution of life; that out of the tangled history of competitive struggle would come progress, harmony, and stability. Now that traditional assumption may have broken down irretrievably. For whatever reason, whether because empirical data suggest it or because extrascientific cultural trends do, the experience of so much rapid social change in our daily lives, scientists are beginning to focus on what they had long managed to avoid seeing. The world is more complex than we ever imagined, they say, and indeed, some would add, ever can imagine.[28]

Despite the obvious complexity of their subject matter, ecologists have been among the slowest to join the cross-disciplinary science of chaos. I suspect that the influence of Clements and Odum, lingering well into the 1970s, worked against the new perspective, encouraging faith in linear regularities and equilibrium in the interaction of species. Nonetheless, eventually there arrived a day of conversion. In 1974 the Princeton mathematical ecologist Robert May published a paper with the title "Biological Populations with Nonoverlapping Generations: Stable Points, Stable Cycles, and Chaos."[29] In it he admitted that the mathematical models he and others had constructed were inadequate approximations of the ragged life histories of organisms. They did not fully explain, for example, the aperiodic outbreaks of gypsy moths in eastern hardwood forests or the Canadian lynx cycles in the subarctic. Wildlife populations do not follow some simple Malthusian pattern of increase, saturation, and crash.

More and more ecologists have followed May and begun to try to bring their subject into line with chaotic theory. William Schaefer is one of them; though a student of Robert MacArthur, a leader of the old equilibrium school, he has been lately struck by the same anomaly of unpredictable fluctuations in populations as May and others. Though taught to believe in "the so-called 'Balance of Nature'," he writes, "...the idea that populations are at or close to equilibrium, things now are beginning to look very different."[30] He describes himself has having to reach far across the disciplines, to make connections with concepts of chaos in the other natural sciences, in order to free himself from his field's restrictive past.

The entire study of chaos began in 1961, with efforts to simulate weather and climate patterns on a computer at MIT. There, meteorologist Edward Lorenz came up with his now famous "Butterfly Effect," the notion that a butterfly stirring the air today in a Beijing park can transform storm systems next month in New York City. Scientists call this phenomenon "sensitive dependence on initial conditions." What it means is that tiny differences in input can quickly become substantial differences in output. A corollary is that we cannot know, even with all our artificial intelligence apparatus, every one of the tiny differences that have occurred or are occurring at any place or point in time; nor can we know which tiny differences will produce which substantial differences in output. Beyond a short range, say, of two or three days from now, our predictions are not worth the paper they are written on.

The implications of this "Butterfly Effect" for ecology are profound. If a single flap of an insect's wings in China can lead to a torrential downpour in New York, then what might it do to the Greater Yellowstone Ecosystem? What can ecologists possibly know about all the forces impinging on, or about to impinge on, any piece of land? What can they safely ignore and what must they pay attention to? What distant, invisible, minuscule events may even now be happening that will change the organization of plant and animal life in our back yards? This is the predicament, and the challenge, presented by the science of chaos, and it is altering the imagination of ecologists dramatically.

John Muir once declared, "When we try to pick out anything by itself, we find it hitched to everything else in the universe."[31] For him, that was a manifestation of an infinitely wise plan in which everything functioned with perfect harmony. The new ecology of chaos, though impressed like Muir with interdependency, does not share his view of "an infinitely wise plan" that controls and shapes everything into order. There is no plan, today's scientists say, no harmony apparent in the events of nature. If there is order in the universe—and there will no longer be any science at all if all faith in order vanishes—it is going to be much more difficult to locate and describe than we thought.

For Muir, the clear lesson of cosmic complexity was that humans ought to love and preserve nature just as it is. The lessons of the new ecology, in contrast, are not at all clear. Does it promote, in Ilya Prigogine and Isabelle Stenger's words, "a renewal of nature," a less hierarchical view of life, and a set of "new relations between man and nature and between man and man"?[32] Or does it increase our alienation from the world, our withdrawal into post-modernist doubt and self-consciousness? What is there to love or preserve in a universe of chaos? How are people supposed to behave in such a universe? If that is the kind of place we inhabit, why not go ahead with all our private ambitions, free of any fear that we may be doing special damage? What, after all, does the phrase "environmental damage" mean in a world of so much natural chaos? Does the tradition of environmentalism to which Muir belonged, along with so many other nature writers and ecologists of the past, people like Paul Sears, Eugene Odum, Aldo Leopold, and Rachel Carson, make sense any longer? I have no space here to attempt to answer those questions, or to make predictions, but only a warning that they are too important to be left for scientists alone to answer. Ecology cannot today, no more than in the past, be assumed to be all-knowing or all-wise or eternally true.

Whether they are true or false, permanent or passingly fashionable, it does seem entirely possible that these changes in scientific thinking toward an emphasis on chaos will not produce any easing of the environmentalist's concern. Though words like ecosystem or climax may fade away, and some new vocabulary take their place, the fear of risk and danger will likely become greater than ever. Most of us are intuitively aware, whether we can put our fears into mathematical formulae or not, that the technological power we have accumulated is *destructively* chaotic; not irrationally, we fear it and fear what it can do to us as well as the rest of nature.[33] It may be that we moderns, after absorbing the lessons of today's science, find we cannot love nature quite so easily as Muir did; but it may also be that we have discovered more reason than ever to respect it—to respect its baffling complexity, its inherent unpredictability, its daily turbulence. And to flap our own wings in it a little more gently.

Footnotes

1. Paul Sears, *Deserts on the March,* 3rd ed. (Norman: University of Oklahoma Press, 1959), p. 162.
2. Ibid., p. 177.
3. Donald Worster, *Nature's Economy: A History of Ecological Ideas* (New York: Cambridge University Press, 1977).
4. This is the theme in particular of Clement's book *Plant Succession* (Washington: Carnegie Institution, 1916).
5. Worster, *Nature's Economy,* p. 210.
6. Clement's major rival for influence in the United States was Henry Chandler Cowles of the University of Chicago, whose first paper on ecological succession appeared in 1899. The best study of Cowles's ideas is J. Ronald Engel, *Sacred Sands: The Struggle for Community in the Indiana Dunes* (Middletown, Conn.: Wesleyan University Press, 1983), pp. 137–159. Engel describes him as having a less deterministic, more pluralistic notion of succession, one that "opened the way to a more creative role for human beings in nature's evolutionary adventure" (p. 150). See also Ronald C. Tobey, *Saving the Prairies: The Life Cycle of the Founding School of American Plant Ecology, 1895–1955* (Berkeley: University of California, 1981).
7. Sears, *Deserts on the March,* p. 142.
8. This book was co-authored with his brother Howard T. Odum, and it went through two more editions, the last appearing in 1971.
9. Eugene P. Odum, *Fundamentals of Ecology* (Philadelphia: W. B. Saunders, 1971), p. 8.
10. Odum, "The Strategy of Ecosystem Development," Science, p. 164 (18 April 1969), p. 266.
11. The terms "K-selection" and "r-selection" came from Robert MacArthur and Edward O. Wilson, *Theory of Island Biogeography* (Princeton: Princeton University Press, 1967). Along with Odum, MacArthur was the leading spokesman for the view of nature as a series of thermodynamically balanced ecosystems during the 1950s and 60s.
12. Odum, "Strategy of Ecosystem Development," p. 266. See also Odum, "Trends Expected in Stressed Ecosystems," BioScience, 35 (July/August 1985), pp. 419–22.
13. A book of that title was published by Earl F. Murphy: Governing Nature (Chicago: Quadrangle Books, 1967). From time to time, Eugene Odum himself seems to have caught that ambition or lent his support to it, and it was certainly central to the work of his brother, Howard T. Odum. On this theme see Peter J. Taylor, "Technocratic Optimism, H. T. Odum, and the Partial Transformation of Ecological Metaphor after World War II," Journal of the History of Biology, 21 (Summer 1988), pp. 213–44.
14. A very influential popularization of Odum's view of nature (though he is never actually referred to in it) is Barry Commoner's *The Closing Circle: Nature, Man, and Technology* (New York: Knopf, 1971). See in particular the discussion of the four "laws" of ecology (pp. 33–46).
15. Communication from Malcolm Gherrett, Ecology, 70 (March 1989), pp. 41–42.
16. See Michael Begon, John L. Harper, and Colin R. Townsend, *Ecology: Individuals, Populations, and Communities* (Sunderland, Mass.: Sinauer, 1986). In another textbook, Odum's views are presented critically as the traditional approach: R. J. Putnam and S. D. Wratten, *Principles of Ecology* (Berkeley: University of California Press, 1984). More loyal to the ecosystem model are Paul Ehrlich and Jonathan Roughgarden, *The Science of Ecology* (New York: Macmillan, 1987); and Robert Leo Smith, *Elements of Ecology,* 2nd ed. (New York: Harper & Row, 1986), though the latter admits that he has shifted from an "ecosystem approach" to more of an "evolutionary approach" (xiii).
17. William H. Drury and lan C. T. Nisbet, "Succession," Journal of the Arnold Arboretum, 54 (July 1973), p. 360.
18. H. A. Gleason, "The Individualistic Concept of the Plant Association," Bulletin of the Torrey Botanical Club, 53 (1926), 25. A later version of the same article appeared in American Midland Naturalist, 21 (1939), pp. 92–110.
19. Joseph H. Connell and Ralph O. Slayter, "Mechanisms of Succession in Natural Communities and Their Role in Community Stability and Organization," The American Naturalist, 111 (Nov.–Dec. 1977), pp. 1119–1144.
20. Margaret Bryan Davis, "Climatic Instability, Time Lags, and Community Disequilibrium," in *Community Ecology,* ed. Jared Diamond and Ted J. Case (New York: Harper & Row, 1986), p. 269.
21. James R. Karr and Kathryn E. Freemark, "Disturbance and Vertebrates: An Integrative Perspective," *The Ecology of Natural Disturbance and Patch Dynamics,* ed. S.T.A. Pickett and P. S. White (Orlando, Fla.: Academic Press, 1985), pp. 154–155. The Odum school of thought is, however, by no means silent. Another recent compilation has been put together in his honor, and many of its authors express a continuing support for his ideas: L. R. Pomeroy and J. J. Alberts, eds., *Concepts of Ecosystem Ecology: A Comparative View* (New York: Springer-Verlag, 1988).
22. Orie L. Loucks, Mary L. Plumb-Mentjes, and Deborah Rogers, "Gap Processes and Large-Scale Disturbances in Sand Prairies," in Pomeroy and Alberts, ibid., pp. 72–85.
23. For the rise of population biology see Sharon E. Kingsland, *Modeling Nature: Episodes in the History of Population Biology* (Chicago: University of Chicago Press, 1985).
24. An influential exception to this tendency is F. H. Bormann and G. E. Likens, *Pattern and Process in a Forested Ecosystem* (New York: Springer-Verlag, 1979), which proposes (in chap. 6) the model of a "shifting mosaic steady-state." See also P. Yodzis, "The Stability of Real Ecosystems," Nature, 289 (19 February 1981), pp. 674–676.
25. Paul Colinvaux, *Why Big Fierce Animals Are Rare: An Ecologist's Perspective* (Princeton: Princeton University Press, 1978), pp. 117, 135.

26. Thomas Soderqvist, *The Ecologists: From Merry Naturalists to Saviours of the Nation. A Sociologically Informed Narrative Survey of the Ecologization of Sweden,* 18951975. (Stockholm: Almqvist & Wiksell International, 1986), p. 281.
27. This argument is made with great intellectual force by Ilya Prigogine and Isabelle Stengers, *Order Out of Chaos: Man's New Dialogue with Nature* (Boulder: Shambala/ New Science Library, 1984). Prigogine won the Nobel Prize in 1977 for his work on the thermodynamics of nonequilibrium systems.
28. An excellent account of the change in thinking is James Gleick, *Chaos: The Making of a New Science* (New York: Viking, 1987). I have drawn on his explanation extensively here. What Gleick does not explore are the striking intellectual parallels between chaotic theory in science and post-modem discourse in literature and philosophy. Post-Modernism is a sensibility that has abandoned the historic search for unity and order in nature, taking an ironic view of existence and debunking all established faiths. According to Todd Gitlin, "Post-Modernism reflects the fact that a new moral structure has not yet been built and our culture has not yet found a language for articulating the new understandings we are trying, haltingly, to live with. It objects to all principles, all commitments, all crusades—in the name of an unconscientious evasion." On the other hand, and put more positively, the new sensibility leads to a new emphasis on democratic coexistence: "a new 'moral ecology'—that in the preservation of the other is a condition for the preservation of the self." Gitlin, "PostModernism: The Stenography of Surfaces," New Perspectives Quarterly, 6 (Spring 1989), pp. 57, 59. See also N. Catherine Hayles, *Chaos Bound: Orderly Disorder in Contemporary Literature and Science* (Ithaca: Cornell University Press, 1990), esp. Chap. 7.
29. The paper was published in Science, 186 (1974), pp. 645–647. See also Robert M. May, "Simple Mathematical Models with Very Complicated Dynamics," Nature, 261 (1976),45967. Gleick discusses May's work on pages 69–80 of Chaos.
30. W. M. Schaeffer, "Chaos in Ecology and Epidemiology," in *Chaos in Biological Systems,* ed., H. Degan, A. V. Holden, and L. F. Olsen (New York: Plenum Press, 1987), p. 233. See also Schaeffer, "Order and Chaos in Ecological Systems," Ecology, 66 (Feb. 1985), pp. 93–106.
31. John Muir, *My First Summer in the Sierra* (1911; Boston: Houghton Mifflin, 1944), p. 157.
32. Prigogine and Stengers, *Order Out of Chaos,* pp. 312–313.
33. Much of the alarm that Sears and Odum, among others, expressed has shifted to a global perspective, and the older equilibrium thinking has been taken up by scientists concerned about the geo- and biochemical condition of the planet as a whole and about human threats, particularly from the burning of fossil fuels, to its stability. One of the most influential texts in this new development is James Lovelock's *Gaia: A New Look at Life on Earth* (Oxford: Oxford University Press, 1979). See also Edward Goldsmith, "Gala: Some Implications for Theoretical Ecology," The Ecologist, 18, nos. 2/3 (1988): pp. 64–74.

Readings for

LANDSCAPE ECOLOGY

Living in a Mosaic

Monica G. Turner

- **Carolyn T. Hunsaker and Daniel A. Levine**
  Hierarchical Approaches to the Study of Water Quality in Rivers
- **David J. Mladenoff and Theodore A. Sickley**
  Assessing Potential Gray Wolf Restoration in the Northeastern United States
- **Steward T. A. Pickett and Mary L. Cadenasso**
  Landscape Ecology: Spatial Heterogeneity in Ecological Systems
- **William H. Romme and Dennis H. Knight**
  Landscape Diversity: The Concept Applied to Yellowstone Park
- **John A. Wiens**
  Spatial Scaling in Ecology

Introduction to

Carolyn T. Hunsaker and Daniel A. Levine

Hierarchical Approaches to the Study of Water Quality in Rivers

by Monica G. Turner

Understanding how water quality is related to the upland areas within a landscape, especially the distribution of land uses within the watershed, remains an important topic in the study of land-water interactions. In this study, Drs. Hunsaker and Levine conduct a series of statistical analyses to determine what attributes about the landscape are best able to predict water quality in streams and rivers, and over what spatial scales these variables are important. This paper illustrates a relationship between landscape pattern and ecosystem processes, as components of water quality include the concentrations or loadings of nutrients such as nitrogen and phosphorus. In addition, it nicely illustrates an approach to the hierarchical study of landscapes.

Hierarchical Approaches to the Study of Water Quality in Rivers

CAROLYN T. HUNSAKER AND DANIEL A. LEVINE

Spatial scale and terrestrial processes are important in developing models to translate research results to management practices. The location of various types of land use in a watershed is critical to modeling.

Hunsaker, C. T. and D. A. Levine. 1995. Hierarchical approaches to the study of water quality in rivers. *Bioscience* 45(3):193–203.

Land-use change may be the single greatest factor affecting ecological resources. Allan and Flecker (1993), who have identified six major factors threatening the destruction of river ecosystems, state that various transformations of the landscape—hydrologic changes to streams and rivers resulting from changes in land use, habitat alteration, and nonpoint source pollution—are probably the most widespread and potent threats to the well-being of lotic ecosystems.

Measures of landscape structure are necessary to monitor change and assess the risks it poses to ecological resources (Graham et al. 1991, Hunsaker et al. 1990). Landscape ecologists seek to better understand the relationships between landscape structure and ecosystem processes at various spatial scales (Forman and Godron 1986, Risser 1987, Turner 1987, 1989). We use the word structure to refer to the spatial relationships of ecosystem characteristics such as vegetation, animal distributions, and soil types. Processes or function refers to the interactions—that is, the flow of energy, materials, and organisms— between the spatial elements. Because landscapes are spatially heterogeneous areas, or environmental mosaics, the structure and function of landscapes are themselves scale-dependent.

Scientists often organize spatial scale in a hierarchical manner such as nested watersheds and ecoregions (Crowley 1967, O'Neill et al. 1989). For example, the U.S. Geological Survey (USGS) has defined the Hydrologic Unit Codes as a four-level hierarchy or logical arrangement of river basins where each larger unit is an aggregate of smaller units (Seaber et al. 1984). We are developing methods to characterize landscape attributes that influence water quality at various spatial scales. Understanding how scale, both data resolution and geographic extent, influences landscape characterization and how terrestrial processes affect water quality are critically important for model development and translation of research results from experimental watersheds to management of large drainage basins.

Landscape Characterization and Water Quality

Streams and rivers serve as integrators of terrestrial landscape characteristics and as recipients of pollutants from both the atmosphere and the landscape; thus, large rivers are especially good indicators of cumulative impacts. Many studies have shown that the proportion of different land uses within a watershed can account for some of the variability in river water quality (DelRegno and Atkinson 1988, Omernik 1977, Reckhow et al. 1980, Sivertun et al. 1988). For example, the water quality in a watershed with 50% agricultural land use and an intact forest riparian zone may be expected to be better (e.g., lower turbidity and nutrients) than that in a similar watershed without any riparian zone. In addition, several researchers have addressed the issue of whether land use close to streams is a better predictor of water quality than land use over the entire watershed (Omernik 1981, Osborne and Wiley 1988, Wilkin and Jackson 1983). Research on nutrient and sediment movement within small watersheds with forest or grass buffer areas between streams and disturbed uplands generally supports such statements (Cooper et al. 1987, Lowrance et al. 1984, Peterjohn and Correll 1984, Schlosser and Karl 1981). However, conclusions by Omernik et al. (1981) for larger watersheds from a wide variety of hydrologic settings suggest that upland land uses are as important as near- stream land uses.

As the issues being addressed by ecological research and resource management have become more complex and integrative, an interesting question has surfaced with regard to the spatial construct we use to characterize regions. Geographers have struggled for a long time with spatial interrelationships and questions of geographic characterization. A regional characterization scheme is especially difficult for aquatic ecosystems.

Two accepted approaches exist, watersheds (Likens et al. 1977, Lotspeich 1980) and ecoregions (Omernik 1987), and both can be hierarchically constructed. A watershed is an area of land draining to a specific point on a stream or to a lake or wetland; watersheds are based on topography and the observation that water flows downslope because of gravity. An ecoregion is an area of relative homogeneity based on one or more attributes such as climate, bedrock geology, or soil properties.

Using nested watersheds as the hierarchical regional characterization scheme, this article addresses three questions relevant to characterizing landscape attributes important to water quality:

- Are both the proportions of land uses and the spatial pattern of land uses important for characterizing and modeling river water quality in watersheds of different areas?
- Can land use near the stream better account for the variability in water quality than land use for the entire watershed?
- Does the size of the watershed influence statistical relationships between landscape characteristics and water quality or model performance ?

The results of this work are likely to aid the understanding and management of nonpoint-source pollution for large geographic areas.

We performed spatial analyses on raster, or cell, data using several different geographic information systems (which are often referred to by the initials GIS). A geographic information system is a computerized mapping system for capture, storage, management, analysis, and display of spatial and descriptive data. In a raster-based system, numeric values for map data are represented in a grid containing rows and columns of cells of a prescribed size. Each cell corresponds to a fixed area on the earth.

Figure 1. The Wabash River study site in Illinois contains three river basins— the Little Wabash, Skillet Fork, and Saline—comprising 1.35 million ha.

To demonstrate hierarchical approaches to the study of large rivers, we used two case studies: the upper Little Wabash River in Illinois and the Lake Ray Roberts drainage of the Trinity River in Texas. All available monitoring and research data were used. Because we relied on data collected by others and used various sampling designs, we were constrained in the type and intensity of statistical analyses that could be applied. (This problem, which is standard in regional analyses of water quality, grows larger with spatial scale.)

We sought to work in two very different landscapes. Although much water-quality monitoring is done in the United States, few large regions have consistent and long-term monitoring networks that can support a hierarchical analysis. We selected the Illinois and Texas study areas because the water-quality monitoring was adequate for our purposes. In addition, they had land-cover and soils data available in a digital format.

The studies were conducted concurrently; preliminary findings were shared between the projects and helped shape the research. Although the studies undertook two different modeling approaches, lumped for Illinois and spatially explicit for Texas, the use of a hydrologically active area (Novotnoy and Chesters 1981) defined by the stream network and/or topography was incorporated into both. Within a watershed the areas that produce surface runoff are called hydrologically active; the rest of the watershed contributes only to interflow and base flow. The hydrologic activity of an area is a stochastic phenomenon depending on the magnitude and intensity of the storm, soil conditions, and surface characteristics of the area.

The Wabash River Study Site in Southeastern Illinois

The Wabash River study site contains three river basins—Little Wabash, Skillet Fork, and Saline— comprising 1.35 million ha (Figure 1). Land-use/land-cover data came from the USGS (Anderson et al. 1976) based on aerial photographs taken from 1974 to 1976. For land cover (Figure 2), we used a cell size of 200 m on a side (an area of 4 ha). Seven land-use classes occur on the site: agriculture, forest, rangeland, barren, wetland, urban, and water. The stream network and digital elevation models were based on topographic maps with a scale of 1:100,000. Each original cell in the digital elevation model represented 90 m x 70 m; these data were resampled to match the land-cover resolution. One cell in the resampled digital elevation models represented approximately four original cells.

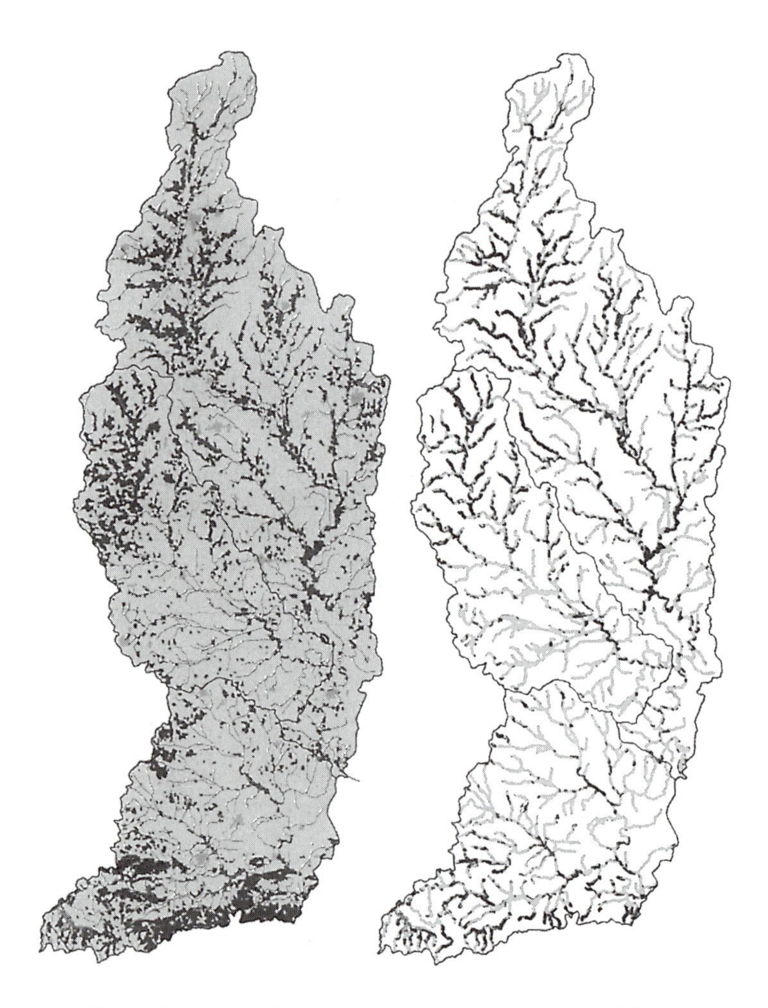

Figure 2. Land-use coverage for the 400-meter equidistant corridor around streams (right) and for the entire Illinois site (left). [Editor's note: see original paper for color image. Here, medium gray indicates rangeland and agriculture; darker gray patches are urban areas (left figure). Other shades indicate mainly forest and wetlands following water channels (right figure).]

Water chemistry data (as concentrations) were retrieved from the U.S. Environmental Protection Agency's STORET (storage and retrieval) database. We identified 47 water-quality monitoring stations with adequate data in the study area (Figure 3). Stations with fewer than three samples in a year or with variances equal to zero were not used. For each station, data were averaged over the period 1974 through 1977. The mean values were weighted by using the number of samples that went into each mean divided by the variance around the mean. In effect, means with larger sample sizes and lower variances had higher weights than means with smaller sample sizes and larger variances. Only data for total nitrogen, total phosphorus, and conductivity were considered extensive enough for analysis (i.e., enough stations and enough samples per station per year). Not all chemistry parameters were monitored at all stations: total nitrogen was measured at 36 stations, total phosphorus at 33, and conductivity at 36. The watersheds of each station included in our study were delineated on USGS 1 :24,000-scale topographic maps, and their boundaries were digitized. The hierarchical structure of the watersheds is illustrated in the network diagram in Figure 3.

Landscape Characterization

Several metrics have been proposed to quantify landscape pattern (Baker and Cai 1992, O'Neill et al. 1988). In our study, we included the proportion of the seven land-use types and the amount of edge between different land uses—forest and agriculture, forest and barren, wetland and agriculture, and wetland and barren. We also used the dominance

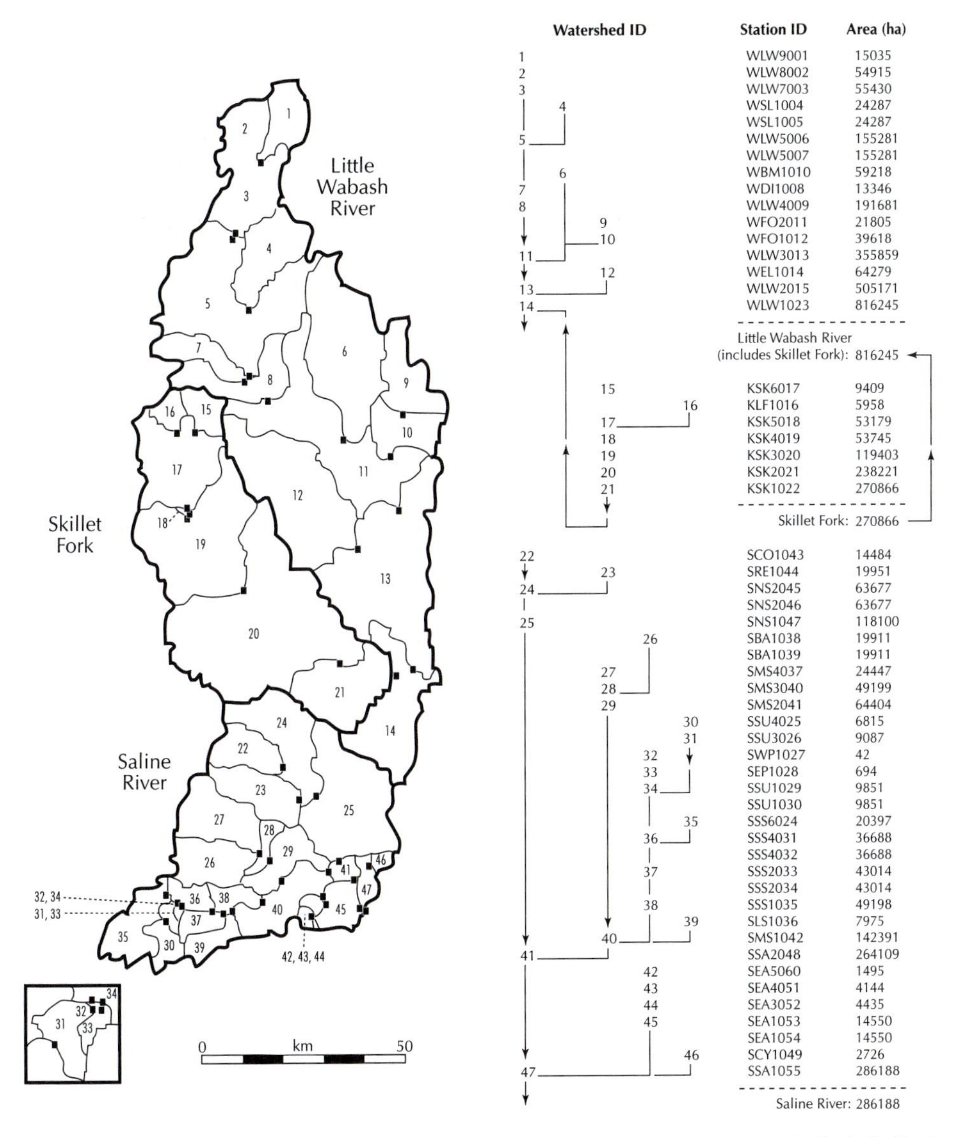

Figure 3. Location of monitoring stations used in the Wabash River study. The chart shows the hydrologic flow between watersheds and the area of each watershed. If two different agencies monitor the same site, only one set of data was used in model development.

metric, which measures the extent to which one or a few land uses dominate the landscape. Another measure employed was a three-cell contagion to measure the extent to which the landscape is fragmented (Hunsaker et al. 1994). The landscape metrics were calculated using a custom program (Timmins and Hunsaker, in press).

Because each of the 13 landscape metrics was needed for each of 47 watersheds and for the corridors around streams, a geographic information system was employed. Individual watershed boundaries, representing the areas draining to each sampling station, were used as an overlay to quantify the total area of each land-cover type within each watershed. Two methods were used to define hydrologically active areas (Figure 4). Equidistance corridors or buffers around the stream network were generated using the ARC/INFO™ geographic information system (ESRI 1987) in widths of 200 m and 400 m (one and two cells) on each side of the stream (Figure 2). These were the smallest areas possible, given the data resolution. A model of the hydrologically active area was also calculated from a digital elevation model using the COUNT program developed by Jensen and Dominque (1988). This program determines the upslope area that drains into each cell. The value in each cell of the resulting file is an actual count of the number of cells that are topographically upslope and therefore contribute hydrologically to that cell. We used an arbitrarily selected COUNT threshold of 3 5 cells (140 ha) to define the hydrologically active area. Every cell with a value of 35 or greater was assumed to be contributing flow and therefore influencing water quality (i.e., a cell had to have 35 or more cells flowing into it to be considered part of the hydrologically active area). Landscape metrics were calculated for the entire watershed and both of the areas estimated to be hydrologically active using different techniques.

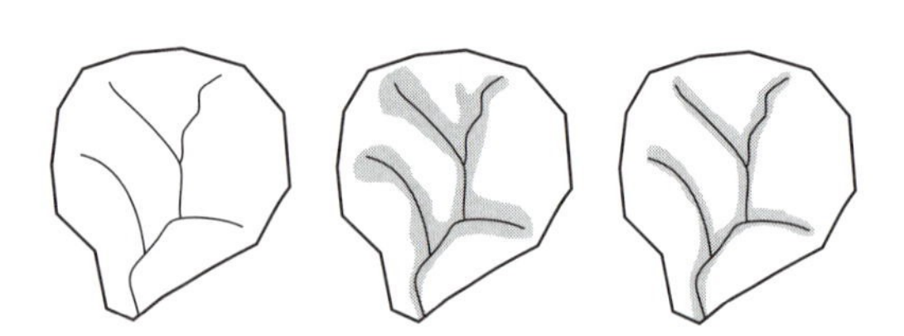

Figure 4. Representation of hydrologically active areas. Left, a stream network within a small watershed. Center, the shaded area represents an irregularly shaped, hydrologically active area surrounding the stream network, which might be produced by an analysis based on topography, soils, and land cover. Right, shaded area represents a regularly shaped hydrologically active area, which would be produced by an equidistance buffer around the stream network.

Modeling Approach

The spatial analysis capabilities of a geographic information system and multivariate statistics were used to develop empirical/statistical models to address our first two questions. Total nitrogen, total phosphorus, and conductivity were treated as the dependent variables, and the 13 landscape metrics were used as explanatory variables. For the Illinois site, the best reduced model (seven or eight variables) was determined using the R-square procedure (SAS Institute 1985). Separate models were developed for the full watersheds, the hydrologically active area based on topography within watersheds (defined by the COUNT program), and the equidistance areas around the stream network within watersheds.

Three generally recommended criteria (Draper and Smith 1981) were used in selecting the best partial model. First, the sum of squared error for the partial model could not differ significantly from the sum of squared error for the full model (13 variables). Second, the amount of variability (R^2) explained by the partial model had to be at least 95% of that explained by the full model. Third, it was preferred that the Cp statistic be positive and reasonably close to the degrees of freedom plus one (Draper and Smith 1981). However, the latter condition was often hard to meet and was relaxed in some cases. Finally, the independent variables in the model had to be scientifically sensible. For example, nutrient levels would logically increase with increased agriculture in a watershed and decrease as the amount of undisturbed forest increases. The effect of collinearity was evaluated using the maximum condition index (Belsley et al. 1980). For all cases, the maximum condition index was less than 13, which indicates only moderate collinearity. Collinearity is a measure of similarity in the variability explained by two parameters and is a measure of independence.

Landscape Pattern and its Relationship to Water Quality

We evaluated three chemistry parameters that have extensive data collection—total nitrogen, total phosphorus, and conductivity—and that are representative of general water quality. Nitrogen and phosphorus are nutrients that in excess can cause eutrophication in the form of algal blooms and oxygen depletion. Conductivity provides a rapid assessment of the dissolved solids content (ion concentration) of water and thus can serve as a general indicator of water purity.

Relationships between fish and phytoplankton productivity and total dissolved solids concentrations have been developed using the morphoedaphic index (Adams et al. 1983). The regression models for total phosphorus, total nitrogen, and conductivity are presented in Table 1. The R^2 values are given for both the reduced models (seven or eight variables) and for the individual contribution to the variance by the most influential variables. In other words, the R^2 values under the "single variable model" headings are for each explanatory variable regressed by itself against the water quality parameter.

Both the proportions of land-cover types and their spatial pattern (i.e., amount of edge, dominance, and contagion) are useful in characterizing water quality; however, proportions of cover types consistently account for the most variance (i.e., large R^2). Such landscape metrics can account for 40% to 86% of the variance in stream quality across a range of watershed sizes (1000 to 1.35 million ha). Landscape metrics were least effective for total nitrogen, often accounting for only half of the variance in this parameter. No strong pattern emerged where a few of the same landscape metrics consistently accounted for a significant amount of variance, either within the models for an individual water-quality parameter or within the type of watershed model (entire watershed or hydrologically active area).

Using landscape data for the entire watershed consistently explains the most variance in water quality. Results for equidistance corridors, both 200 m and 400 m, and hydrologically active area based on topography are similar; R^2 values were usually ten units less for those areas than those for the entire watershed. In general, hydrologically active areas around streams contain a significant proportion of the forest remaining in the watersheds. This case is especially true for the Little Wabash basin (Figure 2). The Skillet Fork basin has a predominance of agriculture, even within the stream corridors, while the Saline basin contains a significant amount of barren land (from mining operations) and wetlands near streams.

Figure 5. Location of the Lake Ray Roberts watershed (arrowhead).

In general, the direction of correlations was logical and consistent between the models used. For example, disturbed land covers like agriculture, barren, and rangeland have positive associations with water-quality parameters; that is, as the proportion of agriculture increases, so does the amount of nitrogen, phosphorus, and conductivity. Contagion and proportion of forest are negatively correlated with water-quality parameters (Hunsaker et al. 1992). Thus, an area that has contiguous land covers (is not fragmented) or that is dominated by forests tends to have better water quality.

The Lake Ray Roberts Study Site in Northern Texas

The Lake Ray Roberts study site comprises 179,821 ha and contains two river systems—the Elm Fork of the Trinity River and the Isle du Bois Creek—which form the headwaters of the Trinity River (Figure 5). These two basins were subdivided into 12 watersheds. The drainage

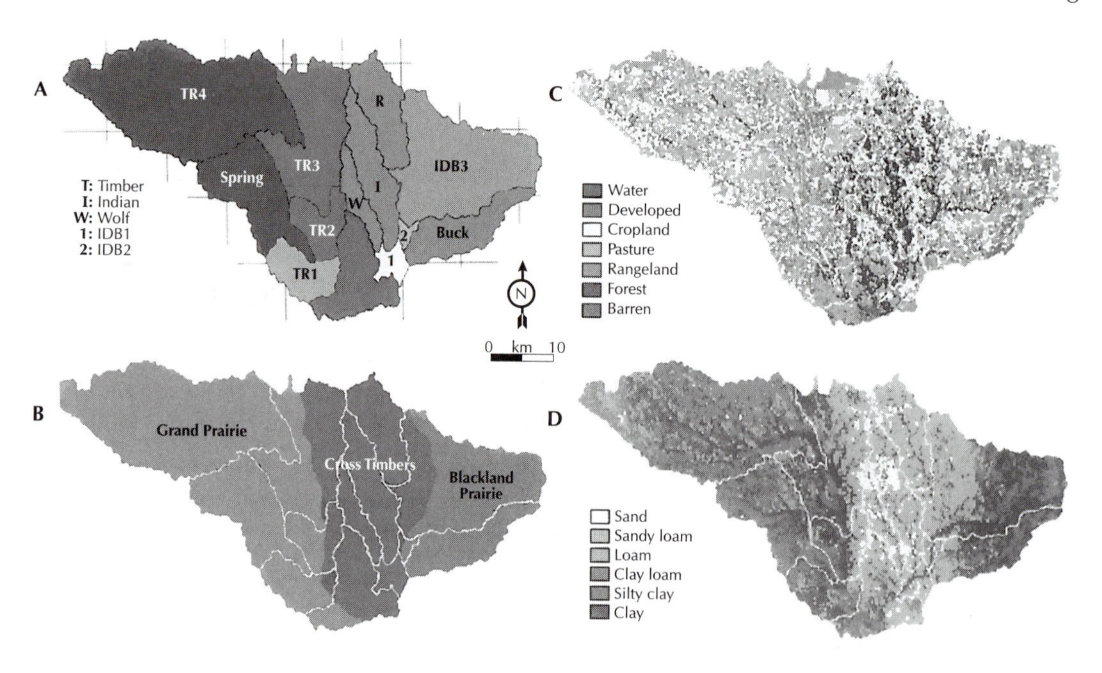

Figure 6. Lake Ray Roberts Basin characteristics. **A.** Delineation of modeling watersheds: watersheds TR1, TR2, TR3, TR4, Spring are drained by the Elm Fork of the Trinity River, and the watersheds W, I, T, Buck, 1, 2 are drained by the Isle du Bois River. **B.** Physiographic regions. **C.**. Land-use/land-cover classification. **D.** Soil textures. [Editor's note: see original paper for color image.]

basins of each river system are different in morphometry, land use, and soil type. These differences are delineated by three physiographic regions. The Grand Prairie region, which is drained by the Elm Fork of the Trinity River, is defined by gentle topography, mostly clay loam soils, and cropland and rangeland as the dominant land cover (Figure 6). The Cross Timbers region, which is drained by the western arm of the Isle du Bois River system, is characterized by slightly more severe topography than the other regions, well-drained sandy soils, and forest and pasture as the dominant landcover types. The Blackland Prairie region, drained by the eastern arm of the Isle du Bois River system, is relatively flat, dominated by clay soils, and has a mixture of cropland and pasture land-cover types.

Water-quality data for this study were obtained from the University of North Texas (Pillard 1988) and were collected biweekly from May 1985 through December 1986. The water-quality parameters used were total phosphorus, total nitrogen, total suspended solids, and instantaneous flow velocity. Total mass loading for a sampling period for each pollutant was determined by multiplying the measured pollutant concentration by the flow for the same date and then multiplying that number by the number of days for the period around each sampling date. Period total loads were then summed to get an annual mass load. Flow and chemistry data were collected during several storm events, and these data dominate the calculations of annual loads.

A cell-based GIS dataset was compiled from various sources. Watershed boundaries, streams, and elevation contours were digitized from 1:24,000 USGS quadrangle maps. Soils data were obtained from the Soil Conservation Service's Map Image Analysis and Display database with a cell resolution of 250 m on a side. The study area contains 105 soil types, and soil attributes were taken from county soil surveys (including mean particle diameter, permeability, and the potential soil erodibility k-factor).

Land-use/land-cover data were obtained from the 1986 Landsat multispectral sensor coverage, classified at 80-meter resolution by the University of North Texas. The landcover classes used for our modeling were: commercial/residential, industrial/transportation (characterized as in Figure 6c), water, cropland, maintained pasture, natural rangeland, forest, and barren land. A digital elevation model, 20-meter resolution, was developed from the digitized contours and used to calculate slope angle, direction of flow, and a COUNT file. Soils and landuse layers were resampled to yield a 20-meter cell size to match the digital elevation model resolution and fit the model requirements.

Modeling Approach

The modeling approach for the Ray Roberts study linked statistical modeling of nutrient and sediment delivery from each cell with the hydrologic flowpath for each cell to determine the total nutrient and sediment load that reached each watershed outlet (Figure 7). Levine et al. (1993) reviewed the literature on vegetated filterstrip research and, using consistent criteria, they selected 13 studies for development of statistical models for delivery ratios. A delivery ratio is determined by dividing the amount of sediment or nutrient that flows out of a plot of land by the amount that flows into it. The physical and chemical parameters within the plot of land, among other parameters, are related to the magnitude of the delivery ratio. Both linear (forward variable selection procedure, R-square) and nonlinear (NLIN,

Table 1. Linear regression models for the Illinois site for entire watershed, contributing area, and 200-meter and 400-meter equidistance corridors. R^2 values are given for both the best seven or eight variable models and for individual independent variables. Variables are a pattern metric, the proportions of watershed in a listed land use or the amounts of edge between two listed land uses.

	Total phosphorus		Total nitrogen		Conductivity	
	Land use	R^2	Land use	R^2	Land use	R^2
Entire watershed						
Single variable model	Urban	25	Urban	17	Barren	55
	Barren	14	Wetland/agriculture	16	Rangeland	37
	Forest/barren*	12	Wetland	11	Forest/barren	24
	Wetland/barren	12	Forest/agriculture	7	Wetland/agriculture	18
	Rangeland	11	Forest	3	Wetland	14
Best complex model	(8 variables)	86	(8 variables)	53	(8 variables)	84
Hydrologically active area						
Single variable model	Contagion†	19	Urban	8	Rangeland	48
	Barren	11	Forest	6	Contagion	11
	Rangeland	11	Barren	5	Barren	7
	Wetland	5	Rangeland	5	Forest	7
	Wetland/barren	4	Wetland/agriculture	4	Forest/barren	5
Best complex model	(8 variables)	76	(8 variables)	41	(8 variables)	76
200-meter and 400-meter corridors‡						
Single variable model	Barren		Contagion		Barren	
	Wetland/agriculture		Dominance†		Forest/barren	
	Forest/barren				Rangeland	
	Contagion				Forest	
Best complex model	(8 variables)	71	(8 variables)	42	(7 variables)	75

*Amount of edge between forest and barren land covers. All variables with a "/" are amount of edge.
†Contagion and dominance have no units.
‡200-meter and 400-meter corridors were analyzed separately. The variables listed under each water-quality parameter were important in both corridor widths. The order of importance changed slightly and the R^2 changed.

Table 2. Total annual loads for each Texas watershed calculated for the entire watershed and for delivery model results.

Location		Total annual loads calculated from entire watersheds			Total annual loads calculated from delivery models		
		Total phosphorus (kg/yr)	Total nitrogen (kg/yr)	Total suspended solids (kg/yr)	Total phosphorus (kg/yr)	Total nitrogen (kg/yr)	Total suspended solids (kg/yr)
Cross Timbers watersheds							
Timber Creek	Observed	2.78×10^3	1.67×10^4	9.37×10^5	2.78×10^3	1.67×10^4	9.37×10^5
	Estimated	1.29×10^4	5.94×10^4	6.23×10^6	2.75×10^3	1.55×10^4	1.31×10^6
	% Difference	364	256	564	–1	–7	39
Indian Creek	Observed	2.68×10^3	1.68×10^4	1.37×10^6	2.68×10^3	1.68×10^4	1.37×10^6
	Estimated	1.13×10^4	5.78×10^4	6.51×10^6	2.64×10^3	1.97×10^4	1.86×10^6
	% Difference	321	244	375	–1	17	36
Wolf Creek	Observed	1.27×10^2	2.69×10^3	7.80×10^4	1.27×10^2	2.69×10^3	7.80×10^4
	Estimated	4.93×10^3	2.23×10^4	2.53×10^6	1.05×10^3	6.73×10^3	6.07×10^5
	% Difference	3782	729	3243	726	150	701
IDBI	Observed	4.83×10^4	3.04×10^5	2.56×10^7	4.83×10^4	3.04×10^5	2.56×10^7
	Estimated	1.08×10^5	4.76×10^5	7.34×10^7	3.82×10^4	1.88×10^5	2.62×10^7
	% Difference	123	57	187	–21	–38	2
Grand Prairie watersheds							
Spring Creek	Observed	8.86×10^3	5.13×10^4	2.49×10^6	8.86×10^3	5.13×10^4	2.49×10^6
	Estimated	2.54×10^4	1.06×10^5	1.88×10^7	8.86×10^3	3.64×10^4	6.42×10^6
	% Difference	187	107	655	0	–29	158
TR4	Observed	2.03×10^4	9.76×10^4	1.73×10^6	2.03×10^4	9.76×10^4	1.73×10^6
	Estimated	6.98×10^4	2.93×10^5	1.07×10^8	2.24×10^4	9.60×10^4	4.62×10^7
	% Difference	244	200	6085	10	–2	167
TR3	Observed	2.72×10^4	1.52×10^5	4.33×10^6	2.72×10^4	1.52×10^5	4.33×10^6
	Estimated	9.93×10^4	4.20×10^5	1.74×10^8	2.67×10^4	1.15×10^5	4.98×10^7
	% Difference	265	176	3918	–2	–24	1050
TR2	Observed	3.36×10^4	2.24×10^5	1.13×10^7	3.36×10^4	2.24×10^5	1.13×10^7
	Estimated	1.08×10^5	4.57×10^5	1.80×10^5	2.95×10^4	1.29×10^5	5.19×10^7
	% Difference	221	104	1493	–12	–43	359
TR1	Observed	3.78×10^4	2.78×10^5	1.30×10^7	3.78×10^4	2.78×10^5	1.30×10^7
	Estimated	1.43×10^5	6.03×10^5	2.07×10^8	4.15×10^4	1.79×10^5	5.83×10^7
	% Difference	278	117	1492	21	–35	348
Blackland Prairie watersheds							
Buck Creek	Observed	8.55×10^3	3.89×10^4	3.64×10^6	8.55×10^3	3.89×10^4	3.64×10^6
	Estimated	1.57×10^4	6.49×10^4	1.16×10^7	8.45×10^3	3.79×10^4	6.07×10^7
	% Difference	84	67	219	–1	–3	1567
IDB3	Observed	2.32×10^4	1.62×10^5	1.44×10^7	2.32×10^4	1.62×10^5	1.44×10^7
	Estimated	5.65×10^4	2.47×10^5	3.92×10^7	2.55×10^4	1.21×10^5	1.72×10^7
	% Difference	144	52	172	9	–25	19
IDB2	Observed	3.33×10^4	2.12×10^5	1.85×10^7	3.33×10^4	2.12×10^5	1.85×10^7
	Estimated	7.35×10^4	3.18×10^5	5.15×10^7	3.41×10^4	1.59×10^5	2.34×10^7
	% Difference	121	50	178	2	–25	26

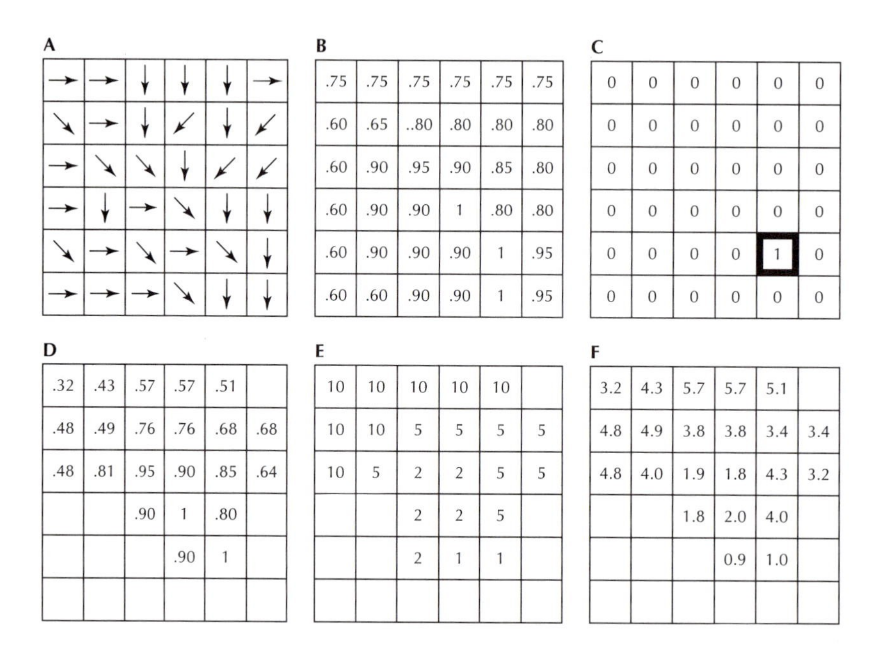

Figure 7. Geographic information system modeling procedure for the Ray Roberts study. **A.** Direction of flow from each cell is calculated based on the digital elevation model. **B.** Cell delivery ratios are calculated using regression equations and GIS layers of soil, slope, and land use. **C.** Location of watershed seed cells are based on location of water-quality sampling locations. **D.** Total flow path delivery is calculated by starting at the seed cell, or outlet, for a watershed (C) and using the flow direction file (A) to walk up the hydrologic flow path for a watershed, multiplying cell delivery ratios (B) together along the way. **E.** Potential nutrient loads are determined by land use and soil types using export coefficients from the literature; potential sediment load is calculated using the Universal Soil Loss Equation. **F.** Actual delivered nutrient and sediment loads from each cell are calculated by multiplying the total flow path delivery ratio file (D) with the potential nutrient and sediment load file (E).

Marquardt method) regression models were developed using data from the 13 studies to determine these statistical relationships (SAS Institute 1985). The models predict the amount (in mass or concentration) of total phosphorus, total nitrogen, and total suspended solids delivered from one side of a plot of land to the other. The amount delivered generally depended on soil type, slope, and vegetative cover.

Although the linear models describe the data used to develop them (R^2 values of 0.78 to 90), when they were applied in the overall modeling approach they did not perform well (Levine et al. 1993). Therefore, nonlinear models were developed and incorporated into a geographic information system (IDRISI™), and delivery ratios for each pollutant and for each cell were calculated. The nonlinear equations developed from the vegetated filter strip data used three or four of the following watershed characteristics: distance of flow, soil permeability, Manning's roughness coefficient (represents land cover), soil mean particle diameter, and slope angle. Manning's roughness coefficient was used in the total phosphorus and total nitrogen models but not in the model for total suspended solids.

Delivery ratios were calculated for all cells in each watershed (Figure 6). Delivery ratios were then accumulated along the hydrologic flow path for each cell, based on the flow direction at each cell, to calculate total flow path delivery ratios. The delivery ratio is assumed to 1.0 (100% delivery) in stream cells.

The models were calibrated by increasing the density of the stream network using the COUNT file. The COUNT threshold was reduced to flow a smaller area to generate beam flow or 100% delivery. Because the soils, topography, and land-cover types in each physiographic region were different, we expected watersheds to behave differently in the hydrologic sense. Therefore, one watershed in each region was used to calibrate the models, resulting in three different calibration COUNT thresholds for each physiographic region. For each pollutant, the data layer for total flow path delivery ratio was multiplied by the data layer for potential nutrient and sediment loading, resulting in a layer representing total pollutant from each cell that reached the watershed outlet. The potential nutrient loads for total phosphorus and total nitrogen were taken from the literature and assigned to each cell based on land use, soil, and precipitation for the area. Potential soil export was estimated using the Universal Soil Loss Equation (Wischmeier and Smith 1978).

The models developed for the Texas site performed well. Total phosphorus and total nitrogen were estimated to within 5% and 15% of observed values respectively (Table 2). However, estimates of total suspended solids were only within 40% of observed values. Calculation of total loads for the entire watershed area resulted in overestimating loads by several orders of magnitude. Models consistently overestimated observed values for the Wolf Creek watershed for each parameter because of an incomplete water-quality dataset; water-quality data were only collected for the final six months in 1986.

The success in calibrating the models by increasing the stream network density demonstrated the importance of this concept for modeling hydrologically active areas and the use of the COUNT program and flow path accumulation of delivery ratios in providing this capability. Calibration steps were initiated with a COUNT threshold of 200, which approximated the blue-line stream network on the 7.5-minute USGS quadrangle maps—a standard input for many nonpoint- source models. Stream network density is related in part to soil texture and is probably not the same for different water-quality parameters, as evidenced by varying model success. (In our application the same COUNT threshold was used for all parameters within watershed for each region.)

Total suspended solids were always overestimated, and it is not necessarily valid to assume that a cell that can transport 100% of soluble nitrogen entering it can also transport 100% of the total suspended solids. Because it takes more energy to transport solids across a cell than soluble nitrogen, the stream network density for total suspended solids would logically be smaller than that for nitrogen. We also say differences in the COUNT values to which the watersheds within different physiographic regions were calibrated. These values fell along a continuum that corresponded to the infiltration characteristics of the soils dominating each region. Thus the clay-dominated soils in the Blackland Prairie region required a COUNT value of 5 (contributing area of 2000 m^2) to generate a stream whereas the sandy soils of the

Cross Timbers region required a COUNT value of 15 (6000 m^2; Table 2).

The application of this model to nested watersheds spanning a range of sizes (4400-100,000 ha) and having different physiographic characteristics allows us to evaluate the influence of scale on spatially distributed modeling. The models appear to be transportable with respect to watershed size and characteristics, although there was a slight decrease in model accuracy in the larger watersheds. The models did perform equally well in each of the physiographic regions.

Conclusions

We compared the two different studies to evaluate the use of land-use data (both the proportions and spatial pattern of land-use types) for modeling water quality and to explore how scale and data resolution influence the type of spatial analyses performed. Both studies indicated that land-use proportions are important for characterizing and modeling water quality in watersheds. However, the spatial pattern—as generalized in the Illinois study by contagion, dominance, and edges—did not greatly influence water-quality models. Additionally, the Illinois study indicates that proximity to streams is not a critical factor in modeling water quality. However, the success of the distributed model in the Texas study indicates that the location of various types of land use in the watershed is critical to modeling.

The large differences in data resolution and the fundamentally different approaches between these studies may explain these seemingly contradictory conclusions. The multivariate regression models developed for Illinois were an effort to create terrestrial metrics of water quality that would provide predictive tools for impact analysis and landscape indicators for monitoring ecological condition. Initial results (Hunsaker et al. 1992) indicated that we could accurately predict water quality using proportions of land cover and/or pattern metrics for entire watersheds (Table 1). The explanatory power of the models lessened when we used the same variables within corridors around the stream network. Omernik (1977) showed similar results by using only the proportions of forest, agriculture, and urban land uses.

For predicting annual nutrient loadings to streams, the Texas study showed to be reasonably accurate the application of empirical equations for delivery of nutrients combined with topographic analysis to define the contributing areas. Wilkin and Jackson (1983) and Osborne and Wiley (1988) also found that land use close to the stream was a better predictor of water quality than was land use for the entire watershed.

Results from the two studies indicate that the use of the COUNT threshold method for defining a hydrologically active area may not have been appropriate at the data resolution used in the Illinois study. The COUNT threshold used in the Illinois study of 35 cells represents an area of at least 140 ha. The COUNT thresholds in the Texas study were between 8 and 15 cells, representing areas between 0.32 to 0.6 ha. Areas slightly larger than this area actually contributed nutrients and sediment to total watershed load, because the threshold technique used in the Texas study identified only areas where delivery was 100%. Nevertheless, the total upslope area contributing nutrients to any one cell was still well within the area of a single cell used in the Illinois study. This result suggests that every cell in the Illinois watershed contributes to the water quality and explains why the models for the entire watershed were as good or better than the models using the stream-based corridors. The data resolution used in Illinois was not appropriate for modeling the hydrological active area. Because the resolution of our two datasets differed by an order of magnitude, we suggest that further work should be done with intermediate data resolutions to determine a breakpoint for the effectiveness of using hydrologically active areas.

The lumped approach employed in the Illinois study did not show any bias between watershed size and model performance, while the Texas study began to show a trend in decreasing model performance with the largest watersheds. The Texas study demonstrated that it is useful to calibrate models within similar physiographic regions (ecoregions). This finding suggests that when employing the lumped modeling approach in an area spanning a number of ecoregions, different models should be developed for each of the ecoregions.

It does not appear useful now to spend the effort (which can be substantial for large geographic areas) to create equidistance corridors or contributing areas from topography in the hope of improving lumped, empirical, nonpoint-source models using coarse-resolution data. However, if the goal is to identify areas critical for management purposes, it is important to identify hydrologically active areas and use a distributed modeling method and fine data resolution.

We conclude that management of nonpoint-source pollution in large river systems could benefit from a two-stage approach. A lumped approach with coarse-resolution data could be used as a screening method to identify watersheds making the most significant pollutant contributions. Then, a high-resolution distributed modeling technique could be used for those smaller watersheds identified as critical for specific management actions.

Acknowledgments

We wish to thank John Beauchamp and Barbara Jackson of Oak Ridge National Laboratory (ORNL) for statistical advice and computer analysis on this research. The Illinois Natural History Survey provided digital data for Illinois. Research was sponsored by the ORNL Exploratory Studies Program, U.S. Department of Energy, and the Office of Research and Development of the U.S. Environmental Protection Agency, under Interagency Agreement DWS9934921-01 with the U.S. Department of Energy, under contract DE-AC05-840R21400 with Martin Marietta Energy Systems, Inc. Although the research described in this article has been funded in part by the U.S. Environmental Protection Agency, it has not been subjected to agency review. Therefore, it does not necessarily reflect the views of the agency. Any mention of trade names or commercial products does not constitute an endorsement or a recommendation for use.

References cited

Adams, S. H., B. L. Kimmel and G. R. Ploskey. 1983. Sources of organic matter for reservoir fish production: a trophic-dynamics analysis. Can. J. Fish. Aquat. Sci. 40: 1480–1495.

Allan, J. D., and A. S. Flecker. 1993. Biodiversity conservation in running waters. BioScience 43:32–43.

Anderson, J. R., E. E. Hardy, J. T. Roach and R. E. Witmer. 1976. A land use and land cover classification system for use with remote sensor data. Geological Survey Professional Paper 964. U.S. Geological Survey, Washington, DC.

Baker, W. L., and Y. Cai. 1992. The r.leprograms for multiscale analysis of landscape structure using the GRASS geographical information system. Landscape Ecol. 7:291–302.

Belsley, D. A., E. Kuh, and R. E. Welsch 1980. Regression Diagnostics. John Wiley & Sons, New York.

Cooper, J. R., J. W. Gilliam, R. B. Daniels and W. P. Robarge. 1987. Riparian areas as filters for agricultural sediment. Soil Sci. Soc. Am. J. 51:416–420.

Crowley, J. M. 1967. Biogeography. Can. Geogr. 11:312–326.

DelRegno, K. J., and S. F. Atkinson. 1988. Nonpoint pollution and watershed management: A remote sensing and geographic information system (GIS) approach. Lake Reservoir Manage. 4:17–25.

Environmental Systems Research Institute (ESRI). 1987. ARC/INFO Users Guide. ESRI, Redlands, CA.

Draper, N. R., and H. Smith Jr. 1981. *Applied Regression Analysis.* 2nd ed. John Wiley & Sons, New York.

Forman, R. T. T., and M. Godron. 1986. *Landscape Ecology.* John Wiley & Sons, New York.

Graham, R. L., C. T. Hunsaker, R. V. O'Neill and B. L. Jackson. 1991. Ecological risk assessment at the regional scale. Ecological Applications 1:196–206.

Hunsaker, C. T., R. L. Graham, G. W. Suter II, R. V. O'Neill, L. W. Barnthouse, and R. H. Gardner. 1990. Assessing ecological risk on a regional scale. Environ. Manage. 14:325–332.

Hunsaker, C. T., D. A. Levine, S. P. Timmins, B. L. Jackson, and R. V. O'Neill. 1992. Landscape characterization for assessing regional water quality. In: *Ecological Indicators Elsevier Applied Science.* D. H. McKenzie, D. E. Hyatt, and V. J. McDonald, eds. New York. Pp. 997–1006

Hunsaker, C. T., R. V. O'Neill, B. L. Jack son, S. P. Timmins, D. A. Levine, and D. J. Norton. 1994. Sampling to characterize landscape pattern. Landscape Ecol. 9:207–226.

Jensen, S. K., and J. O. Dominque. 1988. Extracting topographic structure from digital elevation data for geographic information system analysis. Photogrammetric Engineering and Remote Sensing 54:1593–1600.

Levine, D. A., C. T. Hunsaker, S. P. Timmins, and J. J. Beauchamp. 1993. A geographic information system approach to modeling nutrient and sediment transport. ORNL report 6736. Environmental Sciences Division, Oak Ridge National Laboratory, Oak Ridge, TN.

Likens, G. E., F. H. Bormann, R. S. Pierce, J. S. Eaton, and N. M. Johnson. 1977. *Biogeochemistry of a Forested Ecosystem.* Springer- Verlag, New York.

Lotspeich, F. B. 1980. Watersheds as the basic ecosystem: This conceptual framework provides a basis for a natural classification system. Water Resour. Bull. 16:581–586.

Lowrance, R., R. Todd, J. Fair Jr., O. Hendrickson Jr., R. Leonard, and L. Asmussen. 1984. Riparian forests as nutrient filters in agricultural watersheds. BioScience 34:374–377.

Novotnoy, V. and G. Chesters. 1981. *Handbook of Nonpoint Pollution: Sources and Management.* Van Nostrand Reinhold, Atlanta, GA.

Omernik, J. M. 1977. Nonpoint source stream nutrient level relationships: A nationwide study. EPA-600/3-77-105. U.S. Environmental Protection Agency, Corvallis, OR.

Omernik, J. M. 1987. Ecoregions of the conterminous United States. Annals of the Association of American Geographers 77:118–125.

Omernik, J. M., A. R. Abernathy, and L. M. Male. 1981. Stream nutrient levels and proximity of agricultural and forest land to streams: Some relationships. J. Soil Water Conserv. 36:227–231.

O'Neill, R. V., et al. 1988. Indices of landscape pattern. Landscape Ecol. 1:153–162.

O'Neill, R. V., A. R. Johnson, and A. W. King. 1989. A hierarchical framework for the analysis of scale. Landscape Ecol. 3:193–205.

Osborne, L. L., and M. J. Wiley. 1988 Empirical relationship between land use/cover and stream water quality in an agricultural watershed. J. Environ. Manage. 26:9–27.

Peterjohn, W. T., and D. L. Correll. 1984. Nutrient dynamics in an agricultural watershed: Observations on the role of a riparian forest. Ecology 65:1466–1475.

Pillard, D. A. 1988. Pre-impoundment estimations of nutrient loading to Ray Roberts Lake and prediction of post-inundation trophic status. Ph.D. dissertation. University of North Texas, Denton, TX.

Reckhow, K. H., M. N. Beaulac, and J. T. Simpson. 1980. Modeling phosphorus loading and lake response under uncertainty: A manual and compilation of export coefficients. EPA 440/5-80-011. U.S. Environmental Protection Agency, Washington, DC.

Risser, P. G. 1987. Landscape ecology: State of the art. In: *Landscape Heterogeneity and Disturbance.* M. G. Turner, ed. Springer-Verlag, New York. Pp. 3–14

SAS Institute, Inc. 1985. SAS User's Guide: Statistics. Version S ed. SAS Institute Inc., Cary, NC.

Schlosser, I. J., and J. R. Karr. 1981. Water quality in agricultural watersheds: Impact of riparian vegetation during base flow. Water Resour. Bull. 17:233–240.

Seaber, P. R., F. P. Kapinos, and G. L. Knapp. 1984. State hydrologic unit maps. Open File Report 84-708. U.S. Geological Survey, Reston, VA.

Sivertun, A., L. E. Reinelt, and R. Castensson. 1988. A GIS method to aid in non-point source critical area analysis. International Journal of Geographic Information Systems 2:365–378.

Timmins, S. P., and C. T. Hunsaker. In press. Tools for visualizing landscape pattern for large geographic areas. GIS World.

Turner, M. G., ed. 1987. *Landscape Heterogeneity and Disturbance.* Springer-Verlag, New York.

Turner, M. G. 1989. Landscape ecology: The effect of pattern on process. Annul Rev. Ecol. Syst. 20:171–197.

Wilkin, D. C., and R. W. Jackson. 1983. Nonpoint water quality contributions from land use. J. Environ. Syst. 13:127–136.

Wischmeier, W. H., and D. D. Smith. 1978. Predicting Rainfall Erosion Loss—A Guide to Conservation Planning. Agriculture Handbook 537. U.S. Department of Agriculture, Washington, DC.

Introduction to

David J. Mladenoff and Theodore A. Sickley

Assessing Potential Gray Wolf Restoration in the Northeastern United States: A Spatial Prediction of Favorable Habitat and Potential Population Levels

by Monica G. Turner

In this article, Drs. Mladenoff and Sickley apply some of the techniques typically used in landscape ecology (geographic information systems and spatial modeling) to a question of importance to conservation and restoration ecology. Mladenoff and coworkers have studied the expansion of wolves from their refuges in northern Minnesota to other areas in the Upper Midwest from which they had been extirpated. By analyzing data on the locations of individual wolves over many years and relating this to data describing the landscape, Mladenoff and colleagues developed a spatial model to predict the likelihood of wolf occurrence across northern Minnesota, Wisconsin and Michigan. In the following article, they apply these spatial models developed in the Upper Midwest to the New England states.

Assessing Potential Gray Wolf Restoration in the Northeastern United States: A Spatial Prediction of Favorable Habitat and Potential Population Levels

DAVID J. MLADENOFF AND THEODORE A. SICKLEY

Original publication in *Journal of Wildlife Management* (1998), 62(1):1–10. Reprinted with permission.

Abstract

The northeastern United States was previously identified under the U. S. Endangered Species Act (ESA) as a potential location for restoration of a population of the endangered eastern timber wolf or gray wolf *(Canis lupus)*. The gray wolf has been protected under the ESA since 1974. We used Geographic Information Systems (GIS) and a logistic regression model based on regional road abundance to estimate that the Northeastern states from Upstate New York to Maine contain >77,000 km^2 of habitat suitable for wolves. Using current habitat distribution and available ungulate prey (deer and moose), we estimate the area is capable of sustaining a population of approximately 1,312 wolves (90% CI = 816–1,809). This estimate is equivalent to new, much higher potentials estimated for northern Wisconsin and Upper Michigan, where wolves are rapidly recovering in the U.S. Midwest. Potential wolf densities vary from a low of <12/1,000 km^2 in the Adirondack Region of Upstate New York, where prey densities are lowest, to 20–25/1,000 km^2 in northern Maine and New Hampshire. A contiguous area of favorable habitat from Maine to northeastern Vermont (>53,500 km^2) is capable of supporting approximately 1,070 wolves (90% CI = 702–1,439). Such large areas are increasingly rare and important for wolf recovery if populations large enough to have long-term evolutionary viability are to be maintained within the United States. However, large-scale restoration of a top carnivore like the wolf has other consequences for overall forest biodiversity in eastern forests because wolf recovery is dependent on high levels of ungulate prey, which in turn have other negative effects on the ecosystem. In the United States, planning for wolf restoration in the Northeast should take advantage of experience elsewhere, especially the upper Midwest.

The gray wolf has been protected since 1974 under the U.S. ESA of 1973. However, a recent revision of wolf taxonomy considers the eastern timber wolf to be *Canis lupus lycoan*. This subspecies is now considered extinct in the eastern United States but extant in southeastern Canada (Nowak 1995). A restored population of this subspecies in the Northeast may be more important for its persistence in the United States than previously assumed because the recovering wolf population in the upper Midwest is not *C. l. lycoan* (Nowak 1995).

When protected in 1974, wolves within the contiguous United States were known to exist

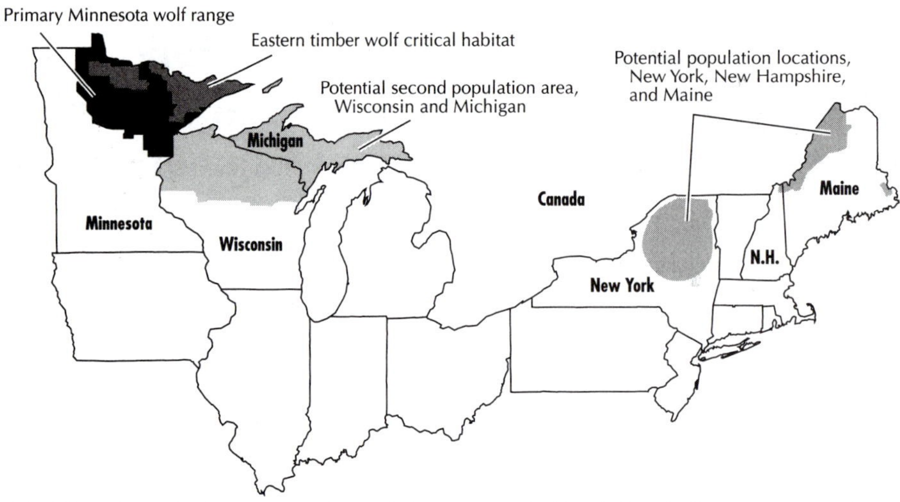

Figure 1. Officially designated primary gray wolf population range and critical habitat in Minnesota, secondary population area in Wisconsin and Minnesota, and potential population locations in New York and Maine (redrawn from U.S. Fish and Wildlife Service 1992).

only in northern Minnesota and on Isle Royale in Lake Superior, but wolves formerly occupied most of what is now the lower 48 states (Mech and Rausch 1976, Mech et al. 1994). Since that time, wolves have increased nearly 3-fold (700–2,000) in Minnesota, and since the late 1970s have naturally recolonized portions of northern Wisconsin and, more recently, Upper Michigan (Fuller 1995). Wolves originating in Canada have also begun recolonizing Montana in the western United States for at least the past decade (Boyd et al. 1995). Wolves also have been actively reintroduced into Yellowstone National Park and Idaho. The initial reintroduction was successful, but active reintroduction is controversial in these western states, where some groups fear livestock depredation and government restrictions (Fritts et al. 1995).

Because wolf recovery elsewhere has progressed successfully, there is increased focus on the Northeast. Attention is especially beginning to focus on the region from Upstate New York to northern Maine, a location identified for a potential second population of gray wolves separate from Minnesota (Figure 1). A second, separate population outside Minnesota is specified for official recovery of the wolf in the East (U.S. Fish and Wildlife Service 1992). If the second population is assumed connected to Minnesota (e.g., located in Wisconsin and Michigan), the population must be >100 animals in a >12,800km^2 area, for at least 5 years. If the second population is more distant from that in Minnesota (e.g., the Northeast), the population must be at least 200 wolves in at least a 25,600-km^2 area.

Potential areas in the northeastern United States were generally identified by the U.S. Fish and Wildlife Service in the revised recovery plan for the endangered eastern timber wolf (Figure 1; U.S. Fish and Wildlife Service 1992). These areas were qualitatively identified based on consensus of biologists and interested groups (U.S. Fish and Wildlife Service 1992).

To assist recovery planning in the Northeast, more precise delineation of areas suitable for wolves is required, as are estimates of the potential populations these areas may eventually contain. Recent work suggests that the several hundred animals previously proposed as constituting a viable wolf population may not be adequate when long-term evolutionary potential and viability are considered (Lance 1995). In light of this evidence, assessing potential habitat area and potential population size is critical for conservation planning.

Previously, we used GIS techniques to derive a predictive spatial habitat model for the Lake States of the Midwest (northern Minnesota, northern Wisconsin, Upper Michigan; Mladenoff et al. 1995). We used telemetry data from wolves in 14 colonizing packs radio-collared in northern Wisconsin from 1979 to 1993. We found that land-cover classes of agriculture and deciduous forest were negatively associated with wolf packs, whereas mixed deciduous-conifer forest and conifer-forested wetlands were positively associated. Public lands and industrial forests were preferred, while smaller-ownership private lands were avoided. Wolves also preferred areas with road densities <0.23 km/km^2; nearly all packs were in areas of <0.45 km/km^2 and human density was <1.52 individual/km^2. Roads per se are not avoided by wolves but serve as indicators of human contact, and thus likelihood of deliberate or accidental human-caused mortality. From these variables, we derived a logistic regression model with road density as the primary variable. This model correctly classified 13 of 14 radio-collared pack territories and 12 of 14 randomly located non-pack areas used in the analysis (Mladenoff et al. 1995).

More recently, we further tested the model by analyzing new data (1993–96) from 23 packs not used in creation of the model. Pack territories were delineated with radiotelemetry (13 packs) and mapped via field methods such as winter track surveys and howling surveys (10 packs). Field-mapped pack territories, although less precise, yielded more conservative (larger) pack territories in terms of our analysis. The original logistic regression model correctly classified 11 of the 13 radio-collared packs and 7 of the 10 field-mapped packs (D. J. Mladenoff et al., unpublished data). We also estimated potential wolf population sizes and spatial density patterns based on our habitat map and relations derived within the Lake States and elsewhere. This information includes data on area requirements for an average wolf pack, wolf prey needs, and prey (ungulate) population estimates (Mladenoff et al. 1997). We estimated that the 1996 wolf population of approximately 100 animals each in northern Wisconsin and the Upper Peninsula of Michigan was capable of reaching 300–400 in northern Wisconsin and >800 in Upper Michigan (Mladenoff et al. 1997). Current (1997) estimates for Wisconsin show a dramatic increase to approximately 150 wolves (A. P. Wydeven, Wisconsin Department of Natural Resources, personal communication). Here we use GIS, our logistic model, and similar population estimation methods to derive regional probability maps of potential wolf habitat and population projections for the northeastern United States from the Adirondack Region of Upstate New York to Maine (Figure 1).

Methods

Habitat Estimation

We followed our previous procedure used for the northern Lake States (Mladenoff et al. 1995) and obtained GIS coverages of roads for the Northeast from the U.S. Census Bureau TIGER/line files (U.S. Census Bureau 1991).

These roads are indicated by solid lines on U.S. Geological Survey 1:100,000 quadrangle maps. The roads include highways and other paved and unpaved roads passable year-round by auto but do not include unimproved trails and forest roads. From these data, we used the ARC/INFO GIS to create a road density (km/km^2) map coverage. Mean road density was calculated in 5- x 5-km cells. We applied 1 of 2 previously generated logistic regression models from our previous work to these mapped data:

$$\text{logit}(p) = -6.5988 + 14.6189R$$

where p is the probability of the response variable (wolf pack presence) and R is road density. We generated a probability map of wolf habitat favorability classes based on the above equation and the relation:

$$p = 1/[1 + e^{(\text{logit}(p))}]$$

where e is the natural exponent.

Wolf Population and Density Estimation

As in the Lake States (Mladenoff et al. 1997) we calculated potential spatial wolf population estimates by using known wolf-prey biomass relations (Fuller 1989, Fuller et al. 1992) for wolf populations in a variety of locations throughout North America. Ungulates constitute 80% of the diet of wolves, and in the eastern United States typically are white-tailed deer *(Odocoileus virginianus)* and, where present,

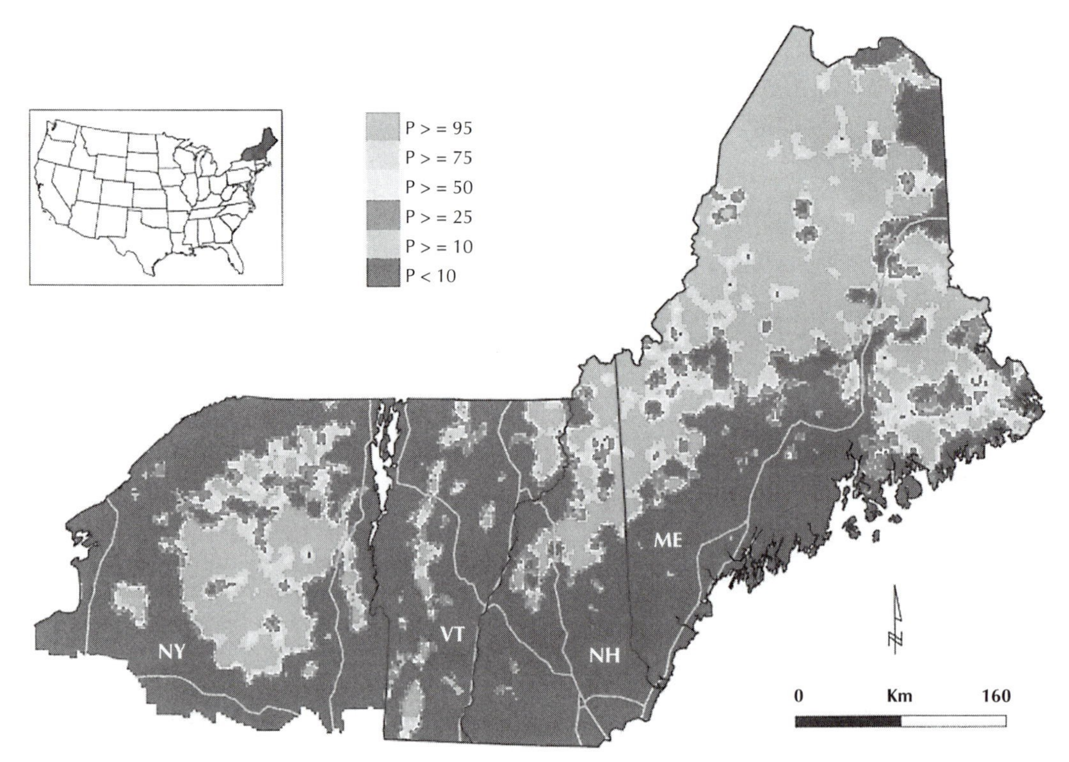

Figure 2. Habitat favorability for the northeastern United States based on the logistic regression model from Mladenoff et al. (1995). Based on the model, classes with p > 0.5 are favorable habitat. [Editor's note: see original paper for color image.]

moose *(Alces alces)*. Wolf density in unexploited populations is largely explained by the prey density (primarily ungulates) on the landscape, which explains 72% of the variation in wolf abundance (Fuller 1989, 1995). In the Lake States, wolves kill the equivalent of 15–18 deer or 5–8 moose $\bullet$wolf$^{-1}\bullet$yr^{-1} (Fuller 1995). This relation is expressed as:

$$N = 4.19X;$$

$R^2 = 0.92$, N is wolf population density (no./km^2), and X is the deer-equivalent prey units (DEPUs; no/km^2) (1 moose = 6 DEPUs). We incorporated a map of prey density and variability (deer and moose) for the Northeast to spatially estimate wolf numbers (Mladenoff et al. 1997).

Results

Potential Habitat Area and Distribution in the Northeast

Applying our logistic model to data from the Northeast produced a spatially explicit map of potential habitat (Figures 2, 3) that differed from the previous, generally estimated areas (U.S. Fish and Wildlife Service 1992; Figures 1, 2), and the model assigns predicted probability levels of habitat favorability (Table 1). In the Adirondack Region of Upstate New York, our favorable habitat projection (16,020 km^2; $p > 0.5$; red to yellow areas on the map) is considerably less (–34%) than suggested previously (24,280 km^2; U.S. Fish and Wildlife Service 1992; Figure 1). Conversely, potential habitat in Maine is considerably greater (32%) by our estimate (47,332 km^2) than suggested previously (35,748 km^2). Our model also identifies additional areas in New Hampshire (5,472 km^2 vs. 377 km^2 previously). We also identified habi-

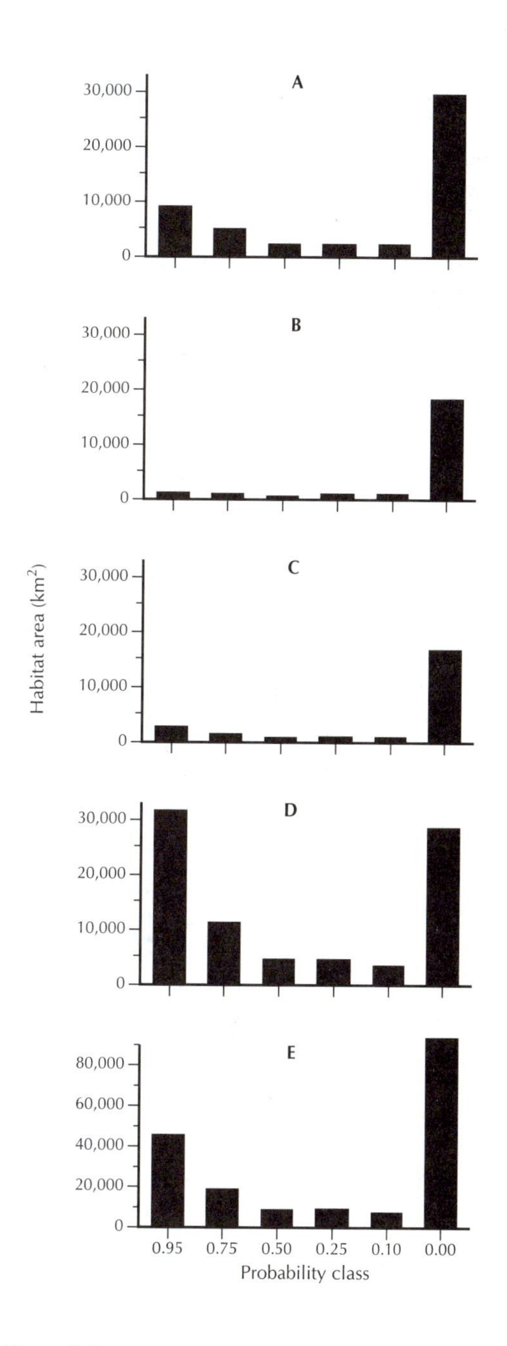

Figure 3. Amount of potential wolf habitat in the probability classes from Figure 2 for **A,** Upstate New York, **B,** Vermont, **C,** New Hampshire, **D,** Maine, and **E,** A–D combined.

Table 1. Habitat probability classes and habitat area (km^2 from the logistic regression model for the 4-state region of the northeastern United States.

Probability (p)	Road density (km/km^2)	Upstate New York Area	Upstate New York (%)	Vermont Area	Vermont (%)	New Hampshire Area	New Hampshire (%)	Maine Area	Maine (%)	Total Area	Total (%)
≥0.95	0–0.25	8,992	18.27	1,340	5.54	3,184	13.25	31,740	37.85	45,256	24.96
0.50.–0.94	0.25–0.38	4,820	9.79	1,416	5.85	1,504	6.26	11,084	13.22	18,824	10.38
0.50–0.74	0.38–0.45	2,208	4.49	868	3.59	784	3.26	4,508	5.38	8,368	4.62
0.25–0.49	0.45–0.53	2,200	4.47	1,212	5.01	976	4.06	4,404	5.25	8,792	4.85
0.10–0.24	0.53–0.60	1,888	3.84	1,200	4.96	708	2.95	3,424	4.08	7,220	3.98
<0.10	<0.53	29,112	59.15	18,172	75.07	16,868	70.21	28,688	34.21	92,840	51.21

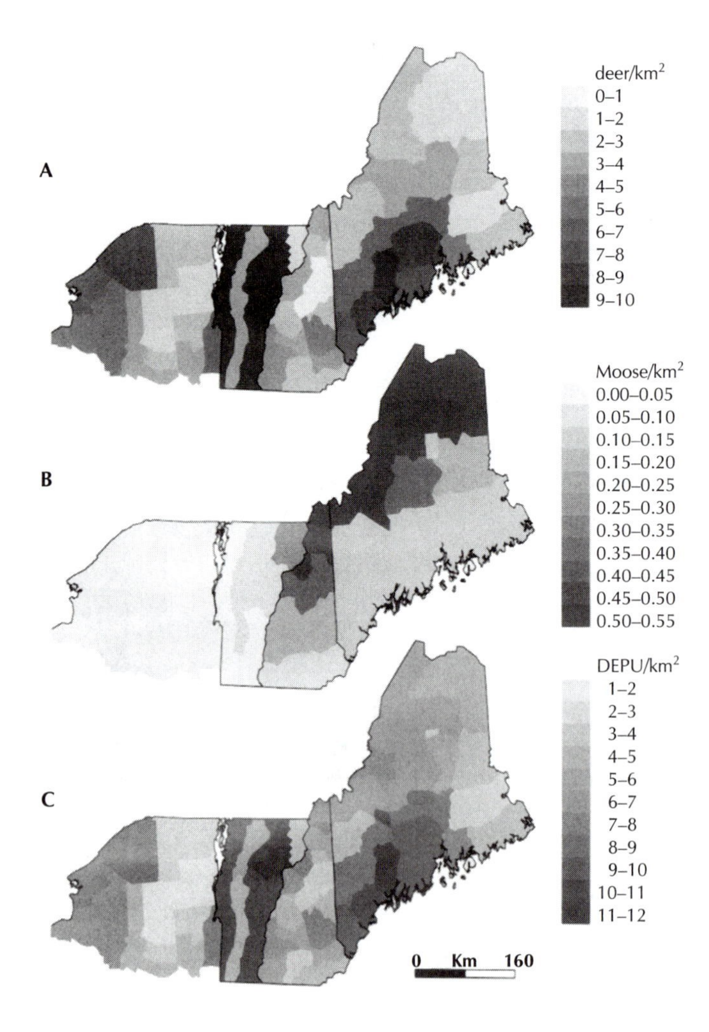

Figure 4. Maps of **A,** deer abundance, **B,** moose abundance, and **C,** combined map of deer-equivalent prey units (DEPU) available for wolves. [Editor's note: see original paper for color image.]

tat in Vermont (3,624 km^2), where potential wolf habitat had not been previously delineated. This habitat occurred primarily along the Green Mountains, which are in the center of the state (Figures 2, 3). The Catskill Mountains in southeastern New York state may also contain a relatively small amount of favorable habitat, but they are not contained within our core study area (Harrison and Chapin 1997). The proportion of habitat in various quality classes ($p > 0.5$ in the model; Figure 2) differs among states in the Northeast (Figure 3). Maine has the most habitat of the 4 states, the greatest proportion of habitat (65%) in the most favorable class ($p > 0.95$), and the next largest amount (23%) in the second-best class ($p > 0.75$; Figure 3).

Potential Wolf Populations

The prey data (Figure 4), combined with the methods of Fuller (1989) and Fuller et al. (1992), yielded a spatially explicit estimate of the potential spatial abundance and distribution of wolves in likely habitat, if the landscape were to be colonized and saturated with wolves at levels known to occur with prey levels that now exist (Figures 5, 6). The overall area we have delineated in the Northeast (77,448 km^2; habitat favorability $p > 0.5$; Figure 2) has a potential wolf population of 1,312 (90% CI = 816–1,809; Figure 6). The potential wolf population in the Adirondack Region could reach approximately 180 (90% CI = 76–284), with potential densities of <13–25 wolves/1,000

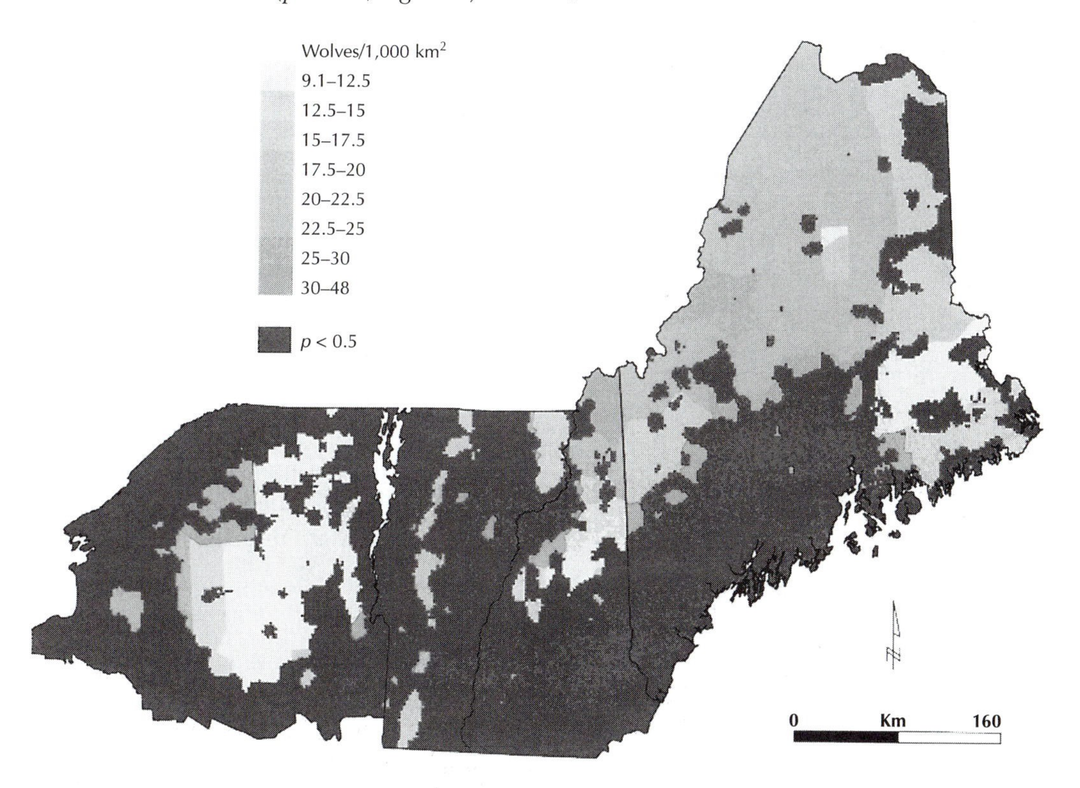

Figure 5. Map of potential wolf density and spatial distribution in the northeastern United States based on a combined analysis in Geographic Information Systems (GIS) of derived favorable habitat (Figure 2) and prey abundance (Figure 4). [Editor's note: see original paper for color image.]

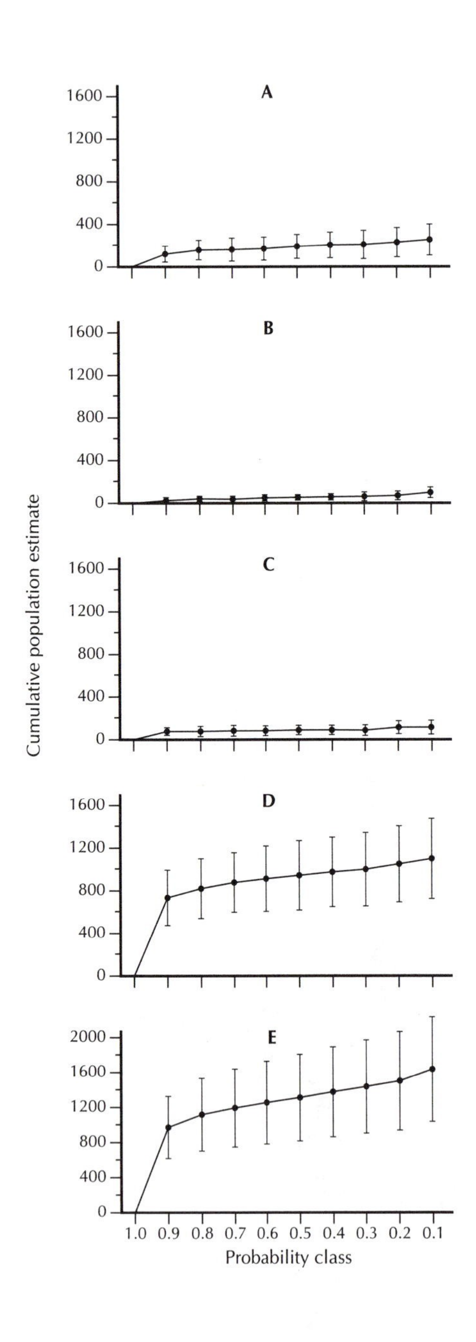

Figure 6. Potential cumulative wolf population for the northeastern United States: **A,** Upstate New York, **B,** Vermont, **C,** New Hampshire, **D,** Maine, and **E,** A–E combined based on habitat-quality classes (Figures 2, 3) saturated at projected wolf density levels (Figures 2, 5).

km^2). In the central Adirondacks, the very low wolf density reflects the low ungulate prey base in that region (<3 deer/km^2). This density is typical of high-elevation areas where lower productivity and high snowfall limit deer abundance (Figure 5). Maine has potential for a total wolf population of approximately 952 (90% CI = 702–1,439). Potential wolf density in Maine is lowest (9–15/1,000 km^2) in the southeast, but northern Maine is a large area (Figure 5) where moose are the primary prey (Figure 4b), and thus the potential wolf density is 20–30/1,000 km^2; Figures 5, 6). A fringe of very high potential wolf density (30–48/km^2) exists where the favorable habitat of northern Maine (low road density) meets the high deer abundance of the southeastern part of the state (Figure 5). New Hampshire has a potential wolf population of 99 (90% CI = 82–136), with densities of 15–30/1,000 km^2) in favorable habitat; lowest densities are in the high-elevation areas of the White Mountains in the central part of the state (Figures 5, 6). The area in Vermont has a potential wolf population of 62 (90% CI = 37–87), and densities similar to New Hampshire (Figures 5, 6). The largest contiguous area of habitat (53,714 km^2) extends from Maine to New Hampshire and into the northeastern corner of Vermont (Figure 2) and has potential for a wolf population of approximately 1,070 (90% CI = 702–1,439). Wolf densities may be lower in areas where prey are migratory or congregate seasonally (Fuller 1995).

Discussion

Comparison of Habitat with the Midwest

The areas of potential wolf habitat we have identified significantly exceed the minimum of 25,000 km^2 specified for a viable population separate from Minnesota (U.S. Fish and

Wildlife Service 1992). The area we have identified in Maine (47,372 km^2) covers more than one-half of the state and nearly equals the area occupied by the large, self-sustaining population in Minnesota (50,200 km^2). The 16,020 km^2 in New York exceeds the favorable habitat in Wisconsin (15,248 km^2) where wolf recolonization has been occurring for nearly 20 years, and the wolf population has increased dramatically from 15 to >150 wolves from 1985 to 1997 (Figures 2, 3; Wydeven et al. 1995; A. P. Wydeven, Wisconsin Department of Natural Resources, personal communication).

The mapped results of the model also clearly indicate important spatial distribution characteristics of the favorable habitat in the Northeast. For example, unlike earlier estimates, the large area of habitat in Maine is shown to be contiguous across the state from west to east (Figures 1, 2). This contiguous habitat also extends into New Hampshire and Vermont, and additional new habitat is also identified as suitable around the White Mountains of New Hampshire (Figures 1, 2). This large contiguous area across the 3 states (53,837 km^2) has important implications for the viability of an eventual wolf population because many potential recovery areas are too small to maintain a wolf population large enough for long-term viability, especially if these areas are isolated from Canada (Fritts and Carbyn 1995, Lande 1995). New areas of favorable habitat identified by our model in Vermont are small (Figure 2). However, these areas in the Green Mountains may be able to contain wolves as the nearby Maine-New Hampshire region and the Adirondacks become colonized (Figure 2). In Wisconsin, wolves have located and colonized a small (1,132 km^2) disjunct area of habitat in central Wisconsin that is separated from the main range in Wisconsin by >100 km of largely agricultural landscape (D. J. Mladenoff et al., unpublished data).

Distribution of habitat among the quality classes also compares favorably with occupied habitat in the Midwest. The total amount of habitat in Maine in the two best classes equals the corresponding amount of best quality habitat in Minnesota where these classes compose 60% in the >0.95 probability class and 25% in the >0.75 probability class (Mladenoff et al. 1995). New York state also has its largest habitat proportion in the >0.95 class, but to a lesser degree than Maine (56%), and 31% is in the >0.75 probability class. In contrast, Wisconsin has slightly less habitat than New York (Figure 3) and only 16% of habitat in the >0.95 probability class, and 46% in the >0.75 probability classes (Mladenoff et al. 1995, Wydeven et al. 1995).

Comparison of Wolf Density Estimates With the Midwest

In the Lake States, we identified potential wolf densities for northern Wisconsin and Upper Michigan of largely 20–40/1,000 km^2, and occasionally more, due to higher prey levels. Prey levels (deer) in the Lake States have recently averaged >8.5/km^2 and not uncommonly reach >15 deer/km^2, particularly in Upper Michigan (Mladenoff et al. 1995). In contrast, potential wolf density estimates for the Northeast states are lower and more spatially variable than those we derived for the Lake States because of significantly lower productivity in the Northeast as represented by prey abundance (Fuller 1995; Figure 4). Prey densities range from 2–3/km^2 in the Adirondack Mountains of New York, 3–4/km^2 in the White Mountains of New Hampshire and southeastern Maine, to an overall biomass index (DEPUs) of 4–6/km^2 in the large habitat region of northern Maine (Figure 4). However, prey are still abundant in the Northeast and, because of the extensive habitat area in the region as a whole (77,448 km^2; $p >$ 0 5 areas in Figure 2), our estimate of a total potential wolf population of 1,312 is near our projected number of 1,424 (90% CI = 829–2,019) for northern Wisconsin and Upper Michigan (Mladenoff et al. 1997).

Wolf Restoration, Forest Biodiversity, and Management

High deer densities are an important positive factor in wolf recovery because prey density influences wolf population size, survival, and reproductive success (Fuller 1995). However, these high deer populations are an artifact of highly altered forest ecosystems that are undergoing extensive harvesting at frequent intervals and provide ideal deer forage and habitat. While wolves are often considered wilderness animals (Peterson 1988), we now know they are also adaptable to semiwild regions with adequate prey, if they are not killed by humans (Mech 1995). Ironically, the high deer populations that assist wolf recovery occur in semideveloped, human-dominated landscapes where wolf mortality is high due to intentional and accidental killing of wolves. However, although assisting wolves, these high deer populations have significant negative effects on other aspects of forest biodiversity, which presents conservationists with complex trade-offs (Mech et al. 1995, Mladenoff et al. 1997). Such very high deer populations would not normally directly limit relatively small wolf populations. However, significant reductions in prey may be spatially limiting to wolves in a landscape where favorable habitat is fragmented and sparse. Because prey reductions cause a corresponding increase in the mean territory size of wolf packs (Fuller 1989, 1995), some habitat patches may no longer be large enough to sustain a wolf pack in a patchy landscape (Mladenoff et al. 1995). This reduction can cause higher mortality if wolves are forced into lower-quality habitat (greater road density and human-caused mortality) where wolf productivity is no longer self-sustaining because more of the landscape becomes a population sink (Mladenoff et al. 1995, 1997).

Public education efforts have had a large role in changing attitudes from negative to positive toward wolves in the upper Midwest, resulting in decreased illegal killing and facilitating wolf recovery (Thiel and Valen 1995, Wydeven et al. 1995). However, with successful restoration in the Midwest, managers must now justify to the public the occasional killing of wolves as they increasingly occupy areas in closer association with humans, and conflicts such as depredation on livestock and pets increase (Mech 1995).

In the Lake States, a region similar to the Northeast in ecosystems, land use, and amount of favorable habitat, the potential wolf population densities we have identified for Wisconsin and Michigan already occur in Minnesota. The recolonizing wolf population has grown from <20 to >250 in northern Wisconsin and Upper Michigan from 1985 to 1996 (Wydeven 1995, Mladenoff et al. 1997). The Northeast has great potential for successful wolf restoration and, given experience elsewhere, wolf recovery may be inevitable whether or not they are actively introduced. Radio-collared wolves in the Lake States have dispersed over 600 km across a largely agricultural landscape (Mech et al. 1995), and may thus find their way into the Northeast over long distances from Canada (Harrison and Chapin 1997). In our analysis, we have assumed that less favorable areas ($P < 0.5$)

will not contain any wolves, as occurred in our Midwest habitat and population projections. Further, we also did not include important secondary prey species (Fuller 1995) such as beaver *(Castor canadensis)* or snowshoe hare *(Lepus americanus)* in our calculations. The population estimates we have derived may therefore be conservative, given wolf reproductive potentials and their adaptability. If a significant source population of wolves eventually establishes in the large contiguous habitat of Maine, less favorable habitat may also become occupied by wolves. The latter situation has occurred in Minnesota (Mech 1995), although such areas may be a regional population sink (Pulliam 1988, Mladenoff et al. 1997). Planning for wolf restoration in the Northeast should take advantage of these experiences in the Lake States but keep in mind the conflicting goals that may occur in restoring a top predator that is dependent on unprecedented levels of deer abundance. Such deer abundance has significant negative effects on the northern forests, a region that remains a highly altered, human-dominated ecosystem (Mladenoff et al. 1997).

Acknowledgments

We benefited from thoughtful reviews of the paper by P. Arcese, M. S. Boyce, R. G. Haight, L. D. Mech, R. O. Peterson, V. C. Radeloff, and A. P. Wydeven. We appreciate assistance from the following state agencies in the northeastern United States that supplied data on prey (deer and moose): New York Department of Environmental Conservation, Wildlife Resources Center; Vermont Agency of Natural Resources, Department of Fish and Wildlife; New Hampshire Fish and Game Department; and Maine Department of Inland Fisheries and Wildlife, Wildlife Division.

Literature Cited

Boyd, D. K., R. C. Paquet, S. Donelon, R. R. Ream, D. H. Pletcher, and C. C. White. 1995. Transboundary movements of a recolonizing wolf population in the Rocky Mountains. In: *Ecology and Conservation of Wolves in a Changing World.* L. N. Carbyn, S. H. Fritts, and D. R. Seip, eds. Canadian Circumpolar Institute, Edmonton, Alberta, Canada. Pp. 141–146.

Fritts, S. H., E. E. Bangs, J. A. Fontaine, W. G. Brewster, and J. F. Gore. 1995. Restoring wolves to the Northern Rocky Mountains of the United States. In: *Ecology and Conservation of Wolves in a Changing World.* L. N. Carbyn, S. H. Fritts, and D. R. seip, eds. Canadian Circumpolar Institute, Edmonton, Alberta, Canada. Pp. 107–126.

Fritts, S. H. and L. N. Carbyn. 1995. Population viability, nature reserves, and the outlook for gray wolf conservation in North America. Restoration Ecology 3:26–38.

Fuller, T. K. 1989 Population dynamics of wolves in north-central Minnesota. Wildlife Monographs 05.

Fuller, T. K. 1995. Guidelines for gray wolf management in the northern Great Lakes region. Technical Publication 271. International Wolf Center, Ely, Minnesota, USA.

Fuller, T. K., W. E. Berg, G. L. Radde, M. S. Lenarz, and G. B. Joselyn. 1992. A history and current estimate of wolf distribution and numbers in Minnesota. Wildlife Society Bulletin 20:42–55.

Harrison, D. J. and T. G. Chapin. 1997. An assessment of potential habitat for eastern timber wolves in the northeastern United States and connectivity with occupied habitat in southeastern Canada. Working Paper 7. Wildlife Conservation Society, New York, New York, USA.

Lande, R. 1995. Mutation and conservation. Conservation Biology 9:782–791.

Mech, L. D. 1995. The challenge and opportunity of recovering wolf populations. Conservation Biology 9:270–278.

Mech, L. D., S. H. Fritts, and D. Wagner. 1995. Minnesota wolf dispersal to Wisconsin and Michigan. American Midland Naturalist 133:368–370.

Mech, L. D., D. H. Pletscher, and C. J. Martinka. 1994. Gray wolves. In: *Our Living Resources 1994.* E. T. LaRoe, G. S. Farris, C. E. Puckett, and P. D. Doran, eds. U. S. National Biological Survey, Washington, D.C., USA. Pp. 6–8.

Mech, L. D, and R. A. Rausch. 1976. Status of the wolf in North America. In: Wolves. Supplemental Paper 43. D. H. Pimlott, editor.World Conservation Union, Morges, Switzerland. Pp. 83–88.

Mladenoff, D. J., R. G. Haight, T. A. Sickley, and A. P. Wydeven. 1997. Causes and implications of species restoration in altered ecosystems: a spatial landscape projection of wolf population recovery. BioScience 47:21–31.

Mladenoff, D. J., T. A. Sickley, R. G. Haight, and A. R Wydeven. 1995. A regional landscape analysis and prediction of favorable gray wolf habitat in the northern Great Lakes region. Conservation Biology 9:279–294.

Nowak, R. M. 1995. Another look at wolf taxonomy. In: *Ecology and Conservation of Wolves in a Changing World*. L. N. Carbyn, S. H. Fritts, and D. R. Seip, eds. Canadian Circumpolar Institute, Edmonton, Alberta, Canada. Pp. 375–397.

Peterson, R. O. 1988. The pit or the pendulum: issues in large carnivore management in natural ecosystems. In: *Ecosystem Management for Parks and Wilderness*. J. K. Agee and D. R. Johnson, eds. University of Washington Press, Seattle, Washington, USA. Pp. 105–117.

Pulliam, H. R. 1988. Sources, sinks, and population regulation. American Naturalist 132:652–661.

Thiel, R. P., and T. Valen. 1995. Developing a state timber wolf recovery plan with public input: the Wisconsin experience. In: *Ecology and Conservation of Wolves in a Changing World. L.* N. Carbyn, S. H. Fritts, and D. R. Seip, eds. Canadian Circumpolar Institute, University of Alberta, Edmonton, Canada. Pp. 169–178.

U. S. Census Bureau. 1991. TIGER/line census files. Technical Documentation, U. S. Census Bureau, Washington, D.C., USA.

U. S. Fish and Wildlife Service. 1992. Revised recovery plan for the eastern timber wolf. U. S. Fish and Wildlife Service, St. Paul, Minnesota, USA.

Wydeven, A. P., R. N. Schultz, and R. P. Thiel. 1995. Monitoring of a recovering gray wolf population in Wisconsin, 1979–1991. In: *Ecology and Conservation of Wolves in a Changing World*. L. N. Carbyn, S. H. Fritts, and D. R. Seip, eds. Canadian Circumpolar Institute University of Alberta, Edmonton, Canada. Pp. 147–156

Introduction to
Steward T. A. Pickett and M. L. Cadenasso
Landscape Ecology: Spatial Heterogeneity in Ecological Systems

by Monica G. Turner

Dr. Steward T. A. Pickett has made important contributions to our understanding of ecological disturbances, developing our understanding of patch dynamics and succession. He has also forged ahead in developing our ecological understanding of how urban systems function and how ecological dynamics change as one moves along a gradient between rural and urban ecosystems. Common threads throughout much of his work include a focus on spatial dynamics at a variety of scales and the linkage of natural and social sciences. In this article, he and his former student, Dr. Cadenasso, provide a concise summary of landscape ecology. The article is especially useful because it highlights the dual, but not exclusive, foci of landscape ecology: the consideration of ecological processes over large areas, and the emphasis on understanding the causes and effects of spatial patterning in ecological systems. In addition, it covers important points regarding spatial mosaics and fluxes of matter and information across landscapes.

Landscape Ecology: Spatial Heterogeneity in Ecological Systems

S. T. A. PICKETT AND M. L. CADENASSO

Many ecological phenomena are sensitive to spatial heterogeneity and fluxes within spatial mosaics. Landscape ecology, which concerns spatial dynamics (including fluxes of organisms, materials, and energy) and the ways in which fluxes are controlled within heterogeneous matrices, has provided new ways to explore aspects of spatial heterogeneity and to discover how spatial pattern controls ecological processes.

Landscape ecology is the study of the reciprocal effects of spatial pattern on ecological processes (1); it promotes the development of models and theories of spatial relations, the collection of new types of data on spatial pattern and dynamics, and the examination of spatial scales rarely addressed in ecology. Throughout much of its history, ecology sought or assumed spatial homogeneity for convenience or simplicity; scales that lent an apparent uniformity to the processes under study were emphasized, and heterogeneity was taken as a necessary evil or an unwelcome complication. In contrast, landscape ecology regards spatial heterogeneity as a central causal factor in ecological systems, and it considers spatial dynamics and ecology's founding concern with the temporal dynamics of systems to be of equal importance. Factors in temporal dynamics include population growth and regulation, community dynamics or succession, and the dynamics of evolutionary change. The spatial effects of these factors were not entirely ignored before the advent of landscape ecology; some of the oldest roots of ecology are in biogeography. Similarly, evolutionary biology contributed the concern with population subdivision and the role of spatial segregation in population differentiation and speciation.

Ecology uses the concept of a landscape in two ways. The first, which considers a landscape as a specific area based on human scales, is intuitive: Landscapes are ecological systems that exist at the scale of kilometers and comprise recognizable elements, such as forest patches, herds and hedgerows, human settlements, and natural ecosystems (2). The second use of landscape is as an abstraction representing spatial heterogeneity at any scale. In this guise, the landscape is an ecological criterion (3) for a spatial approach to any ecological system. Irrespective of the landscape concept used, there are two major approaches to landscape ecology, reflecting differences in scale. The most common approach is the elucidation of the interactions among the elements of a matrix, especially adjacent ones. This focus exposes the relatively fine-scale mechanisms behind the dynamics and struc-

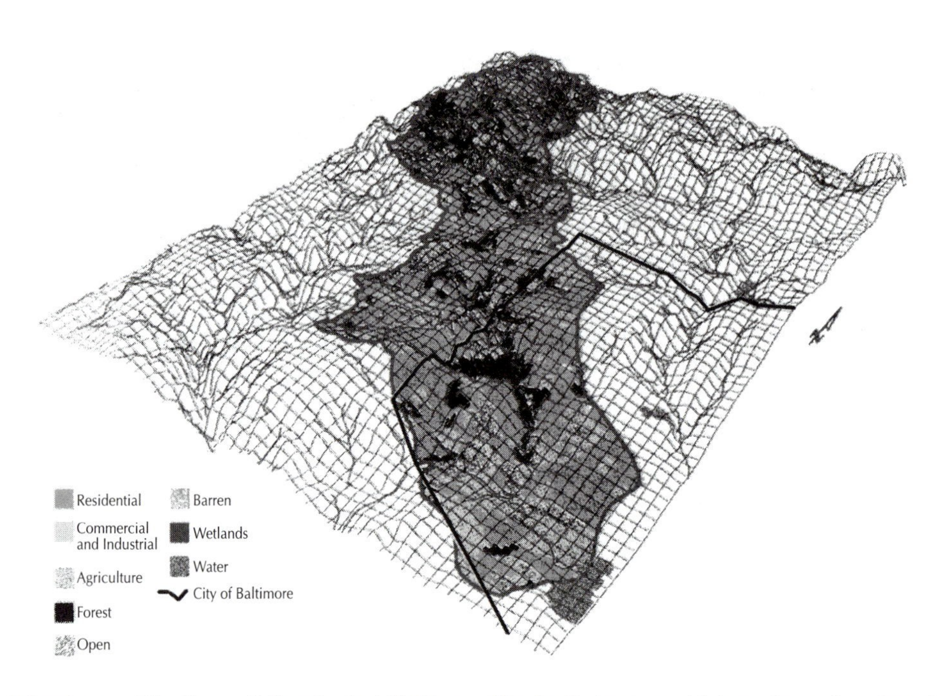

Figure 1. Landscape of the Gwynns Falls watershed (Baltimore, Maryland) showing patch type, size, and configuration, and illustrating structures such as corridors. Finer resolution would expose specific ecological community types. Research on the reciprocal linkages among social, ecological, and hydrological processes uses the spatial configuration of the landscape as the organizing model. Map courtesy of the Revitalizing Baltimore Program and J. M. Grove. [Editor's note: see original paper for color image.]

ture of the entire matrix. The second approach focuses on the coarse-scale dynamics and behaviors of the matrix as a whole. The two approaches are complementary, and both recognize a spatial mosaic with discrete elements. New technologies, such as geographic information systems and electronic databases, integrate these approaches (Figure 1).

Spatial Mosaics

Although all landscapes can be thought of as mosaics, this concept is most easily exemplified by human-dominated landscapes (Figure 1). Landscapes are composed of discrete, bounded patches that are differentiated by biotic and abiotic structure or composition. A predominant, continuous patch or cover type acts as a matrix in which other patch types appear. For example, forest patches are embedded in a matrix of farm fields. There are important correlations between patch characteristics and the ecological parameters within them. For instance, experimental forest fragmentation in the Amazon (4) has revealed that the number of carrion beetles declines with forest fragment size, as does bird diversity. The heterogeneity in landscapes can determine animal population response (5), and landscape matrix variables often explain more of the variation in the abundance and diversity of birds than do within-habitat factors (6). How landscapes are structured is a fundamental question in landscape ecology. In an applica-

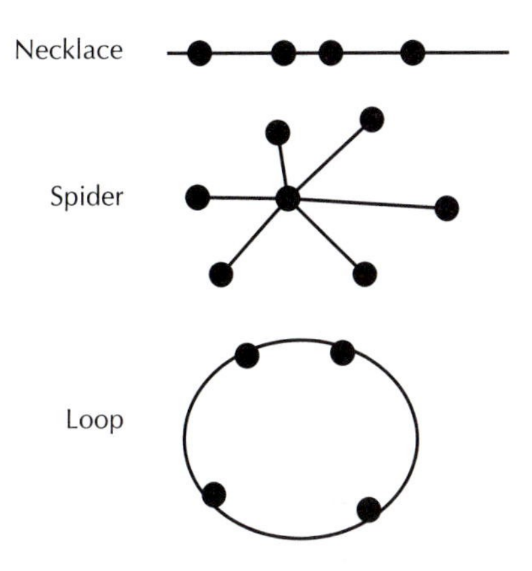

Figure 2. Graph types revealed by analysis of aerial photographs of 25 landscapes (7). Necklaces, spiders, and graph cells or loops were the most common types of connections abstracted.

tion of graph theory to 25 landscapes, in which nodes represented landscape elements identified by aerial photographs and connections between nodes represented contacts between elements (7), 90% of the resulting graphs were "spiders," "necklaces," or loops (Figure 2). This predominance of a few kinds of patterns in different human-dominated landscapes suggests regularities in underlying ecological processes. The patterns suggest comparison with other spatial scales and with landscapes less impacted by humans.

The origins of structure in mosaics are diverse. Patches can arise because of natural or human-caused disturbance, fragmentation of a land cover type, regeneration of a type, persistent differences in environmental resources, or introduction by humans (2). Purely biotic causes of patch formation include the accidents and spatial localization of dispersal of seeds or young, and the spatial segregation resulting from interactions between competitors or between predators and prey. Many such agents of patch formation are episodic. Therefore, patches may form at different times and places in landscapes, leading to a shifting mosaic. After patch formation, environmental conditions or relations among organisms in the patch may change through time. Taken together, the spatial pattern of patch creation and the changes within patches constitute patch dynamics. An example of patch dynamics emerges from the 1988 fires in Yellowstone National Park, which influenced not only patch formation and location but also the subsequent community changes within patches and the response of grazing animals to that shifting mosaic. The disturbance patches in Yellowstone are so large that the mixture of patch types is not at equilibrium over the long term (8). In contrast, disturbances that recur at short time intervals or finer spatial scales can produce an equilibrium patch distribution in a landscape (9). Spatial pattern also exists in marine, freshwater, and wetland environments (10); thus the term "landscape" is not restricted to terrestrial environments.

Landscape pattern can also exist at much finer scales than the examples above. The activities of animals, such as digging and burrowing, can generate spatial heterogeneity (11). Additional examples of fine-scale heterogeneity include the contrasting surface types in deserts, such as rock versus soil or bare soil versus shrub-covered patches (12). Thus, landscape heterogeneity can be expressed at scales that are within the spatial scope of most ecosystem types recognized by ecologists and resource managers.

Flux in Landscapes

Understanding how neighboring elements affect one another, or how they affect a process, is quite different from classical ecological concerns with the structure and function of discrete communities, populations, or ecosystems (13). One of the most well-studied aspects of landscapes is the role of edges (14), such as those of forests. Seed or animal dispersal from forests into fields or clear-cut areas is a widely documented effect on landscape organization. Even the seed stored in some forest soils reflects the current dispersal into the forest more closely than it reflects the identity of the prior occupants of the site (15).

The effects of the forest exterior on processes in the interior are a common topic of concern in landscape ecology. In agricultural landscapes, predation on nests of forest interior birds often decreases with distance from the forest edge (16). The edges of newly isolated tropical wet forest patches experience greater tree mortality and increased recruitment of pioneer species (4). However, simplistic and static views of how edges function can limit understanding of how landscape spatial structure works (17). How these fluxes are actually mediated by the edge is an open question (18) that is beginning to be experimentally examined. Edges can facilitate, inhibit, or remain neutral to crucial fluxes across them as a result of alterations in such mechanisms as wind and water flows, physical limiting factors, habitat availability, animal disperser availability or activity, competition, herbivory, and predation (Figure 3).

Mosaic structures can have a major influence on fluxes of organisms and materials. For example, the species richness and abundance of cavity-nesting birds were associated with patch orientation relative to their migration path but were not affected by patch shape, whereas resident species did not respond to landscape configuration (6). Landscape configuration also affects biogeochemical fluxes. The flux of CO_2 (19), and the movement of various forms of N (20), for example, are determined on the landscape scale by human land uses.

Gene flow and population differentiation are well known to respond to spatial hetero-

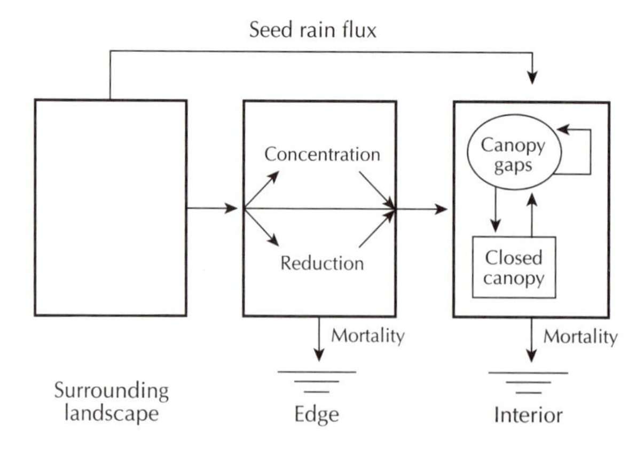

Figure 3. A conceptual model illustrating the possible net effects of the edge of a landscape element on flows from the exterior to the interior of the element. With plants as the motivating organism, the model suggests that the flux from the surrounding landscape is mediated in the edge by concentration (for example, by establishment of seeds and subsequent reproduction of adults) or reduction (for example, by predation of dispersing seeds). Alternatively, the edge may have no net effect, or the flux may avoid the edge, as in dispersal into canopy gaps.

geneity (21). However, the specific ecological processes that link these population factors to their specific landscapes must be determined (22). The concept of a metapopulation—a population that is spatially subdivided yet connected by dispersal—originated in population genetics but has strong links to the landscape approach (23). Theory assumes that populations of a species occur as isolated patches, and that extinction of populations is compensated for by establishment of populations in other patches. The approach has proven valuable in understanding population extinction, establishment, and abundance (24). Various models indicate. that coexistence can result from many specific mechanisms, notably differences in dispersal rates and spatial aggregation of superior competitors (25).

Landscapes and Scale

Landscape ecology is concerned with the causes and effects of heterogeneity rather than with a specific range of spatial scales. However, the degree to which heterogeneity is expressed depends on scale. The basic question about scale in ecology consists of determining whether a given phenomenon appears or applies across a broad range of scales, or whether it is limited to a narrow range of scales (26). Therefore, the search for breaks in scale and the discovery of scales appropriate to different ecological phenomena are critical.

The distribution limits of species of *Quercus* from the arid U.S. southwest constitute an example of how a phenomenon in a landscape is caused by processes that occur at different scales. Seedling establishment is controlled by an interaction of local and regional precipitation, whereas mortality is determined by local elevational changes in moisture relations and the degree of openness of the canopy (27). A second example uses scale to explain patterns of animal body size in contrasting ecosystems. Holling (28) used hierarchical landscape structure as a tool to expose structure in the body size distributions of animals and to link community characteristics with ecosystem productivity. The utility of the approach in contrasting biomes suggests that landscapes can act as a unifying concept in diverse environments.

Humans as Components of Landscapes

Although humans are a conspicuous element of landscapes at the coarse scale, ecologists have struggled to study ecological entities devoid of human influence. However, studies of superficially human-free systems can yield misleading results because the structure and function of systems often reflect human influences that are not obvious (29). There are very few landscapes that do not bear the contemporary, or historical but persistent, stamp of humankind. Landscape ecology has stimulated ecologists to study humans (2) and to interact with other disciplines that are concerned with humans as individuals, societies, and institutions (30).

Human influences on landscapes include altered disturbance types and patterns, addition of new or chronic pollution stresses, widespread alteration of atmospheric chemistry, and introduction of exotic organisms. The spatial extent of these influences is increasing (31); not even apparently remote ecosystems have entirely escaped them. Other large human influences, such as the effects of preindustrial land use on species composition, are subtly hidden in the past. However, ecologists are becoming more knowledgeable about such effects and are incorporating human activities into their concepts and models. Not only does this step enhance basic ecological understanding, it places ecologists in a better position to inform conservation, restoration, and management efforts.

Conclusions

Landscape ecology is a relatively new specialty, which—like all of ecology—is integrative. It consolidates the understanding of the nature, causes, and effects of spatial heterogeneity that ecology had accomplished over many decades. Although that is a useful synthesis in itself, landscape ecology has produced some additional insights. Primary among these is the knowledge that the spatial heterogeneity in ecological systems at various scales often influences important functions, ranging from population structure through community composition to ecosystem processes, and that traditional within-patch explanations were incomplete. Landscape ecology has begun to determine the mechanisms behind the relations of spatial pattern and ecological processes. The heterogeneity of entire matrices as well as the structures of specific boundaries in landscapes have been shown to govern the movement of organisms, materials, and energy. Landscape ecology has become a major stimulus for clarifying the fundamental problem of scale in ecology by showing how processes at various scales interact to shape ecological phenomena and by exposing regularities that have wide explanatory potential. Finally, landscape ecology has focused the attention of ecologists on scales and systems in which human impacts, even subtle and distant ones, are necessary ingredients in ecological models. Together, these advances have brought spatial heterogeneity into ecology to perform valuable explanatory and predictive functions, rather than excluding it as a troublesome source of error.

References

1. Turner, M. G. 1989. Annul Rev Ecol Syst 20:171.
2. Forman, R. T. T. and M. Godron. 1986. *Landscape Ecology.* Wiley, New York.
3. Allen, T. F. H. and T. W. Hoekstra. 1992. *Towards a Unified Ecology.* Columbia Univ. Press, New York.
4. Bienegaard, R. O.Jr., et al. 1992. Bioscience 42:859.
5. Osffeld, R. S. 1992. In: *Effects of Resource Distribution on Animal-Plant Interactions.* Hunter, M. D., T. Ohgushi and P. W. Price, Eds. Academic Press, Orlando, FL. Pp. 43–74.
6. Gurizweiler, K. J. and S. H. Anderson. 1992. Landscape Ecol. 6:293.
7. Cantwell, M. and R. T. T. Forman. 1993. Ibid. 8:239.
8. Turner, M. G. and W. H. Romme. 1994. Ibid. 9:59.
9. Bomann, F. H. and G. E. Likens. 1979. Am. Sci. 67:660.
10. Kolasa, J. and S. T. A. Pickett, Eds. 1991. *Ecological Heterogeneity.* Springer-Verlag, New York.
11. Johnston, C. A. In: (32), pp. 57–80.
12. Jackson, R. B. and M. M. Caldwell. 1993. Ecology 74:612.
13. Malanson, G. P. 1993. *Riparian Landscapes.* Cambridge Univ. Press, New York.
14. Holland, M. M., P. G. Risser and R. J. Naiman, Eds. 1991. *Ecotones: The Role of Landscape Boundaries in the Management and Restoration of Changing Environments.* Chapman & Hall, New York.
15. Beatty, S. W. 1991. J. Biogeogr. 18:553.
16. Andren, H. In: (32), pp. 225–255.
17. Murcia C. 1995. Trends Ecol Evol. 10:58.
18. Forman, R. T. T. and P. N. Moore. 1992) In: *Landscape Boundaries, Consequences for Biotic Diversity and Ecological Flows.* Hansen, A. J. and F. di Castri, Eds. Vol. 92. Springer-Verlag, New York. Pp. 236–258.
19. Dale,V. H. 1994. In: *Effects of Land Use Change on Atmospheric CO_2 Concentrations: Southeast Asia as a Case Study.* Daie, V. H., Ed. Springer-Verlag, New York. Pp.1–14.
20. Groffman, P. M., et al. 1993. Ecology 74:1579.
21. Futuyma, D. J. 1986. *Evolutionary Biology*, 2nd Ed. Sinauer Associates, Sunderland, MA.
22. Young, A. In: (32), pp.153–177.
23. Opdam, P. 1991. Landscape Ecol. 5:93.
24. Harrison, S. 1994. In: *Large-Scale Ecology and Conservation Biology.* Edwards, P. J., R. M. May and N. R. Webb, Eds. Blackwell Scientific, Boston. Pp. 111–128.
25: Hanski, I. In: (32), pp. 203–224.
26. Levin, S. A. 1993. Ecology 73:1943.
27. Neilson, R. P. and L. H. Wulistein. 1983. J. Biogeogr.

10:275.
28. Holling, C. S. 1992. Ecol. Monogr. 62:447.
29. McDonnell, M. J. and S. T A Pickett, Eds. 1993. *Humans as Components of Ecosystems: The Ecology of Subtle Human Effects and Populated Areas.* Springer Verlag, New York.
30. Groffman, P. M. and G. E. Likens, Eds. 1994. *Integrated Regional Models: Interactions Between Humans and Their Environment.* Chapman & Hall, New York.
31. Likens, G. E. 1991. Bioscience 41:130.
32. Hansson, L., L. Fahrig and G. Merriam, Eds. 1995. *Mosaic Landscapes and Ecological Processes.* Chapman & Hall, New York.

Introduction to

William H. Romme and Dennis H. Knight

Landscape Diversity: The Concept Applied to Yellowstone Park

by Monica G. Turner

This article developed from Dr. William H. Romme's doctoral dissertation research, for which Dr. Dennis Knight served as advisor. Romme conducted a detailed study of fire history in a watershed in Yellowstone National Park, well before the major fires of 1988. This work, which was published in *Ecological Monographs* in 1982, laid important groundwork for understanding the frequency and size of fires in northern conifer landscapes, and for placing the 1988 Yellowstone fires in a broader context. The research entailed extensive field studies in which trees that were scarred but not killed by historical fires are identified, and the dates of the fires determined through analysis of the tree rings from a narrow core obtained from the tree. Romme then mapped the spatial patterns of historical fires by date, although only the time period within the lives of the oldest trees (~300 yr) can be studied by this method.

In this *BioScience* article, Romme and Knight discuss the history of fire in Yellowstone over the past several centuries and discuss how this infrequent but large disturbance might have shaped the landscape. This paper presented one of the earliest applications of synthetic metrics to quantify heterogeneity across a landscape, and how it changes through time. Romme and Knight adapted indices of species diversity and evenness that are used frequently by community ecologists to develop indices of landscape diversity and evenness. The landscape indices use the relative abundances of different cover types—in this case, successional stages of the forests in Yellowstone. The authors demonstrate that the composition of the landscape has fluctuated widely over the past few centuries, and they relate these fluctuations to other ecological effects that would ensue.

Landscape Diversity: The Concept Applied to Yellowstone Park

William H. Romme and Dennis H. Knight

Changes in landscape patterns may influence a variety of natural features including wildlife abundance, nutrient flow, and lake productivity. Data suggest that cyclic changes in landscape diversity occur on areas of 100 km² in Yellowstone National Park. When properly managed, large wilderness areas provide the best and probably the only locale for studying the kind of landscape changes that occurred for millennia in presettlement times.

Romme, W. H. and D. H. Knight. 1982. Landscape diversity: The concept applied to Yellowstone Park. *Bioscience* 32:664–670.

Each successive level of biological organization has properties that cannot be predicted from those of less complex levels (Odum 1971). Thus, populations have certain attributes distinct from the characteristics of the individuals of which they are composed, and communities have unique properties beyond the attributes of their component populations. An important level of organization that is now receiving more attention is the landscape, or mosaic of communities that covers a large land unit such as a watershed or a physiographic region (Forman and Godron 1981).

The importance of large-scale landscape patterns has been widely recognized (e.g., Bormann and Likens 1979, Forman 1979, 1982, Forman and Boerner 1981, Forman and Godron 1981, Habeck 1976, Habeck and Mutch 1973, Hansson 1977, Heinselman 1973, Loucks 1970, Luder 1981, Pickett 1976, Reiners and Lang 1979, Rowe 1961, Shugart and West 1981, Sprugel 1976, Sprugel and Bormann 1981, Swain 1980, White 1979, Wright 1974, Zachrisson 1977, and others). A few studies have quantitatively treated changes in landscape patterns (e.g., Hett 1971, Johnson 1977, Johnson and Sharpe 1976, Shugart et al. 1973). We recently made a detailed analysis of landscape composition and diversity in a pristine watershed in Yellowstone National Park in relation to fire and forest regrowth following fire (Romme 1982). In this paper we describe the natural changes that have occurred in landscape pattern over a period of 240 years and the possible consequences of these changes for certain aspects of ecosystem structure and function. Although we focus on Yellowstone in this analysis, the concepts are applicable to other ecosystems as well.

The term *landscape diversity* refers to the diversity of plant communities making up the vegetational mosaic of a land unit. Landscape diversity results from two superimposed vegetation patterns: the distribution of species along gradients of limiting factors, and patterns of disturbance and recovery within the communities at each point along the environmental gradients (Forman and Godron 1981, Reiners and Lang 1979). Both of these patterns contribute to the vegetational diversity of the Yellowstone landscape.

Over the park's 9000 km^2, elevation ranges from about 1800 m along the Yellowstone River in the northern portion to over 3000 m on the high peaks of the east and northwest. As a result, there are pronounced gradients of temperature and moisture, with related patterns in species distribution. The areas at lower elevations in the north support open sagebrush *(Artemisia tridentata)* parks on drier sites and aspen *(Populus tremuloides)* woodlands and Douglas fir *(Pseudotsuga menziesii)* forests in more mesic locations (Despair 1973). On the cooler subalpine plateaus one finds extensive upland coniferous forests of lodgepole pine *(Pinus contorta* var. *latifolia)*, subalpine fir *(Abies lasiocarpa)*, Engelmann spruce *(Picea engelmannii)*, and whitebark pine *(P. albicaulis)*, broken by occasional meadows and sagebrush parks on alluvial and lacustrine soils. The high peaks are covered by forests of spruce, fir, and whitebark pine on sheltered slopes, with alpine or subalpine meadows and boulder fields on the more exposed sites. Pollen analysis of pond sediments indicates that these basic patterns of species distribution have been relatively stable during the last 5000 years (Baker 1970).

However, vegetational patterns related to the second source of landscape diversity—perturbation—have undergone changes during this time. Most of the changes have been natural, as described below, but some aspen and sagebrush communities in northern Yellowstone appear to have been altered somewhat by fire suppression during the last century. Comparisons of 100-year-old photographs with recent photographs of the same sites show that forests today are generally more dense, with an increase in conifers and a decrease in aspen, and that many sagebrush parks now contain more shrubs and fewer grasses and forbes. Streamside thickets of willow *(Salix* spp.) and alder *(Alnus* spp.) also appear less extensive and robust than formerly (Houston 1973)[1] Some have attributed these changes to excessive browsing by elk *(Cervus elaphus)* (Beetle 1974, Peek et al. 1967).

A more common explanation appears to be the virtual elimination of fire in this area from 1886 to 1975. Houston (1973) found that fires formerly recurred at average intervals of 20–25 years in northern Yellowstone, a disturbance frequency that probably was essential for the persistence of plant species and communities representing early stages of secondary succession (notably aspen and herbaceous plants). In the absence of fire, succession has proceeded unchecked and other species such as Douglas fir and sagebrush have become increasingly predominant. Thus fire prevention appears to have modified the overall composition of the northern Yellowstone landscape, reducing landscape diversity by increasing the area covered by late successional plant communities at the expense of early successional communities. The magnitude of this change is relatively small in the context of the entire northern Yellowstone landscape, however, since aspen and herbaceous communities comprised a small fraction of the landscape even in presettlement times (Despain 1973). Similar changes have also been described in several other western parks and wilderness areas following effective fire control (Habeck 1976, Habeck and Mutch 1973, Kilgore and Taylor 1979, Loope and Gruell 1973, Lunan and Habeck 1973). Because a major management goal in the large national parks is to preserve ecosystems in their primeval state (Houston 1971), Yellowstone recently instituted a new fire management policy that allows lightning- caused fires to burn without interference if they do not threaten human life, property, or other values (U.S. National Park Service 1975).

The situation seems to be different on the high subalpine plateaus that dominate most of the central, western, and southern areas of the park. Because of inaccessibility. effective fire control was not accomplished here until about 1950 when fire-fighting equipment and techniques were greatly improved (U.S. National Park Service 1975). Moreover, our research indicated that fire occurs naturally at very long intervals because of very slow forest regrowth and fuel accumulation after fire (Romme 1982). On an average site, 200 years or more are required for a fuel complex to develop that is capable of supporting another destructive fire. On dry or infertile sites, 300–400 years may be necessary. Fires ignited prior to that time are likely to burn a very small area and have a minimal impact on the vegetation (Despair and Sellers 1977, Romme 1982). Recent uncontrolled fires in the park that burned intensely in 300-year-old forests have been observed to stop when they reached a 100-year-old stand, even though weather conditions remained favorable for fire (Despain[2], Despain and Sellers 1977). Thus, in an ecosystem where fire historically occurred at intervals of 200+ years on any particular site, suppression during the last 20–30 years probably has had very little effect on overall landscape pattern. Any major changes that have occurred are largely the result of natural processes that would have taken place even in man's absence.

Although the subalpine landscape apparently has not been substantially altered by man's activities (excluding, of course, those areas of intensive development for visitor use), it has by no means been static during the last 100 years. We found evidence that major fires occur cyclically, i.e., thousands of hectares may burn at intervals of 300–400 years with relatively few major fires in the same area during the intervening periods (Romme 1982). Such a fire cycle can occur because: geologic substrate. soils, and vegetation are very similar over much of the plateau region; forests over large contiguous areas grow and develop a fuel complex at approximately the same rates; and the plateau topography has low relief and few natural barriers to fire spread. Thus one extensive fire tends to be followed by another fire in the same area some 300–400 years later. In other parts of the Rocky Mountains where topographic barriers are more numerous, where succession occurs more rapidly, or where fuel characteristics are different, this particular type of fire cycle may not occur.

Little Firehole River Watershed

We conducted our study in the Little Firehole River watershed, which covers 73 km^2 on the Madison Plateau, a large rhyolite lava flow in west-central Yellowstone. Coniferous forests predominate, with lodgepole pine occurring throughout and subalpine fir, Engelmann spruce, and whitebark pine being found on more mesic sites. Alluvial deposits in the central and northern parts of the watershed support subalpine meadows or open coniferous forests with rich shrub and herbaceous understories. The topography is generally flat or gently sloping, with an average elevation of about 2450 m.

Fire history during the last 350 years was determined using the fire-scar methods developed by Heinselman (1973) and Arno and Sneck (1977). Major fires occurred in 1739, 1755, and 1795 (± 5 years), collectively burning over half of the upland area. Of the forested areas that did not burn at that time, nearly all were located either on topographically protected sites (ravines, lower northeast-facing slopes) that burn rarely (Romme and Knight 1981, Zachrisson 1977), or in places that had been burned by a moderately large fire in 1630, less than 200 years earlier, and were covered by young forests. Since 1795 only three fires >4 ha have occurred, and all three were relatively small (<100 ha). The absence of recent large fires is almost certainly due to a lack of suitable fuel conditions over most of the watershed, not to fire suppression by man. In fact, park records show that only one fire has been controlled in this area, a 90-ha burn in 1949. The fire probably would not have covered a much larger area even without suppression, since it was surrounded by young forests and topographically sheltered sites. Today the areas burned in the 1700s support lodgepole pine forests that are all developing more-or-less synchronously; in another 100–150 years extensive portions of the watershed will again have fuel conditions suitable for a large destructive fire.

Three stages of forest regrowth following fire (early, middle, and late successional) can be recognized on upland sites. Early successional stages are usually present for about the first 40 years and are characterized by an abundant growth of herbs and small shrubs. The large dead stems of the former forest remain standing throughout most of this period, and an even-aged cohort of lodgepole pine becomes established. Middle successional stages are marked by the maturation and dominance of the even-aged pine cohort, beginning with canopy closure around 40 years and lasting until senescence around 250–300 years. Herbaceous biomass and species diversity are lowest during this period (Taylor 1973). During late successional stages (250–300+ years) the even-aged pine canopy deteriorates with heavy mortality and is replaced by trees from the developing understory to produce an all-aged, usually mixed-species stand, which then persists until the next destructive fire.

We used our data on fire history and on the rates and patterns of forest succession after fire to reconstruct the sequence of vegetation mosaics that must have existed in the Little Firehole River watershed during the last 240 years. Past landscape patterns were reproduced by first making a map showing the age (time since the last destructive fire) of all homogeneous forest units in 1978, based on extensive field sampling and aerial photography. Then, to reconstruct the landscape of 1738, for example, we subtracted 240 years from the age of each stand in 1978 and determined in which successional stage a stand of that age would have been. Where a fire had occurred more recently than the date of interest (e.g., areas that burned in 1739 in the reconstruction for 1738) we assumed that the stand was in a fire-susceptible late successional stage (Romme 1982).

Figure 1 shows the proportions of the Little Firehole River watershed covered by early, middle, and late successional stages at different times since 1738. In 1738 most of the area was covered by late successional forests, but fires in 1739, 1755, and 1795 greatly reduced the old-growth forests and replaced them with early successional stages. Middle successional stages became most abundant around 1800 and have dominated the watershed since. The early successional stages that were common in the late 1700s and early 1800s have been very uncommon since the mid-1800s. A decrease in middle successional stages after 1938 and an associated increase in late successional stages reflect forest maturation on areas burned in 1739.

To further describe historic patterns in landscape diversity, we calculated three diversity indices (similar to those used for measur-

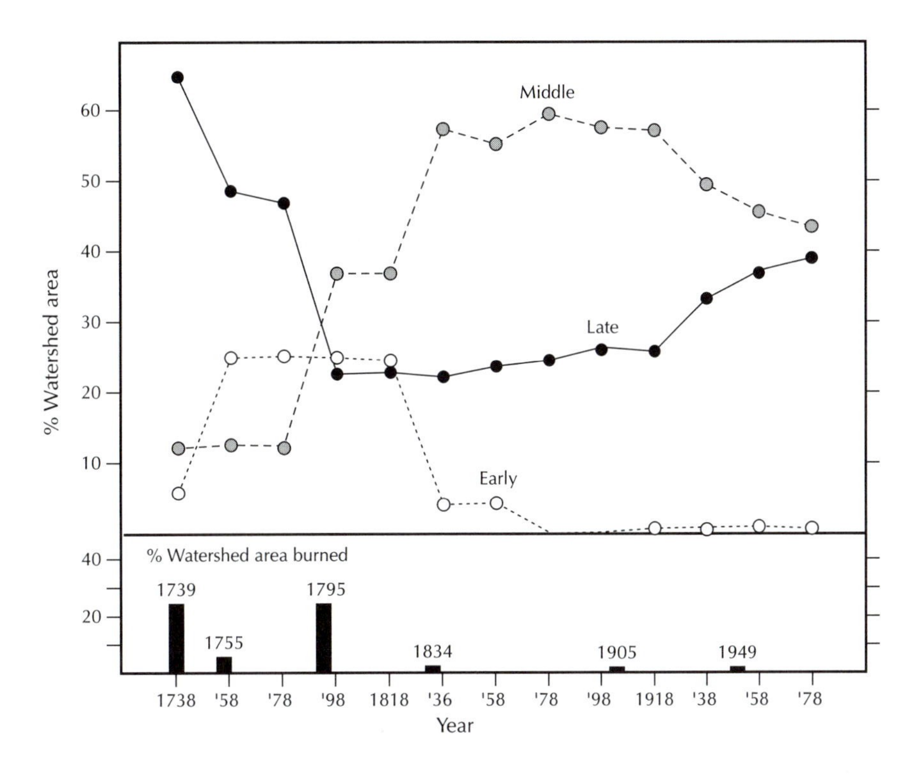

Figure 1. Percent of watershed area covered by early, middle, and late stages of forest succession from 1738–1978 in the 73-km^2 Little Firehole River watershed, Yellowstone National Park.

ing species diversity) and applied them to our landscape reconstructions for 1778–1978. We computed a richness index, based on the number of community types present; an evenness index, reflecting the relative amount of the landscape occupied by each community type; and a patchiness index, indicating the size and interspersion of individual community units as well as the structural contrast between adjacent communities (Romme 1982). Figure 2 shows the results of plotting a weighted average of all three indices, as a measure of overall landscape diversity, and the Shannon index (Pielou 1975), which we calculated by using the proportion of the watershed covered by a community type as a measure of abundance. Both indices reveal a similar pattern: Landscape diversity was high in the late 1700s and early 1800s following the extensive fires of 1739, 1755, and 1795; it fell to a low point in the late 1800s during a 70-year period with no major fires; and it increased again during this century as a result of two small fires plus some variation in the rate of forest maturation in areas burned in 1739 and 1795. This variation in rates of succession is attributable to several factors including localized high densities of the mountain pine beetle *(Dendroctonus ponderosae)* (Romme 1982).

The dramatic changes in landscape composition and diversity in the Little Firehole River watershed during the last 240 years (Fig-

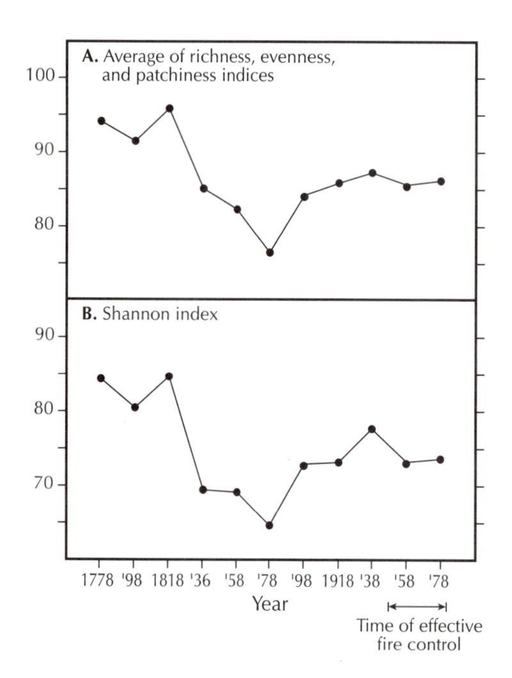

Figure 2. Changes in two measures of landscape diversity in the Little Firehole River watershed from 1778–1978.

ures 1 and 2) must have been associated with significant changes in ecosystem structure and function, including net primary productivity, nutrient cycling, total biomass, species diversity, and population dynamics of individual species. These relationships cannot be fully quantified at this time, but speculation based on existing knowledge is useful.

Implications for Wildlife

Taylor and Barmore (1980) censused breeding birds in a series of lodgepole pine stands representing a gradient from the earliest successional stages after fire through late successional stages in the park. Their data show the pattern of avifaunal succession in a single homogeneous stand. In attempting to answer the question of how breeding bird species and populations change with time in an entire subalpine watershed, we used Taylor and Barmore's (1980) census data to estimate the number of breeding pairs in each stand within our reconstructed vegetation mosaics, summing the estimates for all to arrive at an estimate of breeding pairs in the entire watershed.

Figure 3 shows the results for three representative species and for the total number of breeding pairs of all species. Mountain bluebirds *(Sialia currucoides)* require open habitats with dead trees for nesting. Such habitat was most abundant in the Little Firehole River watershed during the late 1700s and early 1800s when 25–50% of the area was covered by early forest successional stages following the large fires of the 1700s (Figure 1). Consequently, bluebirds may have been very numerous at that time. However, as forests matured, bluebird populations probably dropped dramatically (Figure 3). Today bluebirds are uncommon in the watershed except in the 90-ha area that burned in 1949. Note that this probable population decline was a perfectly natural event. occurring at a time when European man had not yet entered the area.

In contrast to the bluebird, ruby-crowned kinglets *(Regulus calendula)* prefer mature forests. Thus. kinglets were less common when bluebirds were most abundant (Figure 3). The yellow-rumped warbler *(Dendroica auduboni)* breeds successfully in a variety of habitats and as a result the population of this species probably has fluctuated little during the last 240 years despite the major landscape changes that have occurred (Figure 3). Figure 3 also shows that the total number of breeding pairs of all species has probably fluctuated greatly in the last few centuries. The highest numbers apparently were in the late 1700s and early 1800s when landscape diversity was also greatest (Figure 2).

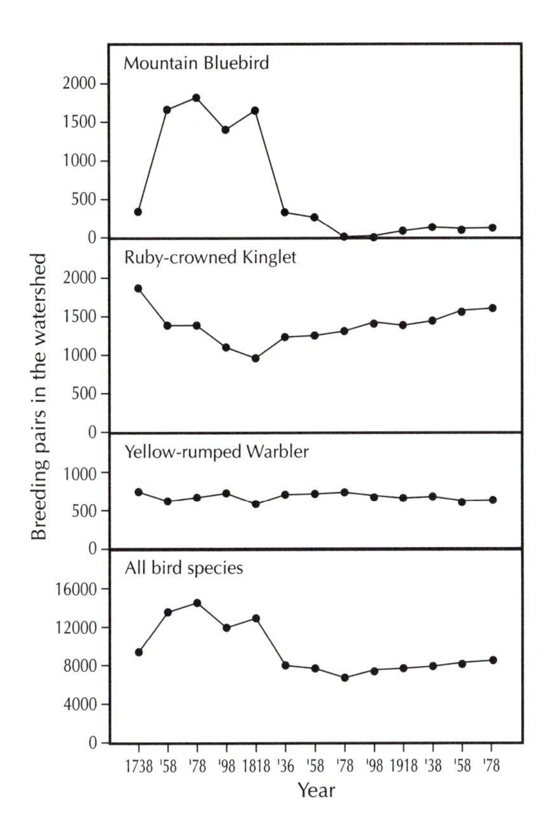

Figure 3. Estimated population sizes of breeding birds in upland forests of the Little Firehole River watershed, based on data from Taylor and Barmore (1980) and the trends shown in Figure 1. Populations in meadows and riparian forests, which cover approximately 16% of the watershed, are not included because appropriate population density data are not available for these habitats.

The population estimates shown in Figure 3 can be challenged easily on the basis that they were derived solely from habitat availability, i.e., the number of hectares of forest present in each age class. Of necessity we have ignored other critical determinants of population density. Nevertheless, the overall patterns are valid to the extent that they show the constraints of habitat on potential populations.

We were also able to consider the effect of landscape change on elk. Using a model much like that developed by Thomas et al. (1976), we examined changes in three critical habitat features during the last 200 years, namely, forage quantity and palatability, shelter (or cover), and water. The forage and shelter provided by an individual forest stand change greatly during postfire succession. Early successional stages usually have the best forage whereas middle and late successional stages provide the best shelter. However, because elk use several different kinds of habitat, the distribution and interspersion of plant communities and successional stages is critical. Thus the center of a large meadow or recently burned area may receive little elk use, despite abundant forage, if it is too distant from shelter or water, and the potential shelter of very extensive tracts of mature forest may be largely ignored if little forage is available (Black et al. 1976, Hershey and Leege 1976, Marcum 1975, Reynolds 1966, Stelfox et al. 1976, Thomas et al. 1976, Winn 1976).

We developed a relative ranking system by which every type of plant community and successional stage in the Little Firehole River watershed was assigned a value from 0–10 to indicate potential forage value (Table 1). These values were subjective, based on published literature and our own observations in the study area. We then divided the watershed into 1429 units of 5 ha each, identified the dominant vegetation type within each unit, and assigned appropriate values to each. Every value was multiplied by a distance coefficient reflecting the distance to the nearest shelter or water if those features were not present within the unit itself (Table 1), the product being our elk habitat index. In this manner we analyzed our reconstructed vegetation mosaics for 1778, 1878, and 1978.

Table 1. Relative values of plant communities and successional stages for elk habitat in the Little Firehole River watershed.*

		Distance Coefficient			
		Distance (m) to cover or water			
Plant community type	**Potential forage value**	**0–320**	**320–800**	**800–1500**	**1500+**
Alluvial woodland adjacent to moist meadow	10	1.0	0.9	0.7	0.5
Meadow	9	1.0	0.9	0.7	0.5
Upland forest:					
early successional stages	7	1.0	0.9	0.7	0.5
late successional stages	3	1.0	0.9	0.7	0.5
middle successional stages	2	1.0	0.9	0.7	0.5

*Based on models and discussions by Asherin 1973, Basile and Jensen 1971, Black et al.1976, Hershey and Leege 1976, Lonner 1976, Lyon 1971, Marcum 1975, 1976, Pengelly 1963, Reynolds 1966, Stelfox et al. 1976, Thomas et al. 1976, and Winn 1976.

Figure 4 (A, B, and C) shows the results for three 5-ha units having different histories of fire and forest regrowth. As a result of changes in stand structure, the quality of elk habitat has varied greatly. However, when we averaged the values for all 1429 individual 5-ha units to obtain an estimate of elk habitat quality for the watershed as a whole, we found much less difference among the landscapes of 1778, 1878, and 1978 (Figure 4D). There are probably two main reasons for this result. First, temporary increases in habitat quality in one part of the watershed (due primarily to the great improvement in forage after fire) have been balanced by decreases resulting from forest maturation on other areas burned earlier. Second, and probably more important, the best habitat is in and around moist meadows where forage, shelter, and water all occur in close proximity. In fact, our model may underestimate the habitat quality of subalpine meadows in the park, since we reduced our elk habitat-index in the centers of large meadows to reflect the distance to shelter. However, the shelter requirement apparently is much less critical for elk populations that are not hunted by man, and elk in the park are frequently observed feeding in the centers of large meadows.[3] We were unable to determine whether the large fires in the surrounding uplands had burned the meadows and adjacent alluvial woodlands. We assumed that the fires in these areas were of low intensity and produced little change in community structure or elk habitat. Although our results suggest that fires may not greatly influence the overall quality of elk summer range on the high plateaus of Yellowstone, elk are attracted to recently burned areas (Davis 1977), and over much of the subalpine zone, moist meadows are less common than in the Little Firehole River watershed. Where meadows are less common, summer elk populations may fluctuate in response to changes in the upland landscape.

Implications for Aquatic Ecosystems

One of the most interesting and attractive features of the park is Yellowstone Lake. This virtually unpolluted subalpine lake covers 354 km^2 and contains populations of the native cutthroat trout *(Salmo clarkii)*. The trout support a complex food chain including pelicans,

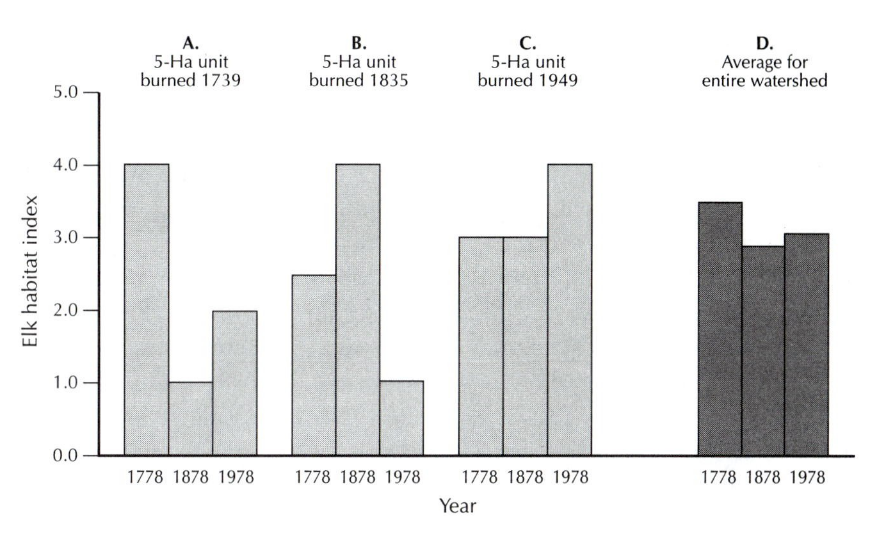

Figure 4. Elk habitat index (see text) for three representative 5-ha units and for the entire Little Firehole River watershed (D) in 1778, 1878, and 1978.

ospreys, otters, and bears. Some evidence indicates that the lake's net primary productivity has declined during the last century, as has its carrying capacity for trout and associated top predators (Shero 1977, U.S. National Park Service 1975). Because the period of apparent decline coincides with attempts at fire control, some have suggested that the cause is reduced nutrient input to the lake due to biotic immobilization by forests. As noted earlier, however, our research indicates that the natural fire regime has not been greatly altered by man's activities in the Yellowstone subalpine zone, particularly in the very remote areas that drain into Yellowstone Lake.

Rather than attribute the cause to fire suppression, we favor the hypothesis that lake productivity is to some extent synchronized with the long-term fire cycle that seems to prevail in the watershed of Yellowstone Lake. A variety of evidence supports this hypothesis. For example, experiments in the Rocky Mountains have shown that removal of mature forest from 40% of a subalpine watershed results in an increase in total water discharge of 25% or more (Leaf 1975). The increase is due to several factors related to the distribution and melting of the winter snowpack. Albin (1979) compared two small tributary streams of Yellowstone Lake; about 20% of one watershed was burned by fires 36 and 45 years previously, whereas the other watershed was unburned. The burned watershed had greater seasonal variation in streamflow and greater total water discharge per hectare. If a large portion of a subalpine watershed burns at intervals of approximately 300 years, as seems to occur in the Little Firehole River watershed, then streamflow also may exhibit a long-term cycle over and above yearly and seasonal fluctuations. During the high-discharge portion of the cycle, especially in years of high snowfall, debris is washed out of stream channels, new channels are cut, and new alluvial deposits are created. Such events influence habitat for fish as well as for floodplain species like willow and alder, which in turn are important browse species for elk and other terrestrial animals (Houston 1973).

But more important to the question of Yellowstone Lake is the nutrient content of stream water. Immediately after deforestation by fire or cutting there often is an increase in dissolved minerals due to erosion, reduced plant uptake, increased microbial activity. increased leaching, and the release of elements from organic matter by fire (Bormann and Likens 1979, McColl and Grigal 1975, Wright 1976). The increase is usually short-lived, lasting several years at most (Albin 1979, Bormann and Likens 1979), but it may be important as a periodic nutrient subsidy (Odum et al. 1979) to oligotrophic aquatic ecosystems. As young forests become established, biotic immobilization is so effective that nutrient concentrations in stream water fall to very low levels (Bormann and Likens 1979, Marks and Bormann 1972, Vitousek and Reiners 1975). Thus a watershed dominated by early and middle forest successional stages (e.g., the Little Firehole River watershed during the 1800s) would produce relatively nutrient-poor water. As forests reach late successional stages. tree growth and net primary productivity decrease, nutrient uptake is less, and consequently the leachate is richer in dissolved minerals (Bormann and Likens 1979, Vitousek and Reiners 1975).[4]

Thus, although the possible connection between fire suppression and reduced productivity in Yellowstone Lake is plausible, an equally attractive alternative hypothesis is that extensive fires in the watershed about 100 years ago replaced many late successional forests with early successional stages. As young forests over much of the watershed began utilizing soil nutrients more efficiently, the total amount leached into stream water feeding the lake was reduced accordingly. If this is true, any recent decline in lake productivity may be a natural phenomenon that has occurred many times in the past and will be alleviated as forests in the watershed mature. Of course, the Yellowstone Lake watershed is very large (ca. 2600 km^2). and landscape patterns over this large area may be in a state of dynamic equilibrium. Or what Bormann and Likens (1979) have referred to as a shifting mosaic steady state. If this is found to be true for the Yellowstone Lake watershed, then total nutrient input to the lake should be about the same from year to year (though the source would vary), and some other explanation for the decline in lake productivity will be required.

Conclusions

After a century of ecological research that focused largely on species or individual communities or ecosystems, there now is a growing interest in still higher levels of organization such as the landscape and biosphere. Changes in landscape patterns influence a variety of natural features including wildlife, water and nutrient flow, and the probability of different kinds of natural disturbances. Given a sufficiently large area and a natural disturbance regime, various measures of landscape pattern may remain fairly constant over time despite dramatic cyclic changes in localized areas such as a small watershed. Such "steady states" have been demonstrated or hypothesized for a Swedish boreal forest (Zachrisson 1977), high-elevation fir forests in New England and elsewhere (Sprugel 1976, Sprugel and Bormann 1981), primeval northern hardwood forests of North America (Bormann and Likens 1979), and mesic deciduous forests of the southern Appalachians (Shugart and West

1981). Our results suggest that strong cyclic changes occur on areas of at least 100 km^2 in Yellowstone National Park, but more research is needed to determine if the landscape patterns in the park as a whole are in a state of equilibrium. Large wilderness areas, when protected from pollutants and managed so that natural perturbations can continue, provide the best and probably the only locale for studying the kind of landscape changes that occurred for millennia in presettlement times.

Acknowledgments

This research was supported by grants from the University of Wyoming–National Park Service Research Center. We thank D. G. Despain and D. B. Houston for sharing their observations and records of natural fires in Yellowstone; L. Irwin for advice on our elk habitat model; N. Stanton and M. Boyce for guidance in estimating diversity; R. Levinson and R. Marrs for assistance with aerial photograph interpretations; M. Cook for computer programming assistance; R. Levinson, L. van Dusen, K. White, and P. White for field assistance; D. G. Despain, L. Irwin, W. H. Martin, W. G. Van der Kloot, and two anonymous reviewers for helpful comments on the manuscript; and K. Diem, M. Meagher, J. Donaldson, and the staff of the Old Faithful Ranger Station for administrative and logistical support.

Footnotes

[1]Houston, D. B. 1976. The Northern Yellowstone Elk. Parts III and IV. Vegetation and habitat relations. Unpublished report. Yellowstone National Park, Wyoming.

[2]D. G. Despain. Personal communication.

[3]L. Irwin. Personal communication.

[4]Pearson, J. A., D. H. Knight and T. J. Fahey. Unpublished ms. Net ecosystem production and nutrient accumulation during stand development in lodgepole pine forest, Wyoming.

References

Albin, D. P. 1979. Fire and stream ecology in some Yellowstone Lake tributaries. Calif. Fish Game 65:216–238.

Arno, S. F. and K. M. Sneck. 1977. A method for determining fire history in coniferous forests of the mountain west. US For. Serv. Gen. Tech. Rep. INT-42.

Asherin, D. A. 1973. Prescribed burning effects on nutrition, production and big game use of key northern Idaho browse species. Doctoral Thesis, University of Idaho, Moscow, ID.

Baker, R. G. 1970. Pollen sequence from late quaternary sediments in Yellowstone Park. Science 168:1449–1450.

Basile, J. V. and C. E. Jensen. 1971. Grazing potential on lodgepole pine clearcuts in Montana. U.S. For. Serv. Res. Pap. INT-98.

Beetle, A. A. 1974. Range survey in Teton County, Wyoming. Part IV: quaking aspen. Univ. of Wyom. Agri. Exp. Station Publ. SM 27.

Black, H., R. J. Scherzinger, and J. W. Thomas. 1976. Relationships of Rocky Mountain elk and Rocky Mountain mule deer habitat to timber management in the Blue Mountains of Oregon and Washington. In: Proceedings of the Elk-Logging–Roads Symposium. Forest, Wildlife, and Range Experiment Station, University of Idaho, Moscow, ID. Pp. 11–31.

Bormann, F. H. and G. E. Likens. 1979. *Pattern and Process in a Forested Ecosystem.* Springer-Verlag, New York.

Davis, P. R. 1977. Cervid response to forest fire and clearcutting in southeastern Wyoming. J. Wildl. Manage. 41:785–788.

Despain, D. G. 1973. Major vegetation zones of Yellowstone National Park. Information Paper Number 19, Yellowstone National Park, WY.

Despain, D. G. and R. E. Sellers. 1977. Natural fire in Yellowstone National Park. West. Wildlands 4:20–24.

Forman, R. T. T., ed. 1979. *Pine Barrens: Ecosystem and Landscape.* Academic Press, New York.

Forman, R. T. T. 1982. Interaction among landscape elements: a core of landscape ecology. In: Perspectives in landscape ecology. Proceedings of the 1981 Symposium of the Netherlands Society for Landscape Ecology. Veldhoven, Pudoc, Wegeningen, the Netherlands, in press.

Forman, R. T. T. and R E. Boerner. 1981. Fire frequency and the Pine barrens of New Jersey. Bull. Torrey Bot. Club 108:34–50.

Forman, R. T. T. and M. Godron. 1981. Patches and structural components for a landscape ecology. BioScience 31:733–740.

Habeck, J. R. 1976. Forests, fuels and fire in the Selway–Bitterroot Wilderness, Idaho. In: Proceedings of the Montana Tall Timbers Fire Ecology Conference and Fire and Land Management Symposium Number 14, 1974. E. V. Komarek, general chairman. Tall Timbers Research Sta-

tion, Tallahassee, FL. Pp. 305–352.
Habeck, J. R. and R. W. Mutch. 1973. Fire-dependent forests in the northern Rocky Mountains. Quat. Res. (ND 3:408–424.
Hansson, L. 1977. Landscape ecology and stability of populations. Landscape Planning 4:85–93.
Heinselman, M. L. 1973. Fire in the virgin forests of the Boundary Waters Canoe Area, Minnesota. Quat. Res. (NY) 3:329–382.
Hershey, T. J., and T. A. Leege. 1976. Influences of logging on elk on summer range in north-central Idaho. In: Proceedings of the Elk-Logging-Roads Symposium; Forest, Wildlife, and Range Experiment Station, University of Idaho, Moscow, ID. Pp. 73–80.
Hett, J. M. 1971. Land-use changes in east Tennessee and a simulation model which describes these changes for three counties. ORNL-IBP-71-8. Oak Ridge National Laboratory, Ecological Sciences Division, Oak Ridge, TN.
Houston, D. B. 1971. Ecosystems of National Parks. Science 172:648–651.
Ibid. 1973. Wildfires in northern Yellowstone National Park. Ecology 54:1111–1117.
Johnson, W. C. 1977. A mathematical model of forest succession and land use for the North Carolina Piedmont. Bull. Torrey Bot. Club 104:334–346.
Johnson, W. C. and D. M. Sharpe. 1976. An analysis of forest dynamics in the northern Georgia Piedmont. Forest Science 22:307–322.
Kilgore, B. M. and D. Taylor. 1979. Fire history of a sequoia-mixed conifer forest. Ecology 60:129–142.
Leaf, C. F. 1975. Watershed management in the central and southern Rocky Mountains: A summary of the status of our knowledge by vegetation types. U.S. For. Serv. Res. Pap. RM-142.
Lonner, T. N. 1976. Elk use-habitat type relationships on summer and fall range in Long Tom Creek, southwestern Montana. In: Proceedings of the Elk–Logging-Roads Symposium. Forest, Wildlife, and Range Experiment Station, University of Idaho, Moscow, ID. Pp. 101–109.
Loope, L. L. and G. E. Gruell. 1973. The ecological role of fire in the Jackson Hole area, northwestern Wyoming. Quat. Res. (NY) 3:425–443.
Loucks, O. L. 1970. Evolution of diversity, efficiency, and community stability. Am. Zool. 10:17–25.
Luder, P. 1981. The diversity of landscape ecology. Definition and attempt at empirical identification. Angew. Botanik 55:321329.
Lunan, J. S. and J. R. Habeck. 1973. The effects of fire exclusion on ponderosa pine communities in Glacier National Park. Montana. Can. J. For. Res. 3:574–579.
Lyon, L. J. 1971. Vegetal development following prescribed burning of Douglas fir in south-central Idaho. U.S. For. Serv. Res. Pap. INT-105.
Marcum, C. L. 1975. Summer-fall habitat selection and use by a western Montana elk herd. Doctoral Thesis. University of Montana, Missoula. MT.
Marcum, C. L. 1976. Habitat selection and use during summer and fall months by a western Montana elk herd. In: Proceedings of the Elk-Logging-Roads Symposium. Forest. Wildlife, and Range Experiment Station, University of Idaho, Moscow, ID. Pp. 91–96.
Marks, P. L. and F. H. Bormann.1972. Revegetation following forest cutting: Mechanisms for return to steady-state nutrient cycling. Science 176:914–915.
McColl, J. G. and D. F. Grigal. 1975. Forest fire: effects on phosphorus movement to lakes. Science 188:1109–1111.
Odum, E. P. 1971. *Fundamentals of Ecology*. Third edition. W. B. Saunders Co. Philadelphia, PA.
Odum, E. P., J. T. Finn. and E. H. Franz. 1979. Perturbation theory and the subsidy stress gradient. BioScience 29:349–352.
Peek, J. M., A. L. Lovaas. and R. A. Rouse. 1967. Population changes within the Gallatin elk herd, 1932–65. J. Wildl. Manage. 31:304–316.
Pickett, S. T. A. 1976. Succession: an evolutionary interpretation. Am. Nar. 110:107–119.
Pengelly, W. L. 1963. Timberlands and deer in the Northern Rockies. J. For. 61:734–740.
Pielou, E. C. 1975. *Ecological Diversity*. Wiley-Interscience. New York.
Reiners, W. A. and G. E. Lang. 1979. Vegetational patterns and processes in the Balsam fir zone, White Mountains. New Hampshire. Ecology 60:403–417.
Reynolds, H. G. 1966. Use of openings in spruce-fir forests of Arizona by elk, deer. and cattle. U.S. For. Serv. Res. Note RM-66.
Romme, W. H. 1982. Fire and landscape diversity in subalpine forests of Yellowstone National Park. Ecol. Monogr. 52:199–221.
Romme, W. H. and D. H. Knight. 1981. Fire frequency and subalpine forest succession along a topographic gradient in Wyoming. Ecology 62:319–326.
Rowe, J. S. 1961. Critique of some vegetational concepts as applied to forests of northwestern Alberta. Can. J. Bot. 39:1007–1017.
Shero, B. R. 1977. An interpretation of temporal and spatial variations in the abundance of diatom taxa in sediments from Yellowstone Lake, Wyoming. Doctoral Dissertation, Univ. of Wyoming, Laramie.
Shugan, H. H.. Jr.. and D. C. West. 1981. Long-term dynamics of forest ecosystems. Am. Sci. 69:647–652.
Shugart, H. H.. Jr., T. R. Crow, and J. M. Hett. 1973. Forest succession models: a rationale and methodology for modeling forest succession over large regions. For. Sci. 19:203–212.
Sprugel, D. G. 1976. Dynamic structure of wave-regenerated *Abies balsamea* forests in the north-eastern United States. J. Ecol. 64:889–911.

Sprugel, D. G. and F. H. Bormann. 1981. Natural disturbance and the steady state in high-altitude balsam fir forests. Science 211:390–393.

Stelfox, J. G., G. M. Lynch, and J. R. McGillis. 1976. Effects of clearcut logging on wild ungulates in the central Albertan foothills. For. Chron. 52:65–70.

Swain, A. M. 1980. Landscape patterns and forest history in the Boundary Waters Canoe Area. Minnesota: A pollen study from Hug Lake. Ecology 61:747–754.

Taylor, D. L. 1973. Some ecological implications of forest fire control in Yellowstone National Park, Wyoming. Ecology 54:1394–1396.

Taylor, D. L. and W. J. Barmore. 1980. Postfire succession of avifauna in coniferous forests of Yellowstone and Grand Teton National Parks, Wyoming. In: Proceedings of the Workshop on Management of Western Forests and Grasslands for Nongame Birds. R. M. DeGraff. Technical Coordinator. U.S. For. Serv. Gen. Tech. Rep. INT–86. Pp. 130–145.

Thomas, J. W., R. J. Miller, H. Black, J. E. Rodiek, and C. Maser. 1976. Guidelines for maintaining and enhancing wildlife habitat in forest management in the Blue Mountains of Oregon and Washington. In: Transactions of the Forty-first North American Wildlife and Natural Resources Conference, Washington, DC. Pp. 452–476.

U.S. National Park Service. 1975. The Natural Role of Fire: A Fire Management Plan for Yellowstone National Park. Unpublished report, Yellowstone National Park, WY.

Vitousek, P. M. and W. A. Reiners. 1975. Ecosystem succession and nutrient retention: a hypothesis. BioScience 25:376–381.

White, P. S. 1979. Pattern, process, and natural disturbance in vegetation. Bot. Rev. 45:229–299.

Winn, D S. 1976. Terrestrial vertebrate fauna and selected coniferous forest habitat types on the north slope of the Uinta Mountains. United States Department of Agriculture, Wasatch National Forest, Region 4.

Wright, H. E.. Jr. 1974. Landscape development, forest fires, and wilderness management. Science 186:487–495.

Wright, R. F. 1976. The impact of forest fire on the nutrient influxes to small lakes in northeastern Minnesota. Ecology 57:649–663.

Zachrisson, O. 1977. Influence of forest fires on the north Swedish boreal forest. Oikos 29:22–32.

Introduction to
John A. Wiens
Spatial Scaling in Ecology

by Monica G. Turner

In its consideration of how populations of organisms respond to the patchiness of their habitats, landscape ecology overlaps considerably with population ecology and conservation biology. Dr. John Wiens has had a long and profound influence on how ecologists think about populations and spatial patterning. Dr. Wiens published a review article on the subject ("Population responses to patchy environments") in the *Annual Review of Ecology and Systematics* back in 1976. Since then, he has continued to be an influential thinker in the field and to make important theoretical and empirical contributions to our understanding of organisms and their responses to landscape pattern. He received the Distinguished Landscape Ecologist award from the U. S. chapter of the International Association for Landscape Ecology (IALE), and he served as president of IALE from 1995 to 1999.

In the following article, Wiens explores the concept of spatial scale, especially as it influences organisms in the environment and their responses to spatial patterning. This article considers many ways in which patterns we observe are dependent upon the scales at which we observe them, and is thus strongly conceptual in flavor. However, it is also replete with a variety of real-world examples. In addition, Wiens briefly discusses several different methods that can be used for determining the scale for observing particular relationships between pattern and process. Ideas relating to scale presented in this article also apply to many other subdisciplines within ecology.

Spatial Scaling in Ecology[1]

J. A. WIENS

The only things that can be universal, in a sense, are scaling things
—*Mitchell Feigenbaum*[2]

Original publication in *Functional Ecology* (1989), 3:385–397. Reprinted with permission.

Introduction

Acts in what Hutchinson (1965) has called the 'ecological theatre' are played out on various scales of space and time. To understand the drama, we must view it on the appropriate scale. Plant ecologists long ago recognized the importance of sampling scale in their descriptions of the dispersion or distribution of species (e.g. Greig-Smith, 1952). However, many ecologists have behaved as if patterns and the processes that produce them are insensitive to differences in scale and have designed their studies with little explicit attention to scale. Kareiva & Andersen (1988) surveyed nearly 100 field experiments in community ecology and found that half were conducted on plots no larger than 1 m in diameter, despite considerable differences in the sizes and types of organisms studied.

Investigators addressing the same questions have often conducted their studies on quite different scales. Not surprisingly, their findings have not always matched, and arguments have ensued. The disagreements among conservation biologists over the optimal design of nature reserves (see Simberloff, 1988) are at least partly due to a failure to appreciate scaling differences among organisms. Controversies about the role of competition in structuring animal communities (Schoener, 1982; Wiens, 1983, 1989) or about the degree of coevolution in communities (Connell, 1980: Roughgarden, 1983) may reflect the imposition of a single scale on all of the species in the community. Current ecological theories do little to resolve such debates, because most of these theories are mute on scale—they can be applied at any scale on which the relevant parameters can be measured.

Recently, however, ecologists studying a wide range of topics have expressed concern about scaling effects (see Dayton & Tegner, 1984; Wiens et al, 1986a; Giller & Gee, 1987; Meetenmeyer & Box, 1987; Frost et al., 1988; Rosswall, Woodmansee & Risser, 1988). 'Scale' is rapidly becoming a new ecological buzzword.

Scientists in other disciplines have recognized scaling issues for some time. The very foundation of geography is scaling. In the atmospheric and earth sciences, the physical processes that determine local and global patterns are clearly linked (e.g. Schumm & Lichty, 1965; Clark, 1985; Dagan, 1986; Ahnert, 1987) and their importance is acknowledged in hierarchies of scale that guide research and define subdisciplines within these sciences. Physical and biological oceanographers often relate their findings to the spectrum of physical processes from circulation patterns in oceanic basins or large gyres to fine-scale eddies or rips (e.g. Haury, McGowan & Wiebe, 1978; Steele, 1978; Legrende & Demers, 1984; Hunt & Schneider, 1987; Platt & Sathyendranath, 1988). Physicists and mathematicians studying fractal geometry, strange attractors, percolation theory, and chaos address scaling as a primary focus of their investigations (Nittman, Daccord & Stanley, 1985; Orbach, 1986; Grebogi, Ott & Yorke, 1987; Gleick, 1987).

Why have ecologists been so slow to recognize scaling? Ecologists deal with phenomena that are intuitively familiar, and we are therefore more likely to perceive and study such phenomena on anthropocentric scales that accord with our own experiences. We have also been somewhat tradition-bound, using quadrats or study plots of a particular size simply because previous workers did. Unlike the physical and earth sciences (and many laboratory disciplines of biology), where our perceptual range has been extended by technology, few tools have been available to expand our view of ecological phenomena (but see Platt & Sathyendranath, 1988; Gosz, Dahm & Risser, 1988).

My thesis in this paper is that scaling issues are fundamental to all ecological investigations, as they are in other sciences. My comments are focused on spatial scaling, but similar arguments may be made about scaling in time.

The Effects of Scale

Some Examples

The scale of an investigation may have profound effects on the patterns one finds. Consider some examples:

- In hardwood forests of the north-eastern United States, Least Flycatchers (*Empidonox minimus* Baird & Baird) negatively influence the distribution of American Redstart (*Setophaga ruticilla* L.) territories at the scale of 4-ha plots. Regionally, however, these species are positively associated (Sherry & Holmes, 1988). Apparently the broad-scale influences of habitat selection override the local effects of interspecific competition. Similar scale-dependency has been found in the habitat relationships of shrub-steppe birds (Wiens, Rotenberry & Van Horne, 1986b), interspecific associations among plant species (e.g. Beals, 1973) or phytoplankton and zooplankton (Carpenter & Kitchell, 1987), and the patterns of coexistence of mosses colonizing moose dung (Merino, 1988) or of ants on mangrove islands (Cole, 1983).
- In the Great Barrier Reef of Australia, the distribution of fish species among coral heads at the scale of patch reefs or a single atoll may be strongly influenced by chance events during recruitment, and the species composition of local communities of fish may be unpredictable (Sale, 1988; Clarke, 1988). At the broader scales of atolls or reef systems, community composition is more predictable, perhaps because of habitat selection, niche diversification, or spatial replacement of species within trophic guilds (Ogden & Ebersole, 1981; Anderson et al., 1981; Green, Bradbury & Reichelt, 1987; Galzin, 1987).
- On the basis of experiments conducted at the scale of individual leaf surfaces, plant physiologists have concluded that stomatal mechanisms regulate transpiration, whereas meteorologists working at the broader scale of vegetation have concluded that climate is the principal con-

trol (Jarvis & McNaughton, 1986; Woodward, 1987). In a similar manner, most of the variation in litter decomposition rates among different species at a local scale is explained by properties of the litter and the decomposers, but at broader regional scales climatic variables account for most of the variation in decomposition rates (Meentemeyer, 1984).

- Domestic cattle grazing in shortgrass prairie use elements of local plant communities quite nonrandomly on the basis of short-term foraging decisions, but use of vegetation types is proportional to their coverage at the broader scale of landscape mosaics (Senft et al., 1987).
- The distribution of phytoplankton in marine systems is dominated by horizontal turbulent diffusion at scales up to roughly 1 km (Platt, 1972; Denman & Platt, 1975). At somewhat broader scales, phytoplankton growth, zooplankton grazing, and vertical mixing override these local effects (Denman & Platt, 1975; Lekan & Wilson, 1978; Therriault & Platt, 1981). At scales of >5 km, phytoplankton patchiness is controlled largely by advection, eddies, and local upwelling occurring over areas of 1–100 km (Gower, Denman & Holyer, 1980; Legrende & Demers, 1984). The same controls operate in lakes, although the transitions occur at finer scales (Powell et al., 1975).

These examples could easily be extended. The salient point is that different patterns emerge at different scales of investigation of virtually any aspect of any ecological system.

Linkages Between Physical and Biological Scales

In the marine phytoplankton and other aquatic systems, physical features may be primary determinants of adaptations of organisms, and physical and biological phenomena may scale in much the same way. However, in many terrestrial environments, atmospheric and geological influences may often be obscured by biological interactions (Clark, 1985). The relationships between climate and vegetation that are evident at broad scales, for example, may disappear at finer scales, overridden by the effects of competition and other biological processes (Greig-Smith, 1979; Woodward, 1987). Local biological interactions have the effect of decoupling systems from direct physical determination of patterns by introducing temporal or spatial lags in system dynamics or creating webs of indirect effects. However, at broader scales, physical processes may dominate or dissipate these biological effects (Levin, 1989). There are exceptions: plant distributions on fine scales may be controlled by edaphic or microtopographic factors, and vegetation may influence climate at regional scales.

System Openness and the Scale of Constraints

Ecological systems become closed when transfer rates among adjacent systems approach zero or when the differences in process rates between adjacent elements are so large that the dynamics of the elements are effectively decoupled from one another. In open systems, transfer rates among elements are relatively high, and the dynamics of patterns at a given scale are influenced by factors at broader scales. However, "openness" is a matter of scale and of the phenomena considered. At the scale of individual habitat patches in a landscape mosaic, for example, population dynamics may be influenced by between-patch dispersal, but at the broader scale of an island containing that landscape, emigration may be nil and the populations closed. The same island, however, may be open with regard to atmospheric flows or broad-scale climatic influences.

The likelihood that measurements made on a system at a particular scale will reveal something about ecological mechanisms is a function of the openness of the system. The species diversity of a local community, for example, is influenced by speciation and extinction, and by range dynamics at regional or biogeographic scales (Ricklefs, 1987). Changes in population size at a location may reflect regional habitat alterations, events elsewhere in a species' range, or regional abundance and distribution rather than local conditions (May, 1981; Vaisanen, Jarvinen & Rauhala, 1986; Roughgarden, Gaines & Pacala, 1987; Wiens, 1989). Habitat selection by individuals may be determined not only by characteristics of a given site but by the densities of populations in other habitats over a larger area (O'Connor & Fuller, 1985). Den Boer (1981) suggested that small local populations may frequently suffer extinction, only to be reconstituted by emigrants from other areas. The fine-scale demographic instability translates into long-term persistence and stability at the scale of the larger metapopulation (Morrison & Barbosa, 1987; DeAngelis & Waterhouse, 1987; Taylor, 1988).

Ecologists generally consider system openness in the context of how broad-scale processes constrain finer-scale phenomena. This is one of the primary messages of hierarchy theory (Allen & Starr, 1982) and of 'supply-side' ecology (Roughgarden et al., 1987) and it is supported by studies of the temporal dynamics of food webs as well (Carpenter, 1988). However, the ways in which fine-scale patterns propagate to larger scales may impose constraints on the broad-scale patterns as well (Huston, DeAngelis & Post, 1988; Milne, 1988). Ecologists dealing with the temporal development of systems (e.g.

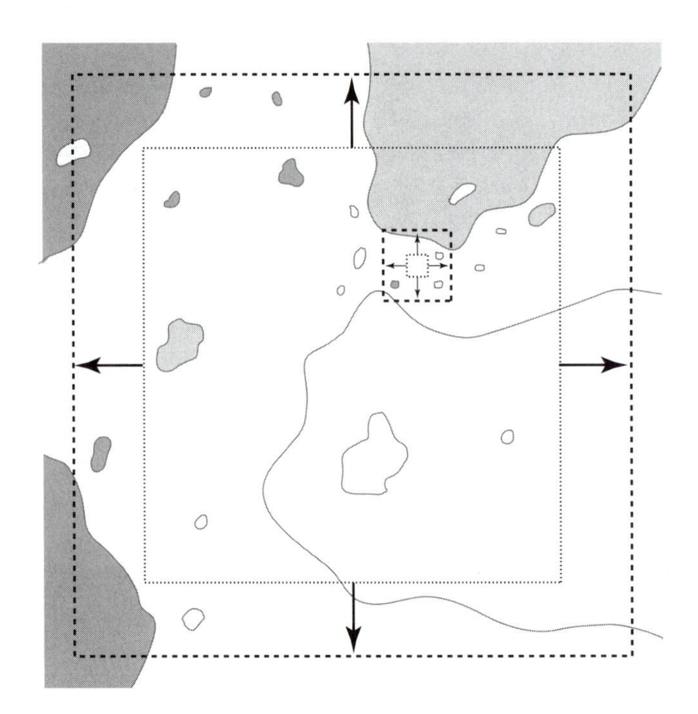

Figure 1. The effects of changing the grain and extent of a study in a patchy landscape. As the extent of the study is increased (large squares), landscape elements that were not present in the original study area are encountered. As the grain of samples is correspondingly increased (small squares), small patches that initially could be differentiated are now included within samples and the differences among them are averaged out.

forest insect epidemics: Barbosa & Schultz, 1987; Rykiel et al., 1988) recognize this sensitivity to small differences in fine-scale initial conditions as the effects of historical events on the subsequent state of the system.

Extent and Grain

Our ability to detect patterns is a function of both the *extent* and the *grain*[3] of an investigation (O'Neill et al., 1986). Extent is the overall area encompassed by a study, what we often think of (imprecisely) as its scale[4] or the population we wish to describe by sampling. Grain is the size of the individual units of observation, the quadrats of a field ecologist or the sample units of a statistician (Figure 1). Extent and grain define the upper and lower limits of resolution of a study; they are analogous to the overall size of a sieve and its mesh size, respectively. Any inferences about scale-dependency in a system are constrained by the extent and grain of investigation—we cannot generalize beyond the extent without accepting the assumption of scale-independent uniformitarianism of patterns and processes (which we know to be false), and we cannot detect any elements of patterns below the grain. For logistical reasons, expanding the extent of a study usually also entails enlarging the grain. The enhanced ability to detect broad-scale patterns carries the cost of a loss of resolution of fine-scale details.

Variance, Equilibrium and Predictability

When the scale of measurement of a variable is changed, the variance of that variable changes. How this happens depends on whether grain or extent is altered. Holding extent constant, an increase in the grain of measurement generally decreases spatial variance. In a perfectly homogeneous area (i.e. no spatial autocorrelation among sample locations), the log-log plot of variance versus grain (or N) has a slope of –1 (Figure 2A). In a heterogeneous area, this slope will generally be between –1 and 0 (O'Neill et al., unpublished), although the relationship may be curvilinear (Figure 2A; Levin, 1989). As grain increases, a greater proportion of the spatial heterogeneity of the system is contained within a sample or grain and is lost to our resolution, while between-grain heterogeneity (= variance) decreases (Figure 2B). If the occurrence of species in quadrats is recorded based on a

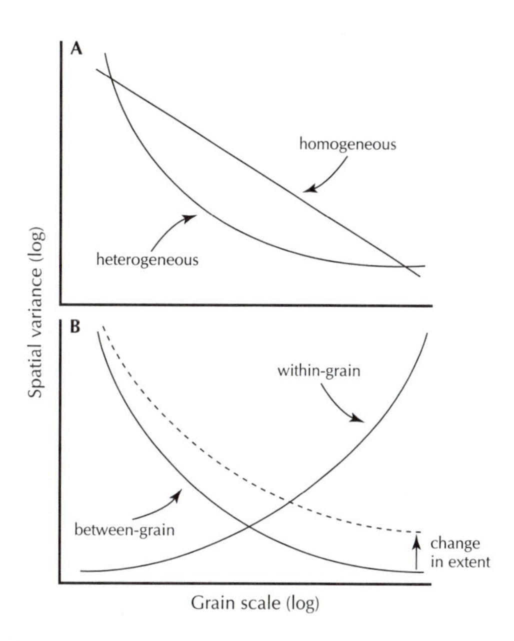

Figure 2. A. As the grain of samples becomes larger, spatial variance in the study system as a whole decreases, albeit differently for homogeneous and heterogeneous areas. This is related to the within- and between-grain (sample) components of variation. **B.** With increasing grain scale, less of the variance is due to differences between samples and more of the overall variation is included within samples (and therefore averaged away). An increase in the extent of the investigation may increase the between-grain component of variance by adding new patch types to the landscape surveyed (Figure 1), but within-grain variance is not noticeably affected.

minimal coverage criterion, rare species will be less likely to be recorded as grain size increases; this effect is more pronounced if the species are widely scattered in small patches than if they are highly aggregated (Turner et al., unpublished). If the measurement criterion is simply the presence or absence of species in quadrats, however, more rare species will be recorded as grain increases, and diversity will increase rather than decrease with increasing grain. Exactly how variance changes with grain scale thus depends on the magnitude and form of the heterogeneity of an area (Milne, 1988, unpublished; Palmer, 1988) and on the type of measurement taken.

Spatial variance is also dependent on the extent of an investigation. Holding grain constant, an increase in extent will incorporate greater spatial heterogeneity, as a greater variety of patch types or landscape elements is included in the area being studied (Figure 1). Between-grain variance increases with a broadening of scale (extent) (Figure 2b).

These considerations also relate to the patterns of temporal variation or equilibrium of ecological systems. Ecologists have often disagreed about whether or not ecological systems are equilibrial (e.g. Wiens, 1984, in press; Chesson & Case, 1986; DeAngelis & Waterhouse, 1987; Sale, 1988). Whether apparent 'equilibrium' or 'nonequilibrium' is perceived in a system clearly depends on the scale of observation. Unfortunately, current theories provide little guidance as to what we might expect: models in population biology (e. g. May & Oster, 1976; Schaffer, 1984; May. 1989) and physics (Gleick, 1987) show that order and stability may be derived at broad scales from finer-scale chaos or that fine-scale determinism may produce broad-scale chaos. depending on circumstances. Perhaps ecological systems follow principles of universality, their final states at broad scales depending on general system properties rather than fine-scale details (cf. Feigenbaum, 1979). Brown (1984) has championed this view, but we still know far too little about the scaling behaviour of ecological systems to consider universality as anything other than an intriguing hypothesis.

Predictability and Space-time Scaling

Because the effects of local heterogeneity are averaged out at broader scales, ecological patterns often appear to be more predictable there. Whether or not the predictions are mechanistically sound depends on the importance of the fine-scale details. The Lotka-Volterra competition equations may predict competitive exclusion of species that in fact are able to coexist because of fine-scale spatial heterogeneity that is averaged away (e.g. Moloney, 1988). These predictions are not really scale-independent but are instead insensitive to important scale-dependent changes.

Our ability to predict ecological phenomena depends on the relationships between spatial and temporal scales of variation (Figure 3). With increased spatial scale, the time scale of important processes also increases because processes operate at slower rates, time lags increase, and indirect effects become increasingly important (Delcourt, Delcourt & Webb, 1983; Clark, 1985). The dynamics of different ecological phenomena in different systems, however, follow different trajectories in space and time. An area of a few square meters of grassland may be exposed to ungulate grazing for only a few seconds or minutes, whereas the temporal scale of microtines in the same area may be minutes to hours and that of soil arthropods days to months or years. There are no standard functions that define the appropriate

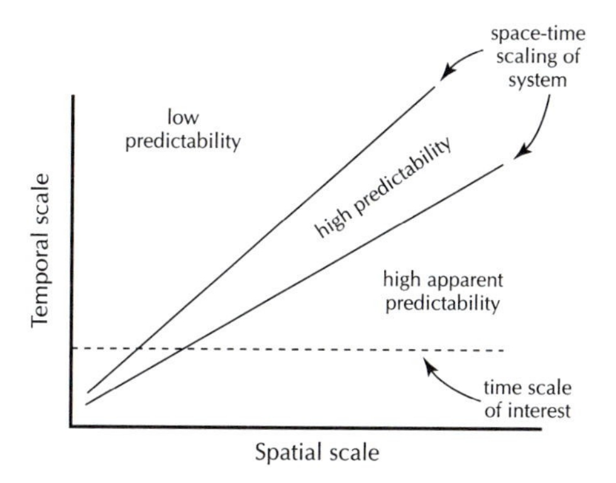

Figure 3. As the spatial scaling of a system increases, so also does its temporal scaling, although these space-time scalings differ for different systems. Studies conducted over a long time at fine spatial scales have low predictive capacity. Investigations located near to the space-time scaling functions have high predictive power. Short-term studies conducted at broad spatial scales generally have high apparent predictability (pseudopredictability) because the natural dynamics of the system are so much longer than the period of study. Often, ecologists and resource managers have been most interested in making and testing predictions on relatively short time scales, regardless of the spatial scale of the investigation.

units for such space-time comparisons in ecology. Moreover, the continuous linear scales we use to measure space and time may not be appropriate for organisms or processes whose dynamics or rates vary discontinuously (e.g. 'physiological time' associated with diapause in insects; Taylor, 1981).

Any predictions of the dynamics of spatially broad-scale systems that do not expand the temporal scale are *pseudopredictions.* The predictions may seem to be quite robust because they are made on a fine time scale relative to the actual dynamics of the system (Figure 3), but the mechanistic linkages will not be seen because the temporal extent of the study is too short. It is as if we were to take two snapshots of a forest a few moments apart and use the first to predict the second. This problem may be particularly severe in resource management disciplines, where the application of policies to large areas is often based on very short-term studies.

Detecting Patterns and Inferring Processes

The characteristics of ecological systems at relatively fine scales differ from those at relatively broad scales (Table 1), and these differences influence the ways ecologists can study the systems. The possibilities for conducting replicated experiments vary inversely with the scale of investigation. The potential for sampling errors of several kinds are greater at finer scales, although the intensity of sampling is generally lower at broader scales. Fine-scale studies may reveal greater detail about the biological mechanisms underlying patterns, but generalizations are more likely to emerge at broader scales. Because the time-frame of ecological processes tends to be longer at broader scales (Figure 3), long-term investigations are more often necessary to reveal the dynamics of the system. The scale of investigation thus determines the range of patterns and processes that can be detected. If we study a system at an inappropriate scale, we may not detect its actual dynamics and patterns but may instead identify patterns that are artifacts of scale. Because we are clever at devising explanations of what we see, we may think we understand the system when we have not even observed it correctly.

Dealing with Scale

Scale Arbitrariness

The most common approach to dealing with scale is to compare patterns among several arbitrarily selected points on a scale spectrum. In his analysis of reef-fish communities, for example, Galzin (1987) compared distributions within a single transect, among several transects

Table 1. General characteristics of various attributes of ecological systems and investigations at fine and broad scales of study. 'Fine' and 'broad' are defined relative to the focus of a particular investigation, and will vary between studies.

Attribute	Scale: Fine	Scale: Broad
Number of variables important in correlations	many	few
Rate of processes or system change	fast	slow
Capacity of system to track short-term environmental variations	high	low
Potential for system openness	high	low
Effects of individual movements on patterns	large	small
Type of heterogeneity	patch	landscape mosaic
Factors influencing species' distribution	resource/habitat distribution, physiological tolerances	barriers, dispersal
Resolution of detail	high	low
Sampling adequacy (intensity)	good	poor
Effects of sampling error	large	small
Experimental manipulations	possible	difficult
Replication	possible	difficult
Empirical rigor	high	low
Potential for deriving generalizations	low	high
Form of models	mechanistic	correlative
Testability of hypotheses	high	low
Surveys	quantitative	qualitative
Appropriate duration of study	short	long

on the same island, and among five islands. Roughgarden et al. (1987) compared the dynamics of rocky intertidal barnacle communities and assemblages of *Anolis* lizards on islands at 'small,' 'medium,' and 'large' spatial scales. Senft et al. (1987) examined herbivore foraging in relation to vegetation patterns at the scales of the local plant community, the landscape, and the region. Multiscale studies of birds have considered patterns at three to five scales, and Wiens et al. (1986a) recognized four scales of general utility in ecological investigations.

In these examples, the definition of the different scales makes intuitive sense and the analyses reveal the scale-dependency of patterns. Casting the relationships in the context of hierarchy theory (Allen & Starr, 1982; O'Neill et al., 1986) may further sharpen our focus on scaling by emphasizing logical and functional linkages among scales. The scales chosen for analysis are still arbitrary, however; they tend to reflect hierarchies of spatial scales that are based on our own perceptions of nature. Just because these particular scales seem 'right' to us is no assurance that they are appropriate to reef fish, barnacles, anoles, cattle, or birds. We need nonarbitrary, operational ways of defining and detecting scales.

Dependence on Objectives and Organisms

What is an 'appropriate' scare depends in part on the questions one asks. Behavioural ecologists, population ecologists, and ecosystem ecologists, for example, all probe the relationship between resources and consumers, but differences in their objectives lead them to focus their investigations at different scales (Pulliam & Dunning, 1987). Conservation of key species or habitats may target particular patches or landscape fragments for management, whereas programmes emphasizing species richness or complexes of communities may concentrate on preserving broader-scale landscape mosaics (Noss, 1987; Scott et al., 1987).

Differences among organisms also affect the scale of investigation. A staphylinid beetle does not relate to its environment on the same scales as a vulture, even though they are both scavengers. What is a resource patch to one is not to the other. The scale on which an oak tree 'perceives' its environment differs from that of an understorey bluebell or a seedling oak (Harper, 1977). Local populations of vagile organisms may be linked together into larger metapopulations and their dynamics may be less sensitive to the spatial configuration of local habitat patches than more sedentary species (Morrison & Barbosa, 1987; Fahrig & Paloheimo, 1988; Taylor, 1988). Chronically rare species may follow different dispersal and scaling functions than persistently common species. Consumers that use sparse or clumped resources are likely to operate at larger spatial scales than those using abundant or uniformly distributed resources, especially if the resources are critical rather than substitutable (Tilman, 1982; O'Neill et al., 1988).

Such scaling differences among organisms may be viewed in terms of 'ecological neighbourhoods' (Addicott et al., 1987) or 'ambits' (Hutchinson, 1953; Haury et al., 1978); areas that are scaled to a particular ecological process, a time period, and an organism's mobility or activity. The ecological neighbourhood of an individual's daily foraging may be quite different from that of its annual reproductive activities. The ecological neighbourhood of the lifetime movements of a tit in a British woodland may comprise an area of a few square kilometers whereas a raptor may move over an area of hundreds or thousands of square kilometers; a nomadic teal of ephemeral desert ponds in Australia may range over the entire continent. Incidence functions (Diamond, 1975) or fragmentation response curves (Opdam, Rijsdijk & Hustings, 1985) depict the ecological neighbourhoods of species with respect to colonization and persistence of populations in areas of different sizes (scales).

To some extent, differences in ecological neighbourhoods among taxa parallel differences in body mass. This raises the possibility of using allometric relationships (e.g. Calder, 1984) to predict scaling functions for organisms of different sizes. On this basis, for example, one might expect the scale of the home range of a 20-g lizard to be approximately 0.3 ha, whereas that of a 20-g bird would be in the order of 4 ha; the parallel scale for a 200-g bird would be 92 ha. Although such an approach ignores variation in allometric relationships associated with diet, age, season, phylogeny, and a host of other factors, it may still provide an approximation of organism-dependent scaling that is less arbitrary than those we usually use.

Because species differ in the scales of their ecological neighbourhoods, studies of interactions among species may be particularly sensitive to scaling. The population dynamics of predators and of their prey, for example, may be influenced by factors operating on different scales (Hengeveld, 1987), and attempts to link these dynamics directly without recognizing the scale differences may lead to greater confusion than enlightenment. The competitive interactions among species sealing the environment in similar ways may be more direct or symmetrical than those between organisms that share resources but operate on quite different scales. If we arbitrarily impose a particular scale

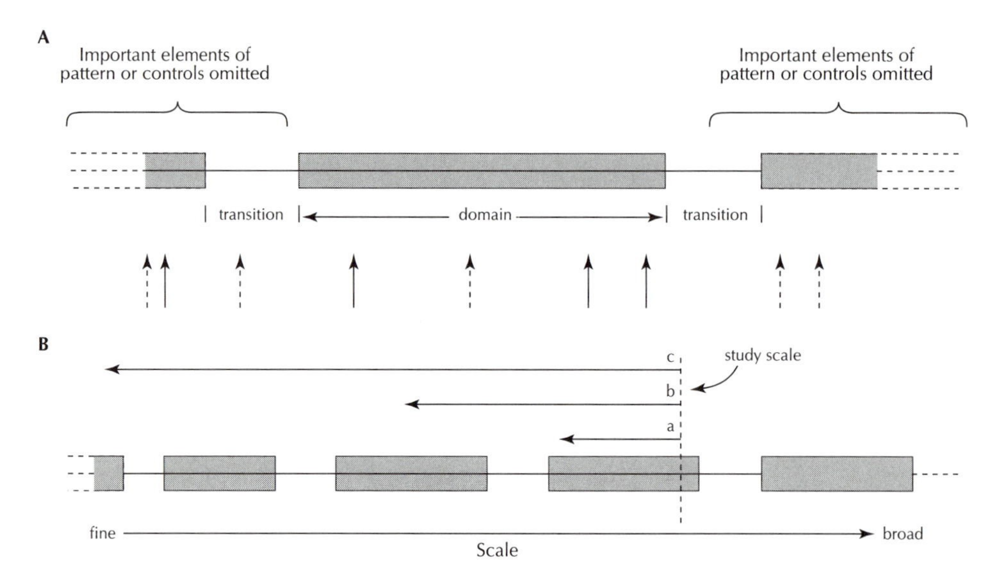

Figure 4. A The domain of scale of a particular ecological phenomenon (i.e. a combination of elements of a natural system, the questions we ask of it, and the way we gather observations) defines a portion of the scale spectrum within which process-pattern relationships are consistent regardless of scale. Adjacent domains are separated by transitions in which system dynamics may appear chaotic. If the focus is on phenomena at a particular scale domain, studies conducted at finer scales will fail to include important features of pattern or causal controls; studies restricted to broader scales will fail to reveal the pattern or mechanistic relationships because such linkages are averaged out or are characteristic only of the particular domain. Comparative investigations based on sampling the scale spectrum at different points in relation to the distribution of scale domains and transitions (solid and dashed vertical arrows) will exhibit different patterns. **B.** If a reductionist approach is adopted to examine patterns found at a particular scale of study, the findings (and inferences about causal mechanisms) will differ depending on how far the reductionism is extended toward finer scales and how many domains are crossed (compare a, b, and c).

(e.g. quadrat size) on a community of organisms that operate on different scales, we truncate the interactions to different degrees for different species.

Domains of Scale

Scale-dependency in ecological systems may be continuous, every change in scale bringing with it changes in patterns and processes. If this is so, generalizations will be hard to find, for the range of extrapolation of studies at a given scale will be severely limited. If the scale spectrum is not continuous, however, there may be *domains* of scale (Figure 4A), regions of the spectrum over which, for a particular phenomenon in a particular ecological system, patterns either do not change or change monotonically with changes in scale. Domains are separated by relatively sharp transitions from dominance by one set of factors to dominance by other sets, like phase transitions in physical systems. Normally well-behaved deterministic systems may exhibit unpredictable behaviour at such transitions (Kitchell et al., 1988), and nonlinear relations may become unstable (O'Neill, personal communication). The resulting chaos makes translation between domains difficult (Heaps & Adam, 1975; May, 1989). The argument over the relative merits of linear versus nonlinear models in ecology (e.g. Patten, 1983) may reflect a failure to recognize the differences in system dynamics within versus between domains.

How may we recognize domains of scale in a way that avoids the arbitrary imposition of preconceived scales or hierarchical levels on natural variation? Several statistical approaches are based on the observation that variance increases as transitions are approached in hierarchical systems (O'Neill et al., 1986). If quadrats in which plant species abundances have been recorded are aggregated into larger and larger groupings, the variance of differences in abundance between pairs of contiguous groups fluctuates as a function of group size (scale). Peaks of unusually high variance indicate scales at which the between-group differences are especially large, suggesting that this may represent the scale of natural aggregation or patchiness of vegetation in the communities (Greig-Smith, 1952, 1979), the boundary of a scale domain. Similar techniques may be used to analyse data gathered on continuous linear transects (Ludwig & Cornelius, 1987). Coincidence in the variance peaks of different features of the system (e.g. plants and soil nutrients, seabirds and their prey) may indicate common spatial scalings and the possibility of direct linkages (Greig-Smith, 1979; Schneider & Piatt, 1986). For a series of point samples, the average squared difference (semivariance) or the spatial autocorrelation between two points may be expressed as a function of the distance between them to estimate the scale of patchiness in a system (Sokal & Oden, 1978; Burrough, 1983). Other investigators have used spectral analysis (Legrende & Demers, 1984) or dimensional analysis (Lewis & Platt, 1982). Obviously, the degree to which any of these methods can reveal scales of spatial patterning is sensitive to grain and extent.

Another approach involves the application of fractal geometry (Mandelbrot, 1983; Peitgen & Saupe, 1988) to ecological patterns. In many physical systems, such as snow crystals, clouds, or flowing fluids, the configuration of patterns differs in detail at different scales but remains statistically 'self-similar' if the changes in pattern measurements are adjusted to the changes in measurement scale (Burrough, 1983; Hentschel & Procaccia, 1984; Nittman et al., 1985). The way in which detail differs with scale is characterized by a fractal dimension, D, which indexes the scale-dependency of the pattern. Statistical self-similarity of patterns (constant D) occurs when processes at fine scales propagate the patterns to broader scales, although self-similar patterns may also arise from the operation of different but complementary processes (Milne, 1988). A change in the fractal dimension of a pattern, on the other hand, is an indication that different processes or constraints are dominant. Regions of fractal self-similarity of pattern may therefore represent domains of scale, whereas rapid changes in fractal dimension with small changes in measurement scale (e.g. the landscape patterns analysed by Krummel et al., 1987 or Palmer, 1988) may indicate transitions between domains. There is a relationship between the sizes and movement patterns of organisms and the fractal dimensions of their habitats (Morse et al., 1985; Weiss & Murphy, 1988, Wiens & Milne, in press), so it may be possible to define ecological neighbourhoods or domains using functions that combine allometry and fractals.

Domains of scale for particular pattern-process combinations define the boundaries of generalizations. Findings at a particular scale may be extrapolated to other scales within a domain, but extension across the transition between domains is difficult because of the instability and chaotic dynamics of the transition zone. Measurements made in different scale domains may therefore not be comparable, and correlations among variables that are evident within a domain may disappear or change sign when the scale is extended beyond the domain (as in the examples of species associations given above). Explanations of a pattern in terms of lower-level mechanisms will differ depending on whether we have reduced to a scale within the same domain, between adjacent domains, or across several domains (Figure 4B). The peril of reductionism in ecology is not so much the prospect that we will be overcome by excessive detail or distracted by local idiosyncrasies but that we will fail to comprehend the extent of our reduction in relation to the arrangement of domains on a scale spectrum.

Of course, not all phenomena in all systems will exhibit the sort of discontinuities in scale-dependence necessary to define domains. Some phenomena may change continuously across broad ranges of scale. The boundaries of even well-defined domains may not be fixed but may vary in time, perhaps in relation to variations in resource levels. The notion of domains, like other conceptual constructs in ecology, may help us to understand nature a bit better, but it should not become axiomatic.

Developing a 'Science of Scale' in Ecology

Recently, Meentemeyer & Box (1987) have called for the development of a 'science of scale' in ecology, in which scale is included as an explicitly stated variable in the analysis. I think that we must go further, to consider scaling issues as a *primary focus* of research efforts. Instead of asking how our results vary as a function of scale, we should begin to search for consistent patterns in these scaling effects. How does heterogeneity affect the size of scale domains? Are the ecological neighbourhoods of organisms in high-contrast landscapes scaled differently from those in areas where the patch boundaries are more gradual? Are there regularities in the transitions between orderly and seemingly chaotic states of ecological systems with changes in scale, in a manner akin to turbulence in fluid flows? Do selective forces influence how organisms scale their environments, so that particular life-history traits are related to responses to particular scales of environmental patchiness? If one adjusts for the size differences between organisms such as a beetle and an antelope that occur in the same prairie, can they then be seen to respond to the patch or fractal structure of the 'landscapes' they occupy in the same way? Are differences between them interpretable in terms of differences in their physiology, reproductive biology, or social organization? Does the spatial heterogeneity of soil patterns at different scales have different effects on how forest ecosystems respond to climatic changes? Is the spread of disturbances a function of the fractal structure of landscapes? Does nutrient redistribution among patches at fine scales lead to instability or stability of nutrient dynamics at broader scales?

To address such questions, we must expand and sharpen the ways we think about scaling. Our ability to detect environmental heterogeneity, for example, depends on the scale of our measurements, whereas the ability of organisms to respond to such patchiness depends on how they scale the environment. Proper analysis requires that the scale of our measurements and that of the organisms' responses fall within the same domain. Because of this, however, the 'proper' scale of investigation is usually not immediately apparent. Moreover, the ecological dynamics within a domain are not closed to the influences of factors at finer or broader domains; they are just different. Ecologists therefore need to adopt a multiscale perspective (Legrende & Demers, 1984; Clark, 1985; Wiens et al., 1986a; Blondel, 1987; Addicott et al., 1987). Studies conducted at several scales or in which grain and extent are systematically varied independently of one another will provide a better resolution of domains, of patterns and their determinants, and of the interrelationships among scales.

We must also develop scaling theory that generates testable hypotheses. One particular focus of such theory must be on the linkages between domains of scale. Our ability to arrange scales in hierarchies does not mean that we understand how to translate pattern-process relationships across the nonlinear spaces between domains of scale, yet we recognize such linkages when we speak of the constraining effects of hierarchical levels on one another or comment on the openness of ecological systems. Perhaps there is a small set of algorithms that can serve to translate across scales. Discovering them requires that we first recognize that ecological patterns and processes are scale-dependent, that this scale-dependency differs for different ecological systems and for different questions that we ask of them, that ecological dynamics and relationships may be well-behaved and orderly within domains of scale but differ from one domain to another and become seemingly chaotic at the boundaries of these domains, and that an arbitrary choice of scales of investigation will do relatively little to define these scaling relationships.

Acknowledgments

A Katharine P. Douglass Distinguished Lectureship at Rocky Mountain Biological Laboratory provided the atmosphere and the impetus to focus my thinking about scaling, and I thank the staff of the laboratory for their hospitality and Mrs. Douglass for endowing the lectureship. Paul Dayton, Kim Hammond, Victor Jaramillo, Natasha Kotliar, Bruce Milne, Bob O'Neill, Eric Peters, LeRoy Poff, Rick Redak, Bill Reiners, David Schimel, Dick Tracy, Monica Turner, James Ward, and Greg Zimmerman commented on the manuscript and agreed on almost nothing. Si Levin, Bob May, Bruce Milne, and Bob O'Neill shared unpublished manuscripts. Ward Watt provided useful editorial suggestions. My investigations of ecological scaling are supported by the United States National Science Foundation (BSR8805829) and Department of Energy (DE-FG0288ER60715).

Notes

[1]Adapted from the first Katharine P. Douglass distinguished Lecture at the Rocky Mountain Biological Laboratory Gothic, Colorado, 23 July 1987.

[2]Quoted in Gleick, 1987, p. 186.

[3]This use of 'grain' differs from that of MacArthur & Levins (1964), who considered grain to be a function of how animals exploit resource patchiness in environments.

[4]Note that what is a fine scale to an ecologist is a large scale to a geographer or cartographer, who express scale as a ratio (e.g. 1:250,000 is a smaller scale than 1:50,000).

References

Addicott, J. F., J. M. Aho, M. F. Antolin, M. F. Padilla, J. S. Richardson and D. A. Soluk. 1987. Ecological neighborhoods: scaling environmental patterns. Oikos:49:340–346.

Ahnert, F. 1987. Process-response models of denudation at different spatial scales. Catena Supplement, 10:31–50.

Allen, T. F. H. and T. B. Starr. 1982. *Hierarchy: Perspectives for Ecological Complexity.* University of Chicago Press, Chicago.

Anderson, G. R. V., A. H. Ehrlich, P. R. Ehrlich, J. D. Roughgarden, B. C. Russell and F. H. Talbot. 1981. The community structure of coral reef fishes. American Naturalist, 117:476–495.

Barbosa, P. and J. C. Schultz. 1987. *Insect Outbreaks.* Academic Press, New York.

Beals, E. W. 1973. Ordination: mathematical elegance and ecological naivete. Journal of Ecology, 61:23–36.

Blondel, J. 1987. From biogeography to life history theory: a multithematic approach illustrated by the biogeography of vertebrates. Journal of Biogeography, 14:405–422

Brown, J. H. 1984. On the relationship between abundance and distribution of species. American Naturalist, 124:255–279.

Burrough, P. A. 1983. Multiscale sources of spatial variation in soil. I. The application of fractal concepts to nested levels of soil variation. Journal of Soil Science, 34:577–597.

Calder, W. A. III 1984. *Size, Function, and Life History.* Harvard University Press, Cambridge, Massachusetts.

Carpenter, S. R. 1988. Transmission of variance through lake food webs. In: *Complex Interactions in Lake Communities.* S. R. Carpenter, ed. Springer-Verlag, New York. Pp. 119–135.

Carpenter, S. R. and J. F. Kitchell. 1987. The temporal scale of variance in limnetic primary production. American Naturalist, 129:417–433.

Chesson, P. L. and T. J. Case. 1 986. Overview: nonequilibrium community theories: chance, variability, history, and coexistence. In: *Community Ecology.* J. Diamond and T. J. Case, eds. Harper and Row. New York. Pp. 229–239.

Clark, W. C. 1985. Scales of climate impacts. Climatic Change, 7:5–27.

Clarke, R. D. 1988. Chance and order in determining fish-species composition on small coral patches. Journal of Experimental Marine Biology and Ecology, 115:197–212.

Cole, B. J. 1983. Assembly of mangrove ant communities: patterns of geographic distribution. Journal of Animal Ecology, 52:339–348.

Connell, J. H. 1980. Diversity and coevolution of competitors, or the ghost of competition past. Oikos, 35:131–138.

Dagan, G. 1986. Statistical theory of groundwater flow and transport: pore to laboratory, laboratory to formation, and formation to regional scale. Water Resources Research. 22:120S–134S.

Dayton, P. K. and M. J. Tegner. 1984. The importance of scale in community ecology: a kelp forest example with terrestrial analogs. In: *A New Ecology. Novel Approaches to Interactive Systems.* P. W. Price, C. N. Slobodchikoff and W. S. Gaud, ed. John Wiley and Sons, New York. Pp. 457–481.

DeAngelis, D. L. and J. C. Waterhouse. 1987. Equilibrium and nonequilibrium concepts in ecological models. Ecological Monographs, 57:1–21.

Delcourt, H. R., P. A. Delcourt, and T. Webb. 1983. Dynamic plant ecology: the spectrum of vegetation change in space and time. Quartenary Science Review, 1:153–175.

den Boer, P. J. 1981. On the survival of populations in a heterogeneous and variable environment. Oecologia, 50:39–53.

Denman, K. L. and T. Platt. 1975. Coherences in the horizontal distributions of phytoplankton and temperature in the upper ocean. Memoirs Societie Royal Science Liege (Series 6), 7:19–36.

Diamond, J. M. 1975. Assembly of species communities. In: *Ecology and Evolution of Communities.* M. L. Cody and J. M. Diamond, eds. Harvard University Press, Cambridge. Massachusetts. Pp. 342–444.

Fahrig, L. and J. Paloheimo. 1988. Effect of spatial arrangement of habitat patches on local population size. Ecology, 69:468–475.

Feigenbaum, M. 1979. The universal metric properties of nonlinear transformations. Journal of Statistical Physics, 21:669–706.

Frost, T. M., D. L. DeAngelis, S. M. Bartell, D. J. Hall, and S. H. Hurlbert. 1988. Scale in the design and interpretation of aquatic community research. In: *Complex Interactions in Lake Communities.* S. R. Carpenter, ed. Springer-Verlag, New York. Pp. 229–258.

Galzin, R. 1987. Structure of fish communities of French Polynesian coral reefs. I. Spatial scale. Marine Ecology-Progress Series, 41:129–136.

Giller, P. S. and J. H. R. Gee. 1987. The analysis of community organization: the influence of equilibrium, scale and terminology. In: *Organization of Communities Past and Present.* J. H. R. Gee and P .S. Giller, eds. Blackwell Scientific Publications, Oxford. Pp. 519–542.

Gleick, J. 1987. *Chaos: Making a New Science.* Viking, New York.

Gosz, J. R., C. N. Dahm and P. G. Risser. 1988. Long-path FTIR measurement of atmospheric trace gas concentrations. Ecology, 69:1326–1330.

Gower, J. F. R., K. L. Denman, and R. J. Holyer. 1980. Phytoplankton patchiness indicates the fluctuation spectrum of mesoscale oceanic structure. Nature, 288:157–159.

Grebogi, C., E. Ott, and J. A. Yorke. 1987. Chaos, strange attractors, and fractal basin boundaries in nonlinear dynamics. Science, 238:632–638.

Green, D. G., R. H. Bradbury, and R. E. Reichelt. 1987. Patterns of predictability in coral reef community structure. Coral Reefs, 6:27–34.

Greig-Smith, P. 1952. The use of random and contiguous quadrats in the study of the structure of plant communities. Annals of Botany, New Series, 16:293–316.

Greig-Smith, P. 1979. Pattern in vegetation. Journal of Ecology, 67:755–779.

Harper, J. L. 1977. *Population Biology of Plants.* Academic Press, New York.

Haury, L. R., J. A. McGowan, and P. H. Wiebe. 1978. Patterns and processes in the time-space scales of plankton distribution. In: *Spatial Pattern in Plankton Communities.* J. H. Steele, ed. Plenum, New York. Pp. 277–327.

Heaps, N. S. and Y. A. Adam. 1975. Non-linearities associated with physical and biochemical processes in the sea. In: *Modeling of Marine Ecosystems.* J. C. P. Nihoul, ed. Elsevier, Amsterdam. Pp. 113–126.

Hengeveld, R. 1987. Scales of variation: their distribution

and ecological importance. Annales Zoologici Fennici, 24:195–202.

Hentschel, H. G. E. and I. Procaccia. 1984. Relative diffusion in turbulent media: the fractal dimension of clouds. Physics Review A, 29:1461–1470.

Hunt, G. L. Jr and D. C. Schneider. 1987. Scale-dependent processes in the physical and biological environment of marine birds. In: *Seabirds: Feeding Ecology and Hole in Marine Ecosystems.* J. P. Croxall, ed. Cambridge University Press, Cambridge. Pp. 7–41.

Huston, M., D. DeAngelis, and W. Post. 1988. New computer models unify ecological theory. BioScience, 38:682–691.

Hutchinson, G. E. 1953. The concept of pattern in ecology. Proceedings of the National Academy of Science of the USA, 105:1–12.

Hutchinson, G. E. 1965. *The Ecological Theater and the Evolutionary Play.* Yale University Press, New Haven, Connecticut.

Jarvis, P. G. and K. G. McNaughton. 1986. Stomatal control of transpiration: scaling up from leaf to region. Advances in Ecological Research, 15:1–49.

Kareiva, P. and M. Andersen. 1988. Spatial aspects of species interactions: the wedding of models and experiments. In: *Community Ecology.* A. Hastings, ed. Springer-Verlag, New York. Pp. 38–54.

Kitchell, J. F., S. M. Bartell, S. R. Carpenter, D. J. Hall, D. J. McQueen, W. E. Neill, D. Scavia and E. E. Werne., 1988. Epistemology, experiments, and pragmatism. In: *Complex Interactions in Lake Communities.* S. R. Carpenter, ed. Springer-Verlag, New York. Pp. 263–280.

Krummel, J. R., R. H. Gardner, G. Sugihara, R. V. O'Neill and P. R. Coleman. 1987. Landscape patterns in a disturbed environment. Oikos, 48:321–324.

Legrende, L. and S. Demers. 1984. Towards dynamic biological oceanography and limnology. Canadian Journal of Fishery and Aquatic Science, 41:2–19.

Lekan, J. F. and R. E. Wilson 1978. Spatial variability of phytoplankton biomass in the surface waters of Long Island. Estuarine and Coastal Marine Science, 6:239–251.

Levin, S. A. 1989. Challenges in the development of a theory of ecosystem structure and function. In: *Perspectives in Ecological Theory.* J. Roughgarden, R. M. May and S. A. Levin, eds. Princeton University Press. Princeton. N. J. Pp. 242–255.

Lewis, M. R. and T. Platt. 1982. Scales of variation in estuarine ecosystems. In: *Estuarine Comparisons.* V. S. Kennedy, ed. Academic Press. New York. Pp. 3–20.

Ludwig, J.A. and J. M. Cornelius. 1987. Locating discontinuities along ecological gradients. Ecology, 68:448–450.

MacArthur, R. H. and R. Levins. 1964. Competition, habitat selection. and character displacement in a patchy environment. Proceedings of the National Academy of Science of the USA, 51:1207–1210.

Mandelbrot. B. 1983. *The Fractal Geometry of Nature.* W.H. Freeman and Company. San Francisco.

Marino, P. C. 1988. Coexistence on divided habitats: mosses in the family Splachnaceae. Annales Zoologici Fennici, 25:89–98.

May, R. M. 1989. Levels of organization in ecology. In: *Ecological Concepts. The Contribution of Ecology to an Understanding of the Natural World.* BES Symposium, No. 29:339–361. Blackwell Scientific Publications, Oxford.

May, R. M. 1981. Modeling recolonization by neotropical migrants in habitats with changing patch structure, with notes on the age structure of populations. In: *Forest Island Dynamics in Man-dominated Landscapes.* R. L. Burgess and D. M. Sharpe, eds. Springer-Verlag. New York. Pp. 207–213.

May, R. M. and G. F. Oster. 1976. Bifurcations and dynamic complexity in simple ecological models. American Naturalist, 110:573–599.

Meentemeyer, V. 1984. The geography of organic decomposition rates. Annals of the Association of American Geographers, 74:551–560.

Meentemeyer, V. and E. O. Box. 1987. Scale effects in landscape studies. In: *Landscape Heterogeneity and Disturbance.* M. G. Turner, ed. Springer-Verlag, New York. Pp. 15–34.

Milne, B. T. 1988. Measuring the fractal geometry of landscapes. Applied Mathematics and Computation, 27:67–79.

Moloney, K. A. 1988. Fine-scale spatial and temporal variation in the demography of a perennial bunchgrass. Ecology, 69:1588–1598.

Morrison, G. and P. Barbosa. 1987. Spatial heterogeneity, population 'regulation' and local extinction in simulated host-parasitoid interactions. Oecologia, 73:609–614.

Morse, D. R., J. H. Lawton, M. M. Dodson, and M. H. Williamson. 1985. Fractal dimension of vegetation and the distribution of arthropod body lengths. Nature, 314:731–733.

Nittman, J., G. Daccord and H .E. Stanley. 1985. Fractal growth of viscous fingers: quantitative characterization of a fluid instability phenomenon. Nature, 314:141–144.

Noss, R. F. (1987) From plant communities to landscapes in conservation inventories: a look at the Nature Conservancy (USA). Biological Conservation, 41:11–37,

O'Connor, R. J. and R. J. Fuller. 1985. Bird population responses to habitat. In: *Bird Census and Atlas Studies: Proceedings of the Vii International Conference on Bird Census Work.* K. Taylor, R. J. Fuller, and P. C. Lack, eds. British Trust for Ornithology, Tring, England. Pp. 197–211.

Ogden. J. C. and J. P. Ebersole. 1981. Scale and community structure of coral reef fishes: a long term study of a large artificial reef. Marine Ecology-Progress Series, 4:97–104.

O'Neill, R.V., D. L. DeAngelis, J. B. Waide and T. F. H. Allen. 1986. *A Hierarchical Concept of Ecosystems.* Princeton University Press. Princeton. N.J.

O'Neill, R.V., B. T. Milne, M. G. Turner and R. H. Gardner.

1988. Resource utilization scales and landscape patterns. Landscape Ecology, 2:63–69.

Opdam, P., G. Rijsdijk and F. Hustings. 1985. Bird communities in small woods in an agricultural landscape: effects of area and isolation. Biological Conservation, 34:333–352.

Orbach, R. 1986. Dynamics of fractal networks. Science, 231:814–819.

Palmer, M. W. 1988. Fractal geometry: a tool for describing spatial patterns of plant communities. Vegetation, 75:91–102.

Patten. B. C. 1983. Linearity enigmas in ecology. Ecological Modeling, 18:155–170.

Peitgen, H. -O. and D. Saupe eds. 1988. *The Science of Fractal Images*. Springer-Verlag. New York.

Platt, T. 1972. Local phytoplankton abundance and turbulence. Deep-Sea Research, 19:183–187.

Platt, T. and S. Sathyendranath. 1988. Oceanic primary production: estimation by remote sensing at local and regional scales. Science, 241:1613–1620.

Powell, T. M., P. J. Richardson, T. M. Dillon, B. A. Agee, B. J. Dozier, D. A. Godden and L. O. Myrup. 1975. Spatial scales of current speed and phytoplankton biomass fluctuations in Lake Tahoe. Science, 189:1088–1090.

Pulliam, H. R. and J. B. Dunning. 1987. The influence of food supply on local density and diversity of sparrows. Ecology, 68:1009–1014.

Ricklefs, R. E. 1987. Community diversity: relative roles of local and regional processes. Science, 235:167–171.

Rosswall, T., R. G. Woodmansee, and P. G. Risser, eds. 1988. *Scales and Global Change*. John Wiley and Sons, New York.

Roughgarden, J. 1983. Competition and theory in community ecology. American Naturalist, 122:583–601.

Roughgarden, J., S. D. Gaines, and S. W. Pacala. 1987. Supply side ecology: the role of physical transport processes. In: *Organization of Communities Past and Present*. J. H. R. Gee and P S. Giller, eds. Blackwell Scientific Publications, Oxford. Pp. 491–518.

Rykiel, E. J. Jr, R. N. Coulson, P. J. H. Sharpe, T. F. H. Allen and R. O. Flamm. 1988. Disturbance propagation by bark beetles as an episodic landscape phenomenon. Landscape Ecology, 1:129–139.

Sale, P. F. 1988. Perception, pattern, chance and the structure of reef fish communities. Environmental Biology of Fishes, 21:3–15.

Schaffer, W. M. 1984. Stretching and folding in Lynx fur returns: evidence for a strange attractor in nature? American Naturalist, 123:798–820.

Schneider, D. C. and J. F. Piatt. 1986. Scale-dependent correlation of seabirds with schooling fish in a coastal ecosystem. Marine Ecology-Progress Series, 32:237–246.

Schoener, T. W. 1982. The controversy over interspecific competition. The American Scientist, 70:586–595.

Schumm, S. A. and R. W. Lichty. (1965) Time, space, and causality in geomorphology. American Journal of Science, 263:110–119.

Scott, J. M., B. Csuti, J. D. Jacobi, & J. E. Estes. 1987. Species richness. BioScience, 37:782–788.

Senft, R. L., M. B. Coughenour, D. W. Bailey, L. R. Rittenhouse, O. K. Sala and D. M. Swift. 1987. Large herbivore foraging and ecological hierarchies. BioScience, 37:789–799.

Sherry, T. W. and R. T. Holmes. 1988. Habitat selection by breeding American Redstarts in response to a dominant competitor, the Least Flycatcher. The Auk, 105:350–364.

Simberloff, D. 1988. The contribution of population and community biology to conservation science. Annual Review of Ecology and Systematics, 19:473–511.

Sokal, R. R. and N. L. Oden. 1978. Spatial autocorrelation in biology. 2. Some biological implications and four applications of evolutionary and ecological interest. Biological journal of the Linnean Society, 10:229–249.

Steele. J., ed. 1978. *Spatial pattern in plankton communities.* NATO Conference Series, Series IV: Marine Sciences, vol. 3. Plenum Press, New York.

Taylor, A. D. 1988. Large-scale spatial structure and population dynamics in arthropod predator-prey systems. Annales Zoologici Fennici, 25:63–74.

Taylor, F. 1981. Ecology and evolution of physiological time in insects. The American Naturalist, 117:1–23.

Therriault, J. -C. and T. Platt. 1981. Environmental control of phytoplankton patchiness. Canadian Journal of Fishery and Aquatic Science, 38, 638–641.

Tilman, D. 1982. *Resource Competition and Community Structure.* Princeton University Press, Princeton, New Jersey.

Vaisanen, R. A., O. Jarvinen, and P. Rauhala. 1986. How are extensive, human-caused habitat alterations expressed on the scale of local bird populations in boreal forests? Ornis Scandinavica, 17:282–292.

Weiss, S. B. and D. D. Murphy. 1988. Fractal geometry and caterpillar dispersal: or how many inches can inchworms inch? Functional Ecology, 2:116–118.

Wiens, J. A. 1983. Avian community ecology: an iconoclastic view. In: *Perspectives in Ornithology* . A. H. Brush and G. A. Clark Jr., eds. Cambridge University Press, Cambridge. Pp. 355–403.

Wiens, J. A. 1984. On understanding a non-equilibrium world: myth and reality in community patterns and processes. In: *Ecological Communities: Conceptual Issues and the Evidence.* D. R. Strong Jr, D. Simberloff, L. G. Abele and A. B. Thistle, eds. Princeton University Press, Princeton, N.J. Pp. 430–457.

Wiens, J. A. 1986. Spatial scale and temporal variation in studies of shrub-steppe birds. In: *Community Ecology*. J. Diamond and T. J. Case, eds. Harper and Row, New York.Pp. 154–172.

Wiens, J. A. 1989. *The Ecology of Bird Communities,* Vol. 2. Processes and Variations. Cambridge University Press. Cambridge.

Wiens, J. A., and B. T. Milne. 1989. Scaling of 'Landscapes' in landscape ecology from a beetle's perspective. Landscape Ecology, 3, in press.

Wiens, J. A., J. F. Addicott, T. J. Case and J. Diamond. 1986a. Overview: The importance of spatial and temporal scale in ecological investigations. In: *Community Ecology*. J. Diamond and T. J. Case, eds. Harper and Row, New York. Pp. 145–153.

Wiens, J. A., J. T. Rotenberry and B. Van Horne. 1986b. A lesson in the limitations of field experiments: shrub-steppe birds and habitat alteration. Ecology, 67:365–376.

Woodward, F. I. 1987. *Climate and Plant Distribution*. Cambridge University Press, Cambridge.

Readings for

Ecosystem Ecology

Integrated Physical, Chemical and Biological Processes

Stephen R. Carpenter

- **Stephen R. Carpenter, James F. Kitchell and James R. Hodgson**
  Cascading Trophic Interactions and Lake Productivity
- **Crawford S. Holling and Gary K. Meffe**
  Command and Control and the Pathology of Natural Resource Management
- **John Pastor, Robert J. Naiman, Bradley Dewey and Pamela McInnes**
  Moose, Microbes, and the Boreal Forest
- **Peter M. Vitousek, Harold A. Mooney, Jane Lubchenco and Jerry M. Melillo**
  Human Domination of Earth's Ecosystems

Introduction to

Stephen R. Carpenter, James F. Kitchell, and James R. Hodgson

Cascading Trophic Interactions and Lake Productivity

by Stephen R. Carpenter

During the 1970s, limnology was dominated by the idea that inputs of nutrients, especially phosphorus, controlled the productivity of lakes. The evidence for the importance of nutrients was convincing. It came from diverse kinds of research, including comparisons of lakes with different nutrient levels, experiments in which lakes were enriched with nutrients, and management experiences. Even though nutrient effects were well supported, there was considerable variability in lake production that could not be explained by nutrients. In some studies, lakes with the same nutrient level varied nearly 100-fold in productivity. Ecologists had long known that ecosystem process rates are controlled by multiple factors. By about 1980, leading limnologists such as Joseph Shapiro, David Schindler and Richard Vollenweider were speculating about other factors that could control lake production. Shapiro, in particular, advanced the idea that grazing was a crucial factor in lake productivity. His ideas were consistent with earlier research on the effects of food web structure on lake plankton. However, it was difficult to understand how food web control could be reconciled with the dominant view of nutrient control, and for many ecologists the alternatives were mutually exclusive. The controversy is represented by many papers from the 1980s and 1990s which contrast "top down versus bottom-up" views of lake productivity.

The trophic cascade theory explains how nutrients and the food web act in combination to control the productivity of lakes. The cascade metaphor was introduced to ecology by Robert Paine in 1980, and adapted by this paper to the problem of understanding lake production. In essence, trophic cascade theory states that nutrients set the potential production of lakes, and food web structure sets the variability around that potential. Fish recruitment (the number of fish that survive and grow to a given size) is highly variable and controlled by factors that are independent of nutrient input. Fluctuating, variable recruitment of fish species, and the predator-prey interactions among the species, leads to fish communities that are dominated by large piscivorous (fish-eating) fishes, or small planktivorous (plankton-eating) fishes, or some mixture of these extremes. In lakes dominated by planktivorous fishes, zooplankton grazers are suppressed and phytoplankton biomass is relatively high. In lakes dominated by piscivorous fishes, planktivorous fishes are suppressed, zooplankton flourish and graze heavily, and phytoplankton biomass is relatively low.

Trophic cascade theory has been tested by comparisons of lakes with contrasting food webs, experimental manipulation of lake food webs, long-term studies of lakes, and management experiences. Predation clearly has major effects on lake productivity, comparable in magnitude to those of nutrients.

Cascading Trophic Interactions and Lake Productivity

STEPHEN R. CARPENTER, JAMES F. KITCHELL, AND JAMES R. HODGSON

Fish predation and herbivory can regulate lake ecosystems; altering food webs by altering consumer populations may be a promising management tool.

Carpenter, S. R., J. F. Kitchell and J. R. Hodgson. 1985. Cascading trophic interactions and lake productivity. *BioScience* 35:634–639.

Limnologists have been studying patterns in lake primary productivity for more than 60 years (Elster 1974). More recently, concern about eutrophication has focused attention on nutrient supply as a regulator of lake productivity. However, nutrient supply cannot explain all the variation in the primary productivity of the world's lakes. Schindler (1978) analyzed a sample of 66 lakes that were likely to be limited in productivity by phosphorus because their nitrogen/phosphorus ratios exceeded five. Phosphorus supply, corrected for hydrologic residence time, explained only 48% of the variance in primary production, and lakes with similar phosphorus supply rates differed nearly a thousandfold in productivity. Phosphorus loading explains 79–95% of the variance in chlorophyll *a* concentration (Dillon and Rigler 1974, Oglesby 1977, Schindler 1978), but chlorophyll *a* concentration is a poor predictor of primary production (Brylinsky and Mann 1973, Oglesby 1977).

The concept of cascading trophic interactions, on the other hand, explains differences in productivity among lakes with similar nutrient supplies but contrasting food webs. The concept reflects an elaboration of long-standing principles of fishery management based on logistic models (Larkin 1978). Simply put, a rise in piscivore biomass brings decreased planktivore biomass, increased herbivore biomass, and decreased phytoplankton biomass (Figure 1). Specific growth rates at each trophic level show

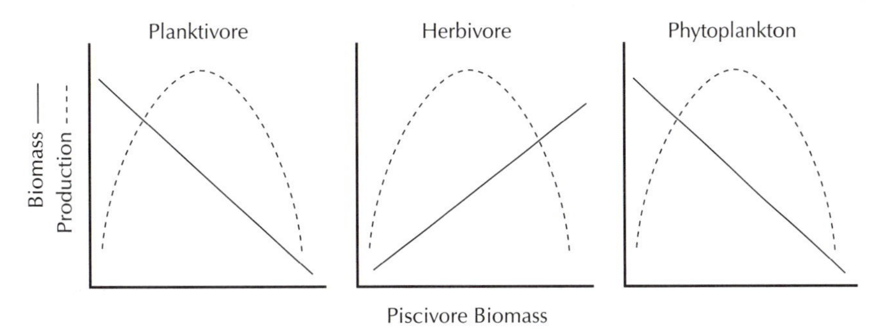

Figure 1. Piscivore biomass in relation to biomass (solid line) and production (dashed line) of vertebrate zooplanktivores, large herbivores, and phytoplankton.

the opposite responses. Productivity at a given trophic level is maximized at an intermediate biomass of its predators. Productivity at all trophic levels, and energy flow through the food web, are highest where intensities of predation are intermediate at all trophic levels (Kitchell 1980). Although this simple conceptual model is heuristically useful, real ecosystems exhibit nonequilibrium dynamics that result from different life histories and variable interactions among the major species.

Cascading trophic interactions and nutrient loading models are complementary, not contradictory. Potential productivity at all trophic levels is set by nutrient supply. Actual productivity depends on the recycling of nutrients and their allocation among populations with different growth rates. The phosphorus availability to phytoplankton, for example, is determined by processes that operate over a wide range of temporal and spatial scales (Harris 1980, Kitchell et al. 1979). Nutrient excretion by zooplankton is a major recycling process (Lehman 1980) that is strongly influenced by selective predation on zooplankton by fishes (Bartell and Kitchell 1978). Thus, by regulating recycling rates, consumers regulate primary production.

The Trophic Cascade

To explain the details of cascading trophic interactions, we consider a lake food web that includes limiting nutrients and four trophic levels: piscivores such as bass, pike, or salmon, zooplanktivores, herbivorous zooplankton, and phytoplankton (Figure 2). Invertebrate planktivores like insect larvae and predacious copepods take smaller prey than vertebrate planktivores like minnows. (Even though the rotifers include herbivores and predators, we will treat them collectively as a size class of zooplankton that includes grazers and is preyed on most heavily by invertebrates.) Small crustacean zooplankton include grazers, such as *Bosmina*, which remain small throughout their life cycle, and the young of large crustacean grazers, such as *Daphnia pulex*, and invertebrate planktivores. We divide the phytoplankton into three functional groups: nannoplankters subject to grazing by all herbivores, edible net phytoplankters like *Scenedesmus* that are grazed only by larger zooplankton, and inedible algae.

Examples of consumers controlling species composition, biomass, and productivity are available for each trophic level. Changes in the density of large piscivorous fishes result in changes in density, species composition, and behavior of zooplanktivorous fishes. In Wisconsin lakes containing bass or pike, spiny-rayed planktivorous fishes replace soft-rayed minnows, which are common in the absence of piscivores (Tonn and Magnuson 1982). The depletion of prey fishes by salmonids stocked in Lake Michigan (Stewart et al. 1981) and in European reservoirs (Benndorf et al. 1984) shows how piscivores can regulate zooplanktivorous fishes. Prey fish biomass declines as their predators increase in density; in contrast, prey fish productivity reaches a maximum at intermediate predator densities (Larkin 1978).

High planktivory by vertebrates is associated with low planktivory by invertebrates as well as high densities of rotifers and small crustaceans. Where planktivorous fishes are absent, invertebrate planktivores and large crustacean zooplankton predominate. Planktivorous fishes select the largest available prey and can rapidly reduce the density of zooplankters larger than about 1 mm (Hall et al. 1976). In contrast, planktivorous invertebrates select and deplete herbivores smaller than 0.5–1 mm. Lynch

(1980) concludes that contrasting planktivore pressures have led to two distinct types of life history in cladoceran herbivores. Heavy planktivory by invertebrates favors large cladocerans that grow rapidly until they cannot be taken by the planktivores. At this size, these cladocerans shift energy allocation from growth to producing many small offspring. Planktivorous fishes, which consume large zooplankton (including invertebrate planktivores), promote dominance of small cladocerans that grow continually, reproduce at an early age, and have small clutches of large offspring.

Differences in size structure among herbivorous zooplankton communities lead to pronounced differences in grazing and nutrient recycling rates. Effects of zooplankton on phytoplankton biomass and productivity are not intuitively clear because they result from countervailing processes (grazing vs. nutrient recycling) and potentially compensatory allometric relationships. Larger zooplankters can ingest larger algae (Burns 1968). Absolute grazing rate (cells • animal^{-1} • t^{-1}) increases with grazer size, but mass-specific grazing rate (cells • mg • animal^{-1} • t^{-1}) declines with grazer size (Peters and Downing 1984). Similarly, absolute excretion rate increases with grazer size, and mass-specific excretion rate decreases with grazer size (Ejsmont-Karabin 1983, Peters and Rigler 1973). Herbivorous zooplankton alter phytoplankton species composition and size structure directly by selective grazing and indirectly through nutrient recycling (Bergquist 1985, Carpenter and Kitchell 1984, Lehman and Sandgren 1985). Changes in phytoplankton size structure imply substantial changes in chlorophyll concentration and productivity because of several allometric relationships. Increasing algal cell size is accompanied by decreases in maximum growth rate, susceptibility to grazing, cell quotas for N and P, and intracellular chlorophyll concentrations and by increases in sinking rate and half-saturation constants for nutrient uptake (Reynolds 1984).

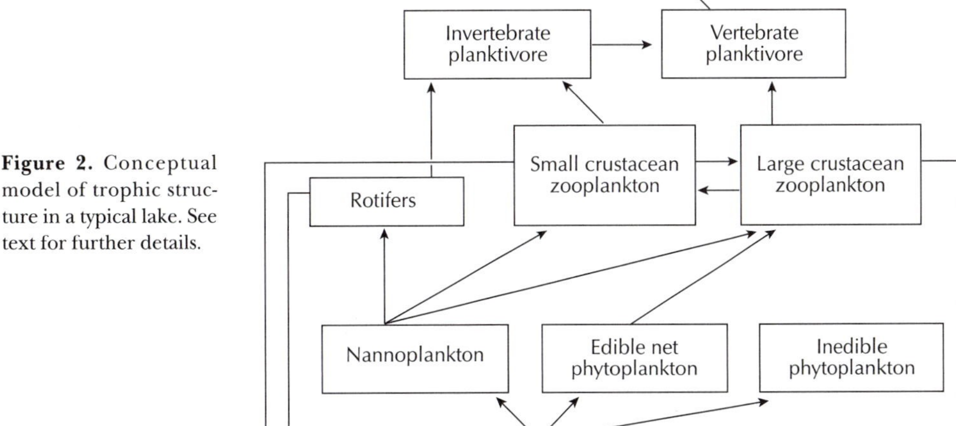

Figure 2. Conceptual model of trophic structure in a typical lake. See text for further details.

We investigated the complex interactions among zooplankton and phytoplankton using simulation models that yielded response surfaces of algal biomass and productivity as functions of zooplankton biomass and zooplankter body size (Carpenter and Kitchell 1984). Chlorophyll *a* concentrations were highest at low biomasses of small herbivores, declining smoothly as both biomass and grazer mass increased. The response of primary production to zooplankton biomass was unimodal, with maximum production at intermediate zooplankton biomass. The zooplankton biomass that maximized primary production declined as herbivore mass increased. At low herbivore biomass, productivity was limited by recycling; it increased as grazer biomass increased. When herbivore biomass was high, productivity was restrained by grazing and declined as grazer biomass rose. Although the model was far more complex than the familiar logistic equation of population biology, the phytoplankton as a whole behaved logistically in two respects: Productivity was related parabolically to chlorophyll *a*, with maximum production at intermediate chlorophyll *a*, and specific productivity decreased as chlorophyll *a* increased.

An increase in piscivore density cascades through the food web in the following way. Vertebrate zooplanktivores are reduced while planktivory by invertebrates increases, shifting the herbivorous zooplankton community toward larger zooplankters and higher biomass. Chlorophyll *a* concentration declines.

A decrease in piscivore density has the reverse effects. Vertebrate zooplanktivory rises at the expense of invertebrate zooplanktivores, and small zooplankters dominate the herbivore assemblage. Chlorophyll *a* concentration rises. A change in piscivore density can increase or decrease primary production, which is a unimodal function of zooplankton biomass.

Rates of Cascading Responses

In natural systems, sequences of cascading trophic interactions will propagate from stochastic fluctuations in piscivore year-class strength and mortality. Fish stocks, reproduction rates, and mortality rates in turn exhibit enormous variance (Peterson and Wroblewski 1984, Steele and Henderson 1984).

Fluctuations in piscivore reproduction do not cascade instantaneously through lake food webs. Rather, lags in ecosystem response occur because generation times differ among trophic levels. In temperate lakes, piscivores and many invertebrate and vertebrate planktivores reproduce annually. Crustacean herbivores and rotifers, which go through a generation in several days, pass through many generations in a summer. Phytoplankton generation times are shorter still, ranging from hours to a few days. Inorganic nutrients turn over in only a few minutes to a few hours. Because of this hierarchy of generation and turnover times, ecosystem components respond at different rates to changes in piscivore abundance.

The longest lags in the trophic cascade result from predatory ontogeny and predatory inertia. Predatory ontogeny occurs when a piscivore cohort develops, and the fish act first as zooplanktivores and then as piscivores (Figure 3, solid lines). As zooplanktivores, the fish drive the ecosystem toward small zooplankton and higher chlorophyll concentrations. These trends reverse as the fish grow and increase the proportion of planktivorous fish in their diet. Predatory inertia refers to the persisting effect of older age classes despite reproductive failure in any one year (Stewart et al. 1981). It takes several consecutive year-class failures to reduce piscivory enough for vertebrate planktivores to increase, with associated shifts in zooplankton and phytoplankton.

Cascading trophic interactions can be reversed by increasing or decreasing the intensity of piscivory. Because of lags, however, responses to increased piscivory involve transitions among food web configurations that do not occur during responses to decreased piscivory. Hysteresis will therefore occur when a change in piscivory is reversed: The sequence of ecosystem states and the rate of transition among states in the reverse pathway will differ from those of the forward pathway. The hysteresis effect is illustrated by two contrasting disturbances (Figure 3). Solid lines show the results of an unusually strong piscivore year class, which could occur naturally or through stocking young fish. Dashed lines show the results of a reduction in piscivores, such as those caused by winter kill or human exploitation. In each case, the system returns to the same state, but the pathways are very different.

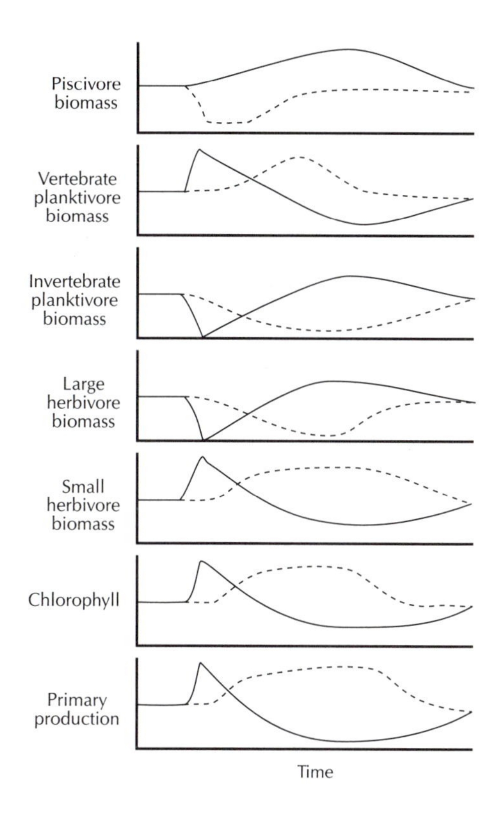

Figure 3. Time course of ecosystem response to a strong piscivore year class (solid line) and a partial winter kill of piscivores (dashed line).

Lake ecosystems are buffeted at irregular intervals by variations in fish recruitment and mortality rates. The system responses are nonequilibrium, transient phenomena that are difficult to detect using long-term averages. Finer-grained time course data are needed.

Correlations vs. Experiments

Ecologists have been urged to develop theories based on multiple regression analyses of data from the literature (Peters 1980). We doubt that this approach can be used successfully to analyze relationships between food web structure and productivity. Correlations among trophic levels reflect nutrient supply effects, which influence biomass at each trophic level in an essentially stoichiometric fashion (cf. McCauley and Kalff 1981). The effects of food web structure are independent of those due to nutrient supply. Therefore, the appropriate statistical procedure is to first remove nutrient effects by regression, and then seek food web effects in the residuals of the regressions. Such a study would be subject to the pitfalls of interpreting regressions pointed out by Box et al. (1978, pp. 487–498). Common statistical problems in data from the literature relevant to cascading trophic interactions are dependencies among predictor variables and lack of control or precise measurements of predictor variables.

Literature data have serious shortcomings, in addition to purely statistical problems, which make them unsuitable for regression analyses of cascading trophic interactions. Frequently, data on biomasses of trophic levels do not distinguish between edible and inedible items or reflect the breadth of predators' diets. Because predators' tastes are catholic, trophic levels are not distinct; they are only a statistical statement about organisms' most prevalent feeding relationships. Typically, researchers look at annual averages; this time scale may not reveal important but transient responses. Finally, the correlation approach does not account for time lags. It is not reasonable to expect today's algal production to correlate with today's biomasses of zooplankton, planktivores, and piscivores. Rather, algal production today may depend on yesterday's zooplankton, which depended on zooplanktivores during the past month, which depended on piscivore recruitment the previous year.

Because of these problems, clear cause-and-effect relationships do not emerge from multiple regression analyses of lake ecosystem data. Experimental manipulations of food webs are a more promising research strategy. The several published accounts of dramatic algal responses to increased grazing pressure in lakes where predation was altered suggest that such experiments will be fruitful.

Case Studies

The idea that fishes can regulate lake ecosystem processes stemmed from the work of Hrbacek et al. (1961). Several experiments have been conducted in which zooplanktivorous fish were removed from lakes, usually with a poison such as rotenone. The consistent conclusions from these studies are that planktivore removal results in greater densities of larger zooplankton, which impose greater grazing pressure on the phytoplankton; increased frequency of grazing-resistant phytoplankters; reduced chlorophyll *a* concentrations and total algal densities; increased Secchi-disk transparency; and reduced total nutrient concentrations in the epilimnion (Benndorf et al. 1984, DiBernardi 1981, Henrikson et al. 1980, Kitchell et al. 1986, Shapiro 1980, Shapiro and Wright 1984). These experiments lack primary productivity data, with one exception. Henrikson et al. (1980) found that primary productivity decreased to 10–20% of baseline levels after zooplanktivorous fish were removed from a Swedish lake.

More extensive productivity data come from experiments using enclosures or microcosms. Korstad (1980) found that increased zooplankton biomass stimulated both productivity and specific productivity of phytoplankton. Bergquist (1985) found that grazed algae (< 22 μm) had maximum productivity at intermediate zooplankton biomass. Production of nongrazed algae was stimulated by increasing zooplankton biomass. Elliott et al. (1983) studied microcosms with no zooplanktivorous fish, those with caged fish that had limited access to the zooplankton, and those with unrestricted fish. Systems with caged fish had intermediate zooplankton biomass and maximum primary production. Studies of periphytic algae have found maximum productivity at intermediate densities of the following herbivores: crayfish (Flint and Goldman 1975), a herbivorous fish (Cooper 1973), tadpoles (Scale 1980), and snails (Gregory 1983). Similar unimodal curves of productivity versus grazer biomass occur in grasslands grazed by ungulates (McNaughton 1979).

Management Implications

In sum, enhanced piscivory can decrease planktivore densities, increase grazer densities, and decrease chlorophyll concentrations. Stocking piscivores therefore has promise as a tool for rehabilitating eutrophic lakes. Shapiro was among the first to recognize the potential of food web alteration as a management tool and has termed the approach biomanipulation (Shapiro and Wright 1986). A recent review has advocated stocking piscivores and/or harvesting zooplanktivores as a practical approach toward enhanced fishery production and mitigation of water quality problems (Kitchell et al. 1986). The approach has been successfully used to control eutrophication in European reservoirs (Benndorf et al. 1984).

Limnology and fisheries biology have developed independently and remain largely separate professions (Larkin 1978, Rigler 1982). An analogous distinction persists between water quality management and fisheries management. The concept of cascading trophic interactions links the principles of limnology with those of fisheries biology and suggests a biological alternative to the engineering techniques that presently dominate lake management. Variation in primary productivity is mechanistically linked to variation in piscivore populations. Piscivore reproduction and mortality control the cascade of trophic interactions that regulate algal dynamics. Through programs of stocking and harvesting, fish populations can be managed to regulate algal biomass and productivity.

Acknowledgments

This article is a contribution from the University of Notre Dame Environmental Research Center, funded by the National Science Foundation through grant BSR 83 08918. We thank David Lodge, Ann Bergquist, and the referees for their constructive comments on the manuscript and Carolyn Robinson for word processing.

References Cited

Bartell, S. M. and J. F. Kitchell. 1978. Seasonal impact of planktivory on phosphorus release by Lake Wingra zooplankton. Verh. Int. Ver. Theoret. Angew. Limnol. 20:466–474.

Benndorf, J., H. Kneschke, K. Kossatz and E. Penz. 1984. Manipulation of the pelagic food web by stocking with predacious fishes. Int. Rev. Gesamten. Hydrobiol. 69:407–428.

Bergquist, A. M. 1985. Effects of herbivory on phytoplankton community composition, size structure, and primary production. Ph.D. dissertation, University of Notre Dame, Notre Dame, IN.

Box, G. E., W. G. Hunter and W. S. Hunter. 1978. *Statistics for Experimenters.* John Wiley & Sons, New York.

Brylinsky, M., and K. H. Mann. 1973. An analysis of factors governing productivity in lakes and reservoirs. Limnol. Oceanogr. 18:1–14.

Burns, C. W. 1968. The relationship between body size of filter-feeding *Cladocera* and the maximum size particle ingested. Limnol. Oceanogr. 13:675–678.

Carpenter, S. R. and J. F. Kitchell. 1984. Plankton community structure and limnetic primary production. Am. Nat. 124:159.

Cooper, D. C. 1973. Enhancement of net primary productivity by herbivore grazing in aquatic laboratory microcosms. Limnol. Oceanogr. 18:31–37.

DiBernardi, R. 1981. Biotic interactions in freshwater and effects on community structure. Boll. Zool. 48:3S3–371.

Dillon, R. J. and E. H. Rigler. 1974. The phosphorus-chlorophyll relationship in lakes. Limnol. Oceanogr. 20:767–773.

Ejsmont-Karabin, J. 1983. Ammonia, nitrogen and inorganic phosphorus excretion by the planktonic rotifers. Hydrobiologia 104:231–236.

Elliott, E. T., L. G. Castanares, D. Perlmutter and K. G. Porter. 1983. Trophic-level control of production and nutrient dynamics in an experimental planktonic community. Oikos 41:7–16.

Elster, H. -J. 1974. History of limnology. Mitt. Int. Ver. Theoret. Angew. Limnol. 20:7–30.

Flint, R. W. and C. R. Goldman. 1975. The effects of a benthic grazer on the primary productivity of the littoral zone of Lake Tahoe. Limnol. Oceanogr. 20:935–944.

Gregory, S. V. 1983. Plant-herbivore interactions in stream systems. In: *Stream Ecology.* Barnes, J. R. and G. W. Minshall,

eds. Plenum Press, New York. Pp. 157–189.
Hall, D. J., S. T. Threlkeld, C. W. Burns and R H. Crowley. 1976. The size-efficiency hypothesis and the size structure of zooplankton communities. Annv. Rev. Ecol. Syst. 7:177–203.
Harris, G. P. 1980. Temporal and spatial scales in phytoplankton ecology:mechanisms methods, models, and management. Can. J. Fish. Aquat. Sci. 37:877–900.
Henrikson, L., H. G. Nyman, H. G. Oscarson and J. A. E. Stenson. 1980. Trophic changes, without changes in the external nutrient loading. Hydrobiologia 68:257–263.
Hrbacek, J., M. Dvorakova, V. Korinek and L. Prochazkova. 1961. Demonstration of the effect of the fish stock on the species composition of zooplankton and the intensity of metabolism of the whole plankton assemblage. Verh. lnt. Ver. Theoret. Angew. Limnol. 14:192–195
Kitchell, J. E. 1980. Fish dynamics and phosphorus cycling in lakes. In: *Nutrient cycling in the Great Lakes: a summarization of the factors regulating cycling of phosphorus*. Scavia, D. and R. Moll, eds. NOAA Spec. Rep. 83. Great Lakes Environmental Research Laboratory, Ann Arbor, Ml. Pp. 81–91
Kitchell, J. E, H. E Henderson, E. Grygierek, J Hrbacek, S. R. Kerr, M. Pedini, T. Petr, J Shapiro, R. A. Stein, J. Stenson and T. Zaret 1986. Management of lakes by food-chain manipulation. FAO Fish. Tech. Pap. UN Food and Agricultural Organization, Rome. In press.
Kitchell, J. E, R. V. O'Neill, D. Webb, G. Gallepp, S. M. Bartell, J. E. Koonce and B. S. Ausmus. 1979. Consumer regulation of nutrient cycling. BioScience 29:28–34.
Korstad, J. E. 1980. Laboratory and field studies of phytoplankton-zooplankton interactions. Ph.D. dissertation, University of Michigan, Ann Arbor.
Larkin, P. A. 1978. Fisheries management—an essay for ecologists. Annul Rev. Ecol. Syst. 9:57–74.
Lehman, J. T. 1980. Release and cycling of nutrients between planktonic algae and herbivores. Limnol. Oceanogr. 25:620–632.
Lehman, J. T. and C. D. Sandgren. 1985. Species-specific rates of growth and grazing loss among freshwater algae. Limnol. Oceanogr. 30:34–46.
Lynch, M. 1980. The evolution of cladoceran life histories. Q. Rev. Biol. 55:23–42.
McCauley, E. and J. Kalff. 1981. Empirical relationships between phytoplankton and zooplankton biomass in lakes. Can. J. Fish. Aquat. Sci. 38:458–463.
McNaughton, S. J. 1979. Grazing as an optimization process: grass-ungulate relationships in the Serengeti. Am. Nat. 113:691–703.
Oglesby, R. T. 1977. Phytoplankton summer standing crop and annual productivity as functions of phosphorus loading and various physical factors. J. Fish. Res. Board Can. 34:2255–2270.
Peters, R. H. 1980. Useful concepts for predictive ecology. Synthèse 43:215–228.
Peters, R. H. and J. A. Downing. 1984. Empirical analysis of zooplankton filtering and feeding rates. Limnol. Oceanogr. 29:763–784.
Peters, R. H. and E. H. Rigler. 1973. Phosphorus release by *Daphnia*. Limnol. Oceanogr. 13:821–839.
Peterson, I. and J. S. Wroblewski. 1984. Mortality rate of fishes in the pelagic ecosystem. Can. J. Fish. Aquat. Sci. 41:1117–1120.
Reynolds, C. S. 1984. *The Ecology of Freshwater Phytoplankton*. Cambridge University Press, London.
Rigler, F. H. 1982. The relation between fisheries management and limnology. Trans. Am Fish. Soc. 111:121–132.
Schindler, D. W. 1978. Factors regulating phytoplankton production and standing crop in the world's lakes. Limnol. Oceanogr. 23:478–486.
Seale, D. B. 1980. Influence of amphibian larvae on primary production, nutrient flux, and competition in a pond ecosystem. Ecology 61:1531–l550.
Shapiro, J. 1980. The importance of trophic-level interactions to the abundance and species composition of algae in lakes. In: *Hypertrophic Ecosystems*. Barica, J. and L. R. Mur, eds. Dr. W. Junk Publishing Co., The Hague, Netherlands. Pp. 105–115
Shapiro, J. and D. I. Wright. 1984. Lake restoration by biomanipulation. Freshwater Biol. 14:371–383.
Steele, J. H. and E W. Henderson. 1984. Modeling long-term fluctuations in fish stocks. Science 224:985–987.
Stewart, D. J., J. E. Kitchell and L. B. Crowder. 1981. Forage fishes and their salmonid predators in Lake Michigan. Trans. Am. Fish. Soc. 110:751–763.
Tonn, W. M. and J. J. Magnuson. 1982. Patterns in the species composition and richness of fish assemblages in northern Wisconsin lakes. Ecology 63:1149–1166

Introduction to

Crawford S. Holling and Gary K. Meffe

Command and Control and the Pathology of Natural Resource Management

by Stephen R. Carpenter

Management of natural resources involves manipulations by people to achieve goals such as food production, resource extraction, or conservation of particular ecological communities. We may attempt to meet our goals by structuring the ecosystem in particular ways, or by suppressing natural variation. For example, if we are interested in maintaining an oak woodland, we may cut down more shade-tolerant trees such as cherries and maples which could replace the oaks, and prevent fires so the oaks are not burned. Ironically, these rigid controls may increase the risk that the oak woodland will be destroyed by pathogens that can spread more easily through a uniform forest, or by catastrophic fires fueled by massive accumulations of branches and leaves during decades of fire suppression. In contrast to the rigid control systems humans might impose, natural oak woodlands are maintained by continual cycles of disturbance, renewal and recovery at several scales, ranging from local windfall of a canopy tree, to ground fires that consume competitors and debris, to the rare and patchy crown fires that lead to new successional cycles. These natural processes give the ecosystem resilience analogous to a tree in the wind, whereas the control approach creates the rigidity of a beam that must eventually break under increasing stress.

This paper contrasts the "command and control" approach to resource management with the adaptive approach to management. Adaptive management involves continual learning about the ecosystem through deliberate experimentation, combined with flexible changes in management tactics as the ecosystem changes over time. The authors suggest a golden rule: "Natural resource management should strive to retain critical types and ranges of natural variation in ecosystems."

The golden rule and associated tenets of adaptive management pose some significant challenges for us. How can small parcels of land best be managed in view of the fact that one rare disturbance could wipe out the dominant vegetation? How can we meet human needs for living resources, such as fish and game, yet accommodate to the natural population cycles of the animals? How can we reconcile the need for efficient agricultural production to feed a growing human population with the need for heterogeneous landscapes, natural predators to suppress pests, and fallowing cycles to renew soil? How can we develop management institutions that adapt flexibly to changes in ecosystems and societal needs?

Command and Control and the Pathology of Natural Resource Management

C. S. Holling and Gary K. Meffe

Holling, C. S. and G. K. Meffe. 1996. Command and control and the pathology of natural resource management. *Conservation Biology* 10(2):328–337. Reprinted with permission of Blackwell Science, Inc.

Abstract

As the human population grows and natural resources decline, there is pressure to apply increasing levels of top-down, command-and-control management to natural resources. This is manifested in attempts to control ecosystems and in socioeconomic institutions that respond to erratic or surprising ecosystem behavior with more control. Command and control, however, usually results in unforeseen consequences for both natural ecosystems and human welfare in the form of collapsing resources, social and economic strife, and losses of biological diversity. We describe the "pathology of natural resource management," defined as *a loss of system resilience when the range of natural variation in the system is reduced,* encapsulates the unsustainable environmental, social, and economic outcomes of command-and-control resource management. If natural levels of variation in system behavior are reduced through command-and-control, then the system becomes less resilient to external perturbations, resulting in crises and surprises. We provide several examples of this pathology in management. An ultimate pathology emerges when resource management agencies, through initial success with command and control, lose sight of their original purposes, eliminate research and monitoring, and focus on efficiency of control. They then become isolated from the managed systems and inflexible in structure. Simultaneously through overcapitalization, society becomes dependent upon command and control, demands it in greater intensity, and ignores the underlying ecological change or collapse that is developing. Solutions to this pathology cannot come from further command and control (regulations) but must come from innovative approaches involving incentives leading to more resilient ecosystems ,more flexible agencies, more self-reliant industries, and a more knowledgeable citizenry. We discuss several aspects of ecosystem pattern and dynamics at large scales that provide insight into ecosystem resilience, and we propose a "Golden Rule" of natural resource management that we believe is necessary for sustainability: management should strive to retain critical types and ranges of natural variation in resource systems in order to maintain their resiliency.

In ecology, we have an incredibly complex system with no central dogma like that of molecular biology to let us even pretend that we have control. Nowhere is this more apparent than in conservation, where we have persuaded ourselves that some degree of control is really necessary.

Ehrenfeld, 1991

An essential paradox of wilderness conservation is that we seek to preserve what must change.
Pickett and White, 1985

Introduction

Control is a deeply entrenched aspect of contemporary human societies: we control human behavior through laws, incentives, threats, contracts, and agreements; we control the effects of environmental variation by constructing safe dwellings; we control variation in our food resources by growing and storing agricultural products; we control human parasites and pathogens through good hygiene and medical technologies. All contribute to stable societies and human health and happiness, and within certain arenas this desire to control is undeniably to our individual and collective benefit. This approach to solving problems may be collectively referred to as "command and control," in which a problem is perceived and a solution for its control is developed and implemented. The expectation is that the solution is direct, appropriate, feasible, and effective over most relevant spatial and temporal scales. Most of all, command and control is expected to solve the problem either through control of the processes that lead to the problem (e.g., good hygiene to prevent disease, or laws that direct human behavior) or through amelioration of the problem after it occurs (e.g., pharmaceuticals to kill disease organisms, or prisons or other punishment of lawbreakers). The command-and-control approach implicitly assumes that the problem is well-bounded, clearly defined, relatively simple, and generally linear with respect to cause and effect. But when these same methods of control are applied to a complex, nonlinear, and poorly understood natural world, and when the same predictable outcomes are expected but rarely obtained, severe ecological, social, and economic repercussions result.

Humanity's contemporary interactions with nature are based on a mix of slowly developed social norms and expectations, and increasingly on more rapidly developed short-term incentives and controls. When the behaviors of people, institutions, or nature violate the norms, desires, or expectations of society, command and control is often sought as the primary solution in an effort to move human or ecosystem behaviors to a predetermined, predictable state. Consequently, much of natural resource management has been an effort to control nature in order to harvest its products, reduce its threats, and establish highly predictable outcomes for the short-term benefit of humanity. Our thesis is that adoption of such command and control has resulted in a pathology that permeates much of natural resource management and precludes long-term sustainability.

Command-and-Control Management

The command-and-control approach, when extended uncritically to treatment of natural resources, often results in unforeseen and undesirable consequences. A frequent, perhaps universal result of command and control as applied to natural resource management is reduction of the range of natural variation of systems—their structure, function, or both—in an attempt to increase their predictability or stability. That is, variation through time or space (such as system behavior over time, or spatial heterogeneity) is reduced. Thus, a common theme of many resource-management efforts is to reduce natural bounds of variation in ecological systems to make them more predictable, and thus more reliable, for human needs. We dampen extremes of ecosystem behavior or change species composition to attain a predictable flow of goods and services or to reduce destructive or undesirable behavior of those systems. For example, we control agricultural pests through herbicides and pesticides: we convert natural, multi-species, variable-aged forests into monoculture, single-aged plantations; we hunt and kill predators to produce a larger, more reliable supply of game species; we suppress fires and pest outbreaks in forests to ensure a steady lumber supply: we clear forests for pasture development and steady cattle production, and so forth.

Such efforts attempt to replace natural ecological controls, which are largely unknown to us and highly complex and variable, with engineered constructs and manipulations that on the surface seem entirely within our control. The purpose is to turn an unpredictable and "inefficient" natural system into one that produces products in a predictable and economically efficient way. When unanticipated environmental problems then arise, the *a priori* expectation of certainty is not met and results in surprise and crisis—chemical pollution and erosion from monocultures, loss of biological diversity from tree farms, irruption of herbivore populations after predator removal, conflagrations and property loss when fires finally erupt, insect pest outbreaks when spraying stops, and pollution and erosion from grazing. Such crises and surprises, we argue here, are the *inevitable* consequences of a command-and-control approach to renewable resource management, where it is (implicitly or explicitly) believed that humans can select one component of a self-sustaining natural system and change it to a fundamentally different configuration in which the adjusted system remains in that new configuration indefinitely without other, related changes in the larger system.

We call the result "the pathology of natural resource management" (Holling 1986; Holling 1995), a simple but far-reaching observation defined here as follows: *when the range of natural variation of a system is reduced, the system loses resilience.* That is, a system in which natural levels of variation have been reduced through command-and-control activities will be less resilient than an unaltered system when subsequently faced with external perturbations, either of a natural (storms, fires, floods) or human-induced (social or institutional) origin. We believe this principle applies beyond ecosystems and is particularly relevant at the intersection of ecological, social, and economic systems.

Because much of our focus here is on loss of resilience, we must explore that concept further. Resilience of a system has been defined in two very different ways in the ecological literature; these differences in definition reflect which of two different aspects of stability are emphasized. The first definition, and the more traditional one, concentrates on stability near an equilibrium steady-state, where resistance to disturbance and speed of return to the equilibrium are used to measure resilience (Pimm 1984 & O'Neill et al. 1986; Tilman & Downing 1994). We call that *equilibrium resilience.* The second definition, and the one of greater relevance here, emphasizes conditions far from any equilibrium in which instabilities can flip a system into another regime of behavior—to another stability domain (Holling 1973, 1994). In this case the measurement of resilience is the magnitude of disturbance that can be absorbed or accommodated before the system changes its structure by changing the variables and processes that control system behavior. We call that *ecosystem resilience* because its significance becomes clearly apparent for large scale systems over long periods. The first definition focuses on efficiency, constancy, and predictability—attributes at the core of command-and-control desires for fail-safe design. The second focuses on persistence, change and unpredictability—attributes embraced by an adaptive management philosophy. Holling (1973) first emphasized the consequences of these different definitions for ecological systems in order to draw attention to the paradoxes between constancy and change or between predictability and unpredictability.

The Pathology

For illustrative purposes we offer several examples of the pathology of natural resource management in which reduction of variation has led to a less resilient system, in the sense of ecosystem resilience:

(1) The loss of genetic variation in small populations is generally thought to result in a less resilient genetic system (Allendorf & Leary 1986; Meffe 1986), possibly resulting in higher probabilities of population extinction. This is particularly true if the environment changes and previously available genotypes that would be appropriate to the new environment no longer exist. Of course there are exceptions; for example, loss of deleterious recessive alleles is not likely to reduce population resilience and may in fact increase it. But overall, loss of genetic variance may lead to lower population resilience in ecological or evolutionary time.

(2) Stabilization of flows by dams in previously wildly flooding or "flashy" southwestern U.S. rivers results in a native fish fauna that is less resilient in the face of invasive fish species (Meffe 1984; Minckley & Meffe 1987). The high flow variation of unregulated rivers inhibits establishment of exotic fishes, and the last remaining strongholds of southwestern native riverine fishes are all in free-flowing rivers (Minckley & Deacon 1991). When flow variation is stabilized by dams and the process of violent flooding is removed, the resulting lentic conditions favorable to many exotic species allow them to flourish and eliminate native fishes that evolved with high flow variation. Stabilization of discharge variation and the presence of invasive species results in unresilient and declining native fish faunas.

(3) Suppression of fire in fire-prone ecosystems is remarkably successful in reducing the short-term probability of fire in the national parks of the United States and in fire-prone suburban regions. But the consequence is an accumulation of fuel over large areas that eventually produces fires of an intensity, extent, and human cost never before encountered (Kilgore 1976; Christensen et al. 1989). Fire suppression in systems that would frequently experience low-intensity fires results in the systems becoming severely affected by the huge fires that finally erupt; that is, the systems are not resilient to the major fires that occur with large fuel loads and may fundamentally change states after the fire. As this and the previous example serve to demonstrate, suppression or removal of a natural disturbance generally reduces system resilience.

(4) Monocultural, energy-intensive farming practices are the epitome of reduction of variation and loss of resilience. Plant species diversity in a natural forest converted to a monoculture may go from dozens or hundreds to one dominant, plus whatever weeds can escape the herbicides. Monocultures are notoriously susceptible to the effects of drought, flooding, insect or pathogen outbreaks, and

market vagaries. They consequently require large inputs of energy (fertilizers, pesticides, herbicides, irrigation) and often large societal subsidies in the form of price supports, guaranteed loans, disaster relief, and surplus buyouts. These monocultures are fundamentally unresilient to natural or social perturbations.

(5) Natural, lateral flow variation (periodic floodplain inundation) throughout much of the Mississippi River drainage has been reduced by channelization and construction of a series of locks and levees to benefit agriculture, shipping, and floodplain development. As a result, the inextricably combined riverine-social system has little resilience during extreme storm events, as witnessed in the massive flooding of 1993. Attempted command-and-control of the river's flows, allowing expansive floodplain development, resulted in an unresilient riverine-social system and unprecedented economic destruction.

The same phenomenon applies equally well beyond natural resource management to many aspects of human existence. For example bureaucracies are an exercise in variance reduction through regulation and control: their purpose is elimination of extreme behavior through regulation to promote conformity to a specific set of standards, which to some degree is certainly desirable in a civilized society. But deeply entrenched bureaucracies are characteristically unresilient to new challenges because the system discourages innovation or other behavioral variance. This is clearly evidenced by merely presenting a unique situation to a clerk who has been narrowly trained in a highly standardized bureaucracy and watching the incredulous reply, or by the typically negative response to and occasional punishment of a government employee who offers an alternative perspective to the standard operating procedure.

The pathology of natural resource management involves not just a contraction of the resilience of ecosystems in response to human control: two other features make for an ultimate pathology. One feature concerns changes that occur in management agencies, and the other involves changes in economic sectors.

First, loss of ecosystem resilience is accompanied by changes in the management agencies. The initial phase of command-and-control is nearly always quite successful: insect pests are reduced by pesticide use; fishing and hunting are enhanced by stocking or predator removal; forest fires are suppressed for years; floods are minimized by levees. As a consequence, agencies responsible for management shift their attention from the original social or economic purpose to an otherwise laudable effort to increase efficiency and reduce costs—better and more efficient ways to kill insects, eliminate wolves, rear hatchery fish, detect and extinguish fires, or control flows. Priorities thus shift from research and monitoring (why "waste" money studying and monitoring apparent success?) to internal agency goals of cost efficiency and institutional survival. The second feature of the pathology thus emerges: growing isolation of agency personnel from the systems being managed and insensitivity to public signals of concern—in short, growing institutional myopia and rigidity.

At the same time, economic activities exploiting the resource benefit from success and expand in the short term, and we witness greater capital investment in activities such as agricultural production, pulp mills, suburban development, and fishing and hunting. That too is laudable within limits: it is the development of human opportunity and enterprise. But the result is increasing dependency on continued success in controlling nature while, unknown to most, nature itself is losing resilience and increasing the likelihood of unexpected events and eventual system failure. With dependency comes denial, demands by economic interests to keep and expand subsidies, and pressure for further command and control. This third feature provides the final element to the ultimate pathology of command-and-control resource and environmental management. The composite result is increasingly less resilient and more vulnerable ecosystems, more myopic and rigid institutions, and more dependent and selfish economic interests all attempting to maintain short-term success.

If the response to this pathology by other interests, such as the environmental community, is exclusively demand for tighter regulation and prohibition, then the pathology is deepened, because this applies a command-and-control solution to a problem initiated by command and control. The result is that lobby groups battle other lobby groups and generate the gridlocks and train wrecks that are now regional issues—from salmon, owls, fishing, and logging in the Pacific Northwest, to cod, poverty, and cultural survival in Newfoundland, to sugar, urbanization, wildlife, and water in the Everglades (Gunderson et al. 1995).

Such problems, with a complex of causes, do not have simple solutions. We know the goal: more resilient ecosystems, more flexible agencies, more self-reliant industries, and more knowledgeable citizens. We also know the ingredients of the solution, if not the specific ways to combine and use those ingredients. First, replace economic subsidies with incentives designed so that restoration and maintenance of ecosystem resilience is to the benefit of economic enterprise. An example is the conservation policies that reward farmers for restoration of habitat and soils. Second, develop ways for agencies to innovate and learn, and allow them to do so. An example is the application of actively-adaptive environment management approaches, where policies become hypotheses and management actions become the experiments to test those hypotheses (Holling 1978; Walters 1986; Lee 1993; Gunderson et al. 1995). Third, engage people as active partners in the process of science and policy. Examples are the various regional and continental monitoring schemes by which people monitor changes in nature—acid rain in the northeast, bird populations along flyways, water quality in bays and rivers. Monitoring of ecological change over time and space is critical to a better understanding of our managed resources, stems and must be a central component of an adaptive management scenario. Monitoring provides the data for the management experiment and the basis for deciding the success or failure of the approach. Fourth, develop local partnerships among broad constituencies that all stand to gain (or lose) together from good (or poor) resource management.

The Behavior of Natural Ecosystems

Our suggestions would be more effective with a better understanding of ecosystem behavior, structure, and dynamics at all spatial scales from the plant to the planet and at all temporal scales from seconds to millennia. The surprises and crises created by the pathology are not only the consequence of incomplete knowledge of how to control nature's variability or improper controls being applied. They also include ignorance of the constructive role that variation plays in maintaining the integrity of ecosystem function in the face of unexpected events. Recently, a group of ecologists working with large-scale terrestrial, freshwater, and marine ecosystems developed a synthesis of their experience with natural, disturbed, and managed ecosystems (Holling et al. 1995). They identified key features of ecosystem structure and dynamics that explain why surprise and crisis are an inevitable outcome of command-and-control approaches. They concluded with the following lessons:

(1) Ecological change is not continuous and gradual; rather, it is episodic, with slow accumulation of natural capital such as biomass or nutrients, punctuated by sudden releases and reorganization of that capital as the result of internal or external nature, processes or of human imposed catastrophes. Rare events, such as hurricanes or the arrival of invading species, can unpredictably shape structure at critical times or at locations of increased vulnerability; the effects of these rare events can persist for very long periods. Therein lies one of the sources of new options that environmental diversity and variation provide. Irreversible or slowly reversible states exist; once the system flips into such a state, only explicit management intention can restore its previous self-sustaining state, and even then success is not assured (Walker 1981). *Conclusion: Critical processes function at radically different rates and at spatial scales covering several orders of magnitude, and these rates and scales cluster around a few dominant frequencies.*

(2) Spatial attributes are not uniform or scale-invariant. Rather, productivity and textures are patchy and discontinuous at all scales, from the leaf to the individual to the vegetation patch to the landscape to the planet. There are several different ranges of scales, each with different attributes of patchiness and texture (Holling 1992). *Conclusion: Scaling up from small to large cannot be a process of simple linear addition: non-linear processes organize the shift from one range of scales to another. Not only do the large and slow control the small and fast, the latter occasionally "revolt" to affect the former.*

(3) Ecosystems do not have single equilibria, with functions controlled to remain near them. Rather, multiple equilibria, destabilizing forces far from equilibria, and absence of equilibria, define functionally different stable states, and movement between states maintains an overall structure and diversity. *Conclusion. On the one hand, destabilizing forces are important in maintaining diversity, resilience, and opportunity. On the other hand, stabilizing forces are important in maintaining productivity and biogeochemical cycles, and, even when these features are perturbed, they recover rapidly if the stability domain is not exceeded (e.g., recovery of lakes from eutrophication or acidification; Schindler 1990; Schindler et al. 1991).*

(4) Policies and management that apply fixed rules for achieving constant yields independent of scale (e.g., constant carrying capacity of cattle or wildlife or constant sustainable yield of fish, wood, or water) lead to systems that gradually lose resilience—systems that suddenly break down in the face of disturbances that previously could be absorbed (Holling 1986). *Conclusion: Ecosystems are moving targets, with multiple potential futures that are uncertain and unpredictable. Therefore management has to be flexible, adaptive, and experimental at scales compatible with the scales of critical ecosystem functions (Holling 1978; Walters 1986; Lee 1993; Gunderson et al. 1995).*

A Closer Look at Resilience

The features we have discussed are the consequences of the stabilizing properties of natural ecosystems. In the ecological literature these properties have been given focus through debates on the meaning and reality of the resilience of ecosystems.

Earlier, we briefly defined resilience in two ways. These two aspects of a system's stability have very different consequences for evaluating, understanding, and managing complexity and change. Ecosystem resilience, our preferred definition, focuses on the interplay between stabilizing and destabilizing properties, which are at the heart of present issues of development and the environment: global change, biodiversity loss, ecosystem restoration, and sustainable development. Nevertheless, much of present ecological theory uses the equilibrium definition of resilience, even though that definition reinforces the pathology of equilibrium-centered command and control. That is because much of that theory draws predominantly from traditions of deductive mathematical theory (Pimm 1984) in which simplified, untouched ecological systems are imagined, or from traditions of engineering in which the motive is to design systems with a single operating objective (Waide & Webster 1976; DeAngelis et al. 1980; O'Neill et al. 1986), or from small-scale quadrat experiments in nature (Tilman & Downing 1994) in which long-term, large-scale successional or episodic transformations are not of concern. That makes the mathematics more tractable, it accommodates the engineer's goal to develop optimal designs, and it provides the ecologist with a rationale for utilizing manageable, small sized, and short-term experiments, all reasonable goals. But these traditional concepts and techniques make the world appear more simple, tractable, and manageable than it really is. They carry an implicit assumption that there is global stability—that there is only one equilibrium steady-state, or, if other operating states exist, they should be avoided with safeguards and regulatory controls. They transfer the command-and-control myopia of exploitive development to similarly myopic demands for environmental regulations and prohibitions.

Those who emphasize ecosystem resilience, on the other hand, come from traditions of applied mathematics and applied resource ecology at the scale of ecosystems, such as the dynamics and management of freshwater systems (Fiering 1982), forests (Clark et al. 1979), fisheries (Walters 1986), semiarid grasslands (Walker et al. 1969), and interacting populations in nature (Dublin et al. 1990; Sinclair et al. 1990). Because these studies are rooted in inductive rather than deductive theory formation and in experience with the effects of large-scale management disturbances, the reality of flips from one stable state to another cannot be avoided (Holling 1986). Indeed, management and resource exploitation can overload waters with nutrients, turn forests into grasslands, trigger collapses in fisheries, and transform savannas into shrub-dominated semideserts.

These two different views of resilience reflect two different traditions of ecological science: that of equilibrium resilience is experimental, analytical, and focuses on small spatial scales and short durations; that of ecosystem resilience is integrative, synthetic, and focuses on multiple scales. The consequences lead not only to opposite views of system behavior but to opposite views of system structure that have major consequences for policy. For example, there is a debate over whether every species is important in ecosystem dynamics and function or whether only a smaller subset is involved in self-organization (Baskin 1994). On the one side is evidence from controlled experiments showing that declining generalized diversity reduces productivity (Naeem et al. 1994), or that reducing numbers of grass species reduces rates of recovery from drought (Tilman &

Downing 1994). In such examples, however, the physical limitations of the experiments limit the conclusions to small-scale interactions (plots ranged 1–4 m on a side) over short periods and to the set of structuring species that happened to be selected at those scales. In contrast, those who argue that a subset of species control dynamics and function draw their evidence from large-scale manipulations of whole ecosystems such as lakes (Schindler 1990), from an understanding of process function at different scales (Holling 1992; Levin 1992), from landscape and ecosystem-scale models (Clark et al. 1979; Costanza et al. 1986; Walters & Gunderson 1994), and from field measures of disturbed and managed ecosystems (Hughes 1994). These observations address boreal, marine, freshwater, and savanna ecosystems and indicate that functional diversity is determined not by all species but by species involved in a set of structuring processes (Schindler 1990; Holling et al. 1995). Examples include the set of grass species and ungulate grazers that maintain the productivity and resilience of savannas (Walker et al. 1969) and the tree species and suite of 35 species of insectivorous birds that mediate budworm outbreak dynamics in the eastern boreal forest (Holling 1988).

Any ecosystem contains hundreds to thousands of species interacting among themselves and their physical and chemical environment. But not all those interactions have the same strength or the same direction. That is, although everything might ultimately be connected to everything else if the web of connections is followed far enough, the first-order interactions that structure the system increasingly seem to be confined to a subset of biotic and abiotic variables whose interactions form the "template" (Southwood 1977) or the niches that allow a great diversity of living things to, in a sense, "go along for the ride" (Carpenter & Leavitt 1991; Cohen 1991; Holling 1992). Those species are affected by the ecosystem but do not, in turn, notably affect the ecosystem, at least in ways that our relatively crude methods of measurement can detect. At the extremes, therefore, species can be regarded either as "drivers" or as "passengers" (Walker 1992), although this distinction needs to be treated cautiously. The driver role of a species may become apparent only every now and then under particular conditions that trigger their key structuring function.

This large-scale view of ecosystems highlights where the priority for resource management, ecosystem restoration, or biodiversity policy should lie. Ecological change is not incremental and local but sudden and extensive. If change does occur, there may be fundamental transformations from one ecosystem type to another—from forest to grassland or grassland to a shrubby semi-desert, for example (Walker et al. 1969; Holling 1973). Then control of structure will shift from one set of organizing processes and variables to another. It is the diversity of overlapping influences within those controls that defines the resilience to those sudden shifts.

The fundamental points are that only a small set of self-organizing processes made up of biotic and physical elements are critical in forming the structure and overall behavior of ecosystems, and that these establish sets of relationships, each of which dominates over a definable range of scales in space and time. Each set includes several species of plants or animals, each species having similar but overlapping influence to give functional redundance. It is that set, operating with abiotic processes, that generates and maintains ecosystem resilience. It provides the focus for identifying the types and sources of variation that are critical for maintaining the integrity of a natural system.

Thus, we suggest that an ecosystem-resilience perspective better reflects the reality of large-scale processes and dynamics and provides the most realistic foundation for addressing the challenging and complex resource management issues of the day. It also provides the conceptual basis necessary to appreciate and understand the paradoxes typically encountered in resource management, as well as the pathology we describe here.

A Golden Rule for Natural Resource Management

The various observations presented herein suggest a "Golden Rule' of natural resource management: *Natural resource management should strive to retain critical types and ranges of natural variation in ecosystems.* That is, management should facilitate existing processes and variabilities rather than changing or controlling them. By so doing, ecosystem resilience and the organizing processes and structures of ecosystems will be maintained, thus better serving not only the natural functions and species diversity of those systems but also the long-term (although not necessarily short-term) interests of humanity. This is a more sophisticated way of stating Aldo Leopold's (1949) famous assertion that "A thing is right when it tends to preserve the integrity, stability, and beauty of the biotic community. It is wrong when it tends otherwise." Because we know more today about the dynamics of ecological communities than Leopold did in the 1940s, we would replace "stability" with "resilience;" otherwise, this remains sound advice, and Leopold clearly anticipated the pathology of natural resource management as elaborated here.

We fully recognize that the particular "rule" we propose has far greater conceptual than prescriptive power. Prescriptions and cookbook approaches generally should be avoided in conservation (Meffe & Carroll 1994), if for no other reason than the systems with which we work are idiosyncratic and endlessly varied. No single, detailed prescription can be of much use for more than a single system. Furthermore, our rule is operationally vague. What is a "critical type" or a "critical range" of variation. That, obviously, is specific to a system and is often not known with any degree of assurance. Ehrenfeld (1992) indicated that "...it is extremely difficult to determine a normal state for communities whose parameters are often in a condition of flux because of natural disturbance." Schindler (1987) further explained that "...we usually do not know the normal range for any variable, at least for any time period greater than a few years."

Thus, our advice to "retain critical types and ranges of natural variation" must remain for the present as a management goal to which to aspire, as a conceptual underpinning for management, rather than an operational dictum. In practice this translates to adopting a conservative approach to changing parameters of systems we understand poorly but that we wish to manage. It means that the default condition, unless clearly proven otherwise, should be retention of the natural state rather than manipulation of system components or dynamics. It argues for humility when managing large systems (Stanley 1995). It shifts the burden of proof from managing by system manipulation to managing by minimal intervention, unless proven otherwise. It also argues strongly for adaptive management rather than command-and-control prescriptions, and development of consistent and dedicated monitoring of systems, both natural and managed. Only through long-term data collection can we begin to close the knowledge gap in understanding normal system behavior, particularly its variance.

How would this Golden Rule, at least in concept, modify resource management practices to take into account the pathology of natural resource management? We revisit our earlier examples and indicate how resource management might be altered to adhere to the Golden Rule:

(1) Genetic diversity of small populations should be retained and not further eroded by management practices (Schonewald-Cox et al. 1983; Falk & Holsinger 1991). This includes maintenance of natural gene flow in the wild (Meffe & Vrigenhoek 1988), reserves large enough to maintain large breeding populations or metapopulations of species of concern, and avoidance of population crashes, bottlenecks, or inbreeding in captive breeding programs.

(2) In riverine systems with naturally high variation in discharge, replace stabilization of flows via dams with watershed restoration and protection. Begin to remove dams to restore the critical ecosystem process of discharge variation. Price water to accurately reflect its ecosystem value in order to stimulate conservation measures, and remove flood-prone lands from development. Combine this with a bioregional perspective that matches development practices to natural, regional ecological constraints. Develop a combination of regional and national incentives and disincentives that would eliminate ecologically disastrous development such as large desert cities that rely on water from far outside of the region and mining of fossil water with a temporally limited productive capacity.

(3) Eliminate policies of fire suppression in naturally fire-prone ecosystems. Eliminate incentives that encourage rebuilding in such ecosystems after fire destruction, and develop incentives such as tax reductions to site new housing and other developments away from such areas, eventually to be designated as wilderness.

(4) Proceed from simple monocultures to more complex agroecosystems with integrated pest management and no-till methods (Carroll et al. 1990). Promote, though education and economic means, ecological complexity in agriculture, eliminating as much as possible energetic and societal subsidies, allowing free ecosystem services (e.g., diversity of predators on pests, soil conservation through no-till methods) to support agriculture.

(5) Relocate communities out of floodplains of the Mississippi River and other large riverine systems: use those areas as wildlife refuges and corridors and as natural buffers and recharge zones for agroecosystems (as is now being promoted in some parts of the Mississippi floodplain). Provide disincentives for further floodplain development.

(6) Examine bureaucracies to identify underlying reasons for their general intransigence and brittleness, and promote incentives for alternative behaviors. Develop incentives and rewards for innovation that place streamlining, local solutions, and concern for customers and sustainability above adherence to a command structure.

Conclusions

Rather than pursuing short-term gain through command and control, effective natural resource management that promotes long-term system viability must be based on an understanding of the key processes that structure and drive ecosystems, and on acceptance of both the natural ranges of ecosystem variation and the constraints of that variation for long-term success and sustainability. This is especially urgent when the growth of the human population and its consumption of resources is added to the picture, as it always must be (Meffe et al. 1993). Despite our penchant to control so many systems through command-and-control techniques, with a few conspicuous exceptions the underlying problem of population growth is often ignored. Ironically, our attempts at command and control are usually directed at complex, poorly understood, and nonlinear natural systems, rather than at the fundamental source of the problem—human population growth and consumption—where control is viable, reasonable, and could be effective. A rapidly increasing human population and increasing consumption is resulting in greater demands on and competition for dwindling and increasingly damaged natural resources. The resource problems we encounter today can only multiply as the human population grows, which means that the errors of command and control will be compounded, which will only lead to calls for more command and control by those who do not fundamentally understand the pathology outlined herein. This highlights the urgency of quickly changing our fundamental approaches to natural resource management and developing solutions and appropriate models of management behavior while time and resources still permit.

Command-and-control management can lead to short-term economic returns, but it also increases the vulnerability of ecosystems to perturbations that otherwise could be absorbed. Any move toward truly sustainable human endeavors must incorporate this principle or it cannot succeed. Our observations are also pertinent to the present move toward ecosystem management in the United States and elsewhere. If ecosystem management is to be more than another buzzword, then there is no substitute for understanding the structure and dynamics of natural ecosystems over spatial and temporal scales covering several orders of magnitude. The role of variation in structuring ecosystems and maintaining their resilience, and managing within the constraints of that structure and dynamics, is critical. We must also modify our institutions and policies to recognize the pathology described herein and to root out similar pathologies in institutional and policy behaviors. To ignore this is to perpetuate the pathology of natural resource management and place ecosystems and humanity at great risk.

Acknowledgments

We dedicate this paper to Aldo Leopold, who clearly anticipated the ideas herein, and who was writing about land pathologies as early as 1935. We thank David Ehrenfeld and Garry Peterson for insightful comments on earlier drafts. G. K. Meffe was supported by contract DE-AC09-76SR00-819 between the U.S. Department of Energy and the University of Georgia, as well as a sabbatical leave at the National Biological Service Laboratory, Gainesville, Florida.

Literature Cited

Allendorf, F. W., and R F. Lean. 1986. Heterozygosity and fitness in natural populations of animals. In: *Conservation Biology: The Science of Scarcity and Diversity.* M. E. Soule, ed. Sinauer Associates, Sunderland. Massachusetts. Pp. 57–76

Baskin, Y. 1994. Ecologists dare to ask: How much does diversity matter? Science 264:202–203.

Carpenter, S. R. and P. R. Leavin. 1991. Temporal variation in paleolimnological record arising from a trophic cascade. Ecology 72: 277–285.

Carroll, C. R., J. H. Vandermeer and P. M. Rosset, eds. 1990. Agroecology. McGraw-Hill, New York.
Christensen, N. L., et al. 1989. Interpreting the Yellowstone fires of 1988. BioScicnce 39:678–685.
Clark, W. C., D. D. Jones and C. S. Holling. 1979. Lessons for ecological policy design: A case study of ecosystem management. Ecological Modeling 7:1–53.
Cohen, J. 1991. Tropic topology. Science 251:686–687.
Costanza, R., F. H. Sklar and J. W. Day, Jr. 1986. Modeling spatial and temporal succession in the Atchafalaya/Terrebonne marsh/estuarine complex in South Louisiana. In: *Estuarine Variability.* D. A. Wolfe, ed. Academic Press, New York. Pp. 387–404
DeAngelis, D. L., W. M. Post and C. C. Travis. 1980. *Positive Feedback in Natural Systems.* Springer-Verlag, New York.
Dublin, H. T., A. R. E. Sinclair and J. McGlade. 1990. Elephants and fire as causes of multiple stable states in the Serengeti-mara woodlands. Journal of Animal Ecology 59:114–1164.
Ehrenfeld, D. 1991. The management of diversity: a conservation paradox. In: *Ecology, Economics, Ethics: the Broken Circle.* F. H. Borman and S. R. Kellert, eds. Yale University Press, New Haven. Connecticut. Pp. 26–39.
Ehrenfeld, D. 1992. Ecosystem health and ecological theories. In: *Ecosystem Health: New Goals for Environmental Management.* R. Costanza, B. G. Norton and B. D. Haskell, eds. Island Press, Covelo, California. Pp. 135–143.
Falk, D. A. and K. E. Holsinger. 1991. *Genetics and Conservation of Rare Plants.* Oxford University Press, New York.
Fiering, M. B. 1982. A screening model to quantify resilience. Water Resources Research 18:27–32.
Gunderson, L., C. S. Holling and S. Light, eds. 1995. *Barriers and Bridges to the Renewal of Ecosystems and Institutions.* Columbia University Press, New York.
Holling, C. S. 1973. Resilience and stability of ecology systems. Annual Review of Ecology and Systematics 4:1–23.
Holling, C. S. 1978. *Adaptive Environmental Assessment and Management.* John Wiley and Sons, New York.
Holling, C. S. 1986. The resilience of terrestrial ecosystems, local surprise and global change. In: *Sustainable Development of the Biosphere.* W. C. Clark and R. E. .Munn, eds. Cambridge University Press, Cambridge. Pp. 292–317.
Holling, C. S. 1988. Temperate forest insect outbreaks, tropical deforestation and migratory birds. Memoirs of the Entomological Society of Canada 146:21–32.
Holling, C. S. 1992. Cross-scale morphology, geometry and dynamics of ecosystems. Ecological Monographs 62:447–502.
Holling, C. S. 1994. New science and new investments for a sustainable biosphere. In: *Investing in Natural Capital.* A. M. Jansson, M. Hannmer, C. Folke and R. Costanza, eds. Island Press, Washington, D.C. Pp. 57–73.
Holling, C. S. 1995. What barriers? What bridges? In: *Barriers and Bridges to the Renewal of Ecosystems and Institutions.* L. H. Gunderson, C. S. Holling and S. S. Light. eds. Columbia University Press, New York. Pp.3–34.
Holling, C. S., D. W. Schindler, B. W. Walker and J. Roughgarden. 1995. Biodiversity in the functioning of ecosystems: An ecological primer and synthesis. In: *Biodiversity Loss: Ecological and Economics Issues.* C. Penings, K. G. Maler, C. Folke, B. O. Jansson and C. S. Holling, eds. Cambridge University Press, New York. Pp. 44–83.
Hughes, T. P. 1994. Catastrophes, phase shifts and large scale degradation of a Caribbean Coral Reef. Science 265:1547–1551.
Kilgore, B. M. 1976. Fire management in the natural parks: an overview. In: *Proceedings of Tall Timbers Fire Ecology Conference.* Florida Stare University Research Council, Tallahassee. Pp. 45–57.
Lee, K. N. 1993. Compass and gyroscope: Integrating science and politics for the environment. Island Press, Covelo. California
Leopold, A. 1949. *A Sand County Almanac and Sketches Here and There.* Oxford University Press, New York.
Levin, S. A. 1992. The problem of pattern and scale in ecology. Ecology 73:1943–l967.
Meffe, G. K. 1984. Effects of abiotic disturbance on coexistence of predator-prey fish species. Ecology 65:1525–1534.
Meffe, G. K. 1986. Conservation genetics and the management of endangered fishes. Fisheries 11:14–23.
Meffe, G. K. and C. R. Carroll. 1994. Principles of conservation biology. Sinauer Associates, Sunderland, Massachusetts.
Meffe, G. K. and R. C. Vrigenhoek. 1988. Conservation genetics in the management of desert fishes. Conservation Biology 2: 157–169
Meffe, G. K., A. H. Ehrlich and D. Ehrenfeld. 1993. Human population control: the missing agenda. Conservation Biology. 7:1–3
Minckley, W. L. and J. E. Deacon, eds. 1991. Battle against extinction: Native fish management in the American West. University of Arizona Press, Tucson.
Minckley, W. L. and G. K. Meffe. 1987. Differential selection by flooding in stream-fish communities of the arid American Southwest. In: *Evolutionary and Community Ecology of North American Stream Fishes.* D. C. Heins and W. J. Matthews, eds. University of Oklahoma Press, Norman. Pp. 93–104.
Naeern S., L. J. Thompson, S. P. Lawler, J. H. Lawton and R. M. Woodfin. 1994. Declining biodiversity can alter the performance of ecosystems. Nature 368:734–737.
O'Neill, R. V., D. L. DeAngelis, J. B. Waide, T. F. H. Allen. 1986. A hierarchical concept of ecosystems. Princeton University Press, Princeton, New Jersey.
Picken, S. T. A. and P. S. White, eds. 1985. The ecology of natural disturbance and patch dynamics. Academic Press, Orlando, Florida.
Pimm, S. L. 1984. The complexity and stability of ecosys-

tems. Nature 307:321–326.

Schindler, D. W. 1987. Detecting ecosystem responses to anthropogenic stress. Canadian Journal of fisheries and Aquatic Sciences 44(suppl. 1):6–25.

Schindler, D. W. 1990. Experimental perturbations of whole lakes as tests of hypotheses concerning ecosystem structure and function. Oikos 57:25–41.

Schindler, D. W., et al. 1991. Freshwater acidification, reversibility and recovery: Comparisons of experimental and atmospherically-acidified lakes. In: *Acidic Deposition: its Nature and Impacts.* F. T. Last and R. Watling, eds. Proceedings of the Royal Society of Edinburgh, Edinburgh. Pp. 193–226.

Schonewald-Cox, C. M., S. M. Chambers, B. MacBride and L. Thomas, eds. 1983. *Genetics and conservation: A Reference for Managing Wild Animal and Plant Populations.* Benjamin-Cummings, Menlo Park, California.

Sinclair, A. R. E., P. D. Olsen and T. D. Redhead. 1990. Can predators regulate small mammal populations? Oikos 59:382–392.

Southwood, T. R. E. 1977. Habitat, the template for ecological strategies? Journal of Animal Ecology 46:337–365.

Stanley, T. R., Jr. 1995. Ecosystem management and the arrogance of humanism. Conservation Biology 9:255–262

Tilman, D. and J. A. Downing. 1994. Biodiversity and stability in grasslands. Nature 367:363–365.

Waide, J. B. and J. R. Webster. 1976. Engineering systems analysis: applicability to ecosystems. In: *Systems Analysis and Simulation in Ecology.* B. C. Parton, ed. Academic Press, New York. Pp. 329–372.

Walker, B. H. 1981. Is succession a viable concept in African savanna ecosystems? In: *Forest Succession: Concepts and Application.* D. C. West, H. H. Shugart and D. B. Botkin, eds. Springer-Verlag, New York. Pp. 431–447.

Walker, B. H., 1992. Biological diversity and ecological redundancy. Conservation Biology 6:18–23.

Walker, B. H., D. Ludwig, C. S. Holling and R.M. Peterman. 1969. Stability of semi-arid savanna grazing systems. Journal of Ecology 69:473–498.

Walters, C. J. 1986. *Adaptive Management of Renewable Resources.* McGraw Hill, New York.

Walters, C. and L. Gundewn. 1994. A screening of water policy alternatives for ecological restoration in the Everglades. In: *Everglades: The Ecosystem and its Restoration.* S. M. Davis and J. Ogden, eds. St. Lucie Press, Delray Beach, Florida. Pp. 757–767

Introduction to
John Pastor, Robert J. Naiman, Bradley Dewey and Pamela McInnes
Moose, Microbes, and the Boreal Forest

by Steven R. Carpenter

Large predators and herbivores in terrestrial systems can affect primary production and soil properties in ways analogous to trophic cascades in aquatic ecosystems. This paper presents data on the effects of moose on boreal forest production, nitrogen cycling and soils. Moose populations are controlled by wolf predation as well as availability of food and habitat. Effects of moose on forest composition, production, and soil characteristics were studied by comparing moose exclosures with areas occupied by moose. Intensive moose browsing leads to fewer deciduous trees, more coniferous trees, and lower levels of nitrogen in the soil. The net result is that the land is less able to support moose populations. Low levels of moose browsing have the opposite effect.

Moose, Microbes, and the Boreal Forest

JOHN PASTOR, ROBERT J. NAIMAN, BRADLEY DEWEY, AND PAMELA MCINNES

Plants and herbivores interact with the soil and its microbes, which are important parts of the overall ecosystem. Through selective browsing, moose change plant communities and ecosystem properties.

Pastor, J., R. J. Naiman, B. Dewey and P. McInnes. 1988. Moose, microbes, and the boreal forest. *BioScience* 38:770–777.

The ecology of moose is intimately tied to that of the boreal forest. Moose first appeared in the fossil record during the glacial advances of the late Pliocene and early Pleistocene, subsequently undergoing adaptive radiation that produced two genera *(Cervalus* and *Alces)* and at least four species (Telfer 1984). The sole extant species, *Alces alces,* radiated into eight subspecies 6500–7000 years ago, of which seven survive (Telfer 1984). The current assemblage of aspen *(Populus tremuloides),* balsam poplar *(Populus balsamifera),* birch *(Betula papyrifera),* spruce *(Picea glauca, Picea mariana),* and balsam fir *(Abies balsamea)* that constitutes the boreal forest arose at about the same time (Larsen 1980).

Both the boreal forest and moose are circumpolar, with nearly identical ranges except for a slight expansion of moose into the northern hardwood forests of eastern North America and Europe. Browsing by moose influences both the plant species present in the forest and the properties of the soil. Interactions between moose and the forest provide a good example of how herbivores influence ecosystem properties over different trophic, organizational, and spatial scales, and how complex feedback loops in these interactions may produce surprising effects.

Herbivores Affect Soil Processes

To simplify ecological complexities, most research on herbivory has focused on direct interactions between plants and herbivores. But the plants and herbivores also interact with the soil and its microbes, which are important parts of the overall ecosystem. When herbivory is considered in this larger context, different and even diametrically opposite conclusions can be obtained compared with those drawn from the more restricted plant-herbivore relationship.

For example, Oksanen et al. (1981) hypothesize that herbivore biomass will be greatest in moderately productive ecosystems and least in both unproductive and highly productive ecosystems. Unproductive ecosystems cannot produce enough forage to sustain herbivores, and very productive ecosystems produce enough food to sustain high predation. This hypothesis assumes that herbivores respond to production of forage but do not affect it, although they may determine the standing crop at any one time. However, if moderate browsing stimulates plant growth by rapid recycling of nutrients through soil microbes (Dyer et al. 1986, Woodmansee 1978), then herbivores can in theory make a moderately productive ecosystem a more productive one. This change should then increase predation, which in turn may cause a decline in herbivore biomass. Thus, studies of the isolated herbivore-plant system predict that herbivore biomass eventually equilibrates along a gradient of plant productivity that is controlled by other factors, but the expanded herbivore-plant-soil system is more dynamic.

Studies of moose and other large mammals demonstrate that soil processes both are influenced by herbivory and indirectly control it. Herbivores, vegetation, and the soil microbes that decompose organic matter are three interacting parts of feedback loops with both positive and negative components (Figure 1). Soil processes affect moose by controlling the supply of browse and the rate at which plants recover from browsing (Bergerud and Manual 1968, Botkin et al. 1981, Peek et al. 1976). Herbivores' excrement, composed of digested organic matter as well as the litter shed by plants, carries organic matter and nutrients into the soil and thus affects microbial processes (Cargill and Jefferies 1984, Hatton and Smart 1984, Kitchell et al. 1979, McKendrick et al. 1980, Mattson and Addy 1975, Schimel et al. 1986, Tiedemann and Berndt 1972, Woodmansee 1978). However, direct shunting of plant materials through the herbivore excretion pathway is minor in forests except during periods of insect outbreak (Swank et al. 1981), when almost the entire nutrient content of foliage is returned to the soil in one pulse.

Another way that moose can affect nutrient cycling is by altering the composition of the plant community through voracious and selective foraging. Moose consume approximately 15 kg of dry food per day or 5–6 metric tons per year (Peterson 1955). In wintering grounds in the Soviet Union, moose annually consume approximately 8% of the biomass of their preferred food species, returning 63% of the biomass consumed to the soil and using 37% for metabolic requirements (Sukachev and Dylis 1964). The feeding requirements of moose are not met uniformly by the species that constitute the boreal forest, which differ sharply in landscape distribution, nutrient cycling properties, and preference by herbivores (Bryant and Kuropat 1980, Larsen 1980, Van Cleve et al. 1983).

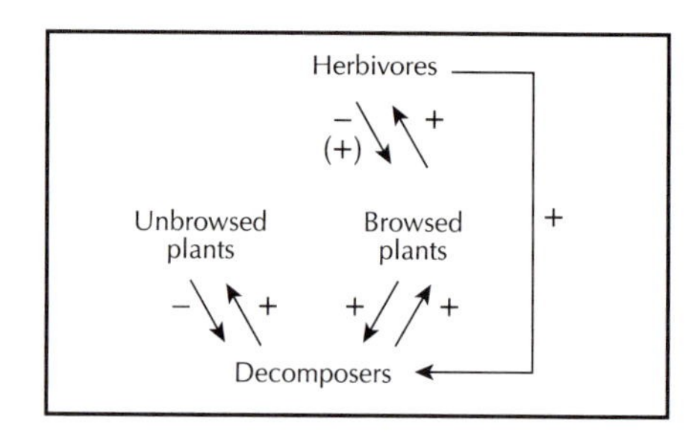

Figure 1. Positive and negative feedbacks in the decomposer-plant-herbivore system. Decomposers positively affect all plants by releasing nutrients from plant litter. Some plants are not browsed and negatively affect decomposers because of high lignin and low nitrogen contents in litter. Other plants are browsed and positively affect decomposers because of low lignin and high nitrogen contents of tissues and litter. These plants support herbivore populations, which in turn negatively affect them by browsing (although light browsing may stimulate growth). Urine, feces, and carcasses of herbivores are rapidly decomposed, and they may enhance nutrient availability in some communities.

Early successional species, such as aspen, balsam poplar, and paper birch grow rapidly and produce easily decomposable and nitrogen-rich litter (Flanagan and Van Cleve 1983, Van Cleve et al. 1983) and are the forage plants preferred by moose (Peterson 1955). They are succeeded by conifers such as spruce and balsam fir, which grow more slowly, have lower nitrogen requirements, and produce slowly decomposing, nitrogen-poor litter (Flanagan and Van Cleve 1983, Van Cleve et al. 1983). Spruce is almost never browsed by moose, apparently because its tissues have high concentrations of resins and lignin and low concentrations of nitrogen (Bryant and Kuropat 1980). Balsam fir is browsed mainly when the preferred foods are not available (Peterson 1955).

The nitrogen, resin, and lignin content of plant tissues affect moose digestion, organic matter decomposition, and soil nitrogen availability, because all three processes are microbially mediated. Small aspen, poplar, and birch, whose easily digestible and decomposable tissues are browsed by moose and other ungulates, are eventually overtopped by unbrowsed spruce and fir, whose litter is low in nutrients and/or high in recalcitrant carbon compounds (Hanley and Taber 1980, Houston 1968, Krefting 1974).

Because of this relationship between feeding preference and litter quality, moose should affect soil nitrogen availability and microbial communities to the extent that they alter plant community composition. This shift in community structure caused by selective browsing may be a negative feedback exerted by moose and other herbivores on the nutrient cycle (Figure 1; Bryant and Chapin 1986, Pastor et al. 1987). Such herbivore-induced changes in plant communities may cascade through the forest ecosystem as they do in lake ecosystems (Carpenter and Kitchell, p. 764 this issue).

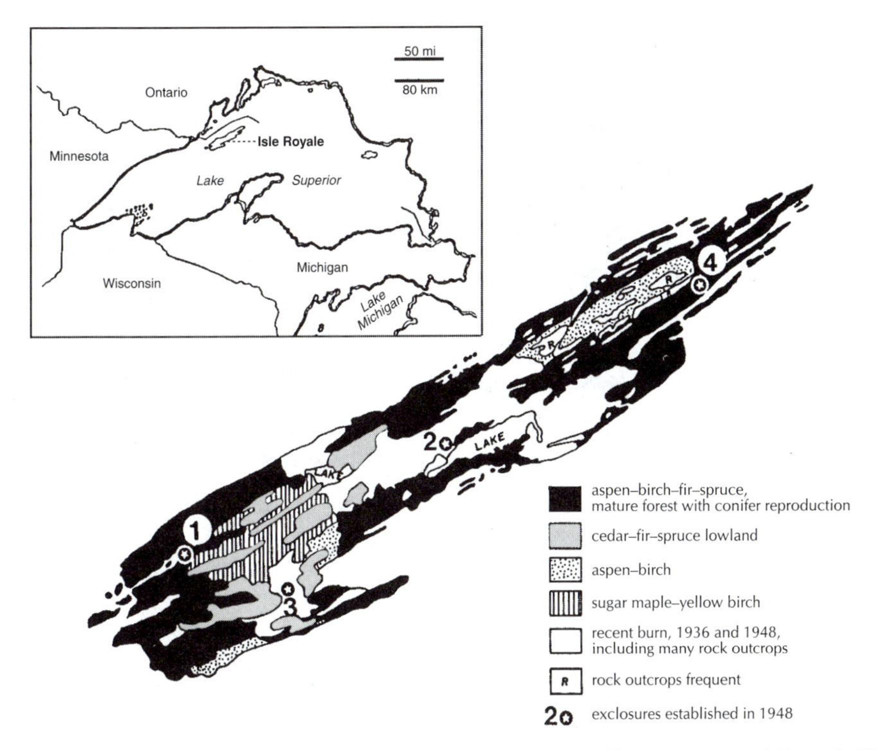

Figure 2. Vegetation map of Isle Royale (from Krefting 1974 and Peterson 1977). Exclosures as follows: 1. Windigo, 2. Siskiwit Lake, 3. Siskiwit Camp, 4. Daisy Farm.

Moose Effects on Isle Royale Forests

We are investigating the relationships between moose and the boreal forests of Isle Royale National Park in Michigan. Here, intensive moose browsing during the past 80 years has greatly altered forest species composition and soil properties. Isle Royale lies in the western arm of Lake Superior (Figure 2). Northern hardwoods are most common on the island's higher and warmer elevations, particularly in the southwestern portion where there are thick till deposits. Boreal species (aspen, paper birch, spruce, and balsam fir) are more common around the cooler lake shore, on the glacially scoured topography of the northeastern end of the island, and (particularly aspen and birch) in fire scars. It is in the boreal communities, particularly those on the southwestern end of the island, where moose densities are greatest (Peterson 1977).

When moose arrived on Isle Royale from Canada or Minnesota near the turn of the century, they found a land of abundant food and no predators. As a consequence, their population expanded rapidly, reaching some 3000 animals on the 574-square-kilometer island in the 1930s (Murie 1934). As their food supply became overbrowsed, the population underwent a severe decline. Several large forest fires in 1936 and 1938 resulted in large amounts of aspen and paper birch regeneration, and the moose population began to recover in the 1940s.

Arrival of wolves *(Canis lupus)* in the late 1940s appeared to stabilize the moose population to levels well below what the food supply could support, and early research concluded that predators or pathogens, not food, are the limiting factors to moose populations (Mech 1966). More recent research indicates that both vegetation and wolves may play important roles in moose population

Figure 3. A. The Windigo exclosure, showing mature forest with closed canopy inside the fence and heavily browsed area outside. **B.** Spruce is the only species growing above browse height on the control plots.

dynamics. Vegetation may set a ceiling on moose biomass and partially determine susceptibility to predation as well as birth rates, with wolves culling young and old moose from the population (Peterson 1977).

Between 1948 and 1950, four exclosures (each 101 m^2) were established on Isle Royale by Laurits Krefting and coworkers from the University of Minnesota. They intended to evaluate the long- term potential of moose to alter plant communities (Krefting 1974, Figure 3). Paired control plots were also established adjacent to each exclosure. Two of these exclosures (Windigo and Daisy Farm) are in upland aspen-birch-spruce-fir forests, one (Siskiwit Camp) is in an upland paper birch stand in the center of a large burn that occurred in 1936, and the final exclosure (Siskiwit Lake) is in a small aspen-willow *(Salix* sp)-alder *(Ahus rugosa)* wetland.

The other herbivores of any abundance on Isle Royale are snowshoe hare *(Lepus americana)* and beaver *(Castor canadensis)*. Hares can enter the exclosures, but population levels rarely are high enough to affect vegetation where the exclosures are located (Peterson 1977). There has never been any recorded instance of beaver affecting the vegetation in these plots, except for an aspen in the Windigo exclosure that may have been cut by a beaver in 1960 (Krefting 1974). Therefore, only moose are affected by the exclosures, and moose are the major herbivore affecting vegetation outside the exclosures.

Species composition, density, and basal area have been determined periodically in all plots since establishment. These are among the oldest mammalian exclosures in continuous existence in North America. We have concentrated on measuring changes in the three upland exclosures (Windigo, Siskiwit Camp, and Daisy Farm).

More than 80% of the aspen, birch, mountain ash *(Sorbus americana)*, and mountain maple *(Acer spicatum)* are continually browsed by moose in the control plots adjacent to the exclosures (Krefting 1974, Risenhoover and Maas 1987). The protection from moose afforded by the exclosures has resulted in consistently greater height growth of aspen, birch, mountain ash, and mountain maple compared with control plots (Krefting 1974). Changes in the plant community are greatest at the Windigo exclosure, where high moose populations have gradually converted the forest into a "spruce-moose savanna" (a term coined by Peter Jordan of the University of Minnesota). Except for a few birch that escaped moose browsing, probably during the population crash of the 1930s, white spruce is the only species that grows above browse height (Figure 3).

Changes in Soil Properties

These changes in plant communities are reflected in the composition of plant litter entering the soil (Figure 4). At Windigo, litter production is significantly greater inside the exclosure compared with outside. Moreover, litter outside the exclosure is dominated more by conifer needles than is litter inside the exclosure. At Siskiwit Camp, moose browsing has not resulted in a change in the plant community, and, although there is greater litter production inside the exclosure compared with outside, litter of both plots is dominated by paper birch. The quantity of litter produced at Daisy Farm has not been significantly affected by moose browsing, although there is significantly greater dominance by conifer litter inside the exclosure compared with outside.

Have these differences in litter inputs inside and outside the exclosures caused cor-

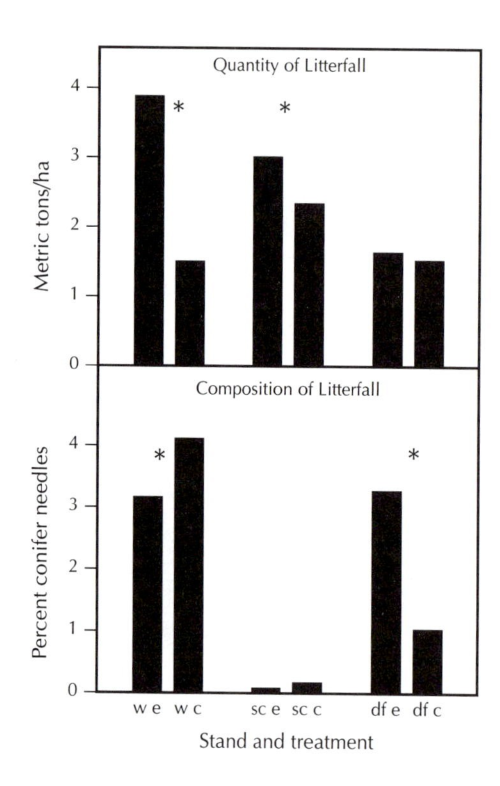

Figure 4. Changes in quantity and composition of autumn litterfall outside and inside exclosures that have excluded moose for 40 years. Explanation of stand and treatment codes: w–Windigo; sc–Siskiwit Camp; df–Daisy Farm, e–exclosure; c–control. *–differences significant at $P < 0.05$.

responding changes in soil properties? Where browsing is most intense at the Windigo site, there are visible differences in the thickness and structure of the forest floor (Figure 5). Outside this exclosure, the floor is thin, being composed mainly of spruce needles, and overlies an A1 horizon. Inside the exclosure, the forest floor thickness is 5–10 cm and overlies a weakly developed A2 horizon. There are no visible differences in forest floor structure at the other two sites.

To examine these apparent changes in soil properties further, soil carbon and nitrogen pools, cation exchange capacity, field soil nitrogen availability (Eno 1960), the pool of potentially available nitrogen and its release rate (Stanford and Smith 1972), microbial respiration rates, and microbial biomass (Jenkinson and Powlson 1976) were measured in the exclosures and control plots. Total carbon and nitrogen pools are the ultimate limit on the amount of nitrogen available for plant growth and the amount of carbon and nitrogen available for microbes.

However, the actual availability of nitrogen for plants is determined by nitrogen mineralization under field conditions, which is the rate at which organic nitrogen is converted to ammonium and nitrate by microbial activity (ammonium is by far the dominant form of mineralized nitrogen in these soils). Only a portion of total nitrogen is usually mineralized, the remainder being recalcitrant to decomposition. Therefore, field mineralization is determined by the proportion of total nitrogen that is potentially mineralizable and the rate at which this pool can be mineralized, a factor that is modified by climatic conditions. Once released, the ammonium must be held on soil-cation-exchange sites until plants can absorb it through root systems, otherwise it will leach beneath the rooting zone.

Microbial activity in the soil is partially determined by fungal, bacterial, and actinomycete biomass. Respiration rates per unit total soil carbon are a useful index of the ability of microbes to decompose soil organic matter.

Soil changes are most apparent at Windigo, where moose browsing has resulted in a statistically significant decrease in six of the eight soil properties measured (Figure 6). Soil carbon, nitrogen, cation exchange capacity, field nitrogen availability, potentially mineralizable nitrogen, and microbial respiration rates were

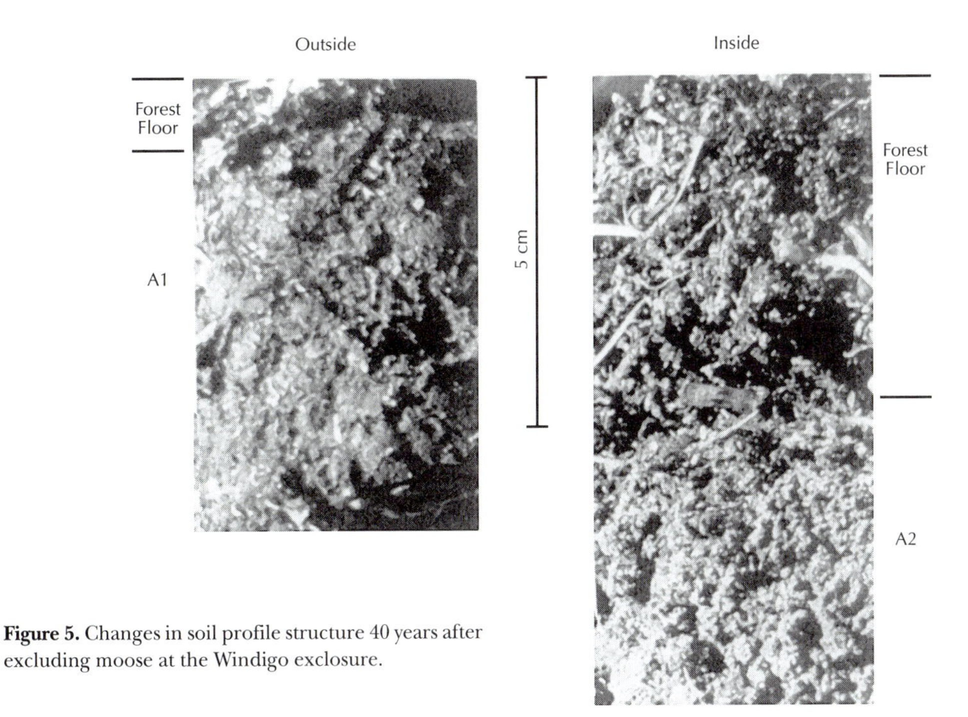

Figure 5. Changes in soil profile structure 40 years after excluding moose at the Windigo exclosure.

significantly greater inside the exclosure compared with outside. The potentially mineralizable nitrogen release rate and microbial biomass were not significantly different.

There were no significant differences in carbon, nitrogen, cation exchange capacity, field mineralization rates, and respiration rates between exclosures and controls at the other two sites. At Siskiwit Camp, potentially mineralizable nitrogen was significantly greater inside the exclosure, but the potential mineralization rate was lower there. This data may account for the lack of difference in the field mineralization measured inside and outside this exclosure.

Similarly, the potential rate of mineralization was significantly greater inside the Daisy Farm exclosure compared with outside, but there is a lower pool of potentially mineralizable nitrogen, resulting in no difference in total field mineralization. Interestingly, microbial biomass was significantly greater inside the Daisy Farm exclosure compared with outside, but it was not different at the other sites. This result is in striking contrast with the opposite pattern seen in the measures of microbial activity, and it may indicate that there are significant changes in microbial communities both among sites and treatments.

We conclude that when moose browsing causes decreases in both quantity and quality of litter through changes in plant community composition, as seen at the Windigo site, soil microbial processes determining nitrogen availability to plants will also decline.

Such changes in ecosystem properties need not be restricted to moose: Hatton and Smart (1984), Tiedemann and Berndt (1972), and Cargill and Jefferies (1984) found larger pool

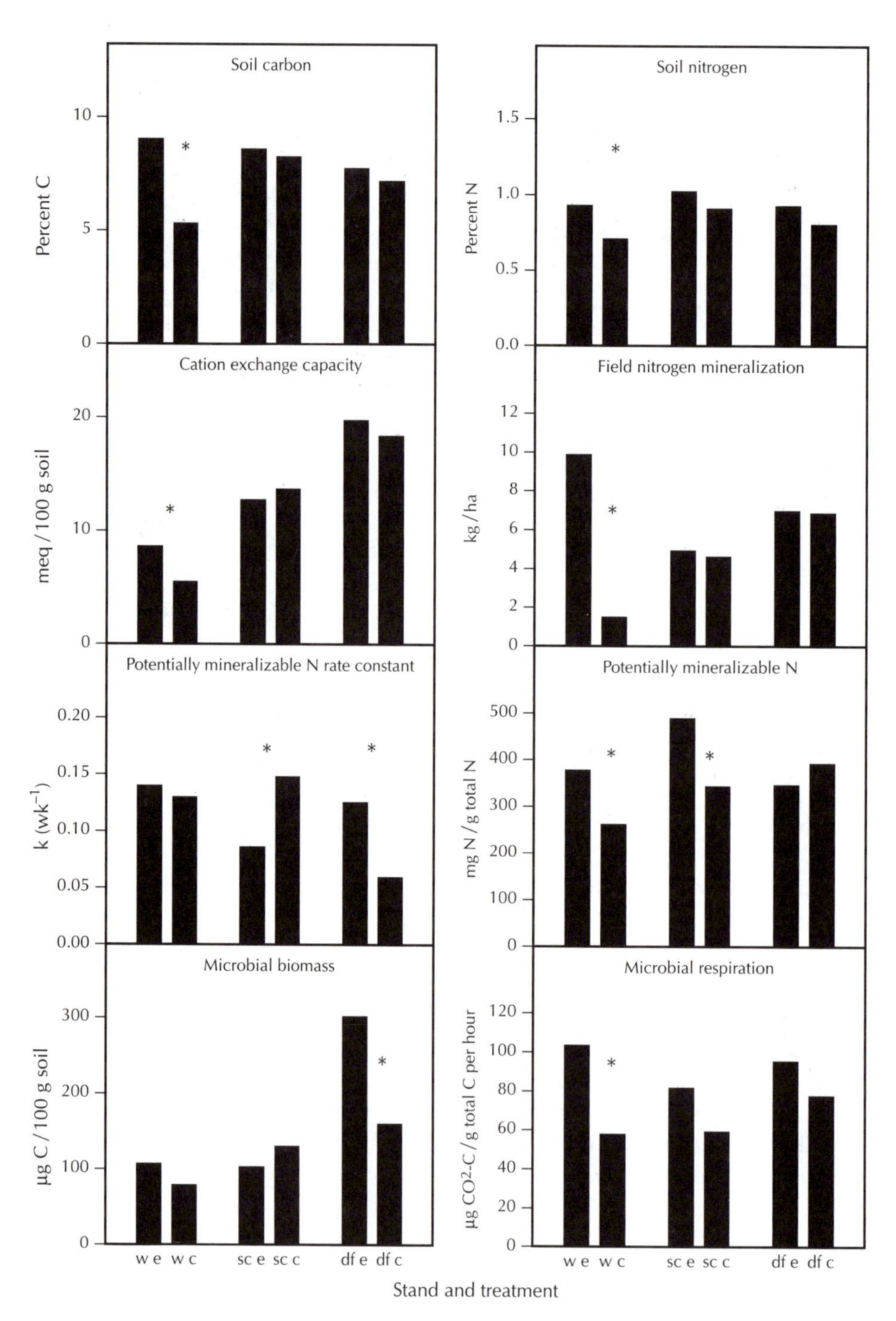

Figure 6. Changes in soil properties 40 years after excluding moose. Stand, treatment, and statistical significance codes as in Figure 4. All properties except cation exchange capacity measured in August 1987.

sizes of soil nutrients or greater cation exchange capacity inside elephant *(Loxodonta africana)*, elk *(Cervus canadensis)*, and snow goose *(Chen hyperborea)* exclosures, respectively, compared with outside. However, we believe this is the first report of changes, upon exclusion of herbivores, in microbial activity commensurate with changes in litter quantity and quality.

These changes in soil nitrogen availability raise several questions. Models of habitat carrying-capacity can explicitly incorporate nutritional constraints on wildlife population growth (Hobbs and Swift 1985), and the results presented here suggest that long-term changes in soil nitrogen availability may affect habitat carrying-capacity in several ways.

The general decline in nitrogen availability may affect how rapidly browsed plants recover from browsing, thus accelerating the invasion of unbrowsed spruce, and the decline may also cause the browsed plants to produce more secondary compounds to deter further herbivory and depress soil microbial activity. These effects in turn may drive down nitrogen availability further (Bryant and Chapin 1986), and they may even provide a selection pressure on evolution of plant defenses and ability to withstand low nutrient availability (Corey et al. 1985).

Increases in abundance of various herbaceous species in heavily browsed areas (Krefting 1974, Risenhoover and Maas 1987) may sometimes compensate for these changes in the composition of tree litter, because most herbaceous plants produce litter that is less lignified than that produced by conifers, and herbaceous plants are an important summer food for moose. Further research on how browsing affects plant community composition will provide insight into how herbivores affect carbon and nitrogen cycles.

Herbivory and the Boreal Landscape

Exclosure experiments are valuable tools for examining how herbivores alter ecosystem processes at particular locations and times. Large herbivores, such as moose with a home range of 300–500 ha or more, must live in a spatially diverse landscape and require spatial diversity for a variety of life's functions and stages. Unless they are very large, exclosures do not capture the spatial dynamics of herbivore-landscape interactions, and these experiments cannot answer such questions as how much forage within a preferred habitat is actually used by moose or how moose select different habitats for different functions (e.g., cover as well as food).

Landscape Dynamics

The spatial patterns of suitable climate, landforms, vegetation, and animal communities (including man and his economic system) all affect the distribution of moose habitat (Telfer 1984). Climate and landforms influence moose movement directly by influencing the thermal regime (Telfer 1984) and the distribution of snow, which limits movement and ability to escape from predators (Peerson 1977), and indirectly by the distribution of vegetation. Moose require a variety of habitats, including wetlands for micronutrients and wet conifer forests for cooling in summer, conifer forests for thermal cover in winter, and young deciduous forests for food (Peek et al. 1976, Peterson 1955, Telfer 1984).

The dynamics of the landscape are equally as important to moose as the heterogeneity at any one time. Geist (1974) hypothesizes that moose populations find refuge in habitats of continuous and usually small disturbances, such as river floodplains, but they expand into transient habitats with an abundance of food created by larger disturbances such as fire. With the advent of fire control programs, timber harvesting is fulfilling the primary role of creating productive, transient habitats (Peek et al. 1976), but the sizes, shapes, distribution, and frequency of harvests are very different from those of natural fires. Therefore, timber harvests and fires may have very different effects on moose populations by differentially affecting the spatial distribution of moose habitat. However, both fire and timber harvesting can stimulate soil nitrogen mineralization and in turn improve forage quality (Hobbs and Spowart 1984).

Thus, the probability that a moose will browse any patch, the amount of time spent browsing that patch, and the subsequent effects on ecosystem processes depends to a large degree on the origin of that patch and the spatial arrangement and dynamics of all required habitat types in the landscape. The extent to which moose can cause changes in ecosystem properties may depend on the scales of ocher disturbances affecting the forest.

Small disturbances, such as windthrows or cuttings around beaver ponds, might be heavily browsed if they are contiguous to conifer cover. Moreover, the adjacent conifers may rapidly invade such small disturbances from the edge. In contrast, large disturbances such as burns or clearcuts may produce so much food that it cannot be overbrowsed by even an expanding moose population (Peek et al. 1976), and conifer seed dispersal may not reach the interior of the disturbance. These disturbance patterns may be the reason for the differences seen between the Windigo site, where disturbances are limited to single tree deaths and windthrows, and the Siskiwit Camp site in the middle of a large burn. Disturbances at the Daisy Farm site are also small, but moose densities do not seem to ever have been as high as on the southwestern end of the island (Krefting 1974), and so browsing intensity has historically been less than at Windigo.

Our observations on Isle Royale indicate that spruce and other unbrowsed conifers may invade heavily browsed areas in a two-step process, whereby a few individual trees become established and then serve as seed sources and secondary foci around which younger individuals become established. The result is a set of spruce groves expanding into heavily browsed areas. Selective moose browsing may initially increase the spatial heterogeneity of forest communities, but, as the unbrowsed groves expand, heterogeneity should decline. There should be corresponding changes in the spatial distribution of soil nitrogen availability, which in turn may influence the rate of expansion, since nitrogen is a limiting factor to plant growth and browse recovery. By selective browsing, moose may therefore contribute to, as well as be affected by, the spatial heterogeneity of forest ecosystems.

Other Herbivores

Herbivory can in itself be a disturbance in boreal landscapes. Yet, moose are but one herbivore that can potentially affect nutrient cycling in boreal forests by selective feeding. Beaver browse on large-diameter aspen and other hardwoods (Neiman et al., page 753 this issue); this can release understory conifers from competition or cause resprouting of the hardwoods from roots and thus perpetuate their occupancy of the site. Spruce budworm *(Choristoneura fumiferana)* and balsam wooley aphid *(Adeiges picea)* can kill large stands of conifers and cause succession to revert to seral stages.

Such changes in plant communities and possible changes in the cycles of nutrients that sustain them, although caused by one herbivore, affect all others indirectly. For example, beaver cutting around ponds could result in heavy regrowth of browse for moose, but intense browsing by moose may prevent the hardwoods from attaining the size preferred by beaver. However, we presently have little data on the relative importance of different herbivores in boreal forests, because most studies, like ours, choose sites where the influence of one herbivore is dominant.

Interaction and feedback among herbivores, plant species, and disturbances may result in the mosaic of nutrient cycling regimes in the boreal landscape. Understanding of the roles of herbivores in ecosystems must account for the origins and consequences of spatial heterogeneity at chemical, population, ecosystem, and landscape scales. Simple long-term exclosure experiments, as the Isle Royale exclosures demonstrate, can provide valuable data for addressing these problems and become increasingly valuable with time.

Computer techniques such as geographic information systems (Neiman et al., page 753 this issue) and models (Corey et al. 1985) can be used to project ecosystem and landscape changes from data collected during exclosure experiments. Increasing sophistication of these techniques, when coupled with data obtained from other field experiments, will greatly expand and alter understanding of the role of herbivores in ecosystems.

Acknowledgments

This research was made possible by a grant from the National Science Foundation's Ecosystem Studies Program and by logistic support from Isle Royale National Park. We thank Yosef Cohen, Elisabeth Holland, Carol Johnston, Ron Moen, and two anonymous reviewers for helpful comments on the manuscript.

References

Bergerud, A. T. and F. Manuel. 1968. Moose damage to balsam fir-white birch forests in Newfoundland. J. Wildl. Manage. 32:729–746.

Botkin, D. B., J. M. Melillo and L. S.-Y. Wu. 1981. How ecosystem processes are linked to large mammal population dynamics. In: *Dynamics of Large Mammal Populations.* C. F. Fowler and T. D. Smith, eds. John Wiley & Sons, New York. Pp. 373–387.

Bryant, J P. and F. S. Chapin. 1986. Browsing-woody plant interactions during plant succession. In: *Forest Ecosystems in the Alaskan Taiga.* K. Van Cleve F. S. Chapin, P. W. Flanagan, L. A. Viereck and C. T. Dyrness, eds. Springer-Verlag, New York. Pp. 213–225.

Bryant, J. P. and P. J. Kuropat. 1980. Selection of winter forage by subarctic browsing invertebrates: the role of plant chemistry. Annul Rev. Ecol. Syst. 11:261–285.

Cargill, S. M. and R. L. Jefferies. 1984. The effects of grazing by lesser snow geese on the vegetation of a subarctic salt marsh. J. Appl. Ecol. 21:669–686.

Carpenter, S. R. and J. F. Kitchell. 1988. Consumer control of lake productivity. BioScience 38:764–769.

Coley, P., J. P. Bryant and F. S. Chapin. 1985. Resource availability and plant antiherbivore defense. Science 230:895–899.

Dyer, M. l., D. L. DeAngelis and W. M. Post. 1986. A model of herbivore feedback on plant productivity. Math. BioSci. 79:171184.

Eno, C. F. 1960. Nitrate production in the field by incubating the soil in polyethylene bags. Soil Sci. Soc. Am. Proc. 24:277–299.

Flanagan, P. W. and K. Van Cleve. 1983. Nutrient cycling in relation to decomposition and organic matter quality in taiga ecosystems. Can. J. For. Res. 13:795–817.

Geist, V. 1974. On the evolution of reproductive potential in moose. Naturaliste Canadien 101:527–S37.

Hanley, T. A. and R. D. Taber. 1980. Selective plant species inhibition by elk and deer in three coniferous communities in western Washington. For. Sci. 26:97–107.

Hatton, J. C. and N. O. E. Smart. 1984. The effect of long-term exclusion of large herbivores on soil nutrient status in Murchison Falls National Park, Uganda. Afr. J. Ecol. 22:23–30.

Hobbs, N. T. and R. A. Spowart.1984. Effects of prescribed fire on nutrition of mountain sheep and mule deer during winter and spring. J. Wildl. Manage. 48:551–560.

Hobbs, N. T. and D. M. Swift. 1985. Estimates of habitat carrying capacity incorporating explicit nutritional constraints. J. Wildl. Manage. 49:814–822.

Houston, D. B. 1968. The shira moose in Jackson Hole, Wyoming. Technical Bulletin No 1, Grand Teton Natural History Association, National Park Service, Jackson Hole, Wyoming

Jenkinson, D. J. and D. F Powlson. 1976. The effects of biocidal treatments on metabolism in soil. V. A method for measuring soil biomass. Soil Biol. Biochem. 8:209–213.

Kitchell, J. F., R. V. O'Neill, D. Webb, G. W. Gallep, S. M. Bartell, J. F. Koonce and B. S. Ausmus. 1979. Consumer regulation of nutrient cycling. BioScience 29:28–34.

Krefting, L.W. 1974. The ecology of the Isle Royale moose, with special reference to the habitat. University of Minnesota Agricultural Experiment Station Technical Bulletin 297, St. Paul.

Larsen, J. A. 1980. *The Boreal Ecosystem.* Academic Press, New York.

McKendrick, J. D., G. O. Batzli, K. R. Everett and J. G. Sanson. 1980. Some effects of mammalian herbivores and fertilization of tundra soils and vegetation. Arct. Alp. Res. 12:565–578.

Mattson, W. J. and N. D. Addy. 1975. Phytophagous insects as regulators of forest primary production. Science 190:515–522.

Mech, L. D. 1966. *The Wolves of Isle Royal.* Fauna of the National Parks of the United States, Fauna Series, Washington, DC.

Murie, A. 1934. *The Moose of Isle Royale.* University of Michigan Museum of Zoology Miscellaneous Publ. 25, Ann Arbor.

Naiman, R. J., C. A. Johnston and J. C. Kelly. 1988. Alteration of North American streams by beaver. BioScience 38: 753–762.

Oksanen L., S. D. Fretwell, J. Arruda and P. Niemeia. 1981. Exploitation ecosystems in gradients of primary productivity. Am. Nat. 118:240–261.

Pastor, J., R. J. Naiman and B. Dewey. 1987. A hypothesis of the effects of moose and beaver foraging on soil carbon and nitrogen cycles, Isle Royale. Alces 23:107–124.

Peek, J. M., D. L. Urich and R. J. Mackie. 1976. Moose habitat selection and relationships to forest management in northeastern Minnesota. Wildlife Monographs No. 48, The Wildlife Society.

Peterson, R. L. 1955. *North American Moose.* University of Toronto Press, Toronto, Canada.

Peterson, R. O. 1977. Wolf Ecology and Prey Relations on Isle Royale. Scientific Monograph Series 7, U.S. Department of Interior, National Park Service, Washington, D.C.

Risenhoover, K. L. and S. A. Maas. 1987. The influence of moose on the composition and structure of Isle Royale forests. Can. J. For. Res. 17:357–366.

Schimel, D. S., W. J. Parton, F. J. Adamsen, R. G. Woodmansee, R. L. Senft and M. A. Stillwell. 1986. The role of cattle in the volatile loss of nitrogen from a shortgrass steppe. Biogeochemistry 2:39–52.

Stanford, G. and S. J. Smith. 1972. Nitrogen mineralization potentials of soils. Soil Sci. Soc. Am. Proc. 36:465–472.

Sukachev, V. and N. Dylis. 1964. Fundamentals of Forest Biogeocoenology. Oliver and Boyd, London.

Swank, W. T., J. B. Waide, D. A. Crossley and R. L. Todd. 1981. Insect defoliation enhances nitrate export from forest ecosystems. Oecologia 51:29.7–299.

Telfer, E. S. 1984. Circumpolar distribution and habitat requirements of moose *(Alces alces).* In: *Northern Ecology and Resource Management.* R. Olson, R. Hastings and F. Geddes, eds. University of Alberta Press, Edmonton, Alberta, Canada. Pp. 145–182.

Tiedemann, A. R. and H. W. Berndt. 1972. Vegetation and soils of a 30-year deer and elk exclosure in Central Washington. Northwest Sci. 46:59–66.

Van Cleve, K., L. Oliver, R. Schlentner, L. A. Viereck and C. T. Dyrness.1983. Productivity and nutrient cycling in taiga forest ecosystems. Can. J. For. Res. 13:747–766.

Woodmansee, R. G. 1978. Additions and losses of nitrogen in grassland ecosystems. BioScience 28:448–453.

Introduction to

Peter M. Vitousek, Harold A. Mooney, Jane Lubchenco and Jerry M. Melillo

Human Domination of Earth's Ecosystems

by Steven R. Carpenter

This article summarizes a growing body of evidence that humans impact virtually all of Earth's ecosystems, and are probably the most dominant species to arise in the history of the planet. Human activities have transformed or degraded nearly half the land surface of Earth, account for more than 20% of the CO_2 in the atmosphere, use about half the accessible fresh water, account for more than half of terrestrial nitrogen fixation, and have fully exploited or depleted about two-thirds of the world's marine fisheries. Many of these effects of humans are accelerating, and are likely to grow because of the expanding human population and the need for further economic development in most of the world. The great spatial extent of the changes, the rates of change, and the combination of kinds of change are unprecedented in Earth's history. The changes are occurring faster than we can accumulate new knowledge of the impacts and their consequences.

The authors suggest three courses of action: (1) reduce the rate at which we alter the Earth system, (2) learn more about how ecosystems respond to human-caused global change and how effects can be mitigated, and (3) accept the fact that direct human management of the planet is inescapable. It is ironic that the maintenance of wild or natural ecosystems now depends entirely on human action. Vitousek and his co-authors address difficult and compelling questions that all students of ecology will one day face: How can the rate of human-caused global change be controlled? What is the role of scientific information in setting priorities and making decisions that affect the rate of human-caused global change?

Human Domination of Earth's Ecosystems

PETER M. VITOUSEK, HAROLD A. MOONEY, JANE LUBCHENCO, JERRY M. MELILLO

Human alteration of Earth is substantial and growing. Between one-third and one-half of the land surface has been transformed by human action, the carbon dioxide concentration in the atmosphere has increased by nearly 30 percent since the beginning of the Industrial Revolution; more atmospheric nitrogen is fixed by humanity than by all natural terrestrial sources combined; more than half of all accessible surface fresh water is put to use by humanity; and about one-quarter of the bird species on Earth have been driven to extinction. By these and other standards, it is clear that we live on a human-dominated planet.

Reprinted with permission from Vitousek, P. M., H. A. Mooney, J. Lubchenco and J. M. Melillo. 1997. Human domination of Earth's ecosystem. *Science* 277:494–499.

All organisms modify their environment, and humans are no exception. As the human population has grown and the power of technology has expanded, the scope and nature of this modification has changed drastically. Until recently, the term "human-dominated ecosystems" would have elicited images of agricultural fields, pastures, or urban landscapes, now it applies with greater or lesser force to all of Earth. Many ecosystems are dominated directly by humanity, and no ecosystem on Earth's surface is free of pervasive human influence.

This article provides an overview of human effects on Earth's ecosystems. It is not intended as a litany of environmental disasters, though some disastrous situations are described; nor is it intended either to downplay or to celebrate environmental successes, of which there have been many. Rather, we explore how large humanity looms as a presence on the globe—how, even on the grandest scale, most aspects of the structure and functioning of Earth's ecosystems cannot be understood without accounting for the strong, often dominant influence of humanity.

We view human alterations to the Earth system as operating through the interacting processes summarized in Figure 1. The growth of the human population, and growth in the resource base used by humanity, is maintained by a suite of human enterprises such as agriculture, industry, fishing, and international commerce. These enterprises transform the land surface (through cropping, forestry, and urbanization), alter the major biogeochemical cycles, and add or remove species and genetically distinct populations in most of Earth's ecosystems. Many of these changes are substantial and reasonably well quantified; all are ongoing. These relatively well-documented changes in turn entrain further alterations to the functioning of the Earth system, most notably by driving global climatic change (1) and causing irreversible losses of biological diversity (2).

Land Transformation

The use of land to yield goods and services represents the most substantial human alteration of the Earth system. Human use of land alters

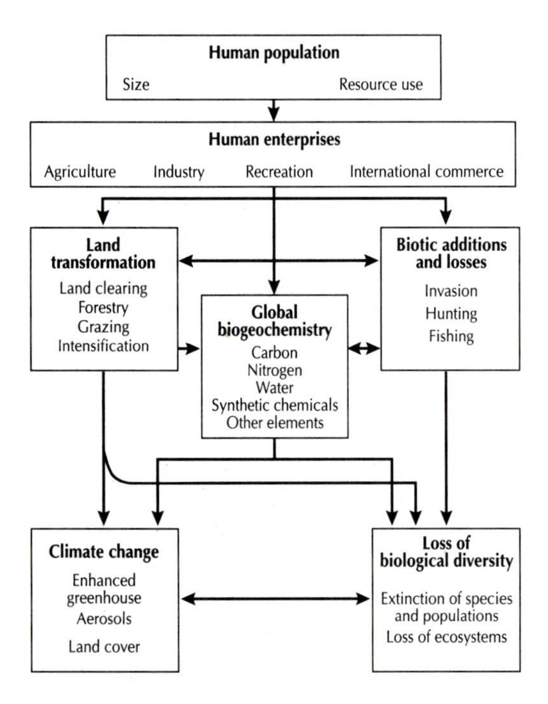

Figure 1. A conceptual model illustrating humanity's direct and indirect effects on the Earth system [modified from (56)].

the structure and functioning of ecosystems, and it alters how ecosystems interact with the atmosphere, with aquatic systems, and with surrounding land. Moreover, land transformation interacts strongly with most other components of global environmental change.

The measurement of land transformation on a global scale is challenging; changes can be measured more or less straightforwardly at a given site, but it is difficult to aggregate these changes regionally and globally. In contrast to analyses of human alteration of the global carbon cycle, we cannot install instruments on a tropical mountain to collect evidence of land transformation. Remote sensing is a most useful technique, but only recently has there been a serious scientific effort to use high-resolution civilian satellite imagery to evaluate even the more visible forms of land transformation, such as deforestation, on continental to global scales (3).

Land transformation encompasses a wide variety of activities that vary substantially in their intensity and consequences. At one extreme, 10 to 15% of Earth's land surface is occupied by row crop agriculture or by urban-industrial areas, and another 6 to 8% has been converted to pastureland (4); these systems are wholly changed by human activity. At the other extreme, every terrestrial ecosystem is affected by increased atmospheric carbon dioxide (CO_2), and most ecosystems have a history of hunting and other low-intensity resource extraction. Between these extremes lie grassland and semiarid ecosystems that are grazed (and sometimes degraded) by domestic animals, and forests and woodlands from which wood products have been harvested; together, these represent the majority of Earth's vegetated surface.

The variety of human effects on land makes any attempt to summarize land transformations globally a matter of semantics as well as substantial uncertainty. Estimates of the fraction of land transformed or degraded by humanity (or its corollary, the fraction of the land's biological production that is used or dominated) fall in the range of 39 to 50% (5) (Figure 2). These numbers have large uncertainties, but the fact that they are large is not at all uncertain. Moreover, if anything these estimates understate the global impact of land transformation, in that land that has not been transformed often has been divided into fragments by human alteration of the surrounding areas. This fragmentation affects the species composition and functioning of otherwise little modified ecosystems (6).

Overall, land transformation represents the primary driving force in the loss of biologi-

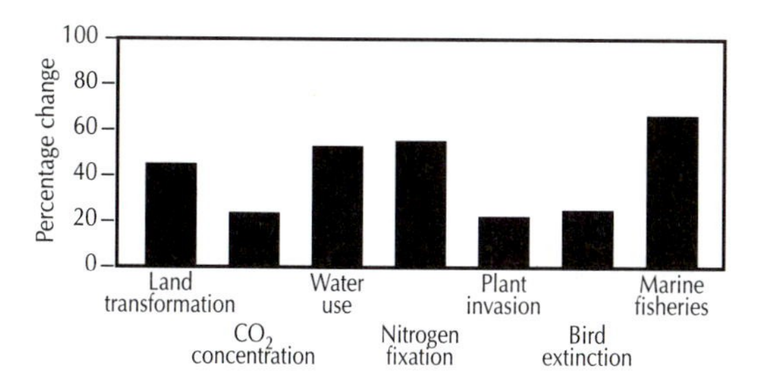

Figure 2. Human dominance or alteration of several major components of the Earth system, expressed as (from left to right) percentage of the land surface transformed (5); percentage of the current atmospheric CO_2 concentration that results from human action (17); percentage of accessible surface fresh water used (20); percentage of terrestrial N fixation that is human-caused (28); percentage of plant species in Canada that humanity has introduced from elsewhere (48); percentage of bird species on Earth that have become extinct in the past two millennia, almost all of them as a consequence of human activity (42); and percentage of major marine fisheries that are fully exploited, overexploited, or depleted (14).

cal diversity worldwide. Moreover, the effects of land transformation extend far beyond the boundaries of transformed lands. Land transformation can affect climate directly at local and even regional scales. It contributes ~20% to current anthropogenic CO_2 emissions, and more substantially to the increasing concentrations of the greenhouse gases methane and nitrous oxide; fires associated with it alter the reactive chemistry of the troposphere, bringing elevated carbon monoxide concentrations and episodes of urban-like photochemical air pollution to remote tropical areas of Africa and South America, and it causes runoff of sediment and nutrients that drive substantial changes in stream, lake, estuarine, and coral reef ecosystems (7–10).

The central importance of land transformation is well recognized within the community of researchers concerned with global environmental change. Several research programs are focused on aspects of it (9, 11); recent and substantial progress toward understanding these aspects has been made (3), and much more progress can be anticipated. Understanding land transformation is a difficult challenge; it requires integrating the social, economic, and cultural causes of land transformation with evaluations of its biophysical nature and consequences. This interdisciplinary approach is essential to predicting the course, and to any hope of affecting the consequences, of human-caused land transformation.

Oceans

Human alterations of marine ecosystems are more difficult to quantify than those of terrestrial ecosystems, but several kinds of information suggest that they are substantial. The human population is concentrated near coasts—about 60% within 100 km—and the oceans' productive coastal margins have been affected strongly by humanity. Coastal wetlands that mediate interactions between land and sea have been altered over large areas; for example, approximately 50% of mangrove ecosystems globally have been transformed or destroyed by human activity (12). Moreover, a recent analysis suggested that although humans use about 8% of the primary production of the oceans, that fraction grows to more than 25% for upwelling areas and to 35% for temperate continental shelf systems (13).

Many of the fisheries that capture marine productivity are focused on top predators, whose removal can alter marine ecosystems out of proportion to their abundance. Moreover, many such fisheries have proved to be unsustainable, at least at our present level of knowledge and control. As of 1995, 22% of recognized marine fisheries were over-

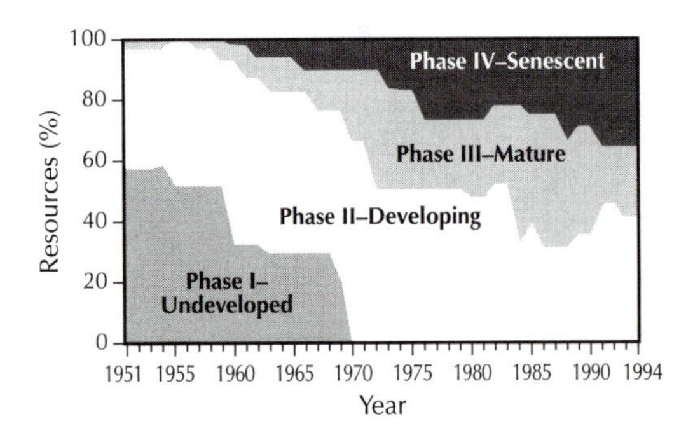

Figure 3. Percentage of major world marine fish resources in different phases of development, 1951 to 1994 [from (57)]. Undeveloped = a low and relatively constant level of catches; developing = rapidly increasing catches; mature = a high and plateauing level of catches; senescent = catches declining from higher levels.

exploited or already depleted, and 44% more were at their limit of exploitation (14) (Figures 2 and 3). The consequences of fisheries are not restricted to their target organisms; commercial marine fisheries around the world discard 27 million tons of nontarget animals annually, a quantity nearly one-third as large as total landings (15). Moreover, the dredges and trawls used in some fisheries damage habitats substantially as they are dragged along the sea floor.

A recent increase in the frequency, extent, and duration of harmful algal blooms in coastal areas (16) suggests that human activity has affected the base as well as the top of marine food chains. Harmful algal blooms are sudden increases in the abundance of marine phytoplankton that produce harmful structures or chemicals. Some but not all of these phytoplankton are strongly pigmented (red or brown tides). Algal blooms usually are correlated with changes in temperature, nutrients, or salinity; nutrients in coastal waters, in particular, are much modified by human activity. Algal blooms can cause extensive fish kills through toxins and by causing anoxia; they also lead to paralytic shellfish poisoning and amnesic shellfish poisoning in humans. Although the existence of harmful algal blooms has long been recognized, they have spread widely in the past two decades (16).

Alterations of the Biogeochemical Cycles

Carbon

Life on Earth is based on carbon and the CO_2 in the atmosphere is the primary resource for photosynthesis. Humanity adds CO_2 to the atmosphere by mining and burning fossil fuels, the residue of life from the distant past, and by converting forests and grasslands to agricultural and other low-biomass ecosystems. The net result of both activities is that organic carbon from rocks, organisms, and soils is released into the atmosphere as CO_2.

The modern increase in CO_2 represents the clearest and best documented signal of human alteration of the Earth system. Thanks to the foresight of Roger Revelle, Charles Keeling, and others who initiated careful and systematic measurements of atmospheric CO_2 in 1957 and sustained them through budget crises and changes in scientific fashions, we have observed the concentration of CO_2 as it has increased steadily from 315 ppm to 362 ppm. Analysis of air bubbles extracted from the Antarctic and Greenland ice caps extends the record back much further; the CO_2 concentration was more or less stable near 280 ppm for thousands of years until about 1800, and has increased exponentially since then (17).

There is no doubt that this increase has been driven by human activity, today primarily by fossil fuel combustion. The sources of CO_2 can be traced isotopically; before the period of extensive nuclear testing in the atmosphere, carbon depleted in ^{14}C was a specific tracer of CO_2 derived from fossil fuel combustion, whereas carbon depleted in ^{13}C characterized CO_2 from both fossil fuels and land transformation. Direct measurements in the atmosphere, and analyses of carbon isotopes in tree rings, show that both ^{13}C and ^{14}C in CO_2 were diluted in the atmosphere relative to ^{12}C as the CO_2 concentration in the atmosphere increased.

Fossil fuel combustion now adds 5.5 ± 0.5 billion metric tons of CO_2–C to the atmosphere annually, mostly in economically developed regions of the temperate zone (18) (Figure 4). The annual accumulation of CO_2–C has averaged 3.2 ± 0.2 billion metric tons recently (17). The other major terms in the atmospheric carbon balance are net ocean-atmosphere flux, net release of carbon during land transformation, and net storage in terrestrial biomass and soil organic matter. All of these terms are smaller and less certain than fossil fuel combustion or annual atmospheric accumulation; they represent rich areas of current research, analysis, and sometimes contention.

The human-caused increase in atmospheric CO_2 already represents nearly a 30% change relative to the pre-industrial era (Figure 2), and CO_2 will continue to increase for the foreseeable future. Increased CO_2 represents the most important human enhancement to the greenhouse effect; the consensus of the climate research community is that it probably already affects climate detectably and will drive substantial climate change in the next century (1). The direct effects of increased CO_2 on plants and ecosystems may be even more important. The growth of most plants is en-

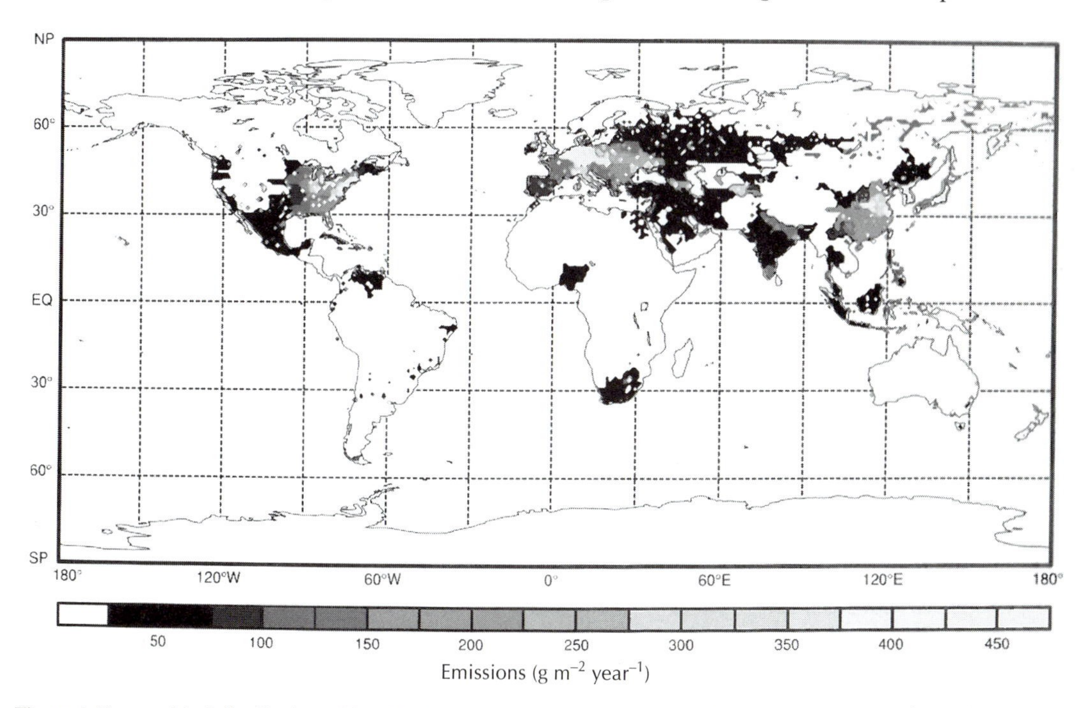

Figure 4. Geographical distribution of fossil fuel sources of CO_2 as of 1990. The global mean is 12.2 g m^{-2} year^{-1}; most emissions occur in economically developed regions of the north temperate zone. EQ, equator; NP, North Pole; SP, South Pole. [Prepared by A. S. Denning, from information in (18)] [Editor's note: see color image in original publication.]

hanced by elevated CO_2, but to very different extents; the tissue chemistry of plants that respond to CO_2 is altered in ways that decrease food quality for animals and microbes; and the water use efficiency of plants and ecosystems generally is increased. The fact that increased CO_2 affects species differentially means that it is likely to drive substantial changes in the species composition and dynamics of all terrestrial ecosystems (19).

Water

Water is essential to all life. Its movement by gravity, and through evaporation and condensation, contributes to driving Earth's biogeochemical cycles and to controlling its climate. Very little of the water on Earth is directly usable by humans; most is either saline or frozen. Globally, humanity now uses more than half of the runoff water that is fresh and reasonably accessible, with about 70% of this use in agriculture (20) (Figure 2). To meet increasing demands for the limited supply of fresh water, humanity has extensively altered river systems through diversions and impoundments. In the United States only 2% of the rivers run unimpeded, and by the end of this century the flow of about two-thirds of all of Earth's rivers will be regulated (21). At present, as much as 6% of Earth's river runoff is evaporated as a consequence of human manipulations (22). Major rivers, including the Colorado, the Nile, and the Ganges, are used so extensively that little water reaches the sea. Massive inland water bodies, including the Aral Sea and Lake Chad, have been greatly reduced in extent by water diversions for agriculture. Reduction in the volume of the Aral Sea resulted in the demise of native fishes and the loss of other biota; the loss of a major fishery; exposure of the salt-laden sea bottom, thereby providing a major source of wind-blown dust; the production of a drier and more continental local climate and a decrease in water quality in the general region; and an increase in human diseases (23).

Impounding and impeding the flow of rivers provides reservoirs of water that can be used for energy generation as well as for agriculture. Waterways also are managed for transport, for flood control, and for the dilution of chemical wastes. Together, these activities have altered Earth's freshwater ecosystems profoundly, to a greater extent than terrestrial ecosystems have been altered. The construction of dams affects biotic habitats indirectly as well; the damming of the Danube River, for example, has altered the silica chemistry of the entire Black Sea. The large number of operational dams (36,000) in the world, in conjunction with the many that are planned, ensure that humanity's effects on aquatic biological systems will continue (24). Where surface water is sparse or overexploited, humans use groundwater—and in many areas the groundwater that is drawn upon is nonrenewable, or fossil, water (25). For example, three-quarters of the water supply of Saudi Arabia currently comes from fossil water (26).

Alterations to the hydrological cycle can affect regional climate. Irrigation increases atmospheric humidity in semiarid areas, often increasing precipitation and thunderstorm frequency (27). In contrast, land transformation from forest to agriculture or pasture increases albedo and decreases surface roughness; simulations suggest that the net effect of this transformation is to increase temperature and decrease precipitation regionally (7, 26).

Conflicts arising from the global use of water will be exacerbated in the years ahead, with a growing human population and with the stresses that global changes will impose on water quality and availability. Of all of the environmental security issues facing nations, an adequate supply of clean water will be the most important.

Nitrogen

Nitrogen (N) is unique among the major elements required for life, in that its cycle includes a vast atmospheric reservoir (N_2) that must be fixed (combined with carbon, hydrogen, or oxygen) before it can be used by most organisms. The supply of this fixed N controls (at least in part) the productivity, carbon storage, and species composition of many ecosystems. Before the extensive human alteration of the N cycle, 90 to 130 million metric tons of N (Tg N) were fixed biologically on land each year; rates of biological fixation in marine systems are less certain, but perhaps as much was fixed there (28).

Human activity has altered the global cycle of N substantially by fixing N_2—deliberately for fertilizer and inadvertently during fossil fuel combustion. Industrial fixation of N fertilizer increased from <10 Tg/year in 1950 to 80 Tg/year in 1990; after a brief dip caused by economic dislocations in the former Soviet Union, it is expected to increase to >135 Tg/year by 2030 (29). Cultivation of soybeans, alfalfa, and other legume crops that fix N symbiotically enhances fixation by another ~40 Tg/year, and fossil fuel combustion puts >20 Tg/year of reactive N into the atmosphere globally—some by fixing N_2, more from the mobilization of N in the fuel. Overall, human activity adds at least as much fixed N to terrestrial ecosystems as do all natural sources combined (Figure 2), and it mobilizes >50 Tg/year more during land transformation (28, 30).

Alteration of the N cycle has multiple consequences. In the atmosphere, these include (i) an increasing concentration of the greenhouse gas nitrous oxide globally; (ii) substantial increases in fluxes of reactive N gases (two-thirds or more of both nitric oxide and ammonia emissions globally are human-caused); and (iii) a substantial contribution to acid rain and to the photochemical smog that afflicts urban and agricultural areas throughout the world (31). Reactive N that is emitted to the atmosphere is deposited downwind, where it can influence the dynamics of recipient ecosystems. In regions where fixed N was in short supply, added N generally increases productivity and C storage within ecosystems, and ultimately increases losses of N and cations from soils, in a set of processes termed "N saturation" (32). Where added N increases the productivity of ecosystems, usually it also decreases their biological diversity (33).

Human-fixed N also can move from agriculture, from sewage systems, and from N-saturated terrestrial systems to streams, rivers, groundwater, and ultimately the oceans. Fluxes of N through streams and rivers have increased markedly as human alteration of the N cycle has accelerated; river nitrate is highly correlated with the human population of river basins and with the sum of human-caused N inputs to those basins (8). Increases in river N drive the eutrophication of most estuaries, causing blooms of nuisance and even toxic algae, and threatening the sustainability of marine fisheries (16, 34).

Other Cycles

The cycles of carbon, water, and nitrogen are not alone in being altered by human activity. Humanity is also the largest source of oxidized sulfur gases in the atmosphere; these affect regional air quality, biogeochemistry, and climate. Moreover, mining and mobilization of phosphorus and of many metals exceed their natural fluxes; some of the metals that are concentrated and mobilized are highly toxic (including lead, cadmium, and mercury) (35). Beyond any doubt, humanity is a major biogeochemical force on Earth.

Synthetic Organic Chemicals

Synthetic organic chemicals have brought humanity many beneficial services. However, many are toxic to humans and other species, and some are hazardous in concentrations as low as 1 part per billion. Many chemicals persist in the environment for decades; some are both toxic and persistent. Long-lived organochlorine compounds provide the clearest examples of environmental consequences of persistent compounds. Insecticides such as DDT and its relatives, and industrial compounds like polychlorinated biphenyls (PCBs), were used widely in North America in the 1950s and 1960s. They were transported globally, accumulated in organisms, and magnified in concentration through food chains; they devastated populations of some predators (notably falcons and eagles) and entered parts of the human food supply in concentrations higher than was prudent. Domestic use of these compounds was phased out in the 1970s in the United States and Canada, and their concentrations declined thereafter. However, PCBs in particular remain readily detectable in many organisms, sometimes approaching thresholds of public health concern (36). They will continue to circulate through organisms for many decades.

Synthetic chemicals need not be toxic to cause environmental problems. The fact that the persistent and volatile chlorofluorocarbons (CFCs) are wholly nontoxic contributed to their widespread use as refrigerants and even aerosol propellants. The subsequent discovery that CFCs drive the breakdown of stratospheric ozone, and especially the later discovery of the Antarctic ozone hole and their role in it, represent great surprises in global environmental science (37). Moreover, the response of the international political system to those discoveries is the best extant illustration that global environmental change can be dealt with effectively (38).

Particular compounds that pose serious health and environmental threats can be and often have been phased out (although PCB production is growing in Asia). Nonetheless, each year the chemical industry produces more than 100 million tons of organic chemicals representing some 70,000 different compounds, with about 1000 new ones being added annually (39). Only a small fraction of the many chemicals produced and released into the environment are tested adequately for health hazards or environmental impact (40).

Biotic Changes

Human modification of Earth's biological resources—its species and genetically distinct populations—is substantial and growing. Extinction is a natural process, but the current rate of loss of genetic variability, of populations, and of species is far above background rates; it is ongoing; and it represents a wholly irreversible global change. At the same time, human transport of species around Earth is homogenizing Earth's biota, introducing many species into new areas where they can disrupt both natural and human systems.

Losses

Rates of extinction are difficult to determine globally, in part because the majority of species on Earth have not yet been identified.

Nevertheless, recent calculations suggest that rates of species extinction are now on the order of 100 to 1000 times those before humanity's dominance of Earth (41). For particular well-known groups, rates of loss are even greater; as many as one-quarter of Earth's bird species have been driven to extinction by human activities over the past two millennia, particularly on oceanic islands (42) (Figure 2). At present, 11% of the remaining birds, 18% of the mammals, 5% of fish, and 8% of plant species on Earth are threatened with extinction (43). There has been a disproportionate loss of large mammal species because of hunting; these species played a dominant role in many ecosystems, and their loss has resulted in a fundamental change in the dynamics of those systems (44), one that could lead to further extinctions. The largest organisms in marine systems have been affected similarly, by fishing and whaling. Land transformation is the single most important cause of extinction, and current rates of land transformation eventually will drive many more species to extinction, although with a time lag that masks the true dimensions of the crisis (45). Moreover, the effects of other components of global environmental change—of altered carbon and nitrogen cycles, and of anthropogenic climate change—are just beginning.

As high as they are, these losses of species understate the magnitude of loss of genetic variation. The loss to land transformation of locally adapted populations within species, and of genetic material within populations, is a human-caused change that reduces the resilience of species and ecosystems while precluding human use of the library of natural products and genetic material that they represent (46).

Although conservation efforts focused on individual endangered species have yielded some successes, they are expensive—and the protection or restoration of whole ecosystems often represents the most effective way to sustain genetic, population, and species diversity. Moreover, ecosystems themselves may play important roles in both natural and human-dominated landscapes. For example, mangrove ecosystems protect coastal areas from erosion and provide nurseries for offshore fisheries, but they are threatened by land transformation in many areas.

Invasions

In addition to extinction, humanity has caused a rearrangement of Earth's biotic systems, through the mixing of floras and faunas that had long been isolated geographically. The magnitude of transport of species, termed "biological invasion," is enormous (47); invading species are present almost everywhere. On many islands, more than half of the plant species are nonindigenous, and in many continental areas the figure is 20% or more (48) (Figure 2).

As with extinction, biological invasion occurs naturally—and as with extinction, human activity has accelerated its rate by orders of magnitude. Land transformation interacts strongly with biological invasion, in that human-altered ecosystems generally provide the primary foci for invasions, while in some cases land transformation itself is driven by biological invasions (49). International commerce is also a primary cause of the breakdown of biogeographic barriers; trade in live organisms is massive and global, and many other organisms are inadvertently taken along for the ride. In freshwater systems, the combination of upstream land transformation, altered hydrology, and numerous deliberate and accidental species introductions has led to particularly widespread invasion, in continental as well as island ecosystems (50).

In some regions, invasions are becoming more frequent. For example, in the San Francisco Bay of California, an average of one new species has been established every 36 weeks since 1850, every 24 weeks since 1970, and every 12 weeks for the last decade (51). Some introduced species quickly become invasive over large areas (for example, the Asian clam in the San Francisco Bay), whereas others become widespread only after a lag of decades, or even over a century (52).

Many biological invasions are effectively irreversible; once replicating biological material is released into the environment and becomes successful there, calling it back is difficult and expensive at best. Moreover, some species introductions have consequences. Some degrade human health and that of other species; after all, most infectious diseases are invaders over most of their range. Others have caused economic losses amounting to billions of dollars; the recent invasion of North America by the zebra mussel is a well-publicized example. Some disrupt ecosystem processes, altering the structure and functioning of whole ecosystems. Finally, some invasions drive losses in the biological diversity of native species and populations; after land transformation, they are the next most important cause of extinction (53).

Conclusions

The global consequences of human activity are not something to face in the future—as Figure 2 illustrates, they are with us now. All of these changes are ongoing, and in many cases accelerating; many of them were entrained long before their importance was recognized. Moreover, all of these seemingly disparate phenomena trace to a single cause—the growing scale of the human enterprise. The rates, scales, kinds, and combinations of changes occurring now are fundamentally different from those at any other time in history; we are changing Earth more rapidly than we are understanding it. We live on a human-dominated planet— and the momentum of human population growth, together with the imperative for further economic development in most of the world, ensures that our dominance will increase.

The papers in this special section summarize our knowledge of and provide specific policy recommendations concerning major human-dominated ecosystems. In addition, we suggest that the rate and extent of human alteration of Earth should affect how we think about Earth. It is clear that we control much of Earth. and that our activities affect the rest. In a very real sense, the world is in our hands—and how we handle it will determine its composition and dynamics, and our fate.

Recognition of the global consequences of the human enterprise suggests three complementary directions. First, we can work to reduce the rate at which we alter the Earth system. Humans and human-dominated systems may be able to adapt to slower change, and ecosystems and the species they support may cope more effectively with the changes we impose, if those changes are slow. Our footprint on the planet (54) might then be stabilized at a point where enough space and resources remain to sustain most of the other species on Earth, for their sake and our own. Reducing the rate of growth in human effects on Earth involves slowing human population growth and using resources as efficiently as is practical. Often it is the waste products and by-products of human activity that drive global environmental change.

Second, we can accelerate our efforts to understand Earth's ecosystems and how they interact with the numerous components of human-caused global change. Ecological research is inherently complex and demanding: It requires measurement and monitoring of populations and ecosystems; experimental studies to elucidate the regulation of ecological processes; the development, testing, and validation of regional and global models; and integration with a broad range of biological, earth, atmospheric, and marine sciences. The challenge of understanding a human-dominated planet further requires that the human dimensions of global change—the social, economic, cultural, and other drivers of human actions—be included within our analyses.

Finally, humanity's dominance of Earth means that we cannot escape responsibility for managing the planet. Our activities are causing rapid, novel, and substantial changes to Earth's ecosystems. Maintaining populations, species, and ecosystems in the face of those changes, and maintaining the flow of goods and services they provide humanity (55), will require active management for the foreseeable future. There is no clearer illustration of the extent of human dominance of Earth than the fact that maintaining the diversity of "wild" species and the functioning of "wild" ecosystems will require increasing human involvement.

References and Notes

1. Intergovernmental Panel on Climate Change. 1996. *Climate Change* 1995. Cambridge Univ. Press, Cambridge. Pp. 9–49.
2. United Nations Environment Program. 1995. *Global Biodiversity Assessment*. V. H. Heywood, Ed. Cambridge Univ. Press, Cambridge.
3. Skole, D. and C. J. Tucker. 1993. Science 260:1905.
4. Olson, J. S., J. A. Watts and L. J. Allison. 1983. *Carbon in Live Vegetation of Major World Ecosystems* (Office of Energy Research, U.S. Department of Energy, Washington, DC).
5. Vitousek, P. M., P. R. Ehrlich, A. H. Ehrlich and P. A. Matson. 1986. Bioscience 36, 368; Kates, R. W., B. L. Turner and W. C. Clark. 1995. In: (35), pp. 1–17; G. C. Daily, Science 269:350.
6. Saunders, D. A., R. J. Hobbs and C. R. Margules. 1991. Conserv. Biol. 5:18.
7. Shukla, J., C. Nobre and P. Sellers. 1990. Science 247: 1322.
8. Howarth, R. W., et al. 1996. Biogeochemistry 35:75.
9. Meyer, W. S. and B. L. Turner II. 1994. Changes in Land Use and Land Cover A Global Perspective. Cambridge Univ. Press, Cambridge.
10. Carpenter, S. R., S. G. Fisher, N. B. Grimm and J. F. Kitchell. 1992. Annu. Rev. Ecol. Syst. 23,119; Smith, S. V. and R. W. Buddemeier, ibid., p. 89; Melillo, J. M., L. C. Prentice, G. D. Farquhar, E. -D. Schuke and O. E. Sala, in (1). Pp. 449–481.
11. Leemans, R. and G. Zuidema. 1995. Trends Ecol. Evol. 10:76.
12. World Resources Institute. 1996. *World Resources 1996–1997*. Oxford Univ. Press, New York.
13. Pauly, D. and V. Christensen. 1995. Nature 374:257.
14. Food and Agricultural Organization (FAO). 1994. FAO Fisheries Tech. Pap. 335.
15. Alverson, D. L., M. H. Freeberg, S. A. Murawski and J. G. Pope. 1994. FAO Fisheries Tech. Pap. 339.
16. O. M. Hallegraeff. 1993. Phycologia 32:79.
17. Schimel, D. S., et al. 1995. In: *Climate Change 1994: Radiative Forcing of Climate Change*. J. T. Houghton et al., Eds. Cambridge Univ. Press, Cambridge. Pp. 39–71.
18. Andres, R. J., G. Marland, L. Y. Fung and E. Matthews. 1996. Global Biogeochem. Cycles 10:419.
19. Koch, G. W. and H. A Mooney. 1996. *Carbon Dioxide and Terrestrial Ecosystems*. Academic Press, San Diego, CA; Komer, C. and F. A. Bazzaz. 1996. *Carbon Dioxide, Populations, and Communities*. Academic Press, San Diego, CA.
20. Postel, S. L., G. C. Daily and P. R. Ehrlich. 1996. Science 271:785.
21. Abramovitz, J. N. 1996. Imperiled Waters, Impoverished Future: The Decline of Freshwater Ecosystems. Worldwatch Institute, Washington, DC.
22. L'vovich, M. I. and G. F. White. 1994. In: (35), pp. 235–252; M. Dynesius and C. Nilsson, Science 266:753.
23. Micklin, P. 1988. Science 241:1170; Kotlyakov, V. 1991. Environment 33:4.
24. Humborg, C., V. Ittekkot, A. Cociasu and B. Bodungen. 1977. Nature 386:385.
25. Gleick, P. H., Ed., 1993. *Water in Crisis*. Oxford Univ. Press, New York.
26. Gornitz, V., C. Rosenzweig, D. Hillel. 1997. Global Planet Change 14:147.
27. Milly, P. C. and K. A. Dunne. 1994. J. Clim. 7:506.
28. Galloway, J. N., W. H. Schlesinger, H. Levy II, A.

Michaels and J. L. Schnoor. 1955. Global Biogeochem. Cycles 9:235.
29. Galloway, J. N., H. Levy II, P. S. Kasibhatta. 1994. Ambio 23:120.
30. Smil, V. In: (35), pp. 423–436.
31. Vitousek, P. M. et al. Ecol. Appl., in press.
32. Aber, J. D., J. M. Melillo, K. J. Nadelhoffer, J. Pastor and R. D. Boone. 1991. ibid. 1:303.
33. Tilman, D. 1987. Ecol. Monogr. 57:189.
34. Nixon, S. W. et al. 1996. Biogeochemistry 35:141.
35. Tumer, B. L. II, et al. 1990. Eds. *The Earth As Transformed by Human Action.* Cambridge Univ. Press, Cambridge.
36. Stow, C. A., S. R. Carpenter, C. P. Madenjian, L. A. Eby and L. J. Jackson. 1995. Bioscience 45:752.
37. Rowland, F. S. 1989. Am. Sci 77:36; Solomon, S. 1990. Nature 347:347.
38. Tolba, M. K. et al. 1992. Eds. *The World Environment 1972–1992.* Chapman & Hall, London.
39. Postel, S. 1987. Defusing the Toxics Threat: Controlling Pesticides and Industrial Waste. Worldwatch Institute, Washington, DC.
40. United Nations Environment Program (UNEP). 1992. Saving Our Planet—Challenges and Hopes (UNEP, Nairobi.
41. Lawton, J. H. and R. M. May, Eds. 1995. *Extinction Rates.* Oxford Univ. Press, Oxford; Pimm, S. L, G. J. Russell, J. L. Gimeman and T. Brooks. 1995. Science 269:347.
42. Olson, S. L. 1989. In: *Conservation for the Twenty-First Century.* D. Western and M. C. Pearl, Eds. Oxford Univ. Press, Oxford, p.50; Steadman, D. W. 1995. Science 267:1123.
43. Barbault, R. and S. Sastrapradja. In: (2), pp. 193–274.
44. Dirzo, R. and A. Miranda. 1991. In: *Plant-Animal Interactions.* P. W. Price, T. M. Lewinsohn, W. Fernandes, W. W. Benson, Eds. Wiley Interscience, New York. P. 273.
45. Tilman, D., R. M. May, C. Lehman and M. A. Nowak. 1994. Nature 371:65.
46. Mooney, H. A., J. Lubchenco, R. Dirzo and O. E. Sala, in (2), pp. 279–325.
47. Elton, C. 1958. *The Ecology of Invasions by Animals and Plants.* Methuen, London; Drake, J. A. et al. 1989. Eds. *Biological Invasions. A Global Perspective.* Wiley, Chichester, UK.
48. Reimanek, M. and J. Randall. 1994. Madrono 41:161.
49. D'Antonio, C. M. and P. M. Vitousek. 1992. Annul Rev. Ecol. Syst. 23:63.
50. Lodge, D. M. 1993. Trends Ecol. Evol. 8:133.
51. Cohen, A. N. and J. T. Carlton. 1995. Biological Study: Nonindigenous Aquatic Species in a United States Estuary: A Case Study of the Biological Invasions of the San Francisco Bay and Delta. U.S. Fish and Wildlife Service, Washington, DC.
52. Kowarik, I. 1995. In: *Plant Invasions-General Aspects and Special Problems.* P. Pysek, K. Prach, M. Rejmánek, M. Wade, Eds. SPB Academic, Amsterdam. P.15.
53. Vitousek, P. M., C. M. D'Antonio, L. L. Loope and R. Westbrooks. 1996. Am. Sci. 84:468.
54. Rees, W. E. and M. Wackemagel. 1994. In: *Investing in Natural Capital: The Ecological Economics Approach to Sustainability.* A M. Jansson, M. Hammer, C. Folke, R. Costanza, Eds. (Island, Washington, DC.
55. Daily, G. C. Ed. 1997. Nature's Services (Island, Washington, DC.
56. Lubchenco, J., et al. 1991. Ecology 72:371; Vitousek, P. M. 1994. ibid. 75, 1861.
57. Garcia, S. M. and R. Grainger. 1996. FAO Fisheries Tech Pap. 359.
58. We thank G. C. Daily, C. B. Field, S. Hobbie, D. Gordon, P. A. Matson, and R. L. Naylor for constructive comments on this paper, A. S. Denning and S. M. Garcia for assistance with illustrations, and C. Nakashima and B. Lilley for preparing text and figures for publication.

Readings for

Physiological Ecology

Tradeoffs for Individuals

James F. Kitchell

Introduction to

William C. Dennison, R. J. Orth, K. A. Moore, et al.
Assessing Water Quality with Submersed Aquatic Vegetation

by James F. Kitchell

This paper presents an example of the value in using the ecological niche as a conceptual framework. The relevance is substantial because submersed aquatic vegetation has declined dramatically in many estuaries and these plants play a vital role as habitat and food for a variety of animals (fishes, shellfish, etc.). Scientists have argued at length about the importance of each of many variables that may be responsible for the decline. Dennison's synthesis provides a means for rising above the squabble and developing an integrated view of the problem.

For plants, light is the most fundamental environmental variable. Light availability can be altered by a host of factors, many of which are influenced by the actions of humans on land and upstream. Nearly all of those can be related to water clarity which, of course, governs light penetration and the growth of vegetation. The authors demonstrate that a field census of maximum depth for healthy vegetation serves as a integrated ecological monitoring tool indicative of such changes as soil erosion from agricultural and urban areas, efficiency of sewage treatment, loading of chemical pollutants, etc. In other words, the plants serve as sentinels of environmental changes that have reduced or increased the dimensions of their realized niche.

Assessing Water Quality with Submersed Aquatic Vegetation

Habitat requirements as barometers of Chesapeake Bay health

WILLIAM C. DENNISON, ROBERT J. ORTH, KENNETH A. MOORE, J. COURT STEVENSON, VIRGINIA CARTER, STAN KOLLAR, PETER W. BERGSTROM, AND RICHARD A. BATIUK

By establishing habitat requirements, scientists, managers, and the public can work toward a clean, healthy bay.

Dennison, W. C., R. J. Orth, K. A. Moore, J. C. Stevenson, V. Carter, S. Kollar, P. W. Bergstrom and R. A. Batiuk. 1993. Assessing water quality with submersed aquatic vegetation. *BioScience* 43(2):86–94.

Worldwide, estuaries are experiencing water quality problems as a result of human population growth in coastal areas. Chesapeake Bay, one of the world's largest estuaries, has experienced deterioration of water quality from nutrient enrichment, sediment inputs, and high levels of contaminants, resulting in anoxic or hypoxic conditions and declines in living resources (Horton and Eichbaum 1991). A mechanism for relating anthropogenic inputs to the health of Chesapeake Bay is through determination of relationships among water quality and various living resources. In particular, the establishment of habitat requirements and restoration goals for critical species living in Chesapeake Bay is a way in which scientists, resource managers, politicians and the public can work toward the goal of obtaining a clean, healthy bay (Funderburk et al. 1991).

We use habitat requirements of submersed aquatic vegetation to characterize the water quality of Chesapeake Bay because of their widespread distribution in the bay, important ecological role, and sensitivity to water quality parameters. Our primary goal is to synthesize information leading to the establishment of quantitative levels of relevant water quality parameters necessary to support submersed aquatic vegetation, a major resource of Chesapeake Bay (Batiuk et al. in press). The development of a habitat requirement approach for Chesapeake Bay could prove useful in other estuaries experiencing water quality degradation.

Submersed Aquatic Vegetation

Submersed aquatic vegetation is comprised of rooted flowering plants that have colonized primarily soft sediment habitats in coastal, estuarine, and freshwater habitats. In Chesapeake Bay, seagrasses in saline regions and freshwater angiosperms that have colonized lower-salinity portions of the estuary constitute a diverse (approximately 20 species) community of submersed aquatic vegetation (collectively known as SAV; Hurley 1990). Seagrasses are typically defined as the approximately 60 species of marine angiosperms (den Hartog, 1970); however, representatives of the several hundred species of freshwater macrophytes are often found in estuarine habitats (Hutchinson 1975). For the purpose of this article, the term submersed aquatic vegetation is used for both marine angiosperms and freshwater macrophytes that are found in Chesapeake Bay. These plants historically have been one of the major factors contributing to the high productivity of Chesapeake Bay (Kemp et al. 1984), especially the abundance of waterfowl.

During the last two decades, there has been an increasing recognition of the importance of submersed aquatic vegetation in coastal and estuarine ecosystems. Submersed aquatic vegetation provides food for waterfowl and critical habitat for shellfish and finfish. This vegetation also affects nutrient cycling, sediment stability, and water turbidity (reviewed in Larkum et al. 1990, McRoy and Helfferich 1977, Phillips and McRoy 1980). However, declines of submersed aquatic vegetation are being documented worldwide (Europe: Giesen et al. 1990; North America: Costa 1988, Orth and Moore 1983; Australia: Cambridge and McComb 1984) because of anthropogenic inputs such as sediments and nutrients that affect the water quality of coastal ecosystems (Thayer et al. 1975).

In Chesapeake Bay, a large-scale decline of submersed aquatic vegetation occurred in the late 1960s and early 1970s (Orth and Moore 1983, 1984). This decline was related to increasing amounts of nutrients and sediments in Chesapeake Bay resulting from development of the bay's shoreline and watershed (Kemp et al. 1983, Twilley et al. 1985). Currently, there are approximately 25,000 ha of submersed aquatic vegetation in Chesapeake Bay (Orth et al. 1991), which is approximately 10% of its historical distribution (Stevenson and Confer 1978).

Submersed aquatic vegetation is particularly crucial as an indicator of water clarity and nutrient levels, because habitat requirements developed for various species of birds, fish, and shellfish in Chesapeake Bay do not incorporate these conditions (Funderburk et al. 1991). Habitat requirements of these other organisms instead focus on chemical parameters (e.g., dissolved oxygen, pH, salinity, toxic compounds, and temperature). Many of the restoration goals for birds, fish, and shellfish involve changes in both environmental quality and management of human harvesting activities. In contrast, submersed aquatic vegetation restoration goals can be linked solely to environmental

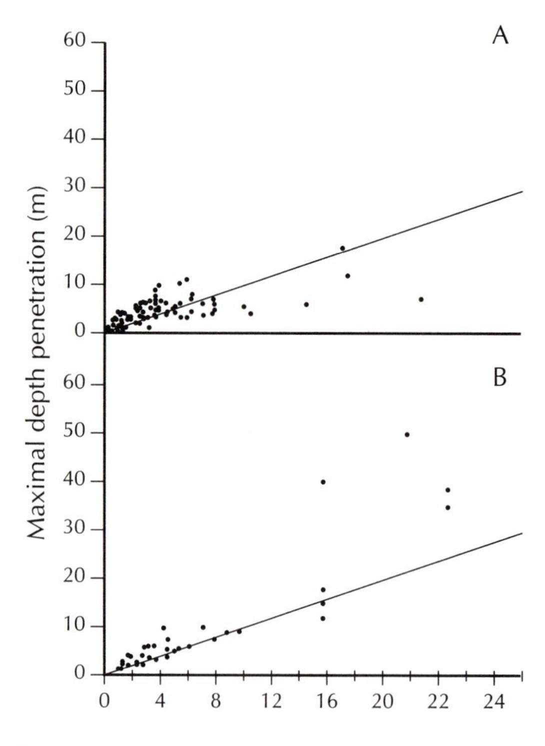

Figure 1. Maximal depth penetration of **A,** freshwater macrophytes and **B,** marine submersed aquatic vegetation plotted as a function of Secchi depth. The 1:1 line of maximal depth penetration and Secchi depth is plotted for reference. Freshwater data from Canfield et al. 1985 and Chambers and Kalff 1985; marine data from references listed in Table 1.

Table 1. Maximal depth limit, light attenuation coefficient (K_d), and minimal light requirements of various species of seagrass. Where Secchi depths were reported, K_d = 1.65/Secchi depth (Giesen et al. 1990). Minimal light requirements were calculated as percent light at the maximal depth limit using $100 \times I_z/I_o = e^{-K_d \cdot Z}$. Range of maximal depth limit and K_d values and means ± SE of minimal light requirement given in locations with multiple data points.

Genus and species	Location	Maximal depth limit (m)	K_d; light attenuation coefficient (m^{-1})	Minimal light requirement (%)
*Amphibolis antaractica**	Waterloo Bay (Australia)	7.0	0.20	24.7
*Cymodocea nodosa**	Ebro Delta (Spain)	4.0	0.57	10.2
*C. nodosa**	Malta	38.5	0.07	7.3
Halodule wrightii†	Florida (US)	1.9	0.93	17.2
Halophila decipiens‡	St. Croix (US)	40.0	0.08	4.4
*H. decipiens**	Northwest Cuba	24.3	0.10	8.8
*Halophila engelmanni**	Northwest Cuba	14.4	0.10	23.7
*Heterozostera tasmanica**	Victoria (Australia)	3.8–9.8	0.36–0.85	5.0 ± 0.6
*H. tasmanica**	Chile	7.0	0.25	17.4
*H. tasmanica**	Spencer Gulf (Australia)	39.0	0.08	4.4
*H. tasmanica**	Waterloo Bay (Australia)	8.0	0.20	20.2
*Posidonia angustifolia**	Waterloo Bay (Australia)	7.0	0.20	24.7
*Posidonia oceanica**	Medas Island (Spain)	15.0	0.17	7.8
*P. oceanica**	Malta	35.0	0.07	9.2
*Posidonia ostenfeldii**	Waterloo Bay (Australia)	7.0	0.20	24.7
*Posidonia sinuosa**	Waterloo Bay (Australia)	7.0	0.20	24.7
*Ruppia maritima**	Brazil	0.7	3.57	8.2
*Syringodium filiforme**	Northwest Cuba	16.5	0.10	19.2
*S. filiforme**	Florida (US)	6.8	0.25	18.3
S. filiforme†	Florida (US)	1.9	0.93	17.2
*Thalassia testudinum**	Northwest Cuba	14.5	0.10	23.5
*T. testudinum**	Puerto Rico	1.0–5.0	0.35–1.50	24.4 ± 4.2
*T. testudinum**	Florida (US)	7.5	0.25	15.3
Zostera marina§	Kattegat (Denmark)	3.7–10.1	0.16–0.36	20.1 ± 2.1
Z. marina□	Roskilde (Denmark)	2.0–5.0	0.32–0.92	19.4 ± 1.3
*Z. marina**	Denmark	1.5–9.0	0.22–1.21	20.6 ± 13.0
*Z. marina**	Woods Hole (US)	6.0	0.28	18.6
*Z. marina**	Netherlands	2.5	0.49	29.4
*Z. marina**	Japan	2.0–5.0	0.38–0.49	18.2 ± 4.5

*Duarte 1991.
†W. J. Kenworthy, personal communication, 1990.
‡Williams and Dennison 1990.
§Ostenfeld 1908.
□Borum 1983.

quality, thus providing for more direct assessment of restoration progress.

The generic nature of submersed aquatic vegetation/light interactions leads to a potential for wider application of submersed aquatic vegetation habitat requirements. Establishment of minimal light requirements for various submersed aquatic vegetation species coupled with water quality monitoring data could be used to establish water clarity and nutrient standards in a variety of coastal environments with the goal of preventing further vegetation declines.

Minimal Light Requirements of Submersed Aquatic Vegetation

Submersed aquatic vegetation requires light for photosynthesis, and its growth, survival, and depth penetration is directly related to light availability (Dennison 1987, Kenworthy and Haunert 1991). The maximal depth at which submersed plants can survive increases with increasing light penetration, as measured with underwater surveys of plant distributions and a Secchi disc (Figure 1). The Secchi depth is the maximal water depth at which a black and white disc (30 centimeter diameter) can be seen from the surface. In spite of the differences between freshwater and marine submersed aquatic vegetation and their habitats (e.g., Stevenson 1988), the general relationships between light availability and depth penetration of submersed aquatic vegetation in various locations are similar in shallow, turbid waters (Secchi depth less than 5 m).

In deeper, clear waters (Secchi depth more than 10 m), a divergence between depth limit and Secchi depth occurs in freshwater versus marine habitats (Duarte 1991). The depth limit for angiosperms in freshwater habitats is 17 m, even at Secchi depths of 23 m (Figure lA). In contrast, submersed aquatic vegetation in marine habitats tend to have depth limits that exceed the higher Secchi depths (Figure 1B). In Chesapeake Bay, Secchi depths are generally 1–2 m, and submersed aquatic vegetation are restricted to shallow water depths (less than 3 m at mean low water [MLW]).

Minimal light requirements for submersed aquatic vegetation are determined from simultaneous measurements of the maximal depth limit for submersed aquatic vegetation and the light attenuation coefficient. A conversion factor between Secchi depth and light attenuation coefficient can be used (Giesen et al. 1990). The percentage of incident light (photosynthetically active radiation [PAR] = 400–700 nm) that corresponds to maximal depth penetration of submersed aquatic vegetation is determined using a negative exponential function according to the Lambert-Beer equation:

$$I_z = I_o \bullet e^{-K_d \bullet Z}$$

where I_z is the PAR light at depth z, I_o is the PAR light just below the water surface, K_d is the light attenuation coefficient, and z is the water depth. Assuming that the minimal light requirement is the light level at the maximal depth penetration (z), percent light can then be determined. In this manner, the average minimal light requirement for freshwater angiosperms from lakes in Canada was determined to be 21.4 ± 2.4% of surface light levels (Chambers and Kalff 1985), and the average minimal light requirement for marine submersed aquatic vegetation was 10.8% (Duarte 1991). However, there is a wide range of minimal light requirements among species (Table 1), likely a result of differences in physiological and morphological adaptations.

Overall, the minimal light requirements of submersed aquatic vegetation (4–29% of incident light measured just below the water surface) are much higher than those of other plants. Terrestrial plants from shade habitats have light requirements on the order of 0.5–2% of incident light measured at the top of the canopy (Hanson et al. 1987, Osmond et al. 1987). Both phytoplankton and benthic algae have minimal light requirements that are significantly less than those of submersed aquatic vegetation charophytes, 2–3% (Sand-Jensen 1988); green algae, 0.05–1.0% (Luning and Dring 1979); brown algae, 0.7–1.5% (Luning and Dring 1979); crustose red algae, 0.0005% (Littler et al. 1985); and lacustrine and marine phytoplankton, 0.5–1.0% (Parsons et al. 1979, Wetzel 1975). Because there is a high minimal light requirement for submersed aquatic vegetation, its survival depends on good water clarity. Therefore, it is important to focus on light attenuation processes to explain the distributions of submersed aquatic vegetation.

The minimal light requirement of a particular species of submersed aquatic vegetation determines the maximal water depth at which it can survive. This relationship is depicted

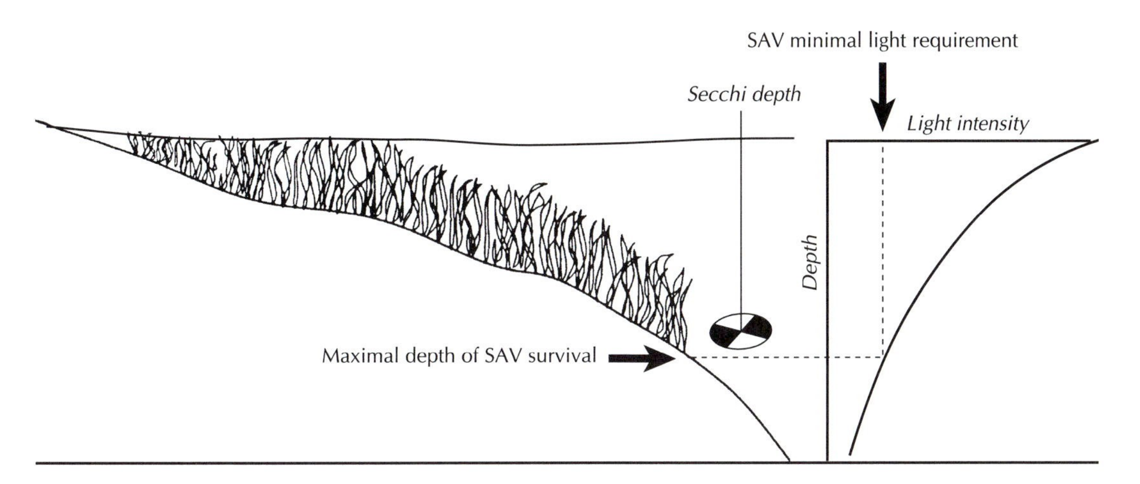

Figure 2. Determination of maximal depth of submersed aquatic vegetation (SAV) survival by the intersection of minimal light requirement and light attenuation curve (% of surface light).

graphically as the intersection of the light intensity versus depth curve with the minimal light requirement (Figure 2). Light intensity is attenuated exponentially with water depth (Figure 2, right side). The minimal light requirement of a particular submersed aquatic vegetation species, as a percentage of incident light, intersects the light curve to give a predicted maximal depth of survival for that species (Figure 2, left side). Light attenuation is temporally and spatially variable, and in the Chesapeake Bay study we used median values taken at monthly intervals during the growing season to characterize light attenuation. Maximal depth limits of submersed aquatic vegetation are less temporally variable, with time intervals of months to years before changes are observed; consequently, annual surveys are generally made.

Minimal light requirements are consistent for each species of submersed aquatic vegetation, with little variation within species (Table 1). The differences in temporal variability of light attenuation, maximal depth limit, and minimal light requirements often results in an imbalance in relative accuracy of these parameters. Knowledge of two of these three unknowns (average light attenuation coefficient [K_d], minimal light requirement, or maximal depth of submersed aquatic vegetation survival) allows determination of the remaining unknown. For example, an assessment of the maximal depth penetration of the seagrass *Zostera marina* with knowledge of its minimal light requirement (Table 1) allows for the determination of an average light attenuation coefficient. In this manner, depth penetration of submersed aquatic vegetation is used as an integrating "light meter" to assess light regimes on the appropriate temporal and spatial scales (with respect to survival) without intensive sampling programs (cf. Kautsky et al. 1986).

Light attenuation within the water column is a function not only of the water itself, but also of its dissolved and particulate components, which serve to reflect, refract, absorb, and scatter the incident radiation (Figure 3). Particulate organic and inorganic particles are washed in from surrounding uplands or resuspended

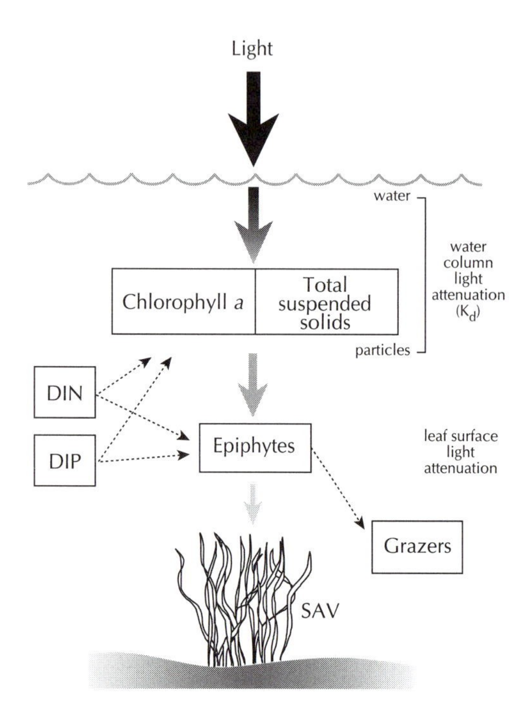

Figure 3. Availability of light for submersed aquatic vegetation (SAV) is determined by light attenuation processes. Water column attenuation, measured as the light attenuation coefficient (K_d), results from absorption and scatter of light by particles in the water (phytoplankton, measured as chlorophyll *a,* and total organic and inorganic particles, measured as total suspended solids) and by absorption of light by water itself. Leaf surface attenuation, largely due to algal epiphytes growing on submersed leaf surfaces, also contributes to light attenuation. Dissolved inorganic nutrients (DIN, dissolved inorganic nitrogen; DIP, dissolved inorganic phosphorus) contribute to phytoplankton and epiphyte components of overall light attenuation, and epiphyte grazers control accumulation of epiphytes.

from bottom deposits and can severely limit light penetration in shallow waters. Inorganic nutrients enhance the growth in the water column of phytoplankton as well as epiphytic algae, which absorb light before it reaches the leaf surface of submersed vegetation. The spectral character of the light may also be changed so that attenuation is greatest in the photosynthetically important blue and red wavelengths of the visible spectrum (Champ et al.1980, Pierce et al. 1986), thereby placing additional stress on submersed aquatic vegetation growth and survival. Thus, light availability is a function of a complex interaction of factors that are directly or indirectly related to water quality, including factors not included in the conceptual model (e.g., dissolved organic matter).

In spite of this complexity, it should be possible to predict submersed aquatic vegetation growth and survival from the known levels of certain key water-quality parameters or, conversely, to predict long-term water-quality levels based on the distribution of submersed aquatic vegetation if the levels of the factors that adversely affect submersed aquatic vegetation are known. This approach does not rely on a complete understanding of the water-quality interactions affecting light attenuation but rather on empirical data on water quality and survival of submersed aquatic vegetation.

Chesapeake Bay Submersed Vegetation

To determine the critical water-quality values that correspond to submersed aquatic vegetation survival, water-quality gradients along subestuaries of Chesapeake Bay were compared to patterns of transplant survival and distributions of submersed aquatic vegetation. We base our analysis of these relationships on case studies of different regions of Chesapeake Bay, by different investigators over several years. Four study areas were used: upper Chesapeake Bay, upper Potomac River, Choptank River, and York River (Figure 4). These areas represent regions of intensive study in the past decade where data on water quality and submersed aquatic vegetation growth were available. The areas span the salinity regimes of

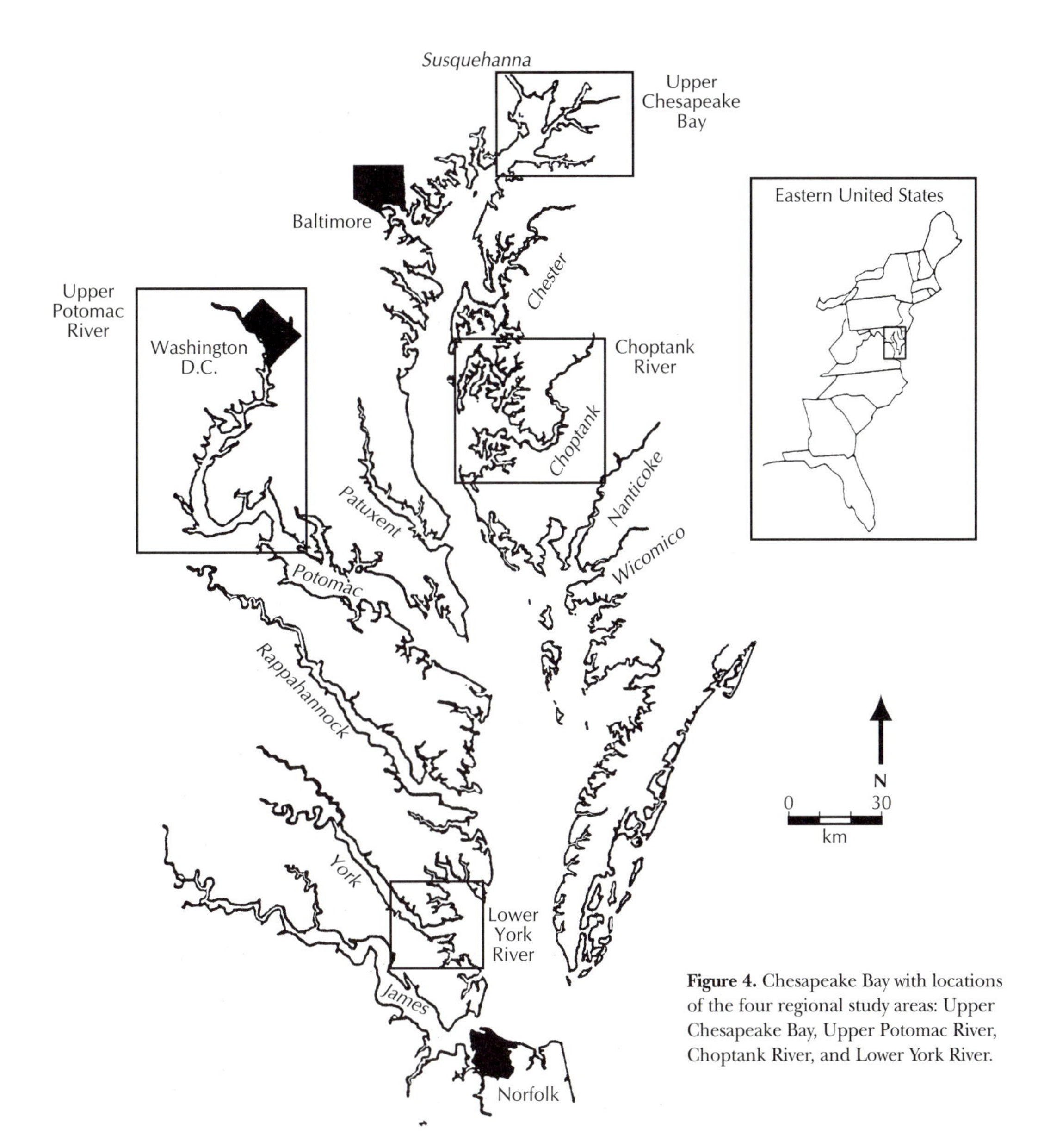

Figure 4. Chesapeake Bay with locations of the four regional study areas: Upper Chesapeake Bay, Upper Potomac River, Choptank River, and Lower York River.

Chesapeake Bay: tidal fresh (0–0.5‰), oligohaline (0.5–5‰), mesohaline (5–18‰), and polyhaline (18–25‰).

The upper Chesapeake Bay, which includes the Susquehanna Flats and the Elk, Sassafras, Northeast and Susquehanna rivers, is a region with tidal freshwater and oligohaline areas. This area historically supported some of the most extensive submersed aquatic vegetation populations in Chesapeake Bay in the 1950s and 1960s (Bayley et al. 1978, Davis 1985). Although there are no precise records on distributions during this period, there are 11,100 ha of bottom that could potentially support submersed aquatic vegetation (less than 2 m water depth MLW). Yet, now only 20% of this area has submersed aquatic vegetation, much of which is sparse (Orth et al. 1991). A variety of species are found in this region, principally *Vallisneria americana, Ceratophyllum demersum, Potamogeton* spp., and *Najas* spp., along with the exotic species *Myriophyllum spicatum.*

The upper Potomac River is also characterized by tidal freshwater and oligohaline waters. Historically, this section had abundant submersed aquatic vegetation through the 1930s (Carter et al. 1985, Cumming 1916, Jaworski et al. 1972). However, subsequent declines left the area nearly devoid of submersed aquatic vegetation until 1983 (Haramis and Carter 1983). Increased water clarity, a result of improvements in sewage treatment and unusual weather conditions, caused a resurgence of submersed aquatic vegetation beginning in 1983. A diverse submersed aquatic vegetation community (13 species) developed, including the exotic species *Hydrilla verticillata* (Carter and Rybicki 1986). Currently, 2500 ha are vegetated, dominated by *H. verticillata, M. spicatum,* and *V. americana* and representing approximately 19% of the river bottom less than 2 m water depth (MLW; Carter and Rybicki 1990, Orth et al. 1991).

The Choptank River is the largest tributary on the eastern shore of the Chesapeake Bay, with mesohaline to tidal freshwater reaches. It was estimated that 15,000 ha of the Choptank River was vegetated with *Ruppia maritima, Zannichellia palustris, M. spicatum,* and several species of *Potamogeton* in the 1960s (Stevenson and Confer 1978). However, in 1990 a single species (*R. maritima*) occupied only 190 ha (Stevenson et al. in press), approximately 1% of Choptank River bottom less than 2 m water depth (MLW; Orth et al. 1991).

The York River is one of the five major tributaries on the western shore of the Chesapeake Bay, with the study area in the polyhaline and mesohaline reaches. Abundant submersed aquatic vegetation consisting of *Zostera marina* and *R. maritima* was found along the shoals of this river in 1971, covering 820 ha (Orth and Gordon 1975). Decline of these species occurred in the 1970s, principally in the upriver and deep water portions of the beds. By 1990, significant regrowth of submersed aquatic vegetation had occurred, primarily in the downriver areas, with upriver areas still unvegetated. Currently, approximately 15% of York River bottom less than 2 m water depth (MLW) is covered with submersed aquatic vegetation (Orth et al. 1991).

The Habitat Requirements Approach

Habitat requirements for submersed aquatic vegetation are defined as the minimal water-quality levels necessary for survival. Survival was defined by the presence of either fluctuating or persistent vegetation beds or the survival of transplants of submersed aquatic vegetation. Areas with persistent beds were defined as areas where submersed aquatic vegetation survived across multiple growing seasons. Areas with fluctuating beds were defined as areas where vegetation was present for one growing season or less or where there appeared to be significant shifts in interannual distribution and abundance patterns.

Water-quality parameters used in the delineation of habitat requirements were chosen because of their availability in water-quality data sets and their relevance to submersed aquatic vegetation survival. Yet, other parameters also

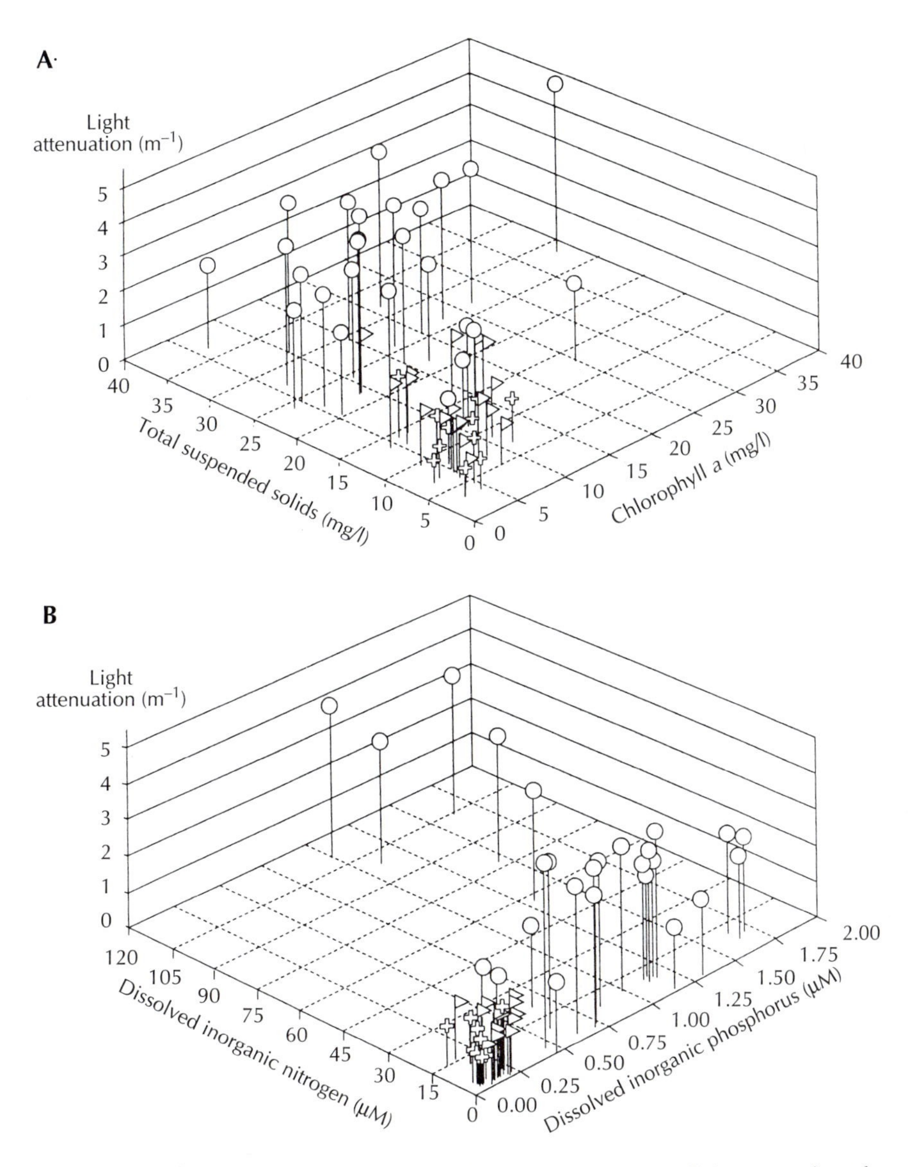

Figure 5. Three-dimensional comparisons of May-October median light attenuation coefficient versus **A,** total suspended solids and chlorophyll *a* concentrations or **B,** dissolved inorganic nitrogen and dissolved inorganic phosphorus at the Choptank River stations for 1986–1989. Plus = persistent submersed aquatic vegetation; flag = fluctuating submersed aquatic vegetation; circle = no submersed aquatic vegetation.

affect survival, and the selected parameters are not independent variables. Some degree of interdependence of these parameters is illustrated by three-dimensional plots of total suspended solids, chlorophyll *a*, dissolved inorganic nitrogen, dissolved inorganic phosphorus, and light attenuation coefficient (Figure 5). But, these parameters were not highly correlated using a Pearson's correlation analysis of all parameters for the data in Figure 5, analyzing separately stations with and without submersed aquatic vegetation (Batiuk et al. in press). Correlations (r) between parameters were all less than 0.5 except for K_d x total suspended solids (r = 0.76 and r = 0.74 in areas with and without seagrass, respectively) and K_d x chlorophyll *a* (r = 0.54 in areas with submersed aquatic vegetation). The lack of appreciable correlation for most of the parameters supports the use of multiple habitat requirements to better predict survival of submersed aquatic vegetation.

Empirical relationships between water-quality characteristics and distributions of submersed aquatic vegetation provided the means of defining habitat requirements for vegetation survival. Habitat requirements were formulated by determining vegetation distributions by transplant survival and bay-wide distributional surveys, measuring water-quality characteristics along large-scale transects that spanned vegetated and unvegetated regions, and combining distributional data and water-quality levels (as in Figure 5) to establish the minimal water quality that supports submersed aquatic vegetation survival.

This correspondence analysis was strengthened by factors common to each of the case studies. Field data were collected over several years (almost a decade in the Potomac River) in varying meteorologic and hydrologic conditions by different investigators. Submersed aquatic vegetation distributions in the four case studies across all salinity regimes were responsive to the five water-quality parameters used to develop habitat requirements. In addition, interannual changes in water quality led to changes in submersed aquatic vegetation distribution and abundance in each region that were consistent with habitat requirements.

Habitat Requirements

Water-quality conditions sufficient to support survival, growth, and reproduction of submersed aquatic vegetation to water depths of one meter below MLW were used as habitat requirements (Table 2). One-meter water depth was chosen because present day Chesapeake Bay submersed aquatic vegetation beds are generally restricted to one meter (MLW) or less. For submersed aquatic vegetation to survive to one meter, light attenuation coefficients less than 2 m^{-1} for tidal fresh and oligohaline regions and less than 1.5 m^{-1} for

Table 2. Chesapeake Bay submersed aquatic vegetation habitat requirements. For each parameter, the maximal growing season median value that correlated with plant survival is given for each salinity regime. Growing season defined as April–October, except for polyhaline (March–November). Salinity regimes are defined as tidal fresh = 0–0.5‰, oligohaline = 0.5–5‰, mesohaline = 5–18‰, polyhaline = more than 18‰.

Salinity regime	Light attenuation coefficient (K_d; m^{-1})	Total suspended solids (mg/l)	Chlorophyll *a* (µg/l)	Dissolved inorganic nitrogen (µM)	Dissolved inorganic phosphorus (µM)
Tidal freshwater	2.0	15	15	–	0.67
Oligohaline	2.0	15	15	–	0.67
Mesohaline	1.5	15	15	10	0.33
Polyhaline	1.5	15	15	10	0.67

mesohaline and polyhaline regions were needed. Measurements of total suspended solids (less than 15 mg/l) and chlorophyll *a* (less than 15 μg/l) were consistent for all regions. The close similarity in habitat requirement values identified for total suspended solids, chlorophyll *a*, and light attenuation coefficient (K_d) for all salinity regimes of Chesapeake Bay suggests that growth and survival of submersed aquatic vegetation, regardless of their location and species at those locations, all respond to these water quality parameters within a small range of values. This correspondence may allow for an overall baywide management strategy for these parameters.

Habitat requirements for dissolved inorganic nitrogen and phosphorus varied substantially between salinity regimes. In tidal freshwater and oligohaline regions, established submersed aquatic vegetation beds survive both episodically and chronically high dissolved inorganic nitrogen; consequently, habitat requirements for dissolved inorganic nitrogen were not determined for these regions. In contrast, maximal dissolved inorganic nitrogen concentrations of 10 μM were established for mesohaline and polyhaline regions. The submersed aquatic vegetation habitat requirement for dissolved inorganic phosphorus was less than 0.67 μM for all regions except for mesohaline regions (less than 0.33 μM).

Differences in nutrient habitat requirements in different regions of Chesapeake Bay are consistent with observations from a variety of estuaries that shifts occur in the relative importance of phosphorus versus nitrogen as limiting factors (e.g., Valiela 1984). Because habitat requirements for nutrient concentrations depended on location (e.g., freshwater versus marine), nutrient reduction strategies could vary depending on the salinity regime. However, nutrient loading in freshwater or oligohaline regions of the estuary affects nutrient concentrations of other salinity regimes, and nutrient reduction strategies may need to be baywide to achieve habitat requirements in each salinity regime.

Habitat requirements can be used to generate distribution and abundance targets for restoration efforts in Chesapeake Bay. Increased water clarity would be required for submersed aquatic vegetation to penetrate to depths greater than one meter. Using a minimal light requirement of 20% (e.g., *Zostera marina*), monthly median light attenuation

Table 3. Application of Chesapeake Bay submersed aquatic vegetation habitat requirements using distributions and water quality data from 1984–1990. Percentages represent the frequency that submersed aquatic vegetation is present when growing season median water quality values are less than the habitat requirements listed in Table 2. Number in parentheses is the number of Chesapeake Bay segments and years with at least 25 ha of submersed aquatic vegetation distribution used to determine percentages. No oligohaline areas had more than 25 ha of vegetation, with the exception of the upper Potomac River, which was not included in this analysis.

Salinity regime	Light attenuation coefficient	Total suspended solids	Chlorophyll *a*	Dissolved inorganic nitrogen	Dissolved inorganic phosphorus
Tidal freshwater	100% (7)	100% (7)	100% (7)	–	100% (7)
Mesohaline	86% (57)	81% (59)	100% (57)	79% (57)	93% (57)
Polyhaline	94% (32)	88% (32)	97% (31)	91% (31)	97% (32)
Total	90% (96)	84% (96	99% (95)	83% (89)	95% (96)

coefficients of 0.80 m^{-1} and 0.54 m^{-1} would be required for revegetation to maximal depths of 2 and 3 m, respectively. Combining these depth limits with Chesapeake Bay hydrography provides estimates of potential vegetation habitat that could be compared to measured distributions, thus providing quantitative method to assess the relative success of Chesapeake Bay restoration efforts.

Chesapeake Bay habitat requirements for submersed aquatic vegetation developed in the four study area were applied to the rest of the bay to test the correspondence of submersed vegetation distributions with the five water quality parameters (Table 3). Chesapeake Bay was divided into 4, segments, and median water-quality values were determined in each segment using Chesapeake Bay Basin Monitoring data for each year between 1984 and 1990. Between 79% and 100% of the segments with at least 25 ha of submersed aquatic vegetation met the habitat requirements for the respective salinity regime giver in Table 2.

The various water quality parameters have differing abilities to predict submersed aquatic vegetation distributions: chlorophyll *a* (99%), dissolved inorganic phosphorus (95%) light attenuation coefficient (90%) total suspended solids (84%), and dissolved inorganic nitrogen (83%); however, the overall average (90%) for all parameters is fairly high and indicates the utility of this approach.

There are few tidal freshwater and no oligohaline stations with more than 25 ha of submersed aquatic vegetation outside the upper Potomac River so testing the habitat requirements in these areas was less intensive. The upper Potomac River was not included in this treatment because the well established submersed aquatic vegetation populations were able to withstand occasional departures from the established distribution (Batiuk et al. in press).

The habitat requirements represent the absolute minimal water-quality characteristics necessary to sustain plants in shallow water. As such, exceeding any of the five water-quality characteristics seriously compromise the chances of submersed aquatic vegetation survival. Improvements in water clarity and nutrient reduction to achieve greater depth penetration of submersed aquatic vegetation would also increase submersed aquatic vegetation density, biomass, and distribution (Carter and Rybicki 1990, Dennison 1987). In addition, improvements of water quality beyond the habitat requirements could lead to the maintenance or reestablishment of a diverse population of native submersed species, with likelihood of long-term survival. Habitat requirements provide a guideline for mitigation efforts involving transplants of submersed aquatic vegetation. If habitat requirements are not met, reestablishment of plant communities via transplant efforts are futile.

Conclusions

The analyses presented here represent a first attempt at linking habitat requirements for a living resource (submersed aquatic vegetation) to water quality standards in an estuarine system. This habitat requirements approach, although deviating from the traditional dose/response measures and direct toxicity studies, provides testable hypotheses concerning water quality/vegetation interactions that can be explored in future studies in other estuaries and perhaps lacustrine systems as well. Additional experimental evidence coming from field and laboratory studies to test the empirical relationships could lead to improved predictive capability of habitat requirements.

The empirical approach used in this study allows for predictive capability without detailed knowledge of the precise nature of vegetation/water quality interactions. Because submersed aquatic vegetation are disappearing rapidly on a global scale, there is a need to provide guidelines on water quality before a more complete understanding of the complex ecological interactions is reached. Submersed aquatic vegetation are convenient natural light meters, integrating water clarity of coastal waters over time scales of weeks to months. Other organisms also possess critical thresholds for a variety of environmental factors that can be used to establish habitat requirements. An important advantage of this approach, requiring only low technology input to achieve a high information yield, is that it can be employed in a variety of settings.

We need to maintain continuous interactions and feedback between the researchers who develop the habitat criteria for individual species and the resource managers who are responsible for regulations that ultimately protect, restore, and enhance living resources. Continued research and monitoring of water quality and living resources, coupled with specific restoration goals, are paramount if these resources are to be part of humanity's future.

Acknowledgments

This article abstracts a technical synthesis (Batiuk et al. in press) that was produced with the help and critical review of a variety of editors, technicians, and program staff from the Chesapeake Bay scientific community. Funding for the research and monitoring data came from a variety of dedicated Chesapeake Bay Program agencies whose commitment to long-term research and monitoring programs made this synthesis possible. Contribution no. 1766 from the Virginia Institute of Marine Science, School of Marine Science, College of William and Mary, and no. 2362 from the University of Maryland Center for Environmental and Estuarine Studies.

References

Batiuk, R. A., R. J. Orth, K. A. Moore, W. C. Dennison, J. C. Stevenson, L. Staver, V. Carter, N. Rybicki, R. E. Hickman, S. Kollar, S. Bieber, P. Heasly and P. Bergstrom. In press. Chesapeake Bay Submerged Aquatic Vegetation Habitat Requirements and Restoration Goals: A Technical Synthesis. US EPA, Chesapeake Bay Program, Annapolis, MD.

Bayley, S., V. D. Stotts, P. F. Springer and J. Steenis. 1978. Changes in submerged aquatic macrophyte populations at the head of the Chesapeake Bay, 1958–1974. Estuaries 1:171–182.

Borum, J. 1983. The quantitative role of macrophytes, epiphytes, and phytoplankton under different nutrient conditions in Roskilde Fjord, Denmark. In: *Proceedings of the International Symposium on Aquatic Macrophytes.* Faculty of Science, Nijmejen, The Netherlands. Pp. 35–40.

Cambridge, M. L., and A. J. McComb. 1984. The loss of seagrass in Cockburn Sound, Western Australia. l. The time course and magnitude of seagrass decline in relation to industrial development. Aquat. Bot. 20:229–243.

Canfield, D. E., K. A. Langeland, S. B. Lindea and W. T. Haller. 1985. Relations between water transparency and maximum depth of macrophyte colonization in lakes. J. Aquat. Plant Manage. 23:25–28.

Carter, V., and N. B. Rybicki. 1986. Resurgence of submersed aquatic macrophytes in the tidal Potomac River, Maryland, Virginia and the District of Columbia. Estuaries 9:368–375.

Carter, V., and N. B. Rybicki. 1990. Light attenuation and submersed macrophyte distribution in the tidal Potomac River and estuary. Estuaries 13:441–452.

Carter, V., J. E. Paschal Jr., and N. Bartow. 1985. Distribution and abundance of submersed aquatic vegetation in the tidal Potomac River and estuary, Maryland and Virginia, May 1976 to November 1981. US Geological Survey Water Supply Paper 2234A.

Chambers, P. A., and J. Kalff. 1985. Depth distribution and biomass of submersed aquatic macrophyte communities in relation to Secchi depth. Can. J. Fish. Aquat. Sci. 42:701–709.

Champ, M. A., G. A. Gold, W. E. Bozzo, S. G. Ackelson and K. C. Vierra. 1980. Characterization of light extinctions and attenuation in Chesapeake Bay, August 1977. In: *Estuarine Perspectives.* V. S. Kennedy, ed. Academic Press, New York. Pp. 263–277.

Costa, J. E. 1988. Distribution, production and historical changes in abundance of eelgrass (*Zostera marina*) in Southeastern Massachusetts. Ph.D. dissertation, Boston University, Boston, MA.

Cumming, H. W. 1916. Investigation of the pollution and sanitary conditions of the Potomac watershed. Treasury Department and U.S. Public Health Service Hygenic Laboratory Bulletin no.104.

Davis, F. W. 1985. Historical changes in submerged macrophyte communities of upper Chesapeake Bay. Ecology 66:981–993.

den Hartog, C. 1970. *Seagrasses of the World.* North-Holland, Amsterdam, The Netherlands.

Dennison, W. C. 1987. Effects of light on seagrass photosynthesis, growth and depth distribution. Aquat. Bot. 27:15– 26.

Duarte, C. M. 1991. Seagrass depth limits.Aquat. Bot. 0:363–377.

Funderburk, S. L., S. J. Jordan, J. A. Mihursky and D. Riley, eds. 1991. Habitat Requirements for Chesapeake Bay Living Resources. Chesapeake Research Consortium, Solomons, MD.

Giesen, W. B. J. T., M. M. van Katwijk and C. den Hartog. 1990. Eelgrass condition and turbidity in the Dutch Wadden Sea. Aquat. Bot. 37:71–85.

Hanson, P. J., J. G. Isebrands and R. E. Dickson. 1987. Carbon budgets of *Quercus rubra* L. seedlings at selected stages of growth: influence of light. In: *Proceedings of the Central Hardwood Forest Conference VI.* R. L. Hay, F. W. Woods and H. DeSelm, eds. University of Tennessee, Knoxville. Pp. 269–276.

Haramis, G. M., and V. Carter. 1983. Distribution of submersed aquatic macrophytes in the tidal Potomac River. Aquat. Bot. 15:65–79.

Horton, T., and W. M. Eichbaum. 1991. *Turning the Tide: Saving the Chesapeake Bay.* Island Press, Washington, DC.

Hurley, L. 1990. *Field Guide to the Submerged Aquatic Vegetation of Chesapeake Bay.* U.S. Fish and Wildlife Service, Chesapeake Bay Program, Annapolis, MD.

Hutchinson, G. E. 1975. *A Treatise on Limnology.* Vol. 3. Limnological Botany. John Wiley & Sons, New York.

Jaworski, N. A., D. W. Lear and O. Villa. 1972. Nutrient management in the Potomac estuary. In: *Nutrients and Eutrophication: The Limiting Nutrient Controversy.* G. E. Likens, ed. American Society of Limnology and Oceanography, Lawrence, KS. Pp. 246–273.

Kautsky, N., H. Kautsky, U. Katusky and M. Waern. 1986. Decreased depth penetration of *Fucus vesiculosus* (L.) since the 1940s indicates eutrophication of the Baltic Sea. Mar. Ecol. Prog. Ser. 28:1–8.

Kemp, W. M., W. R. Boynton, R. R. Twilley, J. C. Stevenson and J. C. Means. 1983. The decline of submerged vascular plants in upper Chesapeake Bay: summary of results concerning possible causes. Mar. Technol. Soc. J. 17:78–89.

Kemp, W. M., W. R. Boynton, R. R. Twilley, J. C. Stevenson and L. G. Ward. 1984. Influences of submerged vascular plants on ecological processes in upper Chesapeake Bay. In: *Estuaries as Filters.* V. S. Kennedy, ed. Academic Press, New York. Pp. 367–394.

Kenworthy, W. J., and D. E. Haunert, eds. 1991. The light requirements of seagrasses: proceedings of a workshop to examine the capability of water quality criteria, standards and monitoring programs to protect seagrasses. NOAA Technical Memorandum NMFS-SERC-287.

Larkum, A. W. D., A. J. McComb and S. A. Shepherd, eds. 1989. *Biology of Seagrasses: A Treatise on the Biology of Seagrasses With Special Reference to the Australian Region.* Elsevier, New York.

Littler, M. M., D. S. Littler, S. M. Blair and J. N. Norris. 1985. Deepest known plant life discovered on an uncharted seamount. Science 227:57–59.

Luning, K., and M. J. Dring. 1979. Continuous underwater light measurement in the sublittoral region. Helgoländer Wissenschaftliche Meeressunters Uchungen 32:403–424.

McRoy, C. P., and C. Helfferich, eds. 1977. *Seagrass Ecosystems: A Scientific Perspective.* Marcel Dekker, New York.

Orth, R. J., and H. Gordon. 1975. Remote sensing of submerged aquatic vegetation in the lower Chesapeake Bay, Virginia. Final Report, NASA- 10720.

Orth, R. J., and K. A. Moore. 1983. Chesapeake Bay: an unprecedented decline in submerged aquatic vegetation. Science 222:51–53.

Orth, R. J., and K. A. Moore. 1984. Distribution and abundance of submerged aquatic vegetation in Chesapeake Bay: an historical perspective. Estuaries 7:531–540.

Orth, R. J., J. F. Nowak, A. A. Frisch, K. Kiley and J. Whiting. 1991. Distribution of submerged aquatic vegetation in the Chesapeake Bay and tributaries and Chincoteague Bay—1990. US EPA, Chesapeake Bay Program, Annapolis, MD.

Osmond, C. B., M. P. Austin, J. A. Berry, W. D. Billings, J. S. Boyer, J. W. H. Dacey, P. S. Nobel, S. D. Smith and W. E. Winner. 1987. Stress physiology and the distribution of plants. BioScience 37:38–48.

Ostenfeld, C. H. 1908. On the ecology and distribution of the grass-wrack (*Zostera marina*) in Danish waters. Report of the Danish Biological Station, Copenhagen, Denmark.

Parsons, T. R., M. Takahashi and B. Hargrave. 1984. *Biological Oceanographic Processes.* Pergamon Press, Oxford, UK.

Phillips, R. C., and C. P. McRoy, eds. 1980. *A Handbook of Seagrass Biology: An Ecosystem Perspective.* Garland, New York.

Pierce, J. W., D. L. Correll, B. Goldberg, M. A. Faust and W. H. Kline. 1986. Response of underwater light transmittance in the Rhode River estuary to changes in water-qual-

ity parameters. Estuaries 9:169–178.
Sand-Jensen, K. 1988. Minimum light requirements for growth in *Ulva lactuca*. Mar. Ecol. Prog. Ser. 50:187– 193.
Stevenson,J. C. 1988. Comparative ecology of submersed grass beds in freshwater, estuarine, and marine environments. Limnol. Oceanogr. 33:867–893.
Stevenson, J. C., and N. M. Confer. 1978. Summary of available information on Chesapeake Bay submerged vegetation. U.S. Fish and Wildlife Service, Office of Biological Services FWS/OBS-78/66.
Stevenson, J. C., L. W. Staver and K. W. Staver. In press. Water quality associated with survival of submersed aquatic vegetation along an estuarine gradient. Estuaries.
Thayer, G. W., D. A. Wolfe and R. B. Williams. 1975. The impact of man on seagrass systems. Am. Sci. 63:288–296.
Twilley, R. R., W. M. Kemp, K. W. Staver, J. C. Stevenson and W. R. Boynton. 1985. Nutrient enrichment of estuarine submersed vascular plant communities. I. Algal growth and associated effects on production of plants and associated communities. Mar. Ecol. Prog. Ser. 23:179–191.
Valiela, I. 1984. *Marine Ecological Processes*. Springer-Verlag, New York.
Wetzel, R. G. 1975. *Limnology*. W. B. Saunders, New York.
Williams, S. L., and W. C. Dennison. 1990. Light availability and diurnal growth of a green macroalga (*Caulerpa cupressoides*) and a seagrass (*Halophila decipiens*). Mar. Biol. 106:437–443.

Introduction to

James J. Elser, Dean R. Dobberfuhl, Neil A. MacKay, and John H. Schampel

Organism Size, Life History, and N:P Stoichiometry

by James F. Kitchell

The living tissues of bacteria and elephants are similar in that they are primarily comprised of molecules containing some mix of carbon, hydrogen, nitrogen, oxygen, phosphorus and sulfur. Organisms can differ in the ratios of those key elements. Elser and his colleagues attempt to build a bridge between the central themes of evolutionary biology (principles of natural selection) and those of ecosystem ecology (principles of thermodynamics) by elaborating an approach known as ecological stoichiometry. This approach emphasizes the relative abundance of nitrogen and phosphorus, the two nutrients most commonly recognized as in limited supply and, therefore, most important in regulating the growth rate of organisms. In developing this argument, the authors challenge the adequacy of dogma based on energy limitation to biological production. They also assert that explanations such as those offered by the trophic cascade concept, which are based on predation and competition interactions, may be necessary, but are not sufficient descriptors of how consumers play a role in the ecosystem context.

Readers will find opportunity to recall their introduction to cell biology because this article starts with the chemical composition of RNA and DNA, then expands to the point where one can imagine the effect that whales have on the growth of phytoplankton. Along the way, readers encounter yet another example of high regard for *Daphnia,* an organism of favor for many ecological investigations.

Organism Size, Life History, and N:P Stoichiometry

Toward a unified view of cellular and ecosystem processes

JAMES J. ELSER, DEAN R. DOBBERFUHL, NEIL A. MACKAY, AND JOHN H. SCHAMPEL

Elemental stoichiometry can provide a new tool to trace the threads of causal mechanisms linking cellular, ecosystem, and evolutionary processes.

Elser, J. J., D. R. Dobberfuhl, N. A. MacKay and J. H. Schampel. 1996. *BioScience* 46(9):674–684.

Ecosystem science and evolutionary biology have long been infrequent and uncomfortable bedfellows (Hagen 1992, Holt 1995, McIntosh 1985). However, the convergence of a global decline in biodiversity and global alterations in biogeochemical cycles provides motivation to overcome past inhibitions. Currently, attempts are being made (Jones and Lawton 1995) to understand relationships between the foci of evolutionary biology (the individual in its species population) and ecosystem science (energy and material flow and storage). Analysis of relationships between species and ecosystems requires a framework appropriate for moving between levels in an imperfect hierarchy of biotic and abiotic components (O'Neill et al. 1986). Although various frameworks are possible, the history of ecology since Lindeman's 1942 paper on the trophic dynamic concept makes it clear that energy has been the currency of choice for ecologists (Hagen 1992).

Although the energetics perspective has had wide application and success, both in studies of individuals and of ecosystems (Brown 1995, Pandian and Vernberg 1987, Wiegert 1988, Wright et al. 1994), critical examination reveals inadequacies in this paradigm. For example, White (1993) argues that, because of disparities between the nitrogen composition of many foods and the nitrogen demands of many consumers, the availability of energy is less important than that of nitrogen in determining the reproductive success and population dynamics of animals. Månsson and McGlade (1993) have also scrutinized energy- based approaches to evolutionary biology and ecosystem dynamics (in particular those proposed by H. T. Odum) and concluded that there are fundamental problems in describing ecosystems using a framework that has a single currency.

Reiners (1986) has presented a more balanced, multidimensional view, proposing elemental stoichiometry as a complementary way to study questions about ecosystems that are unsuited for analysis with energy-based models. Elemental stoichiometry considers relative proportions (ratios) of key elements in organisms in analyzing how characteristics and activities of organisms influence, and are in turn influenced by, the ecosystem in which they are found. In this article we introduce the main concepts and patterns of ecological stoichiometry and synthesize literature from a variety of fields to forge connections, not only between evolutionary and ecosystem sciences but also between the disparate disciplines of cell biology and ecology. Stoichiometry may have a natural advantage in making such connections because it offers an explicit multiple-currency approach that is potentially better suited than a one-currency approach to understanding ecological and evolutionary processes that more closely resemble optimization rather than maximization (Krebs and Houston 1989).

Our approach in this article is as follows. First, we describe recent discoveries that establish the importance of consumer body nitrogen:phosphorus (N:P) ratio in modulating secondary production and consumer-driven nutrient cycling in ecosystems. Second, we review aspects of cellular biochemistry and ultrastructure through the eyes of an ecosystem scientist, focusing on the relative nitrogen and phosphorus contents of important biomolecules and cellular structures. Third, we present examples of how organismal characters such as growth rate and ontogeny are linked with biochemical and cellular investment and thus with body N:P ratio. Finally, we propose a general scenario for allometric variation in body N:P ratio among consumers ranging from bacteria to large vertebrates and use the scenario to predict patterns of consumer-driven nutrient cycling and food quality constraints. In the spirit of Reiners (1986), we employ stoichiometric theory as a complementary approach to the study of biological processes, one that we hope will both reinforce conclusions derived from energetic perspectives as well as provide new insights into biological phenomena that may be puzzling when considered from more traditional single-currency approaches.

Ecological stoichiometry: basic concepts and patterns

Ecological stoichiometry focuses on the relative elemental composition of participants in ecological interactions in ecosystems. Constraints of mass balance must be met both in simple inorganic chemical reactions (Figure lA) and more complex biochemical transformations (Figure 1B); ecological interactions such as competition, predation, or herbivory are also not exempt from thermodynamics (Figure 1C). Thus, in the "ecological play," firm predictions

A. Stoichiometry in chemistry

$$3CaCl_2 + 2Na_3PO_3 \longleftrightarrow Ca_3(PO_4)_2 + 6NaCL$$

B. Stoichiometry in biology (respriation)

$$C_6H_{12}O_6 + 6O_2 \longrightarrow 6CO_2 + 6H_2O$$

C. Stoichiometry in ecology (predator–prey interaction with nutrient recycling)

$$(N_xP_y)_{predator} + (N_aP_b)_{prey} \longrightarrow Q(N_xP_y)_{predator} + (N_{a'}P_{b'})_{waste}$$

Figure 1. The first law of thermodynamics dictates mass balance of multiple elements before and after: **A,** inorganic chemical reactions, **B,** simple biochemical transformations (e.g., respiration of glucose), and **C,** complex ecological - interactions (e.g., predation with nutrient recycling). In the stoichiometry of predator-prey interactions, a prey item of a given elemental composition is consumed by a predator of fixed elemental composition to increase predator biomass by a factor Q, simultaneously producing waste of altered elemental composition. (That is, a':b' may be greater or less than a:b, depending on the relative demands for nitrogen and phosphorus required for producing predator biomass; see Sterner 1990.) The elemental ratio of recycled nutrients (a':b') contributes to the stoichiometry of another ecological interaction—nutrient competition among autotrophs in the ecosystem.

can be made about elemental ratios in the "players" and their "stage" (sensu Lotka 1924) before and after ecological interactions.

One of the best-developed stoichiometric approaches in ecology is resource ratio competition theory (Tilman 1982). This theory, a modification of the graphical approaches of MacArthur (1972), predicts outcomes of competition for inorganic nutrients among autotrophic taxa differing in elemental requirements. In competitive situations, variation in nutrient supply ratios tips the competitive balance in favor of taxa best suited to the supply regime, altering the elemental composition of autotroph community biomass and of the residual chemical environment. Resource ratio theory has been widely supported by studies of competition among autotrophs (Sommer 1989, Tilman 1982). However, stoichiometric approaches have rarely been applied to higher levels in food webs. In this article we highlight studies of elemental ratios in consumers and how they may help in understanding the role of consumers in nutrient cycling and food webs.

Consumer-regulated nutrient cycling is increasingly attracting the attention of ecosystem scientists (DeAngelis 1992, and papers in Naiman 1988) who have traditionally focused on processes mediated by autotrophs and microbes. In recent studies of consumer-driven nutrient recycling in lakes, ecological stoichiometry explains unexpected effects of food web alterations on nitrogen and phosphorus availability (Sterner et al. 1992) and identifies qualitative differences in zooplankton-phytoplankton interactions that occur in marine and freshwater habitats (Elser and Hassett 1994).

These studies have focused on species-specific differences in body N:P ratio of zooplankton that dramatically affect the relative rates of recycling of nitrogen and phosphorus by elementally homeostatic consumers (Sterner 1990). For example, when the food web structure favors dominance by consumers with high body N:P (e.g., calanoid copepods, with body N:P ratio greater than 30:1; all ratios are given as atomic ratios), then the N:P ratio of nutrients recycled by those consumers is low because food items tend to have lower N:P ratios than consumers, which would therefore tend to retain nitrogen and release phosphorus (see Figure 1C). Under such conditions, phytoplankton growth is limited primarily by nitrogen. By contrast, when fish predation on zooplankton is low, permitting dominance by low N:P taxa (especially *Daphnia,* with N:P ratio approximately 12:1), recycling N:P ratio is high and phytoplankton are phosphorus limited (Sterner et al. 1992). Thus, body N:P ratio is critical for understanding nutrient cycling in ecosystems because body N:P ratio directly determines the relative ratios of limiting nutrients recycled by consumers.

Variation in body N:P ratio is also useful in understanding a relatively new aspect of consumer ecology: the role of mineral food quality in influencing consumer growth and reproduction. Mineral nutrition has traditionally been of interest primarily to managers of livestock and game animals (McDowell 1992). However, recent studies of mineral nutrition of freshwater zooplankton indicate that mineral limitation, in particular phosphorus deficiency, may be commonplace in pelagic ecosystems (Elser and Hassett 1994, Sterner and Hessen 1994). In particular, zooplankton taxa with high phosphorus demands for growth (e.g., *Daphnia*) experience reduced growth and reproductive output when feeding on phosphorus-deficient food (Sterner and Hessen 1994). Knowledge of consumer N:P could be critical in identifying taxa most likely to suffer phosphorus limitation in nature and in assessing the extent to which the elemental stoichiometry of available food is likely to affect production of higher trophic levels.

The stoichiometric approaches just described are largely phenomenological, relying on direct measurements of body N:P ratio of dominant consumer taxa (Andersen and Hessen 1991). Observations of strongly contrasting body N:P ratios between taxa raise the question "What causes variation in body N:P?" In biology, answers to that question are of two types (Mayr 1961): proximate (biochemical and physiological) and ultimate (evolutionary). Reiners (1986) addressed both issues by distinguishing between basic "protoplasmic life," which he argues has a standard chemical stoichiometry, and "mechanical structures" (adaptations for specific functions, such as spines for defense or bones for support), which are highly variable in their stoichiometry. Thus, selection for certain mechanical structures in functionally dominant species will alter material cycling, including global ecosystem processes.

However, we believe that there is no characteristic elemental content of "protoplasmic life" and that even unicellular organisms exhibit considerable variation in elemental ratios as a function of their evolved traits. We therefore propose that major changes in organism life history (especially size and growth rate) require substantial changes in the complement of cellular components. Because different cellular components generally have contrasting biochemical constituents that differ strongly in elemental composition, major macroevolutionary patterns must be accompanied by changes in organism stoichiometry. In the following, we focus on the limiting elements nitrogen and phosphorus and the role of heterotrophs (organo-heterotrophic bacteria, protozoa, and multicellular animals) in

Table 1. Major categories, examples, and biological functions of nitrogen- and phosphorous-containing molecules. General information about structure, function, composition, and relative abundance of various molecules is from Lehninger et al. (1993).

Class of molecule	Examples	Functions	Comments
Protein	Collagen, actin communication	Structure, regulation,	Average nitrogen content of the 20 amino acids in proteins is 17.2%
Nucleic acids	DNA, RNA	Storage, transmission, and expression of genetic information	DNA content (as a percentage of cell mass) conservative. RNA: DNA greater than 5:1
Lipids	Phospholipids, glycolipids	Cell membranes	Carbon-rich, minor component of cells (approximately 5% of total cell mass)
Phosphorylated energy storage compounds	ATP, phosphocreatine	High-turnover energy carriers	ATP only approximately 0.05% of invertebrate body mass (DeZwann and Thillart 1985)
Structural carbohydrates	Chitin	Structural support protection	

cycling of those elements. We explore how relationships between major life history traits and cellular organization are reflected in body N:P ratio and thus how evolved characters affect material cycling by consumers at the level of the ecosystem.

How an Ecosystem Scientist Sees Cells

Understanding how selection on major life history traits alters elemental content requires an understanding of the biochemical functions of the various molecules used by organisms and an appreciation of their elemental (especially nitrogen and phosphorus) composition.We review the function and structure of important biomolecules, summarize their relative nitrogen and phosphorus content, and then consider the biochemical and elemental composition of the cellular and subcellular structures constructed from these molecules. As ecosystem scientists, we focus on biomolecules that contain relatively large amounts of nitrogen and phosphorus and also contribute substantively to the cellular and extracellular make-up of organisms. This treatment is therefore a simplification of the actual variety of biochemicals in organisms.

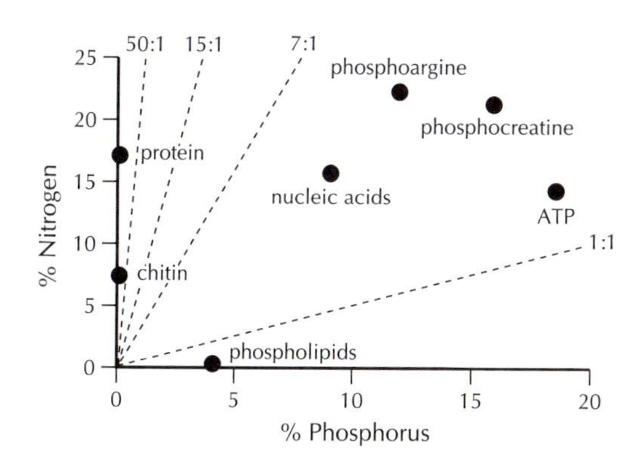

Figure 2. Stoichiometric diagram illustrating the nitrogen and phosphorus composition of biomolecules containing nitrogen and phosphorus. Values for percentage nitrogen and percentage phosphorus are given in terms of weight. Dotted lines depict standard values of atomic (molar) N:P ratio for the purpose of comparing various graphs.

Biochemical Stoichiometry

Major functional classes of organic molecules used in biological systems are listed in Table 1, along with their relative nitrogen and phosphorus composition. For purposes of understanding body N:P ratios, we can neglect most carbohydrates and storage lipids because most of these contain no nitrogen or phosphorus (but see chitin, a structural carbohydrate; Figure 2). Nitrogen-rich molecules include proteins, nucleic acids, and high-energy adenylates (ATP). Note that these three classes of compounds differ little in terms of the percentage of nitrogen (Figure 2); thus, cellular structures or organisms with differing protein:nucleic acid ratios will not differ substantially in their percentage of nitrogen. The main classes of phosphorus-rich molecules (Table 1) are high-energy adenylates (particularly ATP) and nucleic acids (DNA, RNA). Due to their high phosphorus content, these molecules have low N:P ratio (Figure 2). The ATP content of organisms is generally low; thus the contribution of phosphorus from ATP to whole-organism phosphorus is also likely to be low. For example, a study of the ATP content of 22 invertebrate species (DeZwann and Thillart 1985) indicated that phosphorus in ATP as a percentage of dried weight varies from 0.02% to 0.2%, with a mean of 0.05%. Because whole-organism phosphorus content generally ranges from 0.2% to 2%, ATP contributes little more than a fifth of whole organism phosphorus.

The molecules likely to dominate cellular and organismal N:P stoichiometry are thus proteins, due to their high nitrogen content and their substantial contribution to biomass, and nucleic acids, due to their high nitrogen and phosphorus content and their relatively high abundance. More specifically, the strong contrast in N:P stoichiometry between proteins and nucleic acids reflects the high phosphorus content of nucleic acids and indicates that a critical determinant of N:P stoichiometry of cellular structures and whole organisms will be the relative abundance of proteins versus nucleic acids.

Cellular Stoichiometry

Now consider how major biomolecules are deployed in cells, and thus how structures performing specific cellular processes themselves differ in biochemical and elemental composition, with potential consequences for ecosystem processes affected by body N:P stoichiometry.

Cell Membranes

In both prokaryotes and eukaryotes, membranes form selectively permeable physical boundaries of compartments whose composition can be regulated to permit efficient biochemical processing (Evans 1989). Membranes consist primarily of a phospholipid bilayer and associated proteins, with a biochemical composition of 25%–56% lipids, 25%–62% proteins, and 10% carbohydrates (Frausto da Silva and Williams 1991). Thus, membrane percent nitrogen and percent phosphorus are also variable (% N: 5.2–11.1; % P: 1.1–2.4), as is membrane N:P ratio (4.9–23:1; average presented in Figure 3). Although membrane types differ in nitrogen and phosphorus content, increasing membrane contribution to overall cellular biomass would generally drive the N:P ratio down due to the phosphorus-rich phospholipid component. However, the relatively small interspecies differences in contributions of phospholipids to total body mass that have been documented (e.g., Reinhardt and Van Vleet 1986) appear unlikely to contribute greatly to differences in organism N:P ratio.

Nucleus and Chromosomes

The nucleus is the largest organelle in eukaryotic cells and consists of chromosomes, nuclear membrane with externally situated ribosomes, and proteinaceous nucleoplasm matrix. In eukaryotic chromosomes, DNA is complexed with organizing proteins (both histone and nonhistone) into chromatin, which has a protein:DNA ratio of 1–2:1 by weight (Lehninger et al. 1993). Assuming that chromatin proteins are 17.2% N by weight (above) and are present in a 1.5:1 ratio by weight with DNA, eukaryotic genetic material is thus 16.5%

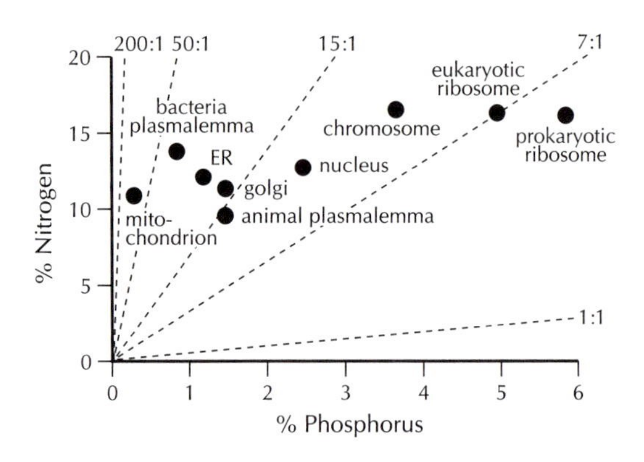

Figure 3. Stoichiometric diagram illustrating the nitrogen and phosphorus composition of major organelles and other cellular structures. Values for percentage nitrogen and percentage phosphorus are given in terms of dry weight percentage. Dotted lines depict standard values of atomic N:P ratio. As described in the text, values for elemental composition were calculated on the basis of reported biochemical composition for each structure, with the exception of nuclei and mitochondria, for which data are based on direct determinations of percent nitrogen and percent phosphorus (Bowen 1979).

N by weight and 3.6% P by weight, with an atomic N:P ratio of approximately 10:1 (Figure 3). The variable complex of materials other than chromosomes in nuclei complicates calculation of the elemental composition of nuclei as a whole. However, cell fracture studies (Bowen 1979) have directly measured the nitrogen and phosphorus composition of nuclei as approximately 12.6% N and 2.5% P (N:P ratio of approximately 11:1; Figure 3).

Ribosomes

Ribosomes are sites of protein synthesis and thus are centrally involved in growth. Ribosomes are composed of ribosomal RNA (rRNA) and protein in ratios of 1.22:1 for eukaryotes and 1.8:1 for bacteria (Campana and Schwartz 1981). Calculations of elemental content based on these data indicate that ribosomes are a particularly phosphorus-rich cellular constituent (eukaryotic ribosomes: 16.3% N, 5.0% P, N:P approximately 7.2:1; prokaryotic ribosomes: 16.1% N, 5.6% P, N:P approximately 6:1; Figure 3). Because ribosomes appear to have lower N:P ratio than other subcellular structures, increasing ribosomal content will tend to lower the N:P ratio.

Mitochondria

Mitochondria are sites of oxidative ATP generation by respiration in eukaryotic cells. Mitochondria are composed of outer and inner membranes and a gel-like inner compartment known as the matrix. The matrix is approximately 50% protein by weight and also contains DNA (approximately 15,000 base pairs in human mitochondria) and ribosomes (Becker 1986). The outer membrane contains nearly equal proportions of lipids and proteins, whereas the inner membrane is nearly 80% protein and only approximately 20% lipid by weight (data for liver mitochondria; Lehninger et al. 1993). Convoluted foldings of the inner membrane increase its surface area, enhancing mitochondrial ATP generation. Not surprisingly, the surface area of the inner membrane correlates with the intensity of tissue respiration (Lehninger et al. 1993). Cells that respire heavily have a larger percentage of inner mitochondrial membrane, which by virtue of its high protein content has a high percent nitrogen and a high N:P ratio, than cells with smaller respiratory demands. Thus, mitochondria probably have an inherently high N:P ratio. Direct observation supports this suggestion: data for percent nitrogen and percent phosphorus from cell fracture studies summarized by Bowen (1979) indicate a mitochondrial N:P ratio of approximately 80:1 in mammalian tissues (Figure 3). Consequently, increased mitochondrial contribution would tend to increase cellular and organismal N:P ratio.

Endoplasmic Reticulum (ER) and the Golgi Complex

ER is a membranous network that constitutes approximately 15% of total cell volume (Mieyal and Blumer 1981). ER is the major component of the intracellular cytocavitary network, is the organizing structure for some ribosomes (rough ER), and is involved in hydroxlation reactions, detoxification, and other metabolic transformations. ER membrane has a high protein:lipid ratio of 2.3 (Becker 1986). Thus, we estimate that ER membrane itself is 12.3% N and 1.2% P by weight and has an N:P ratio of approximately 22:1 (Figure 3). The Golgi complex mediates flows of secretory proteins from the ER to the exterior of the cell and is composed of proteins and lipids intermediate in composition between ER membrane (from which Golgi membrane is thought to arise) and plasma membrane (with which Golgi-derived secretory vesicles eventually fuse in discharging their contents to the outside of the cell; Becker 1986). Assigning the Golgi a protein:lipid ratio of 1:7 (intermediate between the protein:lipid ratios of plasma membrane and ER), we estimate the elemental composition of the Golgi apparatus as 11.2% N and 1.5% P, with an N:P ratio of 17:1 (Figure 3).

Cytoplasm

Cellular cytoplasm is composed of cytosol (the soluble portion of the cytoplasm) and cytoskeleton (an internal framework that gives eukaryotic cells their distinctive shape and internal organization and governs the position and movement of organelles). In typical animal cells, cytoplasm occupies more than half of cell volume. Protein content of the cytosol exceeds 20% (Lehninger et al. 1993), likely reflecting localization of the majority of enzymes of intermediary metabolism. Proteinaceous microtubules dominate the structural components of cytoskeleton (Lehninger et al. 1993). Because cytoskeleton probably contributes more to total cell mass in larger, more differentiated cells than in small, undifferentiated cells, if all else were equal we would expect such cytoskeleton-rich cells to have a higher N:P ratio.

Extracellular Materials

Unicellular organisms release materials to the exterior of the plasma membrane to form cell walls, and in multicellular organisms, extracellular materials are involved in cellular aggregation, tissue organization, maintenance of intercellular spaces, and construction of protective coverings and support structures. Thus, the biochemical and elemental composition of these materials must be considered for a complete view of organism composition. For example, invertebrates commonly use the nitrogenous polysaccharide chitin (Figure 2) for exoskeletons (insects) and carapaces (crustaceans). In vertebrates the fibrous protein collagen is the major extracellular structural protein in connective tissue and bone, making up one-third or more of total body protein in higher vertebrates (Lehninger et al. 1993). Also of critical importance in determining whole organism elemental content in vertebrates is bone itself, which is deposited as apatite [$Ca_5(OH)(PO_4)_3$] within a proteinaceous connective tissue matrix (Frausto da Silva and Williams 1991). The phosphorous content of bone is sufficiently high that bone has a very low N:P ratio (0.8:1) despite the importance of collagen in the bone matrix.

Organismal N:P Ratio and Life History Characters

Primary life history parameters of organisms, especially specific growth rate, are linked to biochemical composition and body N:P stoichiometry. We focus on crustacean zooplankton, the group with which we are most familiar, but data for other groups illustrates how these linkages are general across broad taxonomic and habitat categories.

Growth Rate and N:P Stoichiometry

Studies of nitrogen and phosphorus composition of zooplankton have revealed patterns (both within and between taxa) that are particularly instructive in understanding links between elemental composition, biochemical makeup, and life history strategies. The two dominant groups of crustacean herbivores in freshwater plankton, calanoid copepods and cladocerans, have contrasting life histories. Cladocerans (e.g., *Daphnia*) grow rapidly, reach sexual maturity (for parthenogenetic reproduction) within days of birth under good food conditions, and produce many generations within a growing season. Calanoid copepods, by contrast, grow slowly, reproduce sexually, and generally complete only one or two generations during a year. In addition, copepods undergo complex metamorphosis, whereas cladocerans do not.

Large differences in body N:P ratio accompany these life history contrasts (Figure 4). Both marine and freshwater calanoid copepods have N:P ratios exceeding 30:1 (Andersen and Hessen 1991, Båmstedt 1986, Hessen and Lyche 1991). However, cladocerans, a predominantly freshwater group, have an elemental composition with slightly less nitrogen and considerably more phosphorus, resulting in lower N:P ratios (12–18:1; Andersen and Hessen 1991, Baudouin and Ravera 1972, Hessen and Lyche 1991). For example, *Daphnia* has an N:P ratio of 15:1 (Figure 4). However, carnivorous cladoceran species seem to have higher N:P ratios than herbivorous cladocerans of similar size (Hessen and Lyche 1991).

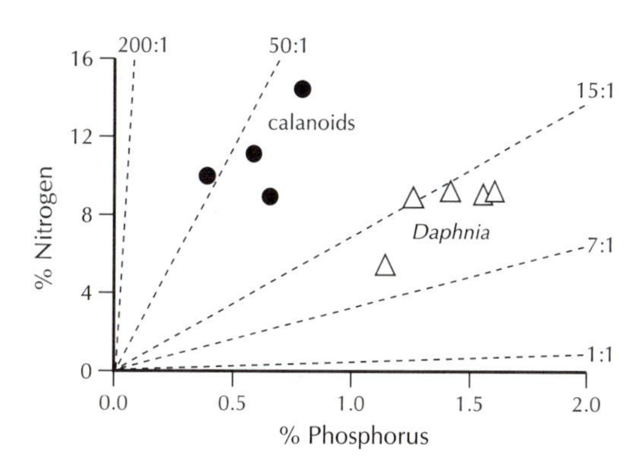

Figure 4. Stoichiometric diagram illustrating the characteristic difference in nitrogen and phosphorus composition for two major herbivorous zooplankton groups: calanoid copepods (circles; four species plotted) and the cladoceran *Daphnia* (triangles; five species plotted). Values for percentage nitrogen and percentage phosphorus are given in terms of dry weight percentage. Dotted lines depict standard values of atomic N:P ratio. Data from Andersen and Hessen (1991) and Hessen and Lyche (1991).

Although data are limited, the biochemical basis of the differences in body N:P ratio between cladocera and copepods is becoming clear (Sterner 1995, Sterner and Hessen 1994). First, nitrogen composition varies from only 8%–10% in both copepods and cladocerans (Andersen and Hessen 1991, Hessen and Lyche 1991). Because nitrogen is a primary constituent of proteins, it is likely that the protein pool varies little across taxa. In contrast, the specific phosphorus content of zooplankton is apparently much more variable: the mean percentage phosphorus values for copepods and cladocerans (especially *Daphnia)* are less than 0.6% and around 1.5%, respectively (Andersen and Hessen 1991, Baudouin and Ravera 1972, Hessen and Lyche 1991). As discussed above, phosphorus is a constituent of several prominent biochemicals in cells: phospholipids, ATP/ADP, and nucleic acids. Phospholipids are a minor constituent in cells, and high-energy adenylates likewise generally contribute less than 1% to dry weight in zooplankters (Båmstedt 1986). This leaves nucleic acids as the remaining candidate to explain variation in percent phosphorus among taxa. In fact, copepods are generally around 2% RNA by weight, whereas *Daphnia* can be as high as 10% RNA (Båmstedt 1986, Baudouin and Scoppa 1975, Dagg and Littlepage 1972, McKee and Knowles 1987). Assuming that RNA is 10% phosphorous (Figure 2), the entire difference in specific phosphorus content of these zooplankton groups (0.8% phosphorous) can be explained by differences in RNA content (Sterner 1995).

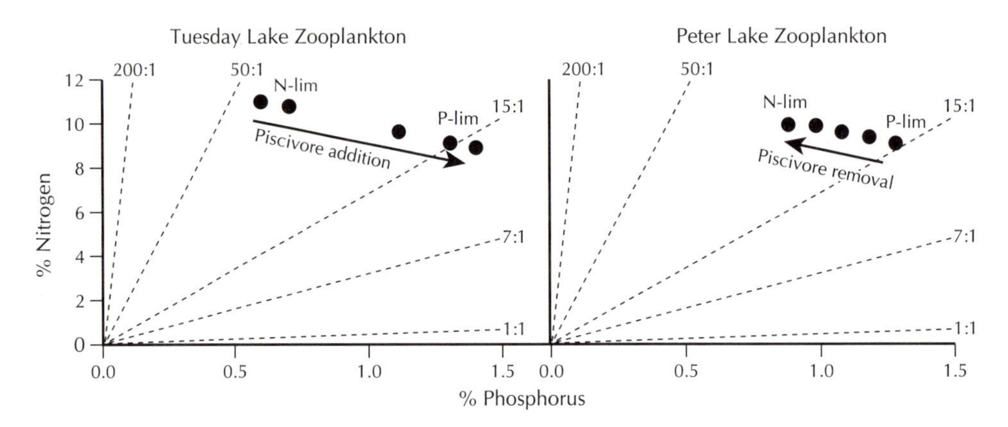

Figure 5. Stoichiometric diagram illustrating changes in estimated nitrogen and phosphorus composition of zooplankton communities during food web manipulations (introduction or removal of piscivorous bass) reported by Wiser et al. (1988). Dotted lines depict standard values of atomic N:P ratio. Values for percentage nitrogen and percentage phosphorus are given in terms of dry weight percentage. Estimates of zooplankton community N:P stoichiometry were made using data for biomass contribution of copepods and *Daphnia* and published values of percentage nitrogen and percentage phosphorus for these groups (see Figure 4). Data points indicate weekly observations during periods of rapid zooplankton change. Shifts in the nature of phytoplankton nutrient limitation (nitrogen versus phosphorus) as a result of changes in zooplankton community N:P stoichiometry are indicated. Tuesday and Peter Lakes are located at the University of Notre Dame Environmental Research Center in Michigan's upper peninsula.

This example provides one of the best illustrations of the fundamental link between organism life history, biochemical investment, and body stoichiometry. Cladocerans in general, and *Daphnia* in particular, have evolved traits that favor rapid growth and high reproductive output, whereas copepods grow more slowly and live longer. *Daphnia's* high growth rate requires a high ribosomal complement for extensive protein synthesis coupled to that high growth rate. Thus, its tissues have high rRNA and therefore high phosphorus contents and low N:P ratio, making *Daphnia* a poor recycler of phosphorus relative to nitrogen.

We can now see the study of Wiser et al. (1988), the first to document qualitative effects of consumers on availability of nitrogen and phosphorus in ecosystems, in a new light.

In Tuesday Lake, predation pressure on zooplankton was reduced by introducing piscivorous bass that reduced zooplankton-feeding minnow populations. *Daphnia* (a superior competitor to copepods by virtue of its high grazing and growth rates) rapidly came to dominate the zooplankton, replacing calanoid copepods (Figure 5). The replacement of high N:P copepods with low N:P *Daphnia* likely produced a high recycling N:P ratio. Thus, the same feature that enables *Daphnia* to achieve dominance under low predation (its rapid growth rate) necessitates investment in biochemical and cellular machinery that lowers body N:P ratio (Figure 3) and elevates recycling N:P ratio in the ecosystem (Figure 5). Conversely, when nearby Peter Lake was manipulated to increase predation intensity on *Daphnia*, slower growing zooplankton species (calanoid copepods and small cladocerans) achieved dominance and, by virtue of their reduced investment in low N:P cellular constituents, generated a low recycling N:P ratio (Figure 5). Such variations in recycling N:P ratio as a function of consumer N:P ratio are now becoming more widely documented (Elser and Hassett 1994, Wiser et al. 1995, Urabe 1993, Urabe et al.1995).

In addition to these nutrient cycling effects, researchers are beginning to explore how the nutritional value of food (specifically food phosphorous content) alters how consumers with different life histories and N:P stoichiometries respond to changes in food web structure. For example, we recently manipulated the food web of a severely phosphorus deficient lake on the Canadian Shield in Ontario by introducing predatory pike to a minnow-dominated food web lacking *Daphnia.* Despite 100-fold reductions in minnow abundance, *Daphnia* increases have been modest and high N:P calanoid copepods remain dominant, suggesting that the poor quality of the lake's severely phosphorus limited phytoplankton prevents dominance by low N:P *Daphnia.* Body stoichiometry thus provides a framework to establish direct mechanistic links between the molecular processes of growth at the level of the cell and the reciprocal interactions between organisms and the ecosystems in which they are found.

Ontogeny and N:P Stoichiometry

Many organisms undergo complex developmental sequences (ontogeny) during which both body size and specific growth rate vary considerably. Thus, we would also expect variation in biochemical and elemental composition to accompany ontogeny. To the extent that N:P stoichiometry changes during development we would also expect variation in an organism's effects on nitrogen and phosphorus cycling and its sensitivity to mineral food quality.

Data on biochemical composition and specific growth rate during ontogeny in a variety of invertebrates show that there are indeed strong ontogenetic shifts in body stoichiometry. For example, daphnids appear to maintain high concentrations of RNA throughout ontogeny, resulting in modestly variable but generally low N:P ratios in all life stages (Baudouin and Ravera 1972, McKee and Knowles 1987; Figure 6A). In contrast, complex metamorphosis in copepods appears to result in more dramatic changes in elemental composition during ontogeny. Protein: RNA ratio in copepod nauplii (stages immediately following hatching) is low (and thus body N:P ratio is low); as development progresses through copepodid (juvenile) stages, growth rate slows and investments in structural proteins increase, causing increases in the protein:RNA ratio (Båmstedt 1986, Dagg and Littlepage 1972) and thus in N:P ratio. Thus, effects of individual copepods on relative nitrogen and phosphorus cycling in nature probably vary strongly during development.

Strong shifts in body biochemical and elemental composition during development are not confined to copepods. Holometabolous terrestrial insects exhibit similar trends in biochemical and elemental composition (Church and Robertson 1966, Lang et al. 1965). For example, body N:P ratio of *Drosophila melanogaster* varies from 9:1 in early larvae to 100:1 immediately before pupation (Church and Robertson 1966; Figure 6A). Variation in body stoichiometry during ontogeny also provides compelling evidence that the main determinant of body N:P in invertebrates is specific growth rate: stage-specific body N:P ratio is strongly ($r^2 > 0.85$) and linearly correlated ($P < 0.01$) with stage-specific growth rate in both *Daphnia* and *Drosophila* (Figure 6B).

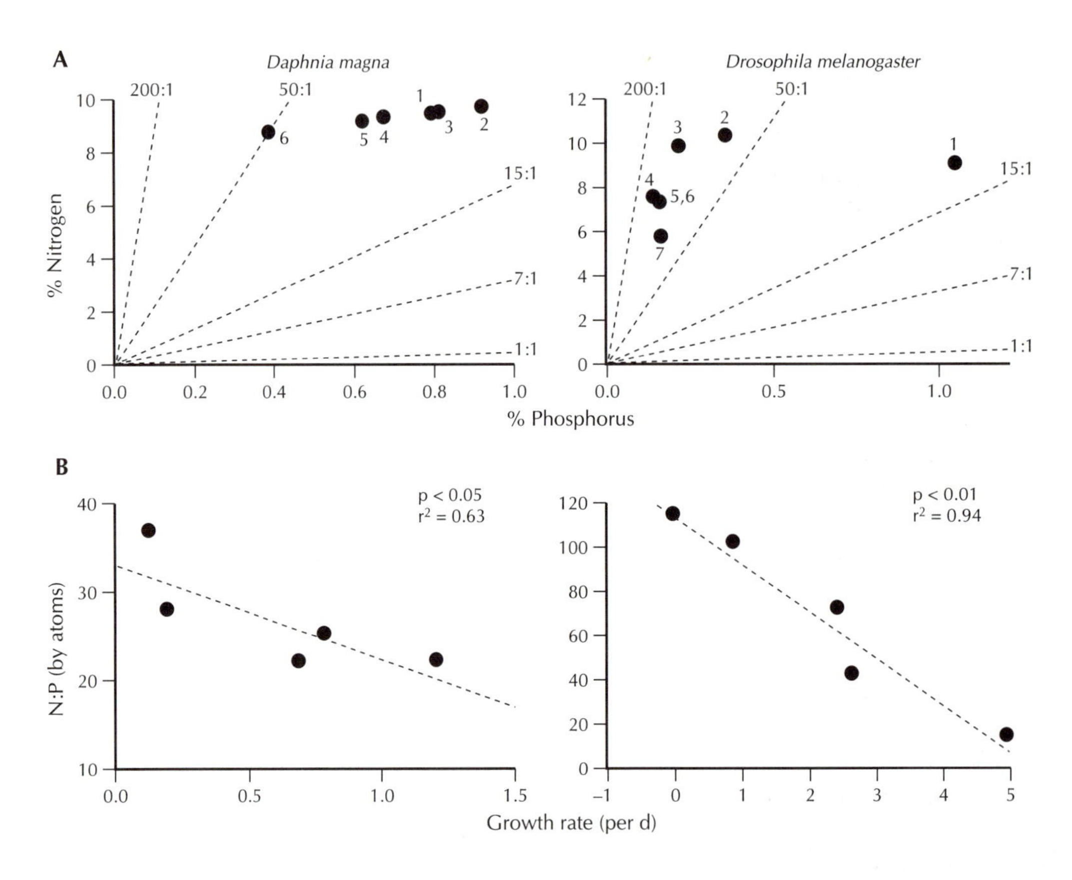

Figure 6. A. Stoichiometric diagrams illustrating ontogenetic variation in percentage nitrogen and percentage phosphorus in *Daphnia magna* and *Drosophila melanogaster.* Dotted lines depict standard values of atomic N:P ratio. Values for percentage nitrogen and percentage phosphorus are given in terms of dry weight percentage. Numbers indicate the ontogenetic sequence of the observations. **B.** Correlation between body N:P ratio and specific growth rate during ontogeny in *D. magna* and *D. melanogaster.* Data for *D. magna* are from McKee and Knowles (1987) and for *D. melanogaster* are from Church and Robertson (1966).

Organism Size and N:P Ratio: a Prediction

We have discussed how evolved characteristics, such as growth rate and developmental sequences, translate into differences in biochemical and elemental composition in consumers. In this section we discuss how macroevolutionary patterns in life history traits also have implications for organismal N:P stoichiometry, focusing on the central parameter of traditional life history theory, organism size. How does body N:P ratio in healthy, actively growing organisms vary in consumer taxa ranging from bacteria (less than 1 picogram [pg]) to large vertebrates (7000 kg or more), a size range of 15 orders of magnitude? The study of allometry in body stoichiometry has considerable potential for new insights into the causes and consequences of evolutionary processes in consumer taxa. These insights may complement those that have resulted from a widespread study of allometry in energetic relations (Schmidt-Nielsen 1984) in consumers.

The question of whether organismal N:P ratio varies significantly and systematically with size distills to whether percent nitrogen and percent phosphorus vary differently with body size. We have already shown that, within certain taxa, percent nitrogen and percent phosphorus are clearly not isometric, because the body N:P ratio varies ontogenetically, and therefore with size, in *Daphnia magna* and, especially, *D. melanogaster* (Figure 6A). But what about allometric variation across taxa ?

Two factors probably influence organism N:P ratio as we proceed from bacteria to large vertebrates. First, the biochemical and elemental composition of protoplasm may vary significantly with size. For example, organisms adapted for rapid growth may contain a greater proportion of RNA, and thus have a lower N:P ratio, than organisms adapted for slower growth. The second factor influencing the size-dependence of body N:P is that increases in size may bring concomitant increases in structural materials, as emphasized by Reiners (1986). For example, the chitinous exoskeletons of arthropods contain more nitrogen than vertebrate structural materials, including bone, which has a very low N:P ratio.

One of the most widely documented allometric patterns is the broad relationship between organism size and specific growth rate (Peters 1983). Allometric declines of specific growth rate (as indexed by production per unit biomass) with size have similar slopes (approximately –0.25) for unicells, poikilotherms, homeotherms, tetrapods, mammals, and fish, although in some cases the intercepts are displaced somewhat and the slope of the relationship for invertebrates is steeper (–0.37). These patterns hold within more specific taxonomic categories as well. For example, in ciliates specific growth rate declines significantly with cell volume (Fenchel 1968).

If the link between organism N:P ratio and specific growth rate (Figure 6) is general, it follows that as organisms increase in size, the organismal N:P ratio should increase (Figure 7). This view is supported by available but extremely limited data. For example, healthy bacterial cells (mass: 1 pg) have extremely rapid growth rates and can have N:P ratio as low as 5:1 (Bratbak 1985). The small flagellate *Paraphysomonas imperforata* (mass: 5 nanogram [ng]) has an N:P ratio of 10:1 (Caron et al. 1985), whereas characteristic N:P ratios for crustacean zooplankton in the 10–100 μg range are generally 12–70:1 (Andersen and Hessen 1991) and those for late instar *Drosophila* larvae (mass: 0.5

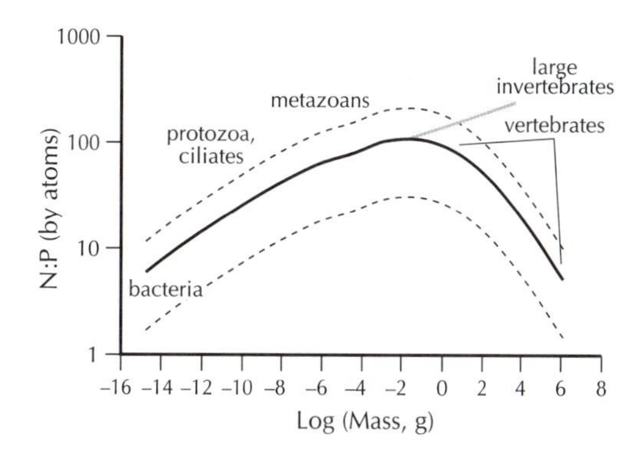

Figure 7. Predicted variation in organism N:P ratio (by atoms) as a function of organism size (g dry weight). Dotted lines above and below the main trend (black line) are meant to indicate that there is likely to be important ecologically or evolutionarily derived variation in organism N:P ratio at any given body size. For large organisms, two trajectories are possible. One trajectory (gray line) corresponds to a continuous increase in organism N:P ratio (which is likely for large invertebrates) and the other to a decline in organism N:P ratio due to increasing bone investment in vertebrates (the main group of animals of large body size).

mg) are 100:1 (Figure 6A). Variation around this general pattern is likely. For example, the different N:P ratios of the similarly sized crustacean zooplankters *Daphnia* and calanoid copepods (12:1 versus greater than 30:1) probably reflect differences in their specific growth rates. Thus, just as allometric patterns of growth rate with body size show considerable scatter, we would also expect to observe substantial deviations from the overall trend of increasing N:P ratio with organism size over this size range. These deviations are likely to have significant evolutionary and ecological causes, as in the case of *Daphnia* versus copepods.

However, the trend of increasing organism N:P ratio with increasing body size is unlikely to continue monotonically throughout the complete range of organism size because other factors, in particular structural investments, come into play at large organism size. In particular, vertebrates, which enter the size continuum at around 100 mg, complicate the picture because contributions of skeletal materials are likely to strongly influence whole organism N:P ratio. Thus, for large organisms, it is necessary to consider how protoplasmic and structural components combine to affect organismal N:P ratio. For both invertebrates and vertebrates of large body size, specific growth rate declines with body size (Peters 1983). Indeed, specific RNA content is known to decline with body size in certain mammalian tissues (Peters 1983). So, N:P ratio of large invertebrate biomass and of "soft-tissue" biomass of vertebrates likely continues to increase with size. However, the structural support investments of vertebrates begin to dominate body N:P ratio and necessitate a different approach to predicting size dependence of N:P ratio within the vertebrates, the dominant group of organisms larger than 10 g.

Evaluating N:P ratio at large body size is facilitated by information on the biochemical and elemental composition of major tissues and organs of vertebrates (Bowen 1966). In addition, the allometry of contributions of these tissues and organs to total body mass is also known (Calder 1984). From these data we can estimate N:P stoichiometry of "theoretical vertebrates" of various body sizes. Various body components vary strongly in elemental content (Figure 8). Tissues such as skin and blood have high protein contents and thus have high N:P ratio (greater than 100:1); in contrast, skeletal material is phosphorus rich due to the deposition of apatite within collagen-based connective tissue (Frausto da Silva and Williams 1991) and has low N:P ratio (0.8:1). The relative contributions of various tissues vary strongly with body size; data for mammals were analyzed by Calder (1984). Of particular interest in this context is an increase in percent skeletal mass with body size from 3.8% of body mass in shrews to 13.6% in elephants (Prange et al. 1979).

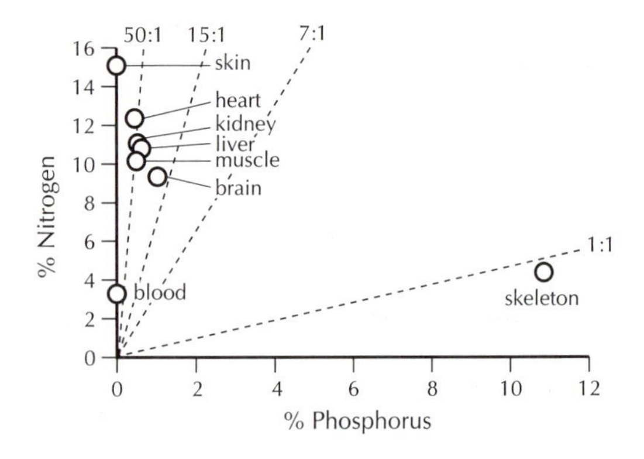

Figure 8. Stoichiometric diagram illustrating nitrogen and phosphorus composition of mammalian organs and organ systems. Dotted lines depict standard values of atomic N:P ratio. Values for percentage nitrogen and percentage phosphorus are given in terms of dry weight percentage. Data from Bowen (1966).

To evaluate N:P variation with body size in vertebrates, we used percent nitrogen and percent phosphorus data for various organ systems compiled by Bowen (1966) as fixed values (although they likely vary with body size as well; see discussion above) and calculated body percent nitrogen and percent phosphorus based strictly on Calder's (1984) equations regarding contributions of tissue and organ systems as a function of body size. These calculations predict a strong size dependence for N:P stoichiometry within vertebrates: we estimate that a 10-g vertebrate is 10.8% N and 0.98% P (N:P ratio: 24:1), whereas a 1000-kg vertebrate is 7.0% N and 1.6% P (N:P ratio: 9.6:1).

The substantial shifts in organism N:P ratio with size (Figure 7) have implications for impacts of consumers on ecosystem nutrient cycling. The increase in organism N:P ratio from bacteria through metazoans implies that larger organisms within this size range will generally be more efficient recyclers of phosphorus than of nitrogen. Relative recycling efficiencies for nitrogen and phosphorus by a consumer depend on the N:P ratios of both prey and consumer (Sterner 1990); thus, the degree to which the consumer differentially recycles nitrogen and phosphorus will be determined by the size dichotomy between predator and prey. Because predators are generally larger than their prey, the increasing N:P ratio with organism size implies that elemental imbalance (food N:P ratio–consumer N:P ratio; Wiser and Hassett 1994) will generally be negative for consumers eating other consumers and thus will result in a low recycling N:P ratio. This is in strong contrast to the interaction between herbivorous consumers and primary producers (which commonly have extremely high N:P ratios); elemental imbalance for the herbivore-producer interaction is frequently strongly positive in lakes, resulting in high recycling N:P ratio (Elser and Hassett 1994).

Decreases in N:P ratio with large body size also probably influence direct effects of large animals on nutrient cycling, an impact that has been increasingly emphasized (Naiman 1988). For example, many studies have shown that migration of anadromous fishes from oceans to lakes, where adults spawn and die, represents a significant source of phosphorus to the lake (Northcote 1988). The low N:P ratio of large vertebrates (Figure 7), including fish (e.g., body N:P ratio of northern pike *[Esox lucius]* is around 12:1 [George 1994]), implies that such fluxes differentially introduce phosphorus (relative to nitrogen) to lakes. Moreover, Carpenter et al. (1992) have shown that fish transport phosphorus from the littoral zone (where they capture prey) to the pelagic zone (where they excrete wastes). However, most studies of fish nutrient cycling have focused exclusively on phosphorus and have not considered nitrogen (but see Vanni 1995).

Our predictions of N:P ratio (Figure 7) indicate that fish have body N:P ratios lower than those of most of their prey (size range: 50 μg–1 g). Thus, stoichiometry predicts that fish should differentially recycle nitrogen relative to phosphorus (Sterner 1990); if fish alter the phosphorus budgets of pelagic ecosystems, then they are also likely to alter the nitrogen budgets even more strongly. Thus, an indirect effect of fish on phytoplankton communities not yet emphasized by aquatic ecologists is an alteration of the N:P supply ratio and thereby a shift in competitive relations among phytoplankton species.

The size dependency of organism N:P ratios also has implications for the role of phosphorus-based food quality in consumer ecology. Organisms with low body N:P ratio, either as adults or at sensitive points in their life history, will have high phosphorus demands for growth and maintenance; these organisms will thus require food that meets not only their energetic demands but also their somatic elemental demand. The recent discovery of effects of phosphorus-deficient food quality on low N:P zooplankton (e.g., *Daphnia*; Sterner and Hessen 1994) provides an intriguing complement to examples of phosphorus-based food quality constraints on vertebrate herbivores (livestock and game populations; McDowell 1992), which also have low N:P ratio. Thus, we suggest that organism N:P ratio provides a tool for identifying taxa, or key life history stages within taxa, that are most; likely to be affected by variation in phosphorus-based food quality in ecosystems. This stoichiometric knowledge may lead to a better understanding of consumer foraging behavior and population regulation that may not be obtained from studies that view production and foraging solely in terms of energetic parameters.

Conclusions

We have shown how the mechanistic bases of a phenomenon occurring at the level of the ecosystem (differential recycling of nitrogen and phosphorus as affected by food web structure) can be traced to the level of the cell and molecule by focusing on N:P stoichiometry. This application of stoichiometric thinking was possible because species that are dominant under contrasting ecological conditions have contrasting life histories requiring different cellular and biochemical investments that necessarily result in differences in body N:P ratio. Elemental stoichiometry can thus provide a new tool to trace the threads of causal mechanisms linking cellular, ecosystem, and evolutionary processes. It offers a means of integrating not only the traditionally disparate disciplines of evolutionary biology and ecosystem science but also ecology and cell biology, two fields that have developed not only independently but often antagonistically. Stoichiometry thus complements energetic perspectives by addressing situations in which energy acquisition and use may not be primary factors dictating fitness or ecological dynamics.

Exploration of the role of stoichiometry in regulation of biological processes is just beginning, even within ecology. Nevertheless, consideration of biochemical and elemental consequences of macroevolutionary trends sets the stage for a unified evolutionary view of the cellular mechanisms that drive ecosystem processes. The power of stoichiometric perspectives arises from the unavoidable demands for chemical elements and the first law of thermodynamics, which applies to all processes by which organisms are born, grow, develop, and die. The essence of this idea was also captured by the Greek philosopher-scientist, Empedocles (quoted by Lotka 1924):

> *There is no coming into being of aught that perishes, nor any end for it... but only mingling, and separation of what has been mingled.*

Acknowledgments

This work was made possible by National Science Foundation grant DEB-9119269 to J. J. Wiser. We are grateful to the members of the Plankton Ecology discussion group, where this project was born. T. H. Chrzanowski, J. F. Harrison, R. P. Hassett, J. Alcock, and three anonymous reviewers provided several important and stimulating comments on the manuscript, and D. E. Chandler helped the authors learn more about cell biology with a gift of several useful texts. We also acknowledge our creative interactions with R. W. Sterner, for whom stoichiometric thinking is now second nature. Wiser is especially grateful to J. P. Collins for providing both encouraging comments and sufficient time for creative thinking while this work was in progress.

References Cited

Andersen, T. and D. O. Hessen. 1991. Carbon, nitrogen, and phosphorus content of freshwater zooplankton. Limnology and Oceanography 36:807–814.

Båmstedt, U. 1986. Chemical composition and energy content. In: *The Biological Chemistry of Marine Copepods*. E. D. S. Corner and S. C. M. O'Hara, eds. Oxford (UK): Oxford University Press. Pp. 1–58.

Baudouin, M. F. and O. Ravera.1972. Weight, size, and chemical composition of some freshwater zooplankton:*Daphnia hyalina* (Leydig). Limnology and Oceanography 17:645–649.

Baudouin, M. F. and P. Scoppa. 1975. The determination of nucleic acids in freshwater plankton and its ecological implications. Freshwater Biology 5:115–120.

Becker, W. M. 1986.*The World of the Cell*. Menlo Park (CA): Benjamin/Cummings Publishing.

Bowen, H. J. M. 1966. *Trace Elements in Biochemistry*. London (UK): Academic Press.

Bowen, H. J. M. 1979. *Environmental chemistry of the elements*. London (UK): Academic Press.

Bratbak, G. 1985. Bacterial biovolume and biomass estimations. Applied and Environmental Microbiology 49:1488–1493.

Brown, J. 1995. *Macroecology*. Chicago (IL): Chicago University Press.

Calder, W. A. 1984. *Size, Function, and Life History*. Cambridge (UK): Harvard University Press.

Campana, T. and L. M. Schwartz. 1981. RNA and associated enzymes. In: *Advanced Cell Biology*. L. M. Schwartz and M. M. Azar, eds. New York: Van Nostrand Reinhold. Pp. 877–944.

Caron, D. A., J. C. Goldman, O. K. Andersen and M. R. Dennett. 1985. Nutrient cycling in a micro-flagellate food chain: II. Population dynamics and carbon cycling. Marine Ecology: Progress Series 24:243–254.

Carpenter, S. R., C. E. Kraft, R. Wright, H. Xi, P. A. Soranno and J. R. Hodgson. 1992. Resilience and resistance of a lake phosphorus cycle before and after food web manipulation. American Naturalist 140:781–798.

Church, R. G.and F. W. Robertson. 1966. A biochemical study of the growth of *Drosophila melanogaster*. Journal of Experimental Zoology 162:337–352.

Dagg, M. J.and J. L. Littlepage. 1972. Relationships between growth rate and RNA, DNA, protein, and dry weight in *Artemia salina* and *Euchaeta elongata*. Marine Biology 17:162–170.

DeAngelis, D. A. 1992. *Dynamics of Nutrient Cycling and Food Webs*. London (UK): Chapman and Hall.

De Zwaan, A. and G. vd Thillart. 1985. Low and high power output modes of anaerobic metabolism: invertebrate and vertebrate strategies. In: *Circulation, Respiration, and Metabolism*. R. Gilles, ed. Berlin (Germany): Springer-Verlag. Pp. 166–192.

Elser, J. J.and R. P. Hassett. 1994. A stoichiometric analysis of the zooplankton-phytoplankton interaction in marine and freshwater ecosystems. Nature 370:211–213.

Elser, J. J., M. M. Elser, N. A. MacKay and S. R. Carpenter. 1988. Zooplankton-mediated transitions between N and P limited algal growth. Limnology and Oceanography 33:1–14.

Elser, J. J., F. S. Lubnow, M. T. Brett, E. R. Marzolf, G. Dion and C. R. Goldman.1995. Factors associated with inter- and intra-annual variation of nutrient limitation of phytoplankton growth in Castle Lake, California. Canadian Journal of Fisheries and Aquatic Sciences 52:93–104.

Evans, W. H. 1989. *Membrane Structure and Function*. Oxford (UK): Oxford University Press.

Fenchel, T. 1968. The ecology of marine microbenthos. III. The reproductive potential of ciliates. Ophelia 5:123–136.

Frausto da Silva, J. J. R. and R. J. P. Williams. 1991. *The Biological Chemistry of the Elements*. Oxford (UK): Clarendon Press.

George, N. B. 1994. Nutrient stoichiometry of piscivore-planktivore interactions in two whole-lake experiments. [M.S. thesis.] University of Texas, Arlington, TX.

Hagen, J. B. 1992. *An Entangled Bank: the Origins of Ecosystem Ecology*. New Brunswick (NJ): Rutgers University Press.

Hessen, D. O. and A. Lyche. 1991. Inter- and intraspecific variations in zooplankton element composition. Archiv fur Hydrobiologie 121:343–353.

Holt, R. D. 1995. Linking species and ecosystems: where's Darwin? In: *Linking Species and Ecosystems*. C. G. Jones and J. H. Lawton, eds. New York: Chapman & Hall. Pp. 273–279

Jones, C. G. and J. H. Lawton. 1995. *Linking Species and Ecosystems*. New York: Chapman & Hall.

Krebs, J. R. and A. I. Houston. 1989. Optimization in ecology. In: *Ecological Concepts*. J. M. Cherret, ed. Oxford (UK): Blackwell Scientific. Pp. 309–338

Lang, C. A., H. Y. Lau and D. J. Jefferson.1965. Protein

and nucleic acid changes during growth and aging in the mosquito. Biochemical Journal 95:372–377.
Lehninger, A. L., D. L. Nelson and M. M. Cox. 1993. *Principles of Biochemistry.* New York: Worth Publishers.
Lindeman, R. L. 1942. The trophic dynamic aspect of ecology. Ecology 23:399–418.
Lotka, A. J. 1924. *Elements of Physical Biology.* Baltimore (MD): Williams and Wilkins.
MacArthur, R. H. 1972. *Geographical Ecology.* Princeton (NJ): Princeton University Press.
Månsson, B. A. and J. M. McGlade. 1993. Ecology, thermodynamics, and H. T. Odum's conjectures. Oecologia 93:582–596.
Mayr, E. 1961. Cause and effect in biology: kinds of causes, predictability, and teleology as viewed by a practicing biologist. Science 134:1501–1506.
McDowell, L. R. 1992. *Minerals in Animal and Human Nutrition.* San Diego (CA): Academic Press.
Mclntosh, R. P. 1985. *The Background of Ecology: Concept and Theory.* Cambridge (UK): Cambridge University Press.
McKee, M. and C. O. Knowles. 1987. Levels of protein, RNA, DNA, glycogen and lipids during growth and development of *Daphnia magna* Straus (Crustacea: Cladocera). Freshwater Biology 18:341–351.
Mieyal, J. J. and J. L. Blumer. 1981. The endoplasmic reticulum. In: *Advanced Cell Biology.* L. M. Schwartz and M. M. Azar, eds. New York: Van Nostrand Reinhold. Pp. 641–688.
Naiman, R. J. 1988. Animal influences on ecosystem dynamics. BioScience 38:750–752.
Northcote, T. G. 1988. Fish in the structure and function of freshwater ecosystems: a "top-down" view. Canadian Journal of Fisheries and Aquatic Sciences 45:361–379.
O'Neill, R. V., D. L. DeAngelis, J. B. Waide and T. F. H. Allen. 1986. *A Hierarchical Concept of Ecosystems.* Princeton (NJ): Princeton University Press.
Pandian, T. J. and F. J. Vernberg. 1987. *Animal Energetics.* San Diego (CA): Academic Press.
Peters, R. H. 1983. *The Ecological Implications of Body Size.* Cambridge (UK): Cambridge University Press.
Prange, H. D., J. F. Andersen and H. Rahn. 1979. Scaling of skeletal mass to body mass in birds and mammals. American Naturalist 113:103–122.
Reiners, W. A. 1986. Complementary models for ecosystems. American Naturalist 127:59–73.
Reinhardt, S. B. and E. S. Van Vleet. 1986. Lipid composition of twenty-two species of Antarctic midwater species and fish. Marine Biology 91:149–159.
Schmidt-Nielsen, K. 1984. *Scaling: Why is Size so Important?* Cambridge (UK): Cambridge University Press.
Sommer, U. 1989. The role of competition for resources in phytoplankton succession. In: *Plankton ecology: Succession in Plankton Communities.* U. Sommer, ed. Berlin (Germany): Springer-Verlag. Pp. 57–106.
Sterner, R. W. 1990. The ratio of nitrogen to phosphorus resupplied by herbivores: zooplankton and the algal competitive arena American Naturalist 136:209–229.
Sterner, R. W. 1995. Elemental stoichiometry of species in ecosystems. In: *Linking Species and Ecosystems.* C. G. Jones and J. H. Lawton, eds. New York: Chapman & Hall. Pp. 240–252.
Sterner, R. W. and D. O. Hessen. 1994. Algal nutrient limitation and the nutrition of aquatic herbivores. Annual Review of Ecology and Systematics 25:1–29.
Sterner, R. W., J. J. Wiser and D. O. Hessen. 1992. Stoichiometric relationships among producers, consumers, and nutrient cycling in pelagic ecosystems. Biogeochemistry 17:49–67.
Tilman, D. 1982. *Resource Competition and Community Structure.* Princeton (NJ): Princeton University Press.
Urabe, J. 1993. N and P cycling coupled by grazers' activities: food quality and nutrient release by zooplankton. Ecology 74:2337–2350.
Urabe, J. M. Nakanishi and K. Kawabata. 1995. Contribution of metazoan plankton to the cycling of N and P in Lake Biwa. Limnology and Oceanography 40:232–241.
Vanni, M. J. 1995. Nutrient transport and recycling by consumers in lake food webs: implications for algal communities. In: *Food Webs: Integration of Patterns and Dynamics.* G. Polis and K. Winemiller, eds. New York: Chapman & Hall. Pp. 81–95.
Wiegert, R. G. 1988. The past, present, and future of ecological energetics. In: *Concepts of Ecosystem Ecology.* L. R. Pomeroy and J. J. Alberts, eds. New York: Springer-Verlag. Pp. 29–55.
White, T. C. R. 1993. *The Inadequate Environment: Nitrogen and the Abundance of Animals.* New York: Springer-Verlag.
Wright, D. H., D. J. Currie and B. A. Maurer. 1994. Energy supply and patterns of species richness on local and regional scales. In: *Species Diversity in Ecological Communities: Historical and Geographic Perspectives.* R. E. Ricklefs and D. Schluter, eds. Chicago (IL): Chicago University Press. Pp. 66–74.

Introduction to

Robert J. Olson and Christofer H. Boggs

Apex Predation by Yellowfin Tuna (Thunnus albacares): *Independent Estimates from Gastric Evacuation and Stomach Contents, Bioenergetics, and Cesium Concentrations*

by James F. Kitchell

How much does an animal eat? How do we measure that? These questions appear in all kinds of ecological contexts. The answers have obvious relevance to behavioral choices made by a forager, to attempts at quantifying predator-prey interactions, evaluating the role of predators in a community context, and their extension to the role of consumers in an ecosystem or landscape perspective. Direct observations allow us to evaluate feeding in many terrestrial environments, but doing so in aquatic habitats is a substantial challenge. The following paper asks the "how much" question in a most challenging way—estimating the rate of feeding by yellowfin tuna, an abundant, highly active and wide-ranging predator of the epipelagic habitat in tropical oceans. Olson and Boggs do an excellent and thorough job of describing the kinds of assumptions and measurements required to accomplish this task.

Three different approaches are compared. In one case, rate of feeding is estimated by measuring rates of input. That involves monitoring stomach contents and multiplying that by the digestion or turnover rate. Although this sounds relatively straightforward, the reader will quickly discover that things are more complex than they would seem at first reckoning.

A second approach involves estimating the amount of energy output. Using the principles of hydrodynamics theory and results of telemetry studies, the authors calculate the energetic cost of swimming, and use that to estimate the amount of energy that must be consumed to balance the overall budget. Exercises of this kind began with Aristotle, have fascinated physicists, mathematicians and engineers for centuries, and once again reveal a mystery known as Gray's Paradox (so named because it is ascribed to Sir James Gray, a noted British zoologist). When the elegant principles and equations of biomechanics are applied to rapidly swimming animals, they usually overestimate the costs of moving between two points. In other words, tunas perform work more efficiently than any machine. We don't know how they do that.

The third approach involves using cesium as a physiological tracer. This approach is representative of the variety of ways that ecologists use the dynamics of individual elements, radioisotopes or stable isotopes to evaluate feeding rates, trophic transfer and bioaccumulation. The relevance of this type of work is growing because fishery exploitation is intensifying in all of the world's waters. Resource managers and conservationists are asking how increased removals of a predator or its prey will affect these ecosystems. In other words, we need to know how much this fish or that actually eats. Olson and Boggs' paper provides a framework for getting those answers. In recognition of its merit, the American Institute of Fishery Research Biologists afforded the W. S. Thompson Award to the authors of this piece. It was deemed the most significant contribution to any of the many fisheries journals published in that year.

Apex Predation by Yellowfin Tuna *(Thunnus albacares)*: Independent Estimates from Gastric Evacuation and Stomach Contents, Bioenergetics, and Cesium Concentrations

ROBERT J. OLSON AND CHRISTOFER H. BOGGS

Original publication in *Canadian Journal of Fisheries and Aquatic Sciences* (1986), 43:1760–1775. Reprinted with permission.

Abstract

Three approaches for estimating predation by yellowfin tuna *(Thunnus albacares)* were compared: (1) stomach analysis adjusted for gastric evacuation; (2) food energy required as a function of swimming speed in yellowfin tracked at sea, and (3) food intake needed to maintain observed cesium concentrations. Gastric evacuation data from captive yellowfin were best fit by linear functions of time for four foods. Fish with high lipid content (mackerel, *Scomber japonicus)* were evacuated at a slower rate (proportion per hour) than smaller fish (smelt, *Hypomesus pretiosus)*, squid *(Loligo opalescens)*, and small fragile fish (nehu, *Stolephorus purpureus)*, all of which had lower lipid contents. Tuna captured in the eastern Pacific had daily rations averaging 3.9% of body mass based on stomach contents and gastric evacuation rates, 5.2% based on bioenergetics estimates, and 6.7% based on the cesium estimate. Swimming costs accounted for one-third to one-half of the energy budget. Annual predation by the eastern Pacific yellowfin population averaged 4.3–6.4 million metric tons during 1970–72, depending on the method used for estimating ration; 34% was frigate tunas *(Auxis* spp.). High growth and turnover rates (*P/B* ratios) of tropical tunas in contrast with low conversion and trophic transfer efficiencies suggest a trophic structure that differs from more productive ecosystems.

Yellowfin tuna *(Thunnus albacares)* are abundant and ubiquitous in tropical regions of the world's oceans. Tunas are opportunistic, generalist predators (Alverson 1963; Blackburn 1968; Magnuson and Heitz 1971). Metabolic rates of tunas greatly exceed those of cold-bodied fishes (Stevens and Dizon 1982) due to their obligate continuous activity (Magnuson 1978) and high "standard" metabolic rates (Brill 1979; Gooding et al. 1981). Estimates of food consumption and an understanding of energy partitioning are needed to explain the ecological strategy of this high-cost mode of living (Stevens and Neill 1978; Kitchell 1983). The large energy requirements of yellowfin, their abundance, and their broad diet make this species useful for estimating pelagic productivity at the intermediate trophic levels occupied by their prey.

Several methods for estimating daily rates of food consumption in fishes have been employed. Direct methods entail quantifying the amount of food in the stomachs of wild-caught fish and adjusting for the gastric evacuation rate determined by laboratory experiments (Elliott and Persson 1978; Jobling 1981a). Indirect methods include constructing energy and nutrient budgets (Davis and Warren 1971; Mann 1978). Both approaches have drawbacks. Direct methods are laborious and provide infrequent point estimates of in situ feeding rate. Bioenergetics modeling of food consumption can be biased by errors in parameter estimation (Bartell et al. 1986) but are useful for evaluating the effect of temperature, body size, and activity of a predator on its food resource (Kitchell 1983). Comparing estimates of food consumption from direct and indirect methods can provide valuable independent validation of laboratory-measured bioenergetics parameters (Rice and Cochran 1984).

In our study, three independent approaches were taken to estimate the flux of energy through a tuna population. In the first, stomach analysis on thousands of field-caught yellowfin tuna was combined with laboratory determinations of gastric evacuation rates for mixed meals of different food items to estimate the daily ration of yellowfin at sea and to estimate predation rates on various prey organisms. The second approach was to employ a model derived from energy expenditures in yellowfin as a function of size and swimming speed in the laboratory (Boggs 1984). This model was applied to swimming speeds measured by acoustic telemetry of yellowfin at sea (Carey and Olson 1982) to estimate typical energy and food requirements. In the third approach, previous determinations of trophic level and cesium concentration in yellowfin and their prey (Mearns et al. 1981) and residence time of cesium in tuna tissue (Folsom et al. 1967) were used to estimate the amounts of prey consumed by yellowfin.

This diversified investigation of food consumption by yellowfin allows for comparison and corroboration of methods, and thus overcomes some of the inadequacies of each approach. The objective was to determine if food consumption by yellowfin in nature was consistent with laboratory results showing a very high energy demand. Validation of laboratory results was sought to increase the credibility of "top-down" trophic models that use bioenergetics parameters such as predation rate (Laevastu and Larkins 1981; Polovina 1984), gross conversion efficiency (Longhurst 1983), or trophic transfer efficiency (Adams et al. 1983) of major predators to describe the trophic system at lower levels. We estimated these parameters for yellowfin tuna, a dominant apex predator in the eastern tropical Pacific Ocean, and used them to estimate the minimum production rate of a variety of prey types.

Materials and Methods

Gastric Evacuation Experiments

Experimental yellowfin were caught by pole-and-line boats using live bait. They were delivered to the Kewalo Research Facility (Nakamura 1972) of the National Marine Fisheries Service in Honolulu, Hawaii, within 2–6 h and placed in 40- or 700-m^3 outdoor tanks. Ambient temperature ranged from 23.5° to 25.5°C. All experiments were conducted within 45 d of capture. Yellowfin exhibit the least degenerative physiological change of any tuna species in captivity. Individuals have been maintained for several years and have gained mass up to 50 kg (C. H. Boggs, pers. obs.).

Sixty-nine yellowfin averaging 36.2 cm in fork length (range = 23.6–45.1 cm) and 973 g (range = 220–1756 g, excluding stomach contents) were individually tagged (with color codes) for easy recognition and were trained to accept mixed meals of dead mackerel

(Scomber japonicus), squid *(Loligo opalescens)*, smelt *(Hypomesus pretiosus)*, and nehu *(Stolephorus purpureus)*. The foods were selected based on their taxonomic and/or gross morphological similarity to important yellowfin prey in the eastern Pacific (Alverson 1963; Anonymous 1984). Red crabs *(Pleuroncodes planipes)*, an important prey, were offered but not eaten by the captive yellowfin.

All food species were acquired frozen, thawed in air, blotted, and weighed to the nearest 0.01 g. Some of the larger mackerel and squid were cut in half to allow ingestion by the yellowfin. Samples of each food type were dried at 60°C to a constant mass to determine water content. Replicate subsamples of dried specimens were analyzed for lipid content using a Soxhlet apparatus (Joslyn 1950) and chloroform-methanol, 2:1 by volume.

The yellowfin were not fed for 24 h prior to experimental feedings, allowing enough time to clear their guts. Although Noble (1973) found that fish may process food more slowly after a moderate period of food deprivation than when feeding continuously, a fairly high frequency of empty stomachs suggests that intervals without food are typical for tunas (Alverson 1963). Mixed meals of the four food organisms were offered, one food particle at a time. The time that each preweighed food particle was eaten by each individually recognized (tagged) fish was recorded to the nearest minute. Feeding continued until the tuna were satiated. The elapsed time from ingestion of each food particle to the time the fish were killed was recorded. Freshly killed fish were weighed to the nearest gram, and fork length was measured to the nearest millimetre. Hemostats were used to ligate the alimentary canal at the esophagus and pyloric sphincter. The stomach was removed from the yellowfin, slit, and the food remains removed, sorted by species, blotted, and weighed to the nearest 0.01 g. The elapsed time from handling the fish to weighing the stomach contents was about 5 min. Stomach contents were oven dried at 60°C to a constant mass at 0.01 g accuracy.

The evacuation data were analyzed separately by food species. Evacuation functions[1] were fitted using wet-mass data, since wet mass was the quantity measured during stomach analysis, and we wished to calculate daily ration on a wet-mass basis. Some data were eliminated prior to curve fitting to correct for a significant bias in this type of data (Olson and Mullen 1986). The problem becomes important when the duration of gastric evacuation experiments is long enough that at least some test individuals empty their stomachs. Empty stomachs must be omitted during data analysis because the exact time they became empty cannot be determined. Prior to the time when the fastest digestors in a sample begin to empty their stomachs, the data include the full range of intraspecific variability expected. But subsequent to that time, an ever-increasing proportion of the sample representing the faster digestors is eliminated from the distribution. Thus, as postprandial time increases, the distribution becomes constricted by the time axis. Olson and Mullen (1986) showed conclusively that "constricted" data distributions can cause serious bias in evacuation rate estimates and a false indication or exaggeration of curvature. Therefore, the data were truncated prior to curve fitting to eliminate those points associated with postprandial times when empty stomachs appear. A simple procedure for choosing points of truncation was explained by Olson and Mullen (1986).

The gastric evacuation data were fitted to linear, square root, and exponential models and the fits evaluated using residual mean squares and statistical procedures of residual analysis, including tests for normality (Filliben 1975), homoscedasticity, and autocorrelation (Wesolowski 1976). The Y_i values (proportion of initial amount recovered from the stomach) were transformed as ln $(Y_i + 1.0)$ and $\sqrt{Y_i + 0.5}$ for the exponential and square root models, respectively (Zar 1974). An arcsine transformation was attempted, but did not improve the approximation of this proportion data to normality. The data were analyzed for the effect of food type, meal size, yellowfin size, mixed meals, partial food particles, and food composition on evacuation rates using analysis of covariance (Dixon and Massey 1957).

Stomach Samples

Yellowfin stomach samples from fish captured by the eastern tropical Pacific purse-seine fishery during 1970–72 (Figure 1) were collected by scientific technicians at tuna canneries in San Diego and San Pedro, California. These yellowfin were captured during daylight hours, and most often in schools associated with dolphins *(Stenella attenuata* and *S. longirostris)*. Fork length of sampled fish was measured to the nearest millimetre and the stomachs were preserved by freezing.

Stomach contents were identified to the lowest possible taxa, counted, and weighed to the nearest 0.1 g. After removing the food, the stomachs were weighed to the nearest

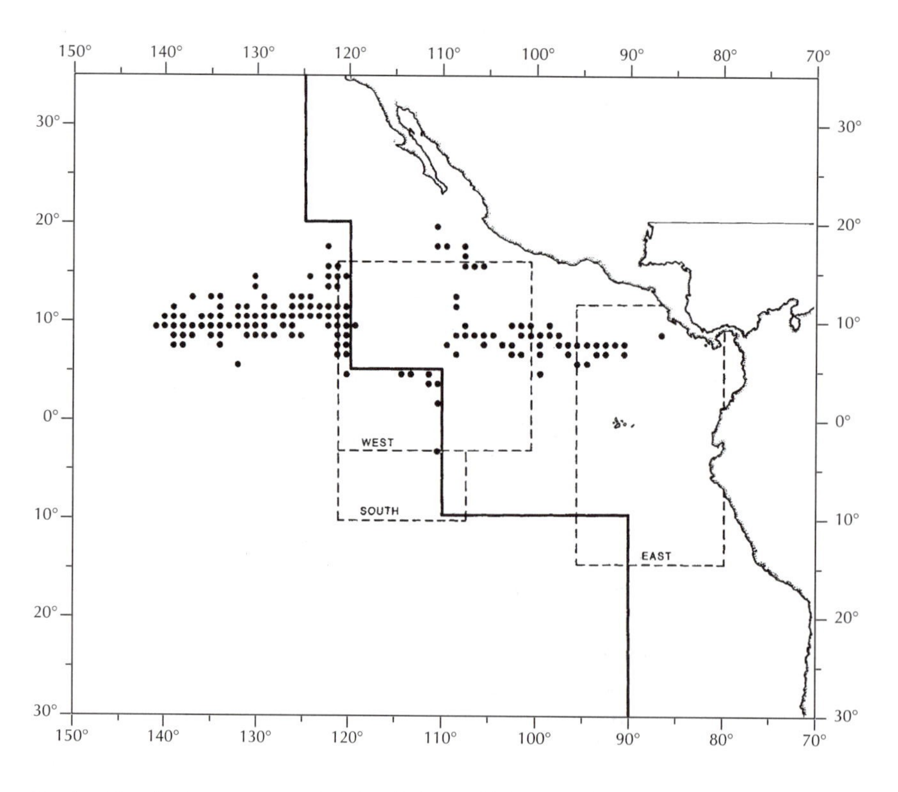

Figure 1. Sampling locations in the eastern tropical Pacific from which stomachs of yellowfin tuna were collected during 1970–72. The western boundary of the regulated fishing area (bold line) encloses 1.696 x 10^{13} m^2 (Sharp and Francis 1976). The broken lines enclose three sampling areas of the EASTROPAC oceanographic expedition (Blackburn et al. 1970).

1.0 g. This was used to adjust for the often missing anteriormost portion of the stomachs due to the canneries' method of eviscerating tuna. A relationship between complete empty stomach mass and yellowfin fork length was determined from an independent sample of 95 yellowfin which we eviscerated. In partially filled complete stomachs the contents were distributed fairly evenly throughout the stomach. Thus, it seemed reasonable that the proportion of each prey category in the stomach contents (if any) missing from each sample was proportional to the weight of the stomach missing.

A second adjustment was made to account for reduced stomach contents due to gear retention and disturbance, since feeding presumably ceases when dolphins and associated yellowfin are chased and while enclosed in a purse seine, but digestion continues at least until death or freezing. The average ration of pursuit and enclosure was determined for fishing operations in the early 1970s and used in conjunction with evacuation rates to adjust upward the weight of stomach contents at time of gut removal.

Stomach contents data were stratified into arbitrary age classes based on yellowfin fork length. Age-class subdivisions were selected to approximate the sizes of yellowfin upon completing their first, second, and third years based on yellowfin otolith increment counts (A. Wild, Inter-American Tropical Tuna Commission, c/o Scripps Institution of Oceanography, La Jolla, CA, pers. comm.).

In representing the relative biomass of prey types consumed, the contribution of each prey to the daily meal was determined by accounting for differences in evacuation rate of each prey. The components of the diet were expressed as proportions of the daily meal.

Ration Estimates

Since preliminary analysis indicated that the evacuation data were poorly fitted by an exponential function, even before data were omitted as described above, an alternative to those methods for calculating daily ration which require an a priori assumption of exponential gastric evacuation (e.g. Elliott and Persson 1978) was needed. A method (Robson 1970) that is appropriate for a variety of evacuation functions was described detail by Olson and Mullen (1986). The model predicts feeding rate ($\hat{r}$, grams per hour) by dividing the mean stomach contents per predator ($\overline{W}_i$, grams) by the integral (A, proportion x hours = hours) of the function that best fits experimental gastric evacuation data. A represents the average amount of time required to evacuate the average proportion of all meals present in the stomach at any instant in time. For a predator that consumes a variety of prey organisms which are evacuated at different rates:

$$\hat{r} = \sum_{i=0}^{I} \frac{\overline{W}_i}{A_i} \quad (1)$$

where subscripts i refer to each of the I prey types. Daily meal is calculated by multiplying $\hat{r}$ by 24 h, and daily ration is daily meal expressed as a percent of body mass.

Bioenergetics Model

Energy requirement calculations were based on a separate study by Boggs (1984) wherein mass and energy content of starved yellowfin were compared with controls to estimate metabolic rates following the approach described by Brett (1973). The method of estimating metabolic rates from starved fish can be used to estimate rates in the field by adjusting for mass, length, and speed and by including the increase due to SDA. Brett (1973) and Boggs (1984) did not find that starvation per se reduced metabolic rate in the highly active fishes studied. The lack of postprandial metabolic increment (SDA), reduced activity, and mass loss during starvation does reduce metabolic rate (Beamish 1964; Glass 1968; Jobling 1980a). However, nonfeeding, mass-specific, speed-specific metabolic rates are not reduced by starvation (Jobling 1980a). In Boggs's (1984) experiments the volitional speeds maintained by tuna differed between individuals, and other individuals were forced to swim fast by altering buoyancy and lift (Magnuson 1970, 1973, 1978). The effects of mass, speed, and length on the metabolic rates of 21 yellowfin were used by Boggs (1984) to fit a model for standard metabolism ($\dot{E}_0$, joules per gram per day) plus swimming work ($\dot{E}_V$, joules per gram per day) where

(2) $$\dot{E}_0 = aM^{\beta}$$

(3) $$\dot{E}_V = fV^{\gamma}L^{\delta}$$

where M is wet mass (grams), V is speed (centimetres per second), L is fork length (centimetres), and a, β, f, γ, and δ are empirically determined constants.

Horizontal and vertical movements of four yellowfin were tracked by Carey and Olson (1982) using acoustic telemetry. The vertical vector of the movements was ignored, since calculations indicated that horizontal movements accounted for almost 100% of distance covered. Swimming speeds were grouped in increments of 0.5 body length • s^{-1}, and the average speed (centimetres per second) and total tracking duration (hours) in each increment were calculated. Then, equation 3 was used to calculate the cost of locomotion ($\dot{E}_V$) for the mean speed in each increment for the four tracked yellowfin.

The sum of locomotory costs at each speed increment was added to standard metabolism ($\dot{E}_0$). Growth rates (grams per day) for eastern Pacific yellowfin of the sizes tracked by Carey and Olson (1982) were calculated based on a size-time relationship determined from recent yellowfin otolith increment counts (A. Wild, pers. comm.) and a previous tetracycline increment validation study (Wild and Foreman 1980). Growth (grams per day) was transformed into joules per day ($\dot{E}_G$) assuming a caloric density of 6.03 x 10^3 J • g^{-1} for wild yellowfin (Boggs 1984). Energy losses due to excretion, egestion, and food assimilation were assumed to account for 35% of the energy consumed in food (Kitchell et al. 1978). Thus, total energy requirements (joules per gram per day) were estimated at

(4) $$\frac{\dot{E}_0 + \dot{E}_V + \dot{E}_G}{1 - 0.35}$$

Cesium Budget

The rate of cesium intake was assumed to balance the rate of cesium loss and to maintain the measured cesium concentration in yellowfin (Mearns et al. 1981) during growth. The rate of loss in yellowfin was assumed to be comparable with the biological half-life of cesium in albacore, *Thunnus alalunga* (Folsom et al. 1967; Young 1970), which is similar to that in other marine fishes (Baptist and Price 1962). A 50% decrease in concentration over 53 d (Folsom et al. 1967) is equivalent to an exponential decline (Cs_{out}) in total cesium content of about 1.3% • d^{-1}. Growth (G) was estimated at 0.63% wet mass • d^{-1} (= percent cesium content per day) for the average-sized yellowfin measured by Mearns et al. (1981). A reasonable value (0.8) (Isaacs 1972) was assumed for cesium assimilation efficiency. Yellowfin cesium concentration (Cs_Y, micrograms per kilogram wet mass) was several times greater than prey cesium concentration (Cs_P) (Mearns et al. 1981). Thus, food intake (F, percent body mass per day) was calculated as

$$F = \frac{(Cs_{out} + G)Cs_Y}{(0.8)Cs_P} \quad (5)$$

The total cesium concentration in the diet was determined from the values given by Mearns et al. (1981) using the proportions of different prey in the diet of age-class 1 and 2 yellowfin in this study. The cesium concentration of squid was used for cephalopods and other invertebrates and that for frigate tuna *(Auxis thazard)* was used for scombrids. The cesium concentration of flyingfish was used for all other fishes. This latter value was assumed, since these other fishes were generally smaller and thus were probably at a trophic level closer to that of flyingfish than that of frigate tuna.

Results and Discussion

Gastric Evacuation

The experimental yellowfin became satiated after about 30 min of feeding. Mean sizes of the four food organisms and the various combinations in which they were consumed are presented in Tables 1 and 2, respectively. Wet mass of the experimental meals averaged 77.1 g ($N = 69$, $\sigma = 36.4$ g, range = 15.8–164.8 g) and was positively correlated ($r = 0.424$, $P < 0.001$) with predator wet body mass (less food mass in the stomach). Relative meal size averaged 8.53% of wet body mass ($\sigma = 3.88\%$, range = 1.31–18.31%) and was negatively correlated ($r = -0.438$, $P < 0.001$) with wet body mass. Subtracting mean water content per food type (Table 1) resulted in a mean experimental meal of approximately 18.9 g ($\sigma = 8.9$ g, range = 3.8–40.2 g) dry mass.

Amounts of the four food types (proportion of initial wet mass) recovered from the yellowfin stomachs as a function of time after feeding are presented in Figure 2. Prior to curve fitting, the data were truncated at postprandial times marked t_B (Figure 2) to avoid a serious bias resulting from distributions constricted by the X-axis ($Y = 0$) (Olson and Mullen 1986). For data with variance comparable with that of Figure 2, Olson and Mullen (1986) found, using Monte Carlo simulations, that data constriction causes the slope to be biased by about 18% ± 9% (mean ± 1 SD). This can cause the illusion of a tail (reduced slope) at later stages of digestion, which may partly explain why curvilinear functions are commonly used with this type of data. The only option to circumvent this problem was to omit some data.

Table 1 Mean size, water content, and lipid content of four experimental foods used in the gastric evacuation experiments (standard deviation in parentheses).

Food species	Mean mass (g)	Mean length (mm)[a]	% water	% lipid: Dry mass	% lipid: Wet mass	N
Mackerel	62.9	188	73.7	30.7 (2.2)	8.1 (0.6)	7
Squid	53.2	160[b]	76.8	19.1 (1.0)	4.4 (0.2)	4
Smelt	17.8	126	76.5	23.6 (1.1)	5.5 (0.3)	5
Nehu	0.4	40	76.2	18.0 (1.2)	4.3 (0.3)	4

[a]Fork length.
[b]Length excludes tentacles.

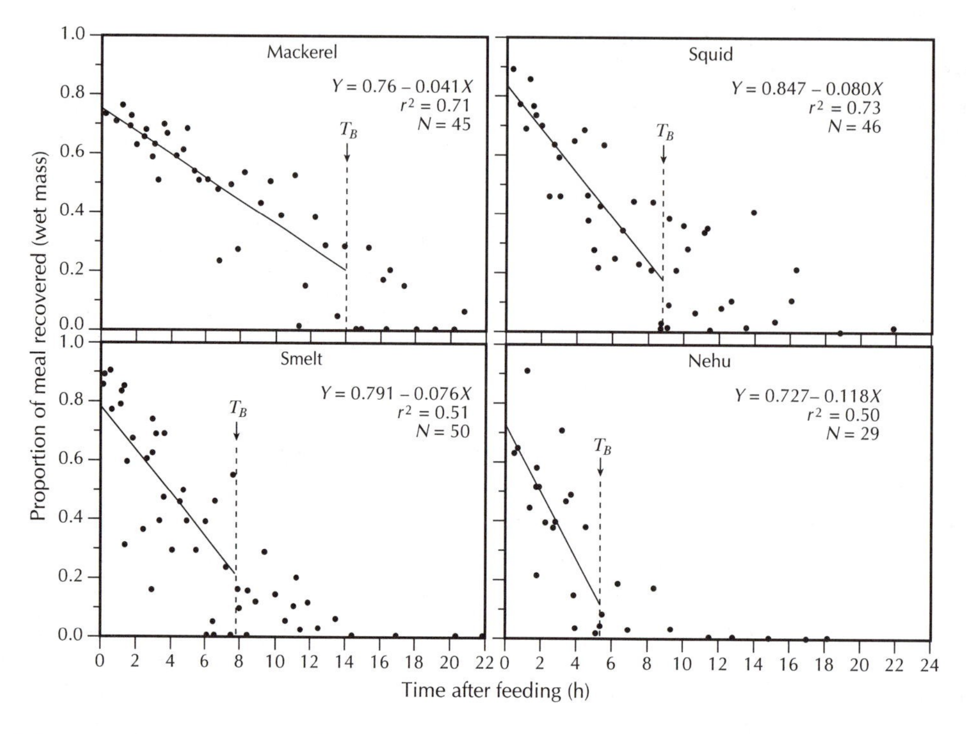

Figure 2. Proportion of initial wet mass of four experimental foods recovered from the stomachs of captive yellowfin versus time after feeding. T_B is the point beyond which data were omitted prior to curve fitting (see text).

Several different mathematical models have been utilized to describe the time course of stomach emptying in fishes, including a linear model (e.g. Swenson and Smith 1973; Jones 1974), an exponential model (e.g. Elliott and Persson 1978), and a square root model (e.g. Jobling 1981a). The four data sets (Figure 2, $X_i < t_B$) were fitted to these three models. For all four food types, residual mean squares (untransformed) of the three fits were homogeneous (F_{max}–test, $P > 0.05$, Sokal and Rohlf 1969). Therefore, residual analysis was used to evaluate the fits. For all four food species, residual analyses indicated that the linear model gave a superior fit, and these regressions (Figure 2) were accepted to represent gastric evacuation rates in yellowfin.[2] The data for dry mass recovered from the stomachs also appeared to be linear. For comparison purposes, Olson and Mullen (1986) fitted the complete unedited data sets (less empty stomachs) to the same three functions. Again, residual mean squares failed to indicate a superior fit to any of the functions (F_{max}–test, $P > 0.05$, Sokal and Rohlf 1969). Residual analysis showed that the linear model gave the best fit for three of the four foods. The squid data were best described by the square root function. Although the exponential model is conceptually more satisfying than the linear model, consensus shows that linear gastric evacuation is plausible for piscivorous fishes that eat large prey (Simenstad and Cailliet 1986). Previous work substantiates this (Swenson and Smith 1973; Diana 1979; Adams et al. 1982).

Table 2. Mixed meal combinations of mackerel (M), squid (Sq), smelt (Sm), and nehu (N) ingested by the experimental yellowfin. Individual food particles were numbered, weighed, and fed in order to individually color-code tagged yellowfin.

Combination No.	Food species	Number that ate meal combination
–	M	2
–	Sq	2
–	Sm	2
–	N	1
1	M, Sq	10
2	M, Sm	3
3	M, N	4
4	Sq, Sm	10
5	Sq, N	1
6	Sm, N	2
7	M, Sq, Sm	11
8	Sq, Sm, N	6
9	M, Sm, N	9
10	M, Sq, Sm, N	6

It is noteworthy that the intercepts of the regression lines fall considerably below 100% (Figure 2). Wet mass of the food in the stomachs a few minutes after feeding was considerably less than the wet mass of food prior to feeding. Magnuson (1969) found the same phenomenon when feeding smelt to skipjack tuna *(Katsuwonus pelamis)*. In Magnuson's experiments, an average of 17.3% of the wet mass of smelt was lost from the food within a few minutes after ingestion. In our experiments, 20.9% of the wet mass of smelt was quickly lost. Magnuson theorized that this loss was equal to the mass of water absorbed by the food during the process of thawing in fresh water. In the present experiments, all food items were air-thawed. Nevertheless, the most plausible explanation for this sudden decrease in food mass is water loss, as there were no visible signs of digestive action on the food particles removed from fish killed within a few minutes after eating. Tuna stomachs may rapidly express water from the food, resulting in a rapid initial loss of mass (Figure 2). Subsequently, food appears to gradually rehydrate with digestive fluids to facilitate mechanical breakdown. Significant positive correlations between percent water in food removed from stomachs and time after feeding were obtained for squid ($P < 0.001$) and nehu ($0.05 > P > 0.02$); correlations were positive but not significant for mackerel and smelt ($0.20 > P > 0.10$).

Factors Affecting Evacuation Rates

The type of prey organism ingested affects significantly the rate at which food is passed from the stomachs of fishes (Windell 1978; Fange and Grove 1979). In yellowfin, significant differences in evacuation rate occurred among the four food types (ANCOVA, $P < 0.0005$). Mackerel were evacuated at a significantly slower rate than squid, smelt, and nehu (*t*-tests for slopes, $P < 0.01$, Zar 1974), which were evacuated at about equal rates (ANCOVA, $P > 0.25$). However, intercepts of the regression lines for squid, smelt, and nehu were significantly different ($P < 0.0005$), and thus a common regression line was not adequate to describe these three data sets.

The evidence for an effect of meal size on gastric evacuation rate of fishes is equivocal. Large meals may take longer to be totally eliminated from the stomach than small meals (Barrington 1957; Steigenberger and Larkin 1974; Jones 1974). However, the amount of food digested during a given time (grams per hour) may increase in proportion to meal size (Hunt 1960; Kitchell and Windell 1968; see review by Windell 1978). In the latter case, the percentage of the meal evacuated per unit time remains constant, and thus the time required to evacuate a meal remains approximately the same regardless of food volume (Magnuson 1969; Tyler 1970; Elliott 1972; Bagge 1977).

The yellowfin gastric evacuation data for each food type were subdivided into three groups according to relative meal size (below 8.0%, between 8.1 and 11.5%, and above 11.6% of body mass). The results of analyses of covariance performed on the subgroups indicated that relative meal size had a significant effect on the evacuation rate of mackerel ($P < 0.0005$), but not on those of squid, smelt, and nehu. Small meals of mackerel were evacuated significantly faster than medium and large meals of mackerel (t-tests for slopes, $P < 0.002$). Therefore, the gastric evacuation rate for mackerel in yellowfin appears to be consistent with the hypothesis of more rapid digestion of small meals, while those for squid, smelt, and nehu seem to follow the alternative hypothesis of a constant percentage per unit time.

Mackerel are larger and appear to be considerably more digestion resistant than the other three food types. The musculature of mackerel is compact, while that of smelt and nehu tends to fall apart in the stomach, providing increased surface area for digestive activity to take place. Squid also have dense musculature; however, the body cavity is more accessible and the body wall is thinner than in mackerel. These results suggest that two or more models of gastric evacuation may be applicable in a single predator, depending on prey type.

Body size is recognized as one of the principal factors which can influence gastric motility in fishes (Windell 1978; Fange and Grove 1979). However, within the size range of yellowfin in this study, analysis of covariance showed that evacuation rates were not significantly influenced by yellowfin size ($P > 0.05$).

Whether foods that are evacuated at distinctly different rates retain their distinct evacuation rates when fed in mixed meals is unknown. Windell (1967) and Elliott (1972) reported that mixed meals of organisms that have similar evacuation rates in bluegill sunfish *(Lepomis macrochirus)* and brown trout *(Salmo trutta)* were evacuated at the same rate as the organisms separately. In the present study, no indication was found to suspect that measuring evacuation rates of food organisms in mixed meals influenced the rate determinations for each food type. For example, evacuation rates (slopes) for mackerel by yellowfin which ate meal combinations 1, 2, 3, 7, 9, and 10 (Table 2) were not significantly different (ANCOVA, $P > 0.05$).

To analyze the effect of cutting some food items into pieces, analysis of covariance was conducted on the data for each food type grouped by whole food specimens, anterior halves, posterior halves, and mixed anterior and posterior halves. Low F-ratios resulted ($P > 0.25$), indicating that the practice of cutting food items did not significantly affect the evacuation rates in these experiments.

Previous workers using prepared foods have demonstrated that low-energy foods are emptied from the stomach more rapidly than foods of high caloric content (Grove et al. 1978; Flowerdew and Grove 1979; Jobling 1981b). Elevated energy levels in natural organisms are commonly related to increased quantities of lipids in the tissues, and the presence of lipids, particularly when in excess of 15% of dry mass (Windell 1978), appears to have a retarding effect on gastric evacuation (Quigley and Meschan 1941; Windell 1967; Fange and Grove 1979). This is generally attributed to a feedback mechanism possibly triggered by a hormone similar to enterogastrone in mammals (Hunt and Knox 1968) produced in response to the presence of lipids or the digestive products of lipids in the duodenum (Windell 1967; Hunt 1975; Jobling 1980b).

In yellowfin, gastric evacuation rates (regression coefficients, Figure 2) were correlated with total lipid content (Table 1) (lipid as percent wet mass: $r = 0.885$, $N = 20$, $P < 0.001$; lipid as percent dry mass: $r = 0.888$, N = 20, P < 0.001) and calories per gram of wet mass ($r = 0.983$, $N = 20$, $P < 0.001$) of the food. It appears, then, that the relationships between caloric content, total lipid content, and evacuation rate found for prepared foods hold true for the natural food organisms tested here. Thus, the high lipid content of mackerel could explain its slow evacuation rate in yellowfin.

Rapid Evacuation Rates in Tunas

This study provides evidence that yellowfin evacuate food from the stomach faster than most other fishes studied, and at about the same rate as skipjack tuna. Magnuson (1969) found that only 1 of 12 species of carnivorous fishes he reviewed had a higher rate of emptying than skipjack. Yellowfin and skipjack emptied their stomachs of smelt in an average of 10.4 and 12 h (Magnuson 1969), respectively. Other piscivores of similar body length reviewed by Magnuson required four to five times longer than skipjack to evacuate a meal. Experiments by Steigenberger and Larkin (1974) on northern squawfish *(Ptychocheilus oregonensis)* were conducted on fish similar in size to our yellowfin, at the same water temperature, and using fish as experimental food. The squawfish required almost 2 h more to empty their stomachs of 3- to 6-cm juvenile rainbow trout *(Salmo gairdneri)* than it took the yellowfin to evacuate nehu, a comparably sized food.

Tunas are a highly specialized group of fishes that must swim continuously to maintain hydrostatic equilibrium and to ventilate their gills (Magnuson 1973; Roberts 1978). Energy expenditures for both "standard" metabolism (Brill 1979) and the metabolic work required for swimming (Gooding et al. 1981; Graham and Laurs 1982; Boggs 1984) are substantially higher than those typical of most other fishes (e.g. Brett 1972). Yet, yellowfin and skipjack are abundant in tropical seas where primary production is purported to be low (but see Kerr 1983) and the food distributions are known to be patchy (Blackburn 1968). It is advantageous for the tropical tunas to have the ability to process large amounts of food in a short time when food is available. The more rapidly food is digested and evacuated from the stomach, the more food a tuna can acquire from what may be very-short-lived aggregations of prey. Any excess energy that is left after meeting metabolic demands is used for growth or stored as body lipids for reproduction and migration (Sharp and Dotson 1977).

Consumption Estimates

STOMACH CONTENTS AND GASTRIC EVACUATION

Total adjustments of prey mass for partial stomachs and gear disturbance by age-class in the 1970–72 stomach samples amounted to increases of 26–46% of the mean observed stomach contents (Table 3). A relationship between complete empty stomach mass (Y, grams) and yellowfin fork length (X, millimetres) was used to adjust for partial stomachs:

$$(6) \quad \ln Y = -12.61 + 2.59 \ln X \ (r^2 = 0.968)$$

The average duration of pursuit and gear enclosure by purse seiners in the early 1970s (2.25 h) and the gastric evacuation functions (Figure 2) were used to calculate adjustments for reduced stomach contents due to fishing

Table 3. Observed and adjusted stomach contents (% of body mass) and size characteristics of four age-classes of yellowfin tuna from which stomach samples were taken in the eastern Pacific Ocean. Adjustments for partial stomachs and gear disturbance are described in the text

		Body mass (g)			Observed stomach contents			Adjusted stomach contents		
Age-class	Fork length (mm)	Min.	Max.	Mean	Min.	Max.	Mean	Min.	Max.	Mean
1	≤ 550	968	3 255	2 629	0	4.98	0.31	0	5.97	0.39
2	551–865	3 310	13 140	8 603	0	3.31	0.48	0	5.55	0.62
3	866–1220	13 187	37 914	22 655	0	4.64	0.38	0	5.50	0.52
4+	>1220	38 010	95 415	53 411	0	2.99	0.39	0	5.46	0.57

gear disturbance. The effect of this adjustment was to increase ration by about 25%.

Mean adjusted stomach contents pooled ranged between 0.39 and 0.62% and maximum adjusted amounts between 5.46 and 5.97% of wet body mass (Table 3). Yellowfin prey were matched with the most similar food organisms tested in the gastric evacuation experiments (Table 4). We assume that the evacuation functions for mackerel, squid, and nehu in small captive yellowfin approximate those of frigate tunas, cephalopods, and gonostomatids/nomeids, respectively, in wild yellowfin. The mean gastric evacuation rate and intercept for the four experimental food organisms (Table 4) were assumed to approximate those of all other prey in yellowfin stomachs. The error created by application of the mean rate to organisms with chitinous exoskeletons is probably not too important to the total ration estimate because crustaceans represented only 3.1–4.8% by mass of the stomach contents (Figure 3). Since the stomach contents indicated that small meals are common, and meal size had an effect on the evacuation rate of mackerel, the regression equation (from ANCOVA above) for mackerel in small meals was used to calculate A, for the Scombridae:

$$(7) \quad Y = 0.856 - 0.0693X \; (r^2 = 0.925).$$

Estimates of average hourly feeding rate, $\hat{r}$, for eastern Pacific yellowfin were calculated

Table 4. Yellowfin prey categories, corresponding gastric evacuation rates (b_1) and intercepts (b_0) from linear regressions fit to evacuation data for mackerel in small meals[a] (M), squid (Sq), and nehu (N) (Figure 2), mean b_1, and b_0, of M, Sq, N, and smelt (Sm) regressions, and A's used in the food consumption model (equation 1). A's were calculated as the area (integral) under the evacuation function: $A = |b_0^2/2b_1|$ (hours).

		Regression values		
Prey	Experimental food species	b_1	b_0	A
Scombridae	M	–0.0693[a]	0.856[a]	5.29
Gonostomatidae	N	–0.1182	0.727	2.24
Nomeidae	N	–0.1182	0.727	2.24
Cephalopoda	Sq	–0.0800	0.847	4.48
Others	Mean of M, Sq, Sm, N	–0.0859	0.805	3.77

[a]From equation 7.

from equation 1 using values for $\overline{W}_i$ and of A_i from data in Figure 3 and Table 4, respectively. Empty stomachs were included for calculating $\overline{W}_i$ since they, as well as full stomachs, reflect the natural feeding condition of the population. Yellowfin are known to feed in the daytime as well as at night (Watanabe 1958; Kume and Morita 1966). Lacking quantitative evidence for diel differences in feeding rate, we have assumed a daily feeding period of 24 h (steady state), recognizing that under some circumstances, nighttime feeding may decline or not occur, which would reduce ration estimates by as much as 50%. Daily meal, then, is assumed to be $24\hat{r}$. Daily rations are the daily meals expressed as a percent of mean mass of yellowfin sampled in each age-class (Table 3).

The initial reduction of wet mass immediately following ingestion in yellowfin is not accounted for in calculating $\hat{r}$ from equation 1. This phenomenon probably occurs with prey in nature as well as in the laboratory. Thus, in calculating daily rations for yellowfin in the eastern Pacific, $\hat{r}$ values were increased by the difference between 100% and the intercept for each food type (Table 4).

Mean daily rations were 2.8, 4.6, 3.6, and 4.5% of body mass per day for age-classes 1, 2, 3, and 4+, respectively. The expected decrease in daily ration with increased size (Kitchell et al. 1978) was not observed. Perhaps the ration of

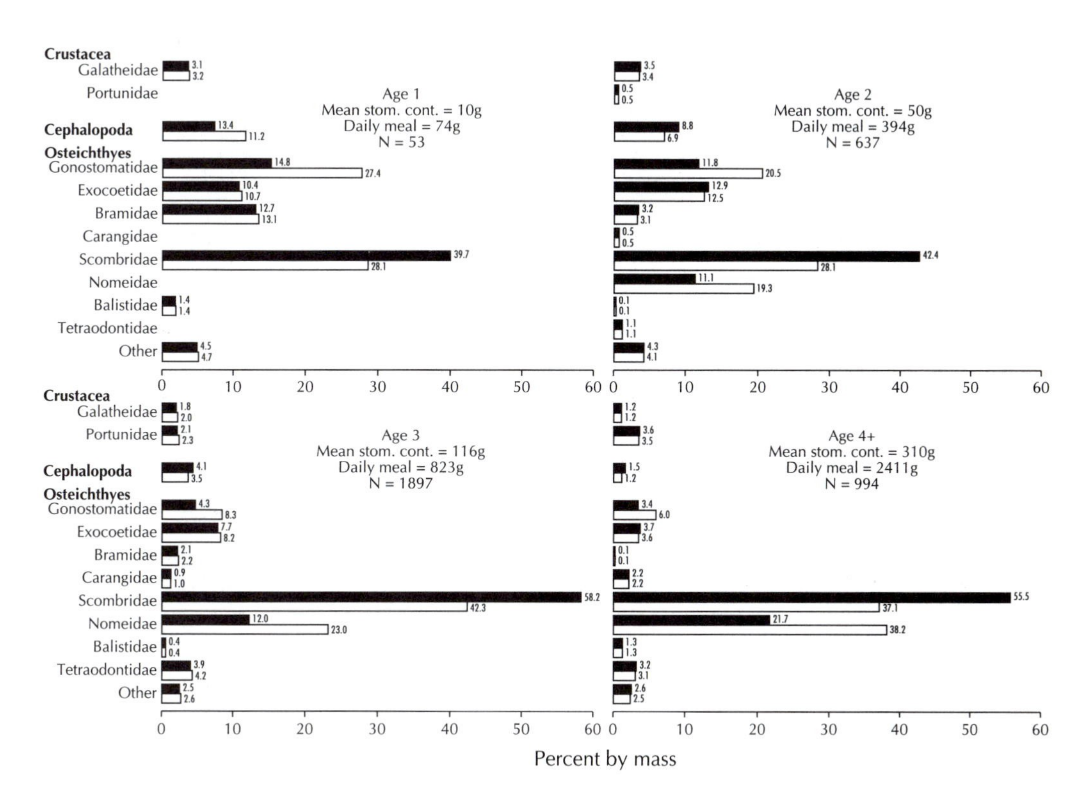

Figure 3. Percent by mass of adjusted stomach contents (black bars) and of daily meal (white bars) for prey found in yellowfin stomachs sampled during 1970–72.

the smallest age-class was poorly represented through some selectivity in the catch. Although the estimated rations are similar to daily rations calculated for small fishes using field data (Seaburg and Moyle 1964; MacKinnon 1973a, 1973b; Swenson and Smith 1973; Lane 1975; Thorpe 1977; Diana 1979; Lane et al. 1979), they are larger than expected for fish of this size.

The tendency to overestimate ration by the assumption of a 24-h feeding period may be offset due to negatively biased stomach samples. Sampling difficulties and an unknown rate of regurgitation during capture would reduce the apparent food mass in stomach samples. Also, gastric motility presumably ceases at death, but digestion is a chemical process that may continue at a reduced rate until the stomach is frozen (Eggers 1977) and after thawing.

Average daily food consumption by eastern Pacific yellowfin appears to be a small fraction

Table 5. Speeds and swimming costs of four yellowfin tuna tracked by Carey and Olson (1982). Observed speeds were grouped by 0.5 body lengths (bl) • s^{-1} increments. The average speed and the total time spent in each increment were used to calculate swimming costs.

Fish ID No. (fork length)	Speed interval (bl • s^{-1})	Mean speed (cm • s^{-1})	Power output (J • g^{-1} h^{-1})		Duration (h)	Swimming cost (J • g^{-1})	Daily swimming cost (J • g^{-1} d^{-1})
2	0–0.5	40	0.8		0.4	0.3	
(87 cm)	0.5–1.0	77	2.4		2.1	5.1	
	1.0–1.5	118	4.9		2.0	9.7	
	1.5–2.0	152	7.4		1.5	11.4	
	2.0–2.5	189	10.6		1.4	14.5	
	2.5–3.0	236	15.3		0.7	11.4	
	3.0–3.5				0	0	
	3.5–4.0	274	19.6		0.3	5.9	
	4.0–4.5	377	33.1		0.4	14.3	
	4.5–5.0	396	36.8		0.4	15.5	
				Total	9.3	88.2	228
3	0–0.5	38	0.7		4.9	3.9	
(89 cm)	0.5–1.0	67	1.9		19.4	36.1	
	1.0–1.5	115	4.6		11.1	50.6	
	1.5–2.0	150	7.1		9.8	69.6	
	2.0–2.5	186	10.1		0.7	7.0	
	2.5–3.5				0	0	
	3.5–4.0	316	24.0		0.2	4.0	
				Total	46.1	170.9	89
4	0–0.5	34	0.6		2.1	1.1	
(98 cm)	0.5–1.0	70	1.8		5.9	10.5	
	1.0–1.5	120	4.3		4.9	21.1	
	1.5–2.0				0	0	
	2.0–2.5	219	11.6		1.4	16.8	
	2.5–3.0	294	18.8		0.5	10.0	
	3.0–3.5	295	19.0		0.2	3.5	
	3.5–4.0	370	27.5		0.5	14.2	
				Total	15.5	77.3	120
5	0–0.5	31	0.5		11.2	5.4	
(96 cm)	0.5–1.0	66	1.6		7.2	11.8	
	1.0–1.5	117	4.3		3.6	15.4	
	1.5–2.0	177	8.4		0.7	6.2	
	2.0–2.5	214	11.5		0.3	3.5	
	2.5–3.5				0	0	
	3.5–4.0	354	26.3		0.6	14.5	
				Total	23.6	56.7	58
				Grand total	94.5	393	

of the maximum possible rate, as evidenced in the laboratory and in the field. Assuming that fish with full stomachs could have continued feeding on the same prey at a rate sufficient to maintain that quantity in the stomachs for 24 h, maximum daily rations would average about 40% of body mass. This is higher, but not double, the maximum rations determined for 1.4-kg skipjack fed small food particles ad libitum for a 12-h daylight period (28–35% of body mass) (Kitchell et al. 1978).

Estimates from Bioenergetics

The mass-specific rate of "standard" metabolism in skipjack tuna was found to decrease significantly with increasing mass in spinalectomized skipjack by Brill (1979). This decrease ($\beta = -0.44$, equation 2) was greater than in other fishes. By extrapolating from swimming metabolism to zero speed, Gooding et al. (1981) found that standard metabolism increased significantly with mass ($\beta = 0.19$). No significant effect of mass on standard metabolism (determined by extrapolation) was found by Boggs (1984). These contradictory results may be related to a small range in mass in the latter two studies. The mean value for standard metabolism ($\dot{E}_0$) determined by Boggs (1984) for yellowfin was 38.4 J • g^{-1} d^{-1} ($a = 38.4$, $\beta = 0$, equation 2).

Boggs (1984) found that mass-specific swimming metabolism ($\dot{E}_V$) of yellowfin had a length exponent (δ) of −1.28, a speed exponent (γ) of 1.64, and a coefficient (f) of 13.7. The speed exponent (γ) was lower than the 2.5–2.8 predicted by hydrodynamic theory (Wu and Yates 1978) but similar to the exponent (1.9) resulting from reanalysis of data on salmon (Wu and Yates 1978; Boggs 1984). In general, the effect of speed on metabolic rate in fishes (Beamish 1978) is less than that predicted by hydrodynamic theory.

Daily swimming costs (Table 5) calculated using equation 3 accounted for one-third to one-half of the energy budget (Table 6) of the four yellowfin tracked by Carey and Olson (1982). The mean swimming cost (weighted by the duration of each track) amounted to 100 J • g^{-1} • d^{-1}. The most frequently. recorded speeds of the four tracked yellowfin were less than 100 • cm s^{-1} (less than one body length per second for the fish, Table 5), but faster

Table 6. Daily ration estimates (% body mass • d^{-1}) for four yellowfin tuna tracked by Carey and Olson (1982) based on energy expenditures for swimming, standard metabolism, growth, excretion, egestion, and food assimilation.

Fish ID No	Fork length (cm)	Mass (kg)	Energy expended (J • g^{-1} • d^{-1}): Swimming cost (Table 5)	Standard metabolism $\dot{E}_0$	Growth ($\dot{E}_G$)	Excretion, egestion, and assimilation	Total consumed	Ration (% • d^{-1})
2	87	13.4	228	38	21	155	441	9.6
3	89	14.3	89	38	20	79	226	4.9
4	98	19.3	120	38	18	95	270	5.9
5	96	18.1	58	38	18	62	175	3.8
Mean			100[a]	38	19		241[b]	5.2[c]

[a]Weighted mean from grand total in Table 5.
[b]From equation 4 using weighted mean swimming costs of 100 J • g^{-1} • d^{-1}
[c]Based on total of 241 J • g^{-1} • d^{-1}

speeds were common, and speeds as high as 300–400 cm s^{-1} were maintained for over half an hour. These short times spent at high speeds contributed substantially to the total estimated cost of swimming (Table 5). Daily swimming costs varied by a factor of 4 among the four tracked yellowfin. There was a trend for speeds and locomotory costs to be low in the two fish that were tracked for a full day or longer (Table 5).

Estimates of total energy consumption ranged from 175 to 441 $J \cdot g^{-1} \cdot d^{-1}$ (Table 6). When total energy requirements were converted into a daily ration using a caloric equivalent of 4.60 kJ • g wet $mass^{-1}$ for food (Kitchell et al. 1978), the mean ration for 87- to 98-cm yellowfin was equivalent to about 5.2% of wet body mass per day. This was about 1.1–1.4 times the estimates from stomach contents and gastric evacuation rates for similarly sized yellowfin (age-classes 2 and 3). This suggests that the energy requirements we estimated were reasonable and that extrapolation of the empirical model for swimming costs (equation 3) did not result in values that were unrealistically low.

Sharp and Francis (1976) estimated a mean energy requirement of 319 $J \cdot g^{-1} \cdot d^{-1}$ for 12.6-kg yellowfin based on a hydrodynamic model for power output and assuming that yellowfin swim at 300 cm s^{-1} for 1.2 h per day and at near minimum speeds (Magnuson 1973) the rest of the day. Sharp (1984) used the same hydrodynamic model to estimate an average energy requirement of 232 $J \cdot g^{-1} \cdot d^{-1}$ for two yellowfin tracked by Carey and Olson (1982). Both of these estimates (Sharp and Francis 1976; Sharp 1984) are within the range of total energy expended estimated using equation 4 (Table 6). Sharp and Francis's (1976) model includes the effects of transition between laminar and turbulent boundary layer conditions. If the velocity exponent does increase radically at Reynold's numbers above about 106 (Sharp and Francis 1976), then the values given in Tables 5 and 6 could be underestimates.

Energy can be supplied from stored reserves of fat as well as from food consumption. Thus, bioenergetics estimates of required energy may exceed actual consumption over a short term. Of course, fat reserves are ultimately derived from surplus energy intake, but the feeding conditions that permit such fat storage are probably not represented by the average stomach contents observed in the field. It is clear from our laboratory studies that yellowfin can eat much more than the overall average rates indicate, and it is clear from field studies that fat content in tunas is highly variable and can increase dramatically in certain areas (Dotson 1978; Vlieg et al. 1983). Tuna bioenergetics may be pictured as "speculative" (Stevens and Neill 1978) in the sense that tuna may exhaust their energy reserves in migrating hundreds of kilometres (Sharp and Dotson 1977) to reach good foraging habitats. This gamble must succeed frequently enough to permit rapid rates of growth among survivors, but the gamble must sometimes fail, resulting in energy depletion, disease, and vulnerability to predation. This would help explain the high natural mortality rates of tuna (Murphy and Sakagawa 1977). Depletion of fat reserves during reproduction would increase the potential for severe energy depletion and might explain reports that sex ratios are skewed strongly towards males in very large yellowfin (Cole 1980; Anonymous 1983). Further investigation of fat dynamics could suggest patterns in foraging behavior, energy utilization, and movements of tunas.

AN ESTIMATE FROM CESIUM CONCENTRATIONS

The proportions of food in the diet (stomach contents adjusted for differential rates of gastric evacuation) of age-class 1 and 2 yellowfin (white bars, Figure 3) were used to represent those of a typical yellowfin in the size range used by Mearns et al. (1981) in their measurements of cesium concentrations.

Thus, approximately 28% of the diet (by wet mass) was frigate tunas and other scombrids, 12% was squid and other invertebrates, 12% was flyingfish, and 48% was other fishes (34% gonostomatids and nomeids). Assigning these proportions (Materials and Methods) to the cesium concentrations of prey given by Mearns et al. (1981) resulted in an estimate of total dietary cesium concentration (Cs_P) of about 21.8 µg • wet kg food^{-1}. Yellowfin cesium concentration (Cs_Y) was given as 60.5 µg • wet kg^{-1}. Thus, from equation 5 the estimated daily ration is 6.7% of body mass per day. This is 2.4 and 1.5 times higher than the estimates from gastric evacuation rates and stomach contents for yellowfin of similar size (age-classes 1 and 2, respectively). The greatest difference is due to the anomalous low ration estimate from stomach contents of age-class 1. The cesium budget estimate is 29% higher than the ration estimate from energy requirements (mean ration, Table 6). The cesium estimate was made for smaller fish than the bioenergetics model estimate, and a larger ration is expected in smaller yellowfin (Kitchell et al. 1978).

The data on cesium concentrations were limited, and the assumptions required to translate these into a ration are crude. This approach may be the least accurate of the three. However, the consistency of results from the cesium approach and the bioenergetics approach implies that stomach analysis under estimates the true amount of food in stomachs of yellowfin at sea, especially in smaller fish. For comparable age-classes, the bioenergetics estimate (5.2%) averaged 1.3 times higher (ages 2 and 3) and the cesium estimate (6.7%) averaged 1.8 times higher (ages 1 and 2) than the direct estimates from stomach contents and gastric evacuation rates. Together, the indirect estimates averaged 1.5 times the results from the direct method.

Apex Predation and Trophic Relations

It is difficult to estimate biomass or production at intermediate trophic levels in pelagic marine ecosystems by direct sampling. As an alternative, bioenergetics models combined with estimates of apex predator biomass from fisheries statistics, tag and recapture studies, or surveys can be used to estimate the rate of prey production required to balance consumption by predators (Laevastu and Larkins 1981; Polovina 1984). This trophic approach can be extended down through the food web to estimate production and biomass at lower levels. The trophic model inputs are predation rates determined from studies like this one, and turnover rates determined from studies of age structure (Allen 1971) and natural mortality (Pauly 1980) or from other estimates of annual production to annual mean biomass (P/B) ratio (Banse and Mosher 1980; Adams et al. 1983; Longhurst 1983; Polovina 1984).

Total predation by yellowfin in the eastern tropical Pacific depends on the number of yellowfin in that region. Cohort analysis has been used to estimate the size of the exploited yellowfin population (Anonymous 1983) in the regulated area shown in Figure 1. The average of estimates for 1970–72 is 4×10^7 individuals (3×10^8 kg). This population was subdivided among the four age-classes used to stratify the stomach contents data (Table 3) according to average size composition of the catch (Anonymous 1983, p. 87). Dividing the numbers and biomass of yellowfin in each age-class by the area of the regulated fishery (Sharp and Francis 1976) (Figure 1) results in estimates of yellowfin density (Figure 4).

Predation rates on each prey type can be estimated from their proportion in the diet. Proportions of prey in the stomach contents (black bars, Figure 3) do not adequately represent relative biomass of prey types consumed due to differential rates of digestion and gastric evacuation (Hess and Rainwater 1939; Macdonald et al. 1982). Although Persson (1984) recommended otherwise, stomach contents were adjusted for differential rates of gastric evacuation (white bars, Figure 3). This adjustment substantially reduced the proportion of Scombridae in the diet compared with the proportion found in stomach samples. Nevertheless, in the offshore areas (Figure 1), scombrids (almost entirely frigate tunas, *Auxis* spp.) were the most important prey by mass in all age-classes except 4+. Small epipelagic and mesopelagic fishes, Nomeidae (mostly *Cubiceps pauciradiatus),* and Gonostomatidae (mostly *Vinciguerria lucetia),* respectively, were either first, second, or third most important prey by mass in the diet of each age-class. Exocoetidae (flyingfish) and cephalopods (mostly squids) were also important by mass. The food habits of yellowfin inhabiting inshore and island areas differ from this pattern (King and Ikehara 1956; Alverson 1963).

Total annual predation by the yellowfin population was divided into predation on each major prey type using the adjusted proportions in the diet (Figure 4). Assuming that a long-term equilibrium exists, the energy passed from prey trophic levels to the predators represents production rather than a decline in standing stock of prey. Under this assumption, predation rates of yellowfin tuna represent minimum rates of production, and in many cases, the only estimates of production for these pelagic ani-

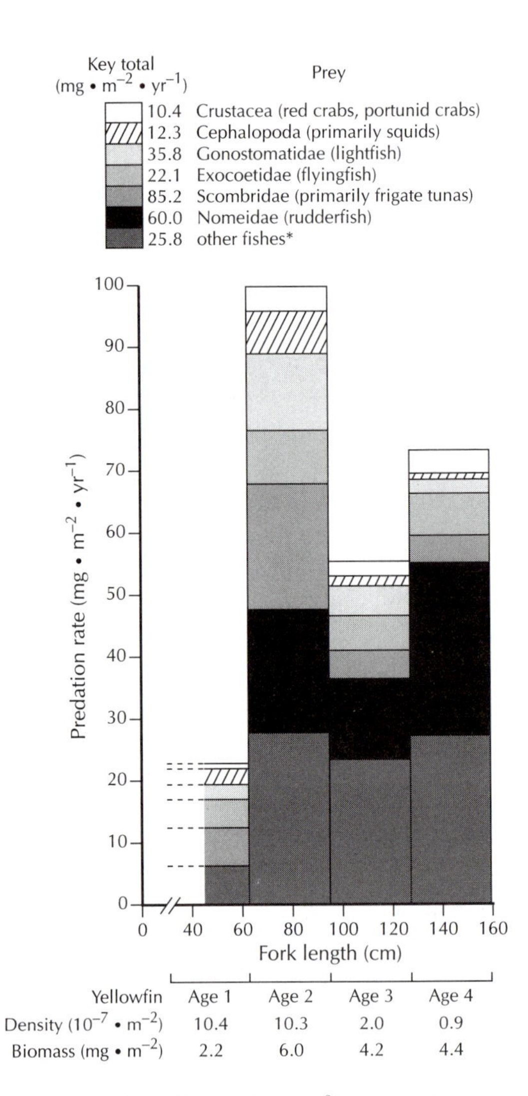

Figure 4. Total prey biomass ($mg \cdot m^{-2}$) consumed per year by the yellowfin population in the eastern tropical Pacific Ocean in 1970–72 based on stomach contents and gastric evacuation rates. Yellowfin density and biomass are given at the bottom for each age-class. *The "other fishes" category was composed primarily of Bramidae, Carangidae, Balistidae, and Tetraodontidae.

mals. Actual predation rates may be about 1–1.5 times higher than these estimates, since the consumption estimates based on energy requirements and cesium concentrations were about 1.5 times the consumption estimates based on stomach contents. This range (1–1.5 times) is incorporated in the following estimates.

The annual production of frigate tunas in the regulated fishing area (Figure 1) during 1970–72 must have amounted to at least 1.4–2.1 million metric tons (*t*). This amount was calculated by multiplying the predation rate on Scombridae (Figure 4) by the area of the regulated fishing region (Figure 1) and the range in the ration estimates (1–1.5 times). This production is more than the average annual world catch of all tunas during those years (1.2 million *t*, FAO 1974). It amounts to about 11–17 times the average biomass of yellowfin caught in this area annually during those years (Anonymous 1983). Frigate tunas are important prey of other apex predators that inhabit the region (Uchida 1981), so these figures underestimate total production of frigate tunas. Possible interaction between frigate tunas and young stages of other tunas could have important implications on the tuna stocks. If frigate tuna prey on larval or postlarval yellowfin, then increasing the commercial harvest of yellowfin, resulting in larger standing stocks of yellowfin prey, might have a deleterious effect on yellowfin recruitment. Previous to this study there were no estimates of standing stock or production for the frigate tunas. These species are largely unexploited in the eastern tropical Pacific.

Turnover rates for frigate tunas can be approximated using an equation relating the *P/B* ratio to body mass (Banse and Mosher 1980) and from the natural mortality rate ($P/B = M$, Allen 1971). Average mass at maturity (400 g) and von Bertalanffy growth parameters ($L_\infty = 53$ cm, $K = 0.4$) were estimated from the sparse data summarized by Uchida (1981). The body mass relationship implies an annual *P/B* ratio of 0.5, but Banse and Mosher's (1980) equation was derived from data on small, temperate fishes. Empirical relationships between *P/B* ratios and body mass should be interpreted cautiously (McLaren and Corkett 1984). Longhurst (1983) discussed the reasons for larger *P/B* ratios for marine fishes in the tropics. Pauly's (1980) formula for natural mortality (*M*) as a function of L_∞, *K*, and mean environmental temperature (25°C) suggests a *P/B* ratio of 0.8 for frigate tunas. Fishing mortality (*F*) in the eastern tropical Pacific is negligible (Anonymous 1983). Based on our production estimate, this would indicate a standing stock of at least 1.7–2.6 million *t* of frigate tunas in the regulated fishing area (Figure 1).

The turnover rate for the yellowfin tuna population can be approximated as the total mortality rate ($P/B = Z = M + F$). Fishing mortality averaged 0.44 during 1970–72, and natural mortality was estimated at 0.8 (Anonymous 1983). Thus, *P/B* was about 1.2 yr^{-1}. In contrast, Banse and Mosher's (1980) equation gives an estimate of only 0.2 for an average mass at maturity of 30 kg (Cole 1980). Some estimates for tropical marine fishes range as high as 3.4–4.5 (Longhurst 1983). The *P/B* for skipjack tuna in the Pacific Ocean is about 5.3 at MSY (R. F. Francis, National Marine Fisheries Service Northwest and Alaska Fisheries Center, 2725 Montlake Blvd East, Seattle, WA 98112, pers. comm.).

High *P/B* ratios could result from an abundant food supply, a high trophic transfer efficiency, or both. Using our estimate of food consumption by yellowfin tuna, we can proceed to make a direct estimate of gross conversion efficiency and trophic transfer efficiency. The average food consumption for an individual yellowfin tuna estimated from bioenergetics was 241 $J \cdot g^{-1} \cdot d^{-1}$ whereas growth was estimated at only 19 $J \cdot g^{-1} \cdot d^{-1}$ (Table 6), indicating a gross conversion efficiency of only 8%. Annual yellowfin production was about 1.2 times the standing stock of yellowfin whereas annual consumption by yellowfin was estimated at 15–22 times the standing stock of yellowfin (Figure 4), indicating a trophic transfer efficiency of only 8–5%. These efficiencies seem low compared with those in more productive ecosystems (Kozlovsky 1968; Adams et al. 1983; Longhurst 1983; Polovina 1984).

The availability of food is difficult to address without a more complete trophic analysis. Primary production and standing stocks of phytoplankton, zooplankton, crustacean micronekton, and fish–cephalopod micronekton in the eastern tropical Pacific were measured by the EASTROPAC oceanographic expedition (Owen and Zeitzschel 1970; Blackburn et al. 1970). These studies examined areal and seasonal variations over a year (1967–68) in a comprehensive survey of a magnitude that is unlikely to be repeated. They found that the average standing stock of fish–cephalopod micronekton in the western region (Figure 1) was about 3 $mL \cdot m^{-2}$ or about 3 $g \cdot m^{-2}$ for a 200-m water column (Blackburn et al. 1970). This is only twice as much as would be consumed annually by the estimated standing stock of frigate tunas if they ate 4% of their body mass per day.

Only 2.8% or 0.08 $g \cdot m^{-2}$ of the fish-cephalopod standing stock measured by Blackburn et al. (1970) consisted of epipelagic fishes whereas 90% (0.23–0.35 $g \cdot m^{-2} \cdot yr^{-1}$) of predation by yellowfin in the offshore regions (Figure 1) was on epipelagic fishes, predominantly frigate tunas, nomeids, and exocoetids (Figure 4). These estimates are not strictly comparable, since some of the prey consumed by yellowfin were larger than micronekton (1–10 cm), but they suggest some degree of scarcity and/or a high annual turnover rate for the epipelagic forage fishes.

Sharp and Francis (1976) estimated that daily consumption by the yellowfin population was small compared with the standing stocks measured by Blackburn et al. (1970). They considered the relative effective productivity at different size-related trophic levels as well as the underestimation of the standing stocks of the more mobile tuna forage species, and argued that yellowfin with fork length greater than 40 cm were not food limited. Of course, the forage available to yellowfin should be based on the productivity of their forage rather than standing stock, but Sharp and Francis were not confident that the available information from the EASTROPAC files was adequate regarding the standing stock figures beyond an absolute minimum estimation (G. D. Sharp, pers. comm.). One should consider that a proportion of the yellowfin forage comprises organisms larger than micronekton, and these prey would be less abundant and less productive than the micronekton. Also, the importance of foraging by other apex predators must be considered.

The population of four dolphin species (*Stenella* spp. and *Delphinus delphis*) is estimated at 7–8 million individuals (R. S. Holt, National Marine Fisheries Service, Southwest Fisheries Center, pers. comm.), and the food consumption of the spotted dolphin (*S. attenuata*) population (3.1–3.5 million individuals) is estimated to be roughly equivalent to that of the yellowfin population (E. F. Vetter, National Marine Fisheries Service, Southwest Fisheries Center, pers. comm.). Thus, food consumption by all four dolphin species could amount to 2 times that of the yellowfin population. Dolphins consume many of the same types of prey as yellowfin (Perrin et al. 1973). We estimate prey consumption by yellowfin to represent less than 33% (100% x (1/(1 + 2))) of the apex predation in this community. The predation rates of other tuna species, billfishes (Istiophoridae and Xiphiidae), dolphinfish (Coryphaenidae), sharks, and seabirds require further investigation before concluding that the food supply is more than adequate.

Conclusions and Implications

The estimates of daily ration for yellowfin tuna derived from the bioenergetics model and tracking study were median among the three independent estimates, lending validity to the use of such models in trophic analyses. The differences between our direct and indirect estimates imply that ration estimates derived from gastric evacuation rates and stomach contents data were too low or that the bioenergetics and cesium budget estimates were too high. The possibility that the bioenergetics model overestimates tuna energy requirements deserves consideration. Metabolic rates of tunas are reported to be 2–3 times higher than in other active fishes swimming at similar speeds (Gooding et al. 1981; Stevens and Dizon 1982; Boggs 1984). Perhaps these findings are due to some bias in the measurement of tuna metabolic rates (due to capture stress, confinement, handling, poor nutrition, starvation, etc.). The independent estimate of energy requirements from stomach contents and gastric evacuation rates indicates that the bioenergetics model could be positively biased by about 30%. This is not enough to account for the major difference between tunas and equally active, cold-bodied fishes. In contrast, the other independent estimate (from cesium concentrations) Suggests that the bioenergetics model is not positively biased.

Ration estimates for dominant apex predators and estimates of trophic transfer efficiency are the raw materials required for "top-down" trophic models that can provide estimates of production and biomass for animals at lower trophic levels. Our independent estimates of ration varied above and below the bioenergetics estimate by less than 30%. Direct sampling of forage biomass at sea may be much less precise than estimates based on the "top-down" approach. Direct standing stock estimates such as those of Blackburn et al. (1970) are hampered by gear inadequacies, time limitations, and increasing costs of ship time. Increased understanding of patchy distributions (Fasham 1978; Wormuth and Roper 1983), gear avoidance and gear selectivity (Kashkin and Parin 1983; Pearcy 1983) suggests that the estimates of Blackburn et al. (1970) may be least accurate for the highly mobile types of prey consumed by yellowfin. Direct sampling must be repetitive to provide more than point estimates whereas trophic models can be dynamic and descriptive of changes at lower levels as a function of the bioenergetics and food habits of the apex predators (Laevastu and Larkins 1981; Stewart et al. 1981). Major apex predators are much more effective than research vessels as samplers of production at intermediate trophic levels.

The "top-down" trophic approach in our study reveals a huge unexploited resource in the form of the frigate tunas in the eastern tropical Pacific. The relationship between frigate tuna production and yellowfin tuna production requires further examination. A more complete trophic analysis could indicate the presence or absence of a food surplus for yellowfin tuna. Such a model will require inputs for the other predators. If adequate estimates for the prey are also found, the analysis can be extended down to the level of primary production and compared with independent estimates, which was done by Adams et al. (1983), Longhurst (1983), and Polovina (1984). At present, there is no compelling evidence for or against food limitation of yellowfin tuna.

The low trophic transfer efficiency between ingestion and production by yellowfin in contrast with their high growth rate, mortality rate, and production to biomass ratio suggests a trophic structure that differs from that of more productive ecosystems. A high production to biomass ratio could be maintained without a high transfer efficiency as long as the ratio between the production of forage and the standing stock of predators is large. In this situation the predator population puts a higher proportion of energy into turnover and maintains a smaller standing stock. This may be characteristic of many tropical pelagic predators that grow rapidly in spite of high metabolic rates in an oligotrophic habitat.

Acknowledgments

Numerous people deserve credit for the success of these investigations. James Joseph, Inter-American Tropical Tuna Commission, and James Kitchell, University of Wisconsin, made this work possible. We sincerely thank the National Marine Fisheries Service, Izadore Barrett, and Richard Shomura for providing us the unique opportunity of working with live tunas at the Kewalo Research Facility (KRF), Honolulu, Hawaii. Valuable expertise was provided by Andrew Dizon, Richard Brill, Randy Chang, Martina Queenth, and Tom Kazama of the KRF. For assistance and advice, we are indebted to Marshall Adams, Roger Carpenter, Lo-chai Chen, David Farris, Sharon Hendrix, John Hunter, Stuart Innes, Witold Klawe, John Magnuson, Ashley Mullen, Jacinta Olson, Jim Rice, Kurt Schaefer, Michael Scott, Gary Sharp, Paul Smith, Patrick Tomlinson, Michael Walker, and George Yates. Support was provided by the Inter-American Tropical Tuna Commission and the National Science Foundation via grant No. PCM-8013473.

Notes

[1]Functions actually represent gastric retention, but we follow convention in calling them gastric evacuation functions.

[2]For calculating daily ration, another equation representing the evacuation rate for mackerel in small meals was used (equation 7). See discussion of the effect of meal size and daily ration based on stomach analysis.

References

Adams, S. M., B. L. Kimmel and G. R. Ploskey. 1983. Sources of organic matter for reservoir fish production: a trophic-dynamic analysis. Can. J. Fish. Aquat. Sci. 40:1480–1495.

Adams, S. M., R. B. McLean and M. M. Hoffman. 1982. Structuring of a predator population through temperature-mediated effects on prey availability. Can. J. Fish. Aquat. Sci. 39:1175–1184.

Allen, K. R. 1971. Relation between production and biomass. J. Fish. Res. Board Can. 28:1573–1581.

Alverson, F. G. 1963. The food of yellowfin and skipjack tunas in the eastern tropical Pacific Ocean. Bull. Inter-Am. Trop. Tuna Comm. 7:293–396. (In English and Spanish)

Anonymous. 1983. Annual report of the Inter-American Tropical Tuna Commission, 1982. Inter-American Tropical Tuna Commission, La Jolla, CA. 294 pp. (In English and Spanish)

Anonymous.1984. Annual report of the Inter-American Tropical Tuna Commission, 1983. Inter-American Tropical Tuna Commission, La Jolla, CA. 272 pp. (In English and Spanish)

Bagge, O. 1977. Meal size and digestion in cod *(Gadus morhua* L) and sea scorpion *(Myoxacephalus scorpius* L). Medd. Dan. Fisk. Havunders. N.S. 7:437–446.

Banse, K. and S. Mosher. 1980. Adult body mass and annual production/biomass relationships of field populations. Ecol. Monogr. 50:355–379.

Baptist, J. P. and T. J. Price. 1962. Accumulation and retention of cesium^{137} by marine fishes. Fish. Bull. U.S. 62:177–187.

Barrington, E. I. W. 1957. The alimentary canal and digestion. In: *The physiology of fishes.* Vol. 1. Metabolism. Brown, M. E., ed. Academic Press, New York, NY. Pp. 109–161.

Bartell, S. M., J. E. Breck, R. H. Gardner and A. L. Brenkert. 1986. Individual parameter perturbation and error analysis of fish bioenergetics models. Can. J. Fish. Aquat. Sci. 43:160–168.

Beamish, F. W. H. 1964. Influence of starvation on standard and routine oxygen consumption. Trans. Am. Fish. Soc. 9:103–107.

Beamish, F. W. H. 1978. Swimming capacity In: *Fish Physiology.* Vol. VII. Hoar, W. S. and D. I. Randall, eds. Locomotion. Academic Press, New York, NY.. Pp. 101–187.

Blackburn, M. 1968. Micronekton of the eastern tropical Pacific Ocean: family composition, distribution, abundance, and relations to tuna. Fish. Bull. U.S. 67:71–115.

Blackburn, M., R. M. Laurs, R. W. Owen and B. Zeitzschel. 1970. Seasonal and areal changes in standing stocks of phytoplankton, zooplankton, and micronekton in the eastern tropical Pacific. Mar. Biol. 7:14–31.

Boggs, C. H. 1984. Tuna bioenergetics and hydrodynamics. Ph.D. thesis, University of Wisconsin, Madison, WI. 198 p.

Brett, J. R. 1972. The metabolic demand for oxygen in fish, particularly salmonids, and a comparison with other vertebrates. Resp. Physiol. 14:151–170.

Brett, J. R. 1973. Energy expenditure of sockeye salmon, *Oncorhynchus nerka,* during sustained performance. J. Fish. Res. Board Can. 30:1799–1809.

Brill, R. W. 1979. The effect of body size on the standard metabolic rate of skipjack tuna, *Katsuwonus pelamis.* Fish. Bull. U.S. 77:494–498.

Carey, F. G. and R. J. Olson. 1982. Sonic tracking experiments with tunas. Int. Comm. Cons. Atl. Tunas, Collect. Vol. Sci. Pap. 17(2):444–457.

Cole, J. S. 1980. Synopsis of biological data on the yellowfin tuna, *Thunnus albacares* (Bonnaterre, 1788), in the Pacific Ocean. In: Synopses of biological data on eight species of scombrids. Inter-Am. Trop. Tuna Comm. Spec. Rep. No. 2. Bayliff ,W. H., ed. Pp. 71–150.

Davis, G. E. and C. E. Warren. 1971. Estimation of food consumption rates. In: *Methods for Assessment of Fish Production in Fresh Waters.* 2nd ed. W. E. Ricker, W. E., ed. I.B.P. Handb. No. 3. Pp. 227–248.

Diana, J. S. 1979. The feeding pattern and daily ration of a top carnivore, the northern pike *(Esox lucius).* Can. J. Zool. 57:2121–2127.

Dixon, W. J. and F. J. Massey, Jr. 1957. *Introduction to Statistical Analysis.* McGraw-Hill, New York, NY. 488 pp.

Dotson, R. C. 1978. Fat deposition and utilization in albacore. In: *The Physiological Ecology of Tunas.* Sharp, G. D. and A. E. Dizon, eds. Academic Press, New York, NY. Pp. 341–355.

Eggers, D. M. 1977. Factors in interpreting data obtained by diel sampling of fish stomachs. J. Fish. Res. Board Can. 34:290–294.

Elliott, J. M. 1972. Rates of gastric evacuation in brown trout, *Salmo trutta* L. Freshwater Biol. 2:1–18.

Elliott, J. M. and L. Persson. 1978. The estimation of daily rates of food consumption for fish. J. Anim. Ecol. 47:977–991.

Fange, R. and D. Grove. 1979. Digestion. In: *Fish Physiology.* Vol. VIII. Bioenergetics and growth. Hoar, W. S., D. J. Randall and J. R. Brett, eds. Academic Press, New York, NY. Pp. 161–260.

FAO. 1974. Yearbook of fishery statistics. Food Agric. Org. U. N. Catches and landings, 1963. Vol. 36: 590 p.

Fasham, M. J. R. 1978. The statistical and mathematical analysis of plankton patchiness. Annul Rev. Oceanogr. Mar. Biol. 16:43–79.

Filliben, J. J. 1975. The probability plot correlation coefficient test for normality. Technometrics 17:111–117.

Flowerdew, M. W. and D. J. Grove. 1979. Some observations of the effects of body weight, temperature, meal size and quality on gastric emptying time in the turbot, *Scophthalmus maximus* L. using radiography. J. Fish Biol. 14:229–238.

Folsom, T. R., D. R. Young and C. Sreekumaran. 1967. An estimate of the response rate of albacore to cesium. In: Symposium on radioecology. Proc. Second Natl. Symp. Nelson, D. J. and F. C. Evans, eds. Ann Arbor, MI. Pp. 337–345.

Glass, N. R. 1968. The effect of time of food deprivation on the routine oxygen consumption of largemouth black bass *(Micropterus salmoides).* Ecology 49:340–343

Gooding, R. M., W. H. Neill and A. E. Dizon. 1981. Respiration rates and low-oxygen tolerance limits in skipjack tuna, *Katsuwonus pelamis.* Fish. Bull. U.S. 79:31–48.

Graham, J. B. and R. M. Laurs. 1982. Metabolic rate of the albacore tuna *Thunnus alalunga.* Mar. Biol. 72:1–6.

Grove, D. J., L. G. Loizides and J. Nott. 1978. Satiation amount, frequency of feeding and gastric emptying rate in *Salmo gairdneri.* J. Fish Biol. 12:507–516.

Hess, A. D. and J. H. Rainwater. 1939. A method for measuring the food preference of trout. Copeia 3:154–157.

Hunt, B. P. 1960. Digestion rate and food consumption of Florida gar, warmouth, and largemouth bass. Trans. Am. Fish. Soc. 89:206–211.

Hunt, J. N. 1975. Interactions of the duodenal receptors which control gastric emptying. Scand. J. Gastroenterol. 10(Suppl. 35):9–21.

Hunt, J. N. and M. T. Knox. 1968. Regulation of gastric emptying. In: Handbook of physiology. Sect. 6, Vol. 4.Code, C. F., ed. American Physiological Society, Washington, DC. Pp. 1917–1935.

Isaacs, J. D. 1972. Unstructured marine food webs and "pollutant analogues." Fish. Bull. U.S. 70:1053–1059.

Jobling, M. 1980a. Effects of starvation on proximate chemical composition and energy utilization of plaice, *Pleuronectes platessa* L. J. Fish Biol. 17:325–334.

Jobling, M. 1980b. Gastric evacuation in plaice, *Pleuronectes platessa* L.: effects of dietary energy level and food composition. J. Fish Biol. 17:187–196.

Jobling, M. 1981a. Mathematical models of gastric emptying and the estimation of daily rates of food consumption of fish. J. Fish Biol. 19:245–257.

Jobling, M. 1981b. Dietary digestibility and the influence of food components on gastric evacuation in plaice, *Pleuronectes platessa* L. J. Fish Biol. 19:29–36.

Jones, R. 1974. The rate of elimination of food from the stomachs of haddock *Melanogrammus aeglefinus*, cod *Cadus morhua* and whiting *Merlangius merlangus*. J. Cons. Int. Explor. Mer 35:225–243.

Joslyn, M. A. 1950. *Methods in Food Analysis Applied to Plant Products*. Academic Press, New York, NY. 525 p.

Kashkin, N. J. and N. V. Parin. 1983. Quantitative assessment of micronektonic fishes by nonclosing gear (a review). Biol. Oceanogr. 2:263–287.

Kerr, R. A. 1983. Are the ocean's deserts blooming? Science (Wash., DC) 220:397–398.

King, J. E. and I. I. Ikehara. 1956. Comparative study of food of bigeye and yellowfin tuna in the central Pacific. Fish. Bull. U.S. 57:61–85.

Kitchell, I. F. 1983. Energetics. In: *Fish Biomechanics*. Webb, P. and D. Weihs , eds. Praeger Publications, New York, NY. Pp. 312–379.

Kitchell, I. F., W. H. Neill, A. E. Dizon and J. J. Magnuson. 1978. Bioenergetic spectra of skipjack and yellowfin tunas. In: *The Physiological Ecology of Tunas*. Sharp, G. D. and A. E. Dizon, eds. Academic Press, New York, NY. Pp. 357–368.

Kitchell, J. F. and J. T. Windell. 1968. Rate of gastric digestion in pumpkinseed sunfish, *Lepomis gibbosus*. Trans. Am. Fish. Soc. 97:489–492.

Kozlovsky, D. G. 1968. A critical evaluation of the trophic level concept. 1. Ecological efficiencies. Ecology 49:48–60.

Kume, S. and Y.. Morita. 1966. Ecological studies of bigeye tuna: III. On bigeye tuna, *Thunnus obesus*, caught by "nighttime longline" in the north Pacific Ocean. Rep. Nankai Reg.Fish. Res. Lab. 24:21–30. (In Japanese; English summary)

Laevastu, T. and H. A. Larkins. 1981. Marine fisheries ecosystem: its quantitative evaluation and management. Fishing News Books, Farnham, Surrey, England. 162 p.

Lane, E. D. 1975. Quantitative aspects of the life history of the diamond turbot, *Hypsopselia glunulala* (Girard), in Anaheim Bay. In: T*he Marine Resources of Anaheim Bay*. Lane, E. D. and C. W. Hill, eds. Calif. Dep. Fish Game Fish Bull. 165:153–173. Pp. 153–173.

Lane, E. D., M. C. S. Kingsley and D. E. Thornton. 1979. Daily feeding and food conversion efficiency of the diamond turbot: an analysis based on field data. Trans. Am. Fish. Soc. 108:530–535.

Longhurst, A. 1983. Benthic–pelagic coupling and export of organic carbon from a tropical Atlantic continental shelf–Sierra Leone. Estuarine Coastal Shelf Sci. 17:261–285.

MacDonald, J. S., K. G. Waiwood and R. H. Green. 1982. Rates of digestion of different prey in Atlantic cod *(Gadus morhua)*, ocean pout *(Macrozoarces americanus)*, winter flounder *(Pseudopleuronecles americanus)*, and American plaice *(Hippoglossoides plaressoides)*. Can. J. Fish. Aquat. Sci. 39:651–659.

MacKinnon, J. C. 1973a. Analysis of energy flow and production in an unexploited marine flatfish population. J. Fish. Res. Board Can. 30:1717–1728.

MacKinnon, J. C. 1973b. Metabolism and its relationship with growth rate of American plaice, *Hippoglossoides plalessoides* Fabr. J. Exp. Mar. Biol. Ecol. 11:297–310.

Magnuson, J. J. 1969. Digestion and food consumption by skipjack tuna *(Kalsawonus pelamis)*. Trans. Am. Fish. Soc. 98:379–392.

Magnuson, J. J. 1970. Hydrostatic equilibrium of *Eulhynnus affinis*, a pelagic teleost without a gas bladder. Copeia 1970:56–85.

Magnuson, J. J. 1973. Comparative study of adaptations for continuous swimming and hydrostatic equilibrium of scombroid and xiphoid fishes. Fish. Bull. U.S. 71:337–356.

Magnuson, J. J. 1978. Locomotion by scombrid fishes: hydromechanics, morphology, and behavior. In: *Fish Physiology*. Vol. VII Locomotion. Hoar, W. S. and D. J. Randall, eds. Academic Press, New York, NY. Pp. 239–313.

Magnuson, J. J. and J. G. Heitz. 1971. Gill raker apparatus and food selectivity among mackerels, tunas, and dolphins. Fish. Bull. U.S. 69:361–370.

Mann, K. H. 1978. Estimating the food consumption of fish in nature.. In: *Ecology of Freshwater Fish Production*. Gerking, S. D., ed. John Wiley & Sons, New York, NY. Pp. 250–273

McLaren, I. A. and C. J. Corkett. 1984. Singular, mass-specific *P/B* ratios cannot be used to estimate copepod production. Can. J. Fish. Aquat. Sci. 41:828–830.

Mearns, A. J., D. R. Young, R. J. Olson and H. A. Schafer. 1981. Trophic structure and the cesium-potassium ratio in pelagic food webs. Calif. Coop. Oceanic Fish. Invest. Rep. 22:99–110.

Murphy, T. C. and G. T. Sakagawa. 1977. A review and evaluation of estimates of natural mortality rates of tunas. Int. Comm. Cons. Atl. Tunas, Collect. Vol. Sci. Pap. 6(1):117–123.

Nakamura, E. L. 1972. Development and uses of facilities for studying tuna behavior. In: *Behavior of Marine Animals: current perspectives in research*. Vol. 2. Winn, H. E. and B. L. Olla, eds. Vertebrates. Plenum Press, New York, NY.

Pp. 245–277.
Nobel, R. L. 1973. Evacuation rates of young yellow perch, *Perca flavescens* (Mitchill). Trans. Am. Fish. Soc. 4:759–763.
Olson, R. J. and A. J. Mullen. 1986. Recent developments for making gastric evacuation and daily ration determinations. Environ. Biol. Fishes 16:183–191.
Owen, R. W. and B. Zeitschel. 1970. Phytoplankton production: seasonal change in the oceanic eastern tropical Pacific. Mar. Biol. 7:32–36.
Pauly, D. 1980. On the interrelationships between natural mortality, growth parameters and mean environmental temperature in 175 fish stocks. J. Cons. Int. Explor. Mer 39:195–212.
Pearcy, W. G. 1983. Quantitative assessment of the vertical distributions of micronektonic fishes with opening/closing midwater trawls. Biol. Oceanogr. 2:289–310.
Perrin, W. F., R. R. Warner, C. H. Fiscus and D. B. Molts. 1973. Stomach contents of porpoise, *Stenella* spp., and yellowfin tuna *Thunnus albacares,* in mixed-species aggregations. Fish. Bull. U.S. 71 10771092.
Persson, L. 1984. Food evacuation and models for multiple meals in fishes. Environ. Biol. Fishes 10:305–309.
Polovina, J. J. 1984. Model of a coral reef ecosystem, Part 1: ECOPATH and its application to French Frigate Shoals. Coral Reefs 3:1–11.
Quigley, J. P. and I. Meschan. 1941. Inhibition of the pyrloric sphincter region by the digestion products of fat. Am. J. Physiol. 134:803–807
Rice, J. A. and P. A. Cochran. 1984. Independent evaluation of a bioenergetics model for largemouth bass. Ecology 65:732–739.
Roberts, J. R. 1978. Ram gill ventilation in fish.In: *The Physiological Ecology of Tunas.* Sharp, G. D. and A. E. Dizon, eds. Academic Press, New York, NY. Pp. 83–88.
Robson, D. S. 1970. On the relation between feeding rate and stomach content in fishes. Biometrics Unit Mimeo. Ser. BU-328-M, Cornell University, Ithaca, NY.
Seaburg, K. G. and J. B. Moyle. 1964. Feeding habits, digestive rates and growth of some Minnesota warmwater fishes. Trans. Am. Fish. Soc. 93:269–285.
Sharp, G. D. 1984. Ecological efficiency and activity metabolism. In: *Flows of Energy and Materials in Marine Ecosystems:* theory and practice. Fasham, M. J. R., ed. NATO Advanced Research Institute, May 13–19, 1982, Carcans, France. Plenum Press, New York, NY. Pp. 459–474.
Sharp, G. D. and R. C. Dotson. 1977. Energy for migration in albacore, *Thunnus alalunga.* Fish. Bull. U.S. 75:447–450.
Sharp, G. D. and R. C. Francis. 1976. An energetics model for the exploited yellowfin tuna, *Thunnus albacares,* population in the eastern Pacific Ocean. Fish. Bull. U.S. 74:36–50.
Simenstad, C. A. and G. M. Cailliet. 1986. Contemporary studies of fish feeding: proceedings of GUTSHOP '84. Dev. Environ. Biol. Fishes 7. (In press)
Sokal, R. R. and F. J. Rohlf. 1969. Biometry: the principles and practice of statistics in biological research. W. H. Freeman, San Francisco, CA. 776 p.
Steigenberger, L. W. and P. A. Larkin. 1974. Feeding activity and rates of digestion of northern squawfish *(Ptychocheilus oregonensis).* J. Fish. Res. Board Can. 31:411–420.
Stevens, E. D. and A. E. Dizon. 1982. Energetics of locomotion in warmbodied fish. Annul Rev. Physiol. 44:121–131.
Stevens, E. D. and W. H. Neill. 1978. Body temperature relations of tunas, especially skipjack. In:*Fish Physiology.* Vol. Vii. Locomotion. Hoar, W. S. and D. J. Randall, eds. Academic Press, New York, NY. Pp. 315–359.
Stewart, D. J., J. F. Kitchell and L. B. Crowder. 1981. Forage fishes and their salmonid predators in Lake Michigan. Trans. Am. Fish. Soc. 110:751–763.
Swenson, W. A. and L. L. Smith, Jr. 1973. Gastric digestion, food consumption, feeding periodicity, and food conversion efficiency in walleye *(Stizostedion vitreum vitreum).* J. Fish. Res. Board Can. 30:1327–1336.
Thorpe, J. E. 1977. Daily ration of adult perch, *Perca fluviatilis* L. during summer in Loch Leven, Scotland. J. Fish Biol. 11:55–68.
Tyler, A. V. 1970. Rates of gastric emptying in young cod. J. Fish. Res. Board Can. 27:1177–1189.
Uchida, R. N. 1981. Synopsis of biological data on frigate tuna, *Auxis thazard,* and bullet tuna, *A. rochei.* U.S. Dep. Commer. NOAA Tech. Rep. NMFS Circ. 436:63 p. (FAO Fish. Synop. 124)
Vlieg, P., G. Habib and G. I. T. Clement. 1983. Proximate composition of skipjack tuna *Kalsawomus pelamis* from New Zealand and New Caledonian waters. N. Z. J. Sci. 26:243–250.
Watanabe, H. 1958. On the difference of the stomach contents of the yellowfin and bigeye tunas from the western equatorial Pacific. Rep. Nankai Reg. Fish. Res. Lab. 7:72–81. (In Japanese; English summary)
Wesolowski, G. O. 1976. *Multiple regression and analysis of variance: an introduction for computer users in management and economics.* John Wiley & Sons, New York, NY. 292 p.
Wild, A. and T. J. Foreman. 1980. The relationship between otolith increments and time for yellowfin and skipjack tuna marked with tetracycline. Bull. Inter-Am. Trop. Tuna Comm. 17:509–560. (In English and Spanish)
Windell, J. T. 1967. Rates of digestion in fishes. In: *The Biological Basis of Freshwater Fish Production.* Gerking, S. D., ed. John Wiley & Sons, New York, NY. Pp. 151–173.
Windell, J. T. 1978. Digestion and the daily ration of fishes. In: *Ecology of Freshwater Fish Production.* Gerking, S. D., ed. John Wiley & Sons, New York, NY. Pp. 159–183.
Wormuth, J . H. and C. F. E. Roper. 1983. Quantitative sampling of oceanic cephalopods by nets: problems and recommendations. Biol. Oceanogr. 2:357–377.
Wu, T. Y. and G. T. Yates. 1978. A comparative mechano-

physiological study of fish locomotion with implications for tuna-like swimming mode. In: *The Physiological Ecology of Tunas*. Sharp, G. D. and A. E. Dizon, eds. Academic Press, New York, NY. Pp. 313–337.

Young, D. R. 1970. The distribution of cesium, rubidium and potassium in the quasi-marine ecosystem for the Salton Sea. Ph.D. thesis, Scripps Institution of Oceanography, University of California, San Diego, CA. 213 p.

Zar, J. H. 1974. *Biostatistical analysis*. Prentice-Hall, Englewood Cliffs, NJ. 620 pp.

Introduction to
Michael G. Ryan and Barbara J. Yoder
Hydraulic Limits to Tree Height and Tree Growth

by James F. Kitchell

Unlike animal cells that have a genetically programmed and limited capacity for cell division, most cell types in plants are capable of virtually unlimited and continuous division. Therefore, forest ecologists and plant physiologists have long wondered why trees stop growing. The following paper reviews the four alternative hypotheses offered to explain this mystery. The basic ideas are:

- bigger trees must allocate more carbon to respiration and, therefore, have little left for growth,
- nutrient uptake by roots reaches the maximum for a given soil type and tree size,
- growth tissues do, in fact, mature and, like animal cells, become inefficient or senescent, and
- as suggested by the title, tree height is limited by the ability to move water from roots to leaves.

The authors do an excellent job of reviewing these alternatives, evaluating the evidence for and against each, and focusing on the plausible feedback mechanisms that account for limits to tree growth. This study will help students understand why some of the oldest trees are not the largest trees. It is an excellent example of application of the hypothetico-deductive method in physiological ecology.

Hydraulic Limits to Tree Height and Tree Growth

MICHAEL G. RYAN AND BARBARA J. YODER

What keeps trees from growing beyond a certain height? A mechanism that can explain differences in maximum tree height at different locations and patterns in height growth with age has eluded ecologists and plant physiologists.

Ryan, M. G. and B. J. Yoder. 1997. Hydraulic limits to tree height and tree growth. *BioScience* 47(4):235–242.

Why do old trees stop growing in height? Trees seem to have mechanisms that slow their growth as they age and prevent them from growing beyond a certain height. For example, a young mountain ash *(Eucalyptus regnans)* east of Melbourne, Australia, may grow 2–3 m per year in height. By 90 years of age, height growth has slowed to 50 cm per year. By 150 years, height growth has virtually ceased, although the tree may live for another century or more. As they stop growing taller, trees also undergo a change in form. For many species, at the same time as the treetops become flattened, primary branches become thickened and smaller branches become more twisted and gnarled because the annual extension of all branches is also reduced (Figure 1).

Why do trees grow to different heights in different places? The mechanism that determines maximum tree height varies among sites. For example, in the front range of the Colorado Rocky Mountains, a seed from a 25 m tall ponderosa pine *(Pinus ponderosa)* may fall into a rocky crevice and never grow more than 1–2 m. On the eastern slope of the Cascade Mountains in Oregon, ponderosa pine soar to 50 m. Thirty kilometers farther east, in a drier climate, the same species struggles to attain 10 m. Similar examples can be found for all tree species. For every species—in fact, for every genotype within a species—maximum height and the rate of height growth vary remarkably from place to place (Figure 1).

Early foresters noticed that the maximum height a tree reaches is correlated with the speed at which it grew in height when young. They also noticed that total wood production was highest at locations where trees achieved the greatest maximum height. Foresters continue to use this relationship to predict growth (Figure 2), and they classify the productivity of sites based on the tree height expected for a particular species at a certain age.

What mechanism causes site-to-site differences in height growth, maximum height, and productivity, even for trees with identical genetic potential for growth? What mechanism causes the slowing of height growth with age and height for all species at all sites? We understand why trees might grow faster at one site than another (trees, like most other plants, grow most rapidly in moist, fertile sites, and slowly in dry, nutrient-poor sites). A mechanism that can explain differences in maximum tree height at different locations and patterns in height growth with age has eluded ecologists and plant physiologists. Such a mechanistic explanation would allow

Figure 1. Ponderosa pine on the eastern slope of the Cascade Mountains in Oregon. (left) Tree near Sisters, Oregon, is approximately 250 years old and 37 m tall. (right) Tree near Bend, Oregon, on an old lava flow. Tree is approximately 250 years old and 7 m tall. Nearby trees that are not on the lava flow reach a height of 30 m. The same woman is standing next to both trees.

better modeling of tree growth and growth response to the environment (Ryan et al. 1996). It would also help us to understand whether conversion of old-growth forests to young plantations would increase rates of carbon storage (Harmon et al. 1990) and whether atmospheric inputs of nitrogen would increase carbon storage in temperate forests (Kauppi et al. 1992).

In this article, we discuss four possible mechanisms to explain the patterns of maximum tree height and age-related changes in height growth within a species. Three of these—respiration, nutrient limitation, and genetic changes in meristem tissue—have been discussed in the literature. (The respiration and nutrient limitation mechanisms were developed to explain patterns of whole tree and stand growth with age, but not tree height specifically.) The fourth, hydraulic limitation, is newer, but we believe that it is the most promising. Accordingly, we examine the hydraulic limitation hypothesis in the most detail.

Proposed Explanations for Tree Growth Pattern

Several mechanisms have been proposed to explain the decline in growth with age and the differences in height growth and maximum height within a species:

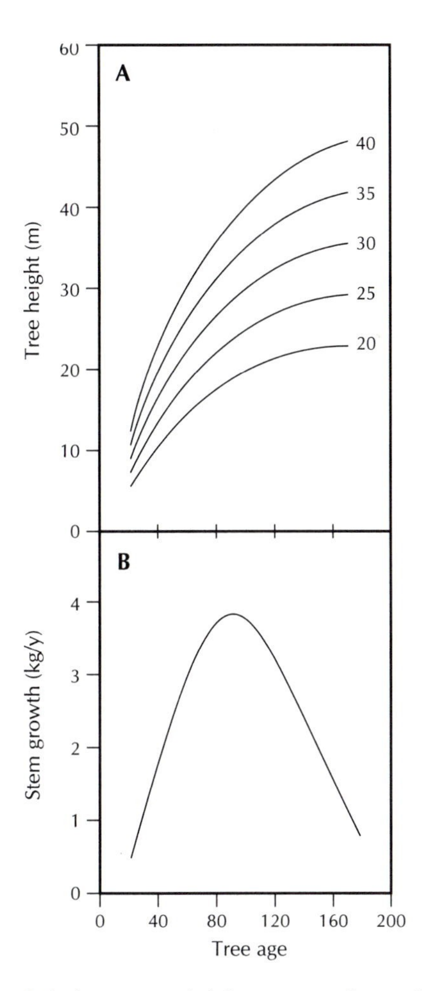

Figure 2. A. Average tree height versus age for ponderosa pine grown on sites with different growing conditions. Numbers on right are "site index" classes (site index is the height, in meters, of a dominant tree at age 100). Curves plotted from equations given in Barrett (1978). Height growth, maximum height, and growth are clearly linked and differ dramatically within a species. **B.** Average stem growth versus age for a ponderosa pine in site index of 30.

- As trees grow, increased autotrophic respiration of woody tissues consumes carbon, so less is available for growth. This is the respiration hypothesis.
- In large trees, nutrients become sequestered in living biomass, and as whole forests age, nutrients become sequestered in woody debris and decaying wood on the forest floor. The subsequent decline in nutrient availability may reduce tree growth in two ways. First, lower nutrient availability may lead to increased allocation of photosynthetic products to root production in older stands and therefore to decreased allocation to aboveground structures. Second, the foliage of older trees with reduced nutrient supply may have lower photosynthetic capacity. This is the *nutrient limitation hypothesis.*
- Growth of the undifferentiated tissue at the growing tips (meristematic tissue) may slow as the number of cell divisions increases. This is the *maturation hypothesis.*
- Total resistance to water flow in trees varies with both the path length and the conductivity of wood in a given cross-section. Hydraulic resistance increases as trees grow taller, because water must travel a longer path and, perhaps, because slower-growing wood is less permeable to water flow. To move the same amount of water through a path with higher resistance requires a higher tension (i.e., lower water potential) in the xylem water column. However, at higher tensions, air bubbles are more likely to form in the xylem water column (cavitation). To prevent cavitation and temporary loss of the water conducting system, leaf stomata in trees with higher resistance (taller trees) close earlier in the day or earlier in a drought cycle. This closure lowers stomatal conductance and photosynthesis, reducing carbon available for wood growth. This is the *hydraulic limitation hypothesis.*

Respiration Hypothesis

Yoda et al. (1965) developed the respiration hypothesis by reasoning that respiration will increase with tree size because the living portion of woody biomass (the sapwood) respires. They also observed that the ratio of woody biomass to leaf area increases as trees grow. Because photosynthesis is proportional to leaf area, higher wood-to-leaf ratios will yield greater autotrophic respiration costs relative to photosynthesis. The increased respiration costs use carbon that, in a smaller, younger tree, would contribute to wood production. Whittaker and Woodwell (1967) used the same reasoning, but with stem surface area instead of biomass.

Although textbooks (e.g., Kramer and Kozlowski 1979, Waring and Schlesinger 1985) generally favor the respiration hypothesis as an explanation for declining growth in older trees, a direct test of this hypothesis (Ryan and Waring 1992) failed to support it. Respiration rates for the living portion of the woody biomass of lodgepole pine *(Pinus contorta* var. *latifolia)* were low. In addition, much of the woody respiration was related to growth processes, and it declined as growth declined. Subsequently, Mencuccini and Grace (1996) showed that in Scots pine *(Pinus sylvestris)*, increases in woody respiration with tree size could not explain the decreased growth of older trees. In general, maintenance respiration rates of woody tissues (the portion of respiration not related to tissue growth) appear to be low, using only 5% to 12% of annual carbon fixation (Ryan et al. 1995).

Other, indirect evidence also contradicts the respiration hypothesis. For example, trees at wider spacings (and therefore with higher wood-to-leaf ratios) maintained high growth after canopy closure, whereas more narrowly spaced trees experienced a decline in growth rates (Fownes and Harington 1990). The respiration hypothesis also cannot account for the variation within a species in maximum tree height for trees growing in similar climates. Finally, it does not explain why growth in diameter continues, tapering off much more gradually, long after growth in height has ceased.

Nutrient Limitation Hypothesis

Several observations have contributed to the nutrient limitation hypothesis. Grier et al. (1981) found that the allocation to fine roots was dramatically greater in an old Pacific silver fir *(Abies amabilis)* forest than in an adjacent, young forest of the same species. Sequestration of nutrients in biomass and detritus in the old stand lowered nutrient availability in soil, and the trees responded to the lower nutrient supply by increasing production of fine roots. Similarly, fine-root biomass was greater in a mature slash pine *(Pinus elliottii)* stand than in a young stand (Gholz et al. 1982). Although fine-root turnover was lower in the mature stand, below-ground respiration costs were higher to maintain the greater root biomass. In both studies, increased allocation below ground in the older stand used carbon that would have gone to wood production in the younger stand, thus slowing the growth of the older trees relative to the younger trees. These observations, together with the finding that nitrogen is sequestered in biomass of a larch *(Larix gmelinii)* forest (Schulze et al. 1995) and with the results of model simulations, which showed a tendency for mature forests to sequester nitrogen in woody biomass and decaying wood (Murty et al. 1996), led to the development of the nutrient limitation hypothesis (Gower et al. 1996).

Although the nutrient limitation hypothesis may sometimes account for slower growth in older trees, it is unlikely to be generally applicable. Nutrient availability in the soil of some old forests can be as much as or greater than that of younger forests (Olsson 1996, Ryan et al. 1996). Below-ground allocation in lodgepole pine is greatest soon after canopy closure and declines as rates of both tree height and diameter growth decrease.[1] The nutrient limitation hypothesis also does not explain why, in mixed-aged stands, tree height limitations occur only in older trees (e.g., Yoder et al. 1994), or why isolated old trees growing in parks with ample nutrients and water do not continue to grow in height indefinitely. Although increased below-ground allocation may sometimes slow the total growth of older trees—possibly to a greater degree than increased autotrophic respiration—the nutrient limitation hypothesis cannot serve as a general explanation for height growth patterns.

Maturation Hypothesis

All higher organisms show maturational changes with age. In woody plants, maturational changes include differences in branching and in foliage biochemistry, physiology, and morphology (Greenwood and Hutchinson 1993). Shoots from older trees are less branchy than those from younger trees; they also have lower photosynthetic rates, are less likely to root, and show less height and diameter growth when grafted onto juvenile rootstock (Greenwood and Hutchinson 1993). When mature branches are grafted onto juvenile rootstock, mature characteristics persist; thus, maturation may involve genetic changes in the meristematic tissue of plants (Greenwood and Hutchinson 1993). The older the tree from which the grafts are derived, the lower the growth rate of these grafted branches. Genetically programmed slowing of growth is thus a potential explanation for the decrease in height growth with age (Greenwood 1989).

However, grafting experiments also provide evidence that maturation does not limit height or slow height growth. The transition from juvenile to mature characteristics occurs early in the life of a tree, generally well before height growth slows. For example, the greatest maturational changes occur between 1 and 4 years for Douglas fir (Ritchie and Keeley 1994) and between 1 and 20 years in larch (Greenwood 1989). Slowing of height growth occurs later in these species (at approximately 30–70 years), suggesting that maturation is not a factor in height growth patterns—trees that are slowing in growth rate are already "mature."

Another difficulty with the maturation hypothesis is the young age of the living portion of the tree: The leaves and fine roots are replaced frequently, making them much younger than the chronological age of the tree as a whole. Ray parenchyma cells (found in the water-conducting xylem) are the oldest cells in a tree, and these rarely exceed 80 years of age, even in very old trees (e.g., Connor and Lanner 1990). It is therefore unlikely that the age of an individual tree influences plant growth (Nooden 1988). Unlike animal cells, which have a mechanism that slows cell division as cells age and limits the total number of cell divisions that can occur (to a few dozen), plant cells undergo thousands of divisions, to judge by the number of vessels or tracheids in a stem, all of which derive from the same precursor cell.

Hydraulic Limitation Hypothesis

Yoder et al. (1994) observed differences in the diurnal patterns of photosynthesis in needles of the same age from young and old trees in two conifer species (Figure 3): old trees

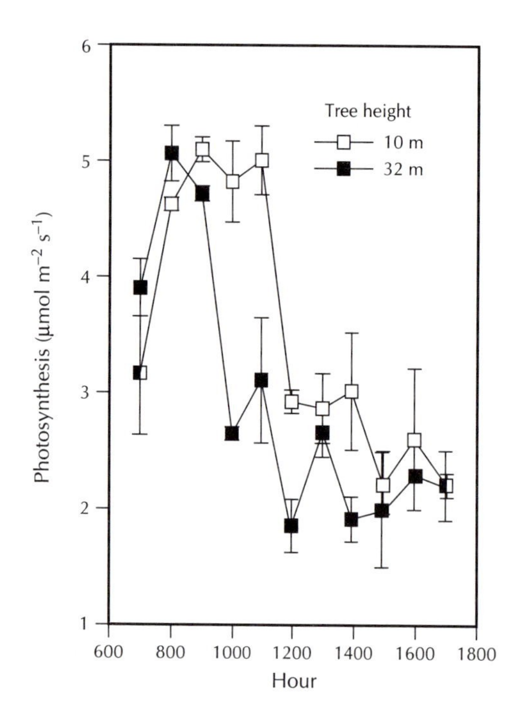

Figure 3. Diurnal course of photosynthesis for one-year-old foliage in the upper one-third of the canopy on young (10 m; open symbols) and old (32 m; closed symbols) ponderosa pine trees. Earlier closure of stomata for the old trees resulted in 25% lower total diurnal photosynthesis. Adapted from Yoder et al. 1994 with permission of the Society of American Foresters.

assimilated less carbon overall. The differences in photosynthesis did not appear to be caused by differences in photosynthetic capacity, because foliar nitrogen contents were similar, as were peak photosynthesis rates. However, stomata on the needles from old trees closed earlier in the day. Yoder et al. (1994) speculated that the earlier stomatal closure might be caused by increased hydraulic resistance resulting from increased xylem path length in taller trees with longer branches.

To understand how hydraulic resistance of stems and branches might be coupled with stomatal behavior and photosynthesis, it is necessary to consider the limitations and controls on water movement through plants. Evaporation of water from leaves provides the force to pull water from the soil, through the roots and xylem, and to the leaves. Because the water column from root to leaf is connected and under tension, this mechanism of sap ascent is called the "cohesion tension theory." Leaf water potentials in trees are generally 1 MPa (megapascal) lower than that of the water in soil, and resistance to water flow in the xylem causes most of the difference (Tyree and Ewers 1991). The high negative pressures theoretically required to move water up a large tree have been repeatedly measured with a pressure bomb and independently verified by measuring the flow of water through stems subjected to increasing water tension by centrifugation (Pockman et al. 1995).

The rate of water flow from the roots to a leaf is equal to the potential difference between the roots and the leaf divided by the total hydraulic resistance along the hydraulic pathway. This relationship is analogous to Ohm's law, which describes electrical current as voltage divided by electrical resistance. In the early morning, transpiration from leaves is low, so a small potential gradient from roots to leaves is required to supply transpirational water. As radiation and temperature increase, so does the atmospheric "pull" on leaf water. The water potential in leaves and stems falls, so that the water flux up the plant can keep pace with increased transpirational losses. But there is a limit to the water tension that leaves and stems can endure. Eventually stomata begin to close, limiting transpirational water loss as well as the root-to-leaf water potential gradient that is required to supply the water. Because stomatal closure also limits the carbon dioxide that can enter leaves for photosynthesis, any reduction in stomatal opening caused by a hydraulic limitation in the vascular system would also decrease daily carbon assimilation by the canopy.

Hydraulic resistance varies with the permeability of the sapwood and the overall path length (height of tree and length of branches). Therefore, hydraulic resistance will increase as the tree gains height and branches lengthen. In addition, the force of gravity results in a gradient of 0.01 MPa/m, even when there is no transpiration. The slow increase of hydraulic resistance as a tree grows will increase water stress in leaves, forcing stomata to close and lowering photosynthesis compared with a shorter tree.

To some extent, trees compensate for their increased size by producing xylem elements with increased permeability (Pothier et al. 1989a), but total resistance still increases with size and age (Mattson-Djos 1981, Mencuccini and Grace 1996). The links between hydraulic resistance and stomatal opening, and between stomatal opening and photosynthesis, mean that a larger hydraulic resistance will likely reduce diurnal and seasonal carbon assimilation. If foliage on older trees produces less photosynthate, then less wood growth will result, because the other carbon costs (respiration, foliage production, root growth) are similar to or slightly greater than those in younger trees (Ryan and Waring 1992). The hydraulic limitation hypothesis predicts that new stem growth in old trees should be most severely limited where the hydraulic resistance is greatest: at the treetop and the tips of very long branches. The flattened tops and gnarled branches of old trees are, thus, an expected result of hydraulic limitations.

The hydraulic limitation hypothesis can also explain why height growth rates and maximum height of trees vary with resource availability. On a nutrient-poor site, trees tend to have low rates of photosynthesis and stomatal conductance (Schulze et al. 1994) and, therefore, slower wood growth (the production of new xylem). The xylem that does form contains a larger proportion of "late-wood" xylem, with very narrow tracheids and, thus, low permeability (Pothier et al. 1989b). Therefore, in resource-poor environments, hydraulic resistance for a given length (i.e., the tree height) will be greater than for a tree growing in a resource-rich environment. In addition, on a dry site, any given leaf water potential has less "pulling power" because the driving force for water movement is the soil-to-leaf potential difference. Similarly, in cold soils a particular leaf water potential has less pulling power than in warmer soils because water becomes more viscous at near-freezing temperatures. Therefore, increased hydraulic resistance from smaller and fewer xylem cells and reduced driving force act together to limit the stature of trees in stressful locations.

The hydraulic limitation hypothesis can also explain why different species or individuals may attain different maximum heights in the same environment. Species that can endure very low xylem tensions or that have limited hydraulic restrictions (e.g., branch nodes) in their architecture are likely to achieve greater maximum heights. Individuals that have been exposed to mechanical stresses, and therefore produce reaction or compression wood with low permeability, are likely to achieve reduced maximum heights.

Evidence for the Hydraulic Limitation Hypothesis

For hydraulic resistance to limit tree growth, three elements are necessary. First, stomata (and, consequently, transpiration and photosynthesis) must respond to changes in hydraulic resistance. Second, hydraulic resistance must increase with tree height or tree age. Third, photosynthesis must be lower on the foliage of older trees.

Do Stomata Respond to Changes in Hydraulic Resistance?

Tyree and Sperry (1988) proposed that stomata may close in response to decreasing bulk leaf water potentials to prevent catastrophic loss of xylem function through cavitation. Cavitation results from tiny air bubbles that form in the xylem water column at high tensions (Zimmermann 1983). If cavitation occurs, the air bubbles interrupt the continuous column of water in the xylem, increasing total hydraulic resistance. Thus, even steeper gradients of water potential are required to maintain the same flux of water up the plant. Unless the plant limits transpirational water losses through stomatal closure, a catastrophic cycle of ever-worsening cavitation will result.

Sperry and colleagues (Sperry 1995, Sperry and Pockman 1993, Sperry et al. 1993) have shown that stomata do behave in a manner consistent with Tyree and Sperry's (1988) proposal: Experimental increases in hydraulic resistance (by cutting partially through stems) induced stomatal closure without measurably affecting leaf water potential (Figure 4). Other results suggest that the ability of the xylem to carry water may control the response of stomata to low humidity (Meinzer and Grantz 1991). Therefore, although the precise signals that induce stomatal closure are still a topic of debate, it is clear that stomatal behavior changes when hydraulic resistance changes.

Several other studies provide evidence that stomata respond to xylem resistance, independent of their well-known response to vapor

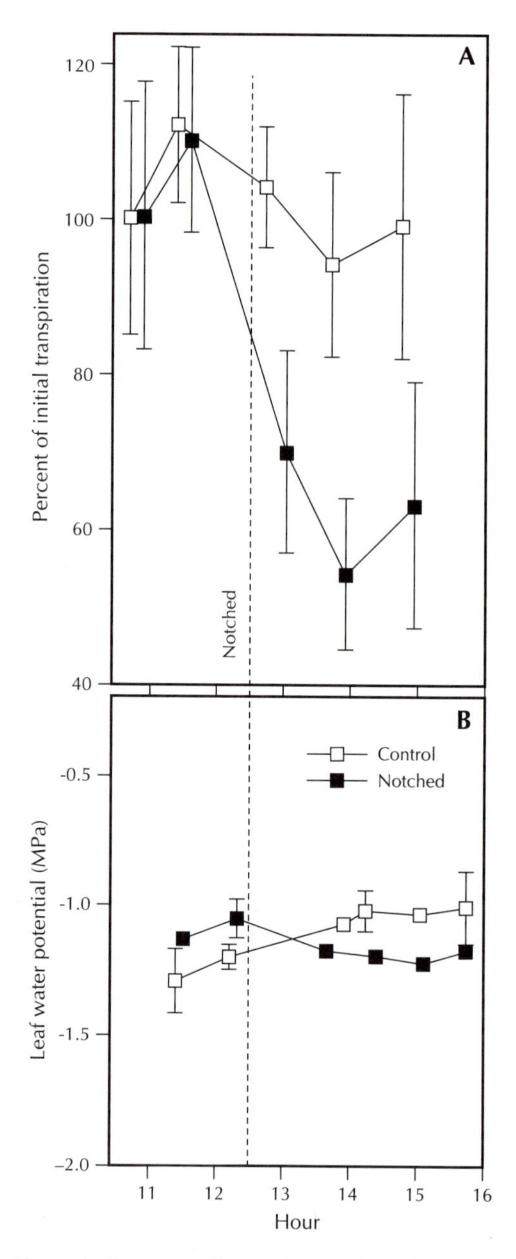

Figure 4. Response of transpiration **A,** and leaf water potential **B,** in *Betula occidentalis* branches before and after notching to increase resistance to water flow. Notching was done at 12.5 hours. After the hydraulic resistance increased (as a consequence of notching), stomata closed to reduce transpiration. Reprinted from Sperry et al. 1993 with permission of Oxford University Press.

pressure differences between the leaf and the air. In these studies, hydraulic resistance was either increased, by root pruning or by notching xylem (Sperry and Pockman 1993, Sperry et al. 1993, Teskey et al. 1983), or reduced (for individual leaves) by removing leaf area (Meinzer and Grantz 1990). In all cases, stomata closed when hydraulic resistance increased, and they opened when hydraulic resistance for individual leaves was reduced. Also, in all cases, leaf water potential remained stable through large changes in stomatal opening (e.g., Figure 4A, B). Additional evidence that stomata respond to changes in hydraulic resistance comes from the studies of Saliendra et al. (1995), who found that stomata that had closed in response to a high leaf-to-air vapor pressure gradient opened immediately in response to root pressurization. The response was also reversible: Releasing the root pressure resulted in stomatal closure.

Studies with stable isotopes of carbon provide further evidence of the effects of hydraulic conductance on stomatal behavior (Waring and Silvester 1994, Yoder et al. 1994). Carbon dioxide in air is mostly $^{12}CO_2$ but contains a small, constant amount of $^{13}CO_2$. When stomata are open wide, the sugar products of photosynthesis have low amounts of ^{13}C because the lighter $^{12}CO_2$ diffuses more quickly into the leaf and binds more readily with the photosynthetic enzymes (Farquhar et al. 1989). When the stomata close, in response to dry air or internal drought, more $^{13}CO_2$ is used for photosynthesis; thus, the amount of ^{13}C in sugars rises. In a recent stable isotope study of well-ventilated, open-grown ponderosa pine trees with no shading from adjacent trees[2] the relative abundance of ^{13}C in one-year-old needles was found to increase from the base of the live crown to the treetops. This finding indicated that stomatal opening was reduced with tree height within individuals. Similar differences were found for one-year-old needles in the upper canopy of young and old ponderosa pine (Yoder et al.1994), Douglas fir, and western hemlock.[3] The ratio of ^{13}C to ^{12}C in foliar cellulose also increases with branch length (Waring and Sylvester 1994). The increased hydraulic resistance with tree height and branch length produces an isotopic signature consistent with increased stomatal closure.

Does Hydraulic Resistance Increase as Trees Age?

Hydraulic resistance increases with tree height and age in several species. For example, transpiration was approximately 20% lower in a 140-year-old spruce stand (*Picea abies*) compared with a 40-year-old spruce stand.[4] These differences in transpiration between young and old spruce forests translate into lower average stomatal conductances to water vapor for the older stand.

In Scots pine, whole-tree hydraulic conductance (leaf transpiration divided by the water potential difference between leaves and soil; the inverse of hydraulic resistance) in 40-year-old trees was estimated to be approximately 20% what it was in seedlings (Mattson-Djos 1981). Similar dramatic reductions in hydraulic conductance were also found for Scots pine in a different location (Mencuccini and Grace 1996). In this study, wood growth per unit leaf area decreased linearly as hydraulic conductance (per unit leaf area) decreased. Finally, canopy hydraulic conductance in 30 m tall, 270-year-old ponderosa pine trees was 50% lower than that of adjacent 10 m tall, 40-year-old trees.[5]

Additional evidence for increasing hydraulic resistance with tree height comes from a study of leaf water potential gradients in a tall tree. The force of gravity on a water column will change xylem water potential approximately –0.01 MPa per meter of tree height (Zimmermann 1983). Steeper gradients in xylem water potential with height (–0.024 MPa/m) indicate the increased resistance of the xylem to water flow in tall trees.[6]

The evidence for increased hydraulic resistance in older trees is not conclusive. Zimmermann et al. (1994) did not find that water potential gradient varied with tree height, end water flow in the sapwood of older Eucalyptus trees was similar to that in young trees, although tree height differed by 35 m (Dunn and Connor 1993). These contrary findings do have an explanation. The pressure probe measurements of Zimmermann et al. (1994) are suspect because insertion of the probe may cause cavitation (Pockman et al. 1995). Older trees often have lower leaf area per unit sapwood area, so that the resistance per unit leaf area could have been higher for old trees in the Dunn and Connor (1993) study. Where leaf water potential gradients have been measured with a pressure bomb, and where hydraulic resistance has been expressed on a leaf area basis, hydraulic resistance was found to be greater in older trees.

Is Photosynthesis Reduced in Old Trees?

In the few available studies, carbon assimilation and stomatal conductance are lower for foliage of old trees than of younger trees. For example, diurnal photosynthesis and stomatal conductance in two gymnosperms, ponderosa pine and lodgepole pine, were lower in old trees than in nearby young ones (Yoder et al. 1994); the same was true for cherry, an angiosperm (Fredericksen et al. 1996). In both studies, rates of photosynthesis in old and young trees were similar in the moist morning air, but they differed as relative humidity decreased (Figure 3). Other studies have shown that photosynthesis and stomatal conductance were uniformly lower in older conifers than in younger ones (Grulke and Miller 1994, Kull and Koppel 1987, Schoettle 1994); these findings suggest that aging and height growth lower photosynthetic capacity. Two explanations could account for the decrease in photosynthetic capacity. Older trees may experience nutrient limitations in addition to hydraulic limitations; alternatively, a chronic water deficit and lower stomatal conductance near the tops and branch tips of older trees may be accompanied by reduced allocation of nitrogen to that foliage.

Perhaps the most interesting and convincing documentation of reduced stomatal conductance and photosynthesis with age comes from measurements of whole stands rather than individual trees. As mentioned above, canopy transpiration was lower (and hydraulic resistance higher) in older spruce stands than in younger stands.[7] Another recent study compared carbon dioxide and energy fluxes between young and old jack pine *(Pinus banksiana)* forests in northern Canada (Sellers et al. 1995). Preliminary results from this study indicate that midday net carbon dioxide uptake and transpiration were consistently higher for the young trees, even though the old trees had more leaf area. These studies support the hypothesis that transpiration and carbon assimilation are constrained by hydraulic resistance in older stands.

A Negative Feedback Loop?

The link between stomatal opening and hydraulic resistance suggests a feedback mechanism that may limit tree height and height growth. As a tree grows, gravitational force and hydraulic resistance increase simultaneously, resulting in limits to carbon assimilation. Because wood production apparently has a low priority for carbon allocation (Waring and Pitman 1985) and other carbon costs remain fixed, decreased assimilation will result in less new xylem. The slower growth also promotes the formation of a higher proportion of "latewood" cells in the xylem—cells that have low permeability to water. Trees may also lose water-conducting xylem to heartwood formation or to the effects of irreversible cavitation. The loss of water-carrying xylem and the decrease in xylem production ultimately combine to constrain the conducting system and maximum height of the tree.

The negative feedback described above suggests that new stem growth in old trees should be most severely limited where hydraulic resistance is greatest: at the tops of tall trees, the tips of very long branches, and the tips of twigs with very small xylem cells. The flattened tops and gnarled branches of old trees are, thus, an expected result of hydraulic limitations.

Evidence from fertilization and thinning studies suggests that this negative feedback loop can be broken if more resources are made available to the stand or to individual trees (e.g., by removing competitors for light, water, and nutrients, or by adding nutrients). For example, in 120-year-old Scots pine in Sweden, fertilization increased xylem production.[8] The trees also initiated new height growth, although they were near their maximum height. Presumably, fertilization either increased photosynthetic capacity or reduced allocation to root production (Gower et al. 1996), releasing carbon for xylem production.

Conclusions

A maximum height that varies with resource availability and slower height growth in older individuals appear to be universal for trees. Old trees are different, both physiologically and morphologically, from younger trees: They have lower rates of photosynthesis, reduced height and diameter growth rates, and a distinctive architecture. Nutrition, carbon allocation (including respiration), meristematic activity, and the tree's hydraulic architecture can all potentially change with tree growth and promote slower growth in older trees. In fact, these processes may interact.

Respiration or costs of producing additional roots become more important if photosynthesis is reduced as a result of hydraulic limitation. Additionally, slow xylem growth, whatever the cause, may lead to smaller tracheids or vessels and, therefore, to increased hydraulic resistance. However, if hydraulic resistance changes with tree height and branch length, and if stomata respond to the increasing resistance by closing to avoid cavitation, then all trees will face hydraulic limitation to photosynthesis—whether as isolated individuals, in uniform stands, or in stands of mixed age classes. Regardless of other constraints, trees will have to accommodate increased hydraulic resistance and its consequences as they grow.

Assessing the physiology of large trees has been a difficult logistical problem, but new techniques and resources are now available. Construction cranes now provide access to trees more than 60 m tall in both conifer and tropical forest sites, and sap flow instrumentation can now be used to measure the whole-tree water flux and whole-tree canopy resistance. Stable isotopes of carbon allow an assessment of stomatal behavior integrated over a year or more. With these techniques, we may be close to answering some of our oldest questions about tree height.

Acknowledgments

Research was supported by USDA (NRICGP #9401021) and NSF (DEB93–06356). We gratefully acknowledge the help of the Sisters Ranger District, USDA Forest Service, for access to study sites.

Footnotes

[1]S. Resh and F. W. Smith, manuscript in preparation. Colorado State University, Fort Collins, CO.

[2]B. J. Yoder and J. A. Panek, 1996, submitted manuscript. Oregon State University, Corvallis, OR.

[3]See footnote 2.

[4]M. Alsheimer, B. M. Kostner, E. Falge, and J. D. Tenhunen, manuscript in prep. University of Bayreuth, Germany.

[5]M. G. Ryan et al., manuscript in preparation.

[6]N. M. Holbrook et al., 1996, submitted manuscript. Harvard University, Cambridge, MA.

[7]See footnote 4.

[8]S. Linder, 1996, personal communication. Swedish University of Agricultural Sciences, Uppsala, Sweden

References Cited

Barrett, J. W. 1978. Height growth and site index curves for managed, even-aged stands of ponderosa pine in the Pacific Northwest. USDA Forest Service Research Paper nr PNW-232. Portland (OR): USDA Forest Service.

Connor, K. F. and R. M. Lanner. 1990. Effects of tree age on secondary xylem and phloem anatomy in stems of Great Basin bristlecone pine *(Pinus longaeva)*. American Journal of Botany 77:1070–1077.

Dunn, G. M. and D. J. Connor. 1993. An analysis of sap flow in mountain ash (*Eucalyptus regnans*) forests of different age. Tree Physiology 13:321– 336.

Farquhar, G. D., K. T. Hubick, A. G. Condon and R. A. Richards. 1989. Carbon isotope fractionation and plant water-use efficiency. Pp. 21–40. In: *Stable Isotopes in Ecological Research*. P. W. Rundel, J. R. Ehleringer and K. W. Nagy, eds. New York: Springer-Verlag.

Fownes, J. H. and R. A. Harington. 1990. Modeling growth and optimal rotations of tropical multipurpose trees using unit leaf rate and leaf area index. Journal of Applied Ecology 27:886–896.

Fredericksen, T. S., K. C. Steiner, J. M. Skelly, B. J. Joyce, T. E. Kolb, K. B. Kouterick and J. A. Ferdinand. 1996. Diel and seasonal patterns of leaf gas exchange and xylem water potentials of different-sized *Prunus serotina* Ehrh. trees. Forest Science 42:359–365.

Gholz, H. L.and R. F. Fisher. 1982. Organic matter production and distribution in slash pine *Pinus elliottii* plantations. Ecology 63:1827–1839.

Gower, S. T., R. E. McMurtrie and D. Murty. 1996. Aboveground net primary production decline with stand age: potential causes. Trends in Ecology and Evolution 11:378–382.

Greenwood, M. S. 1989. The effect of phase change on annual growth increment in eastern larch *(Larix laricina* [Du Roi] K. Koch). Annales des Sciences Forestieres 46 (supplement):171s–177s.

Greenwood, M. S. and K. W. Hutchinson. 1993. Maturation as a developmental process. In: *Clonal Forestry 1: Genetics and Biotechnology*. M. R. Ahuja, W. J. Libby, eds. Berlin (Germany): Springer-Verlag. Pp. 14–33.

Grier, C. C., K. A. Vogt, M. R. Keyes and R. L. Edmonds. 1981. Biomass distribution and above- and below-ground production in young and mature *Abies amabilis* zone ecosystems of the Washington Cascades. Canadian Journal of Forest Research 11:155–167.

Grulke, N. E. and P. R. Miller. 1994. Changes in gas exchange characteristics during the life span of giant sequoia: implications for response to current and future concentrations of atmospheric ozone. Tree Physiology 14:659–668.

Harmon, M. E., W. K. Ferrell and J. F. Franklin. 1990. Effects on carbon storage of conversion of old-growth forests to young forests. Science 247:699–702.

Kauppi, P. E., K. Mielikainen and K. Kuusela. 1992. Biomass and carbon budget of European forests, 1971 to 1990. Science 256:70–74.

Kramer, P. J. and T. T. Kozlowski. 1979. *Physiology of Woody Plants*. New York: Academic Press.

Kull, O., A. Koppel. 1987. Net photosynthetic response to light intensity of shoots from different crown positions and age in *Picea abies* (L.) Karst. Scandinavian Journal of Forest Research 2:157–166.

Mattson-Djos, E. 1981. The use of pressure bomb and porometer for describing plant water status in tree seedlings. In: Vitality and quality of nursery stock—proceedings of a Nordic symposium, Helsinki (Finland). P. Puttonen, ed. Department of Silviculture, University of Helsinki. Pp. 45–57.

Meinzer, F. C. and D. A. Grantz. 1990. Stomatal and hydraulic conductance in growing sugarcane: stomatal adjustment to water transport. Plant Cell and Environment 13:383–388.

Meinzer, F. C. and D. A. Grantz. 1991. Coordination of stomatal, hydraulic, and canopy boundary layer properties: do stomata balance conductances by measuring transpiration? Physiologia Plantarum 83:324–329.

Mencuccini, M. and J. Grace. 1996. Hydraulic conductance, light interception and needle nutrient concentration in Scots pine stands and their relation with net primary productivity. Tree Physiology 16:459– 468.

Murty, D., R. E. McMurtrie and M. G. Ryan.1996. Declining forest productivity in ageing forest stands—a modeling analysis of alternative hypotheses. Tree Physiology 16:187–200.

Noodén, L. D. 1988. Whole plant senescence. In: *Senescence and Aging in Plants.* L. D. Noodén, and A. D. Leopold, eds. San Diego (CA): Academic Press. Pp. 391–439.

Olsson, U. R. 1996. Nitrogen cost of production along a lodgepole pine chronosequence. [Master's thesis.] Colorado State University, Fort Collins, CO.

Pockman, W. T., J. S. Sperry and J. W. O'Leary. 1995. Sustained and significant negative water pressure in xylem. Nature 378:715–716.

Pothier, D., H. A. Margolis and R. H. Waring. 1989a. Patterns of change in saturated sapwood permeability and sapwood conductance with stand development. Canadian Journal of Forest Research 19:432–439.

Pothier, D., H. A. Margolis and R. H. Waring. 1989b. Relation between the permeability and the anatomy of jack pine sapwood with stand development. Canadian Journal of Forest Research 19:1564–1570.

Ritchie, G. A., J. W. Keeley. 1994. Maturation in Douglas-fir: 1. Changes in stem, branch and foliage characteristics associated with ontogenetic ageing. Tree Physiology 14:1245–1259.

Ryan, M. G., R. H. Waring. 1992. Maintenance respiration and stand development in a subalpine lodgepole pine forest. Ecology 73:2100–2108.

Ryan, M. G., S. T. Gower, R. M. Hubbard, R. H. Waring, H. L. Gholz, W. P. Cropper and S. W. Running.1995. Woody tissue maintenance respiration of four conifers in contrasting climates. Oecologia 101:133–140.

Ryan, M. G., D. Binkley and J. H. Fownes. 1996. Age related decline in forest productivity: pattern and process. Advances in Ecological Research 27:213–262.

Saliendra, N. Z., J. S. Sperry and J. P. Comstock. 1995. Influence of leaf water status on stomatal response to humidity, hydraulic conductance, and soil drought in *Betula occidentalis.* Planta 196:357–366.

Schoettle, A. W. 1994. Influence of tree size on shoot structure and physiology of *Pinus contorta* and *Pinus aristata.* Tree Physiology 14:1055–1068.

Schulze, E. -D., F. M. Kelliher, C. Korner, J. Lloyd and R. Leuning. 1994. Relationships among maximum conductance, ecosystem surface conductance, carbon assimilation rate, and plant nitrogen nutrition: a global ecology scaling exercise. Annual Review of Ecology and Systematics 25:629–660.

Schulze, E. -D., et al. 1995. Above-ground biomass and nitrogen nutrition in a chronosequence of pristine Dahurian *Larix* stands in Eastern Siberia. Canadian Journal of Forest Research 25:943–960.

Sellers, P., et al. 1995. The Boreal Ecosystem Atmosphere study (BOREAS): an overview and early results from the 1994 field year. Bulletin of the American Meteorological Society 76:1549–1577.

Sperry, J. S. 1995. Limitations on stem water transport and their consequences. In: Plant Stems: *Physiology and Functional Morphology.* B. L. Gartner, ed. San Diego (CA): Academic Press. Pp. 105–124.

Sperry, J. S. and W. T. Pockman. 1993. Limitation of transpiration by hydraulic conductance and xylem cavitation in *Betula occidentalis.* Plant Cell and Environment 16:279–287.

Sperry, J. S., N. N. Alder and S. E. Eastlack. 1993. The effect of reduced hydraulic conductance on stomatal conductance and xylem cavitation. Journal of Experimental Botany 44:1075–1082.

Teskey, R. O., T. M. Hinckley and C. C. Grier. 1983. Effect of interruption of flow path on stomatal conductance of *Abies amabilis.* Journal of Experimental Botany 34:1251–1259.

Tyree, M. T. and J. S. Sperry. 1988. Do woody plants operate near the point of catastrophic xylem disfunction caused by dynamic water stress? Plant Physiology 88:574–580.

Tyree, M. T. and F. W. Ewers. 1991. The hydraulic architecture of trees and other woody plants (Tansley Review nr 34). New Phytologist 119:345–360.

Waring, R. H. and G. B. Pitman. 1985. Modifying lodgepole pine stands to change susceptibility to mountain pine beetle attack. Ecology 66:889–897.

Waring, R. H. and W. H. Schlesinger. 1985. *Forest Ecosystems: Concepts and Management.* Orlando (FL): Academic Press.

Waring, R. H. and W. B. Silvester. 1994. Variation in foliar ^{13}C values within the crowns of *Pinus radiata* trees. Tree Physiology 14:1203–1213.

Whittaker, R. H. and G. M. Woodwell. 1967. Surface area relations of woody plants and forest communities. American Journal of Botany 54:931–939.

Yoda, K., K. Shinozaki, H. Ogawa, K. Hozumi and T. Kira. 1965. Estimation of the total amount of respiration in woody organs of trees and forest communities. Journal of Biology, Osaka City University 16:15–26.

Yoder, B. J., M. G. Ryan, R. H. Waring, A. W. Schoettle and M. R. Kaufmann. 1994. Evidence of reduced photosynthetic rates in old trees. Forest Science 40:513–527.

Zimmermann, M. H. 1983. *Xylem Structure and the Ascent of Sap.* Berlin (Germany): Springer-Verlag.

Zimmermann, U., F. C. Meinzer and A. Haase. 1994. Xylem water transport: is the available evidence consistent with the cohesion theory? Plant, Cell and Environment 17:1169–1181.

Readings for

Behavioral Ecology

Investigating the Adaptive Value of Behavior

Robert L. Jeanne

■ **H. Jane Brockmann, Alan Grafen and Richard Dawkins**
Evolutionary Stable Nesting Strategy in a Digger Wasp

■ **Ronald L. Mumme**
Do Helpers Increase Reproductive Success?

■ **Paul Schmid-Hempel, Alejandro Kacelnik and Alasdair I. Houston**
Honeybees Maximize Efficiency by Not Filling Their Crop

Introduction to

H. Jane Brockmann, Alan Grafen and Richard Dawkins

Evolutionarily Stable Nesting Strategy in a Digger Wasp

by Robert L. Jeanne

For her Ph.D. thesis at the University of Wisconsin, Jane Brockmann studied the nesting behavior of two populations of great golden digger wasp, *Sphex ichneumoneus*. The females of these wasps dig burrows in the soil, then go out in search of katydids, which they sting into paralysis and carry back to their burrow. After several paralyzed prey are stashed in the nest, the female lays an egg on one of them and seals the burrow. After the wasp larva hatches, it feeds on the provisioned food. It eventually pupates, spends the winter protected in the burrow, and emerges as an adult wasp the following summer. Brockmann found, however, that not all females follow this scenario. Some females occasionally enter burrows partially dug or provisioned by other females and adopt them as their own. A given female, in the course of her short summer nesting season, can be both a "digger" and an "enterer." In other words, a female digger wasp can pursue alternative nesting "strategies."

In the following paper Brockmann and her colleagues analyze her data on the frequency of each alternative nesting behavior in the two populations she studied. They use the data to test between two alternative models of mixed evolutionarily stable strategies (ESS). This represents one of the first successful quantitative tests of an ESS model. Data from one of the two populations Brockmann studied support one of the models, while those of the other population do not support it. Even for the first population, however, as so often happens with field studies, the results do not fit the favored model precisely. After reading the paper, one is left with several new questions about the behavior of the wasps in these two populations. For example, why does each individual adopt a mixed strategy rather than specializing on either digging or entering? What environmental (or behavioral) change would cause the frequency of the two strategies to shift toward "digger"?

This is how science proceeds, from question, to proposed explanation, to new questions, to revised explanation, and so on. As a final bonus, in the discussion at the end of the paper the authors offer a rare glimpse into the evolution of their thinking as they struggled with how to explain the behavior.

Evolutionarily Stable Nesting Strategy in a Digger Wasp

H. Jane Brockmann, Alan Grafen and Richard Dawkins

Reprinted from *Journal of Theoretical Biology*, Vol. 77, Brockmann, H. J., Grafen, A., and Dawkins, R., Evolutionarily stable nesting strategy in a digger wasp, pp. 473–496, 1979, by permission of the publisher Academic Press Limited.

Abstract

Two alternative "strategies" will not coexist in a population unless on average they are equally successful. The most likely way for such an equilibrium to be maintained is through something equivalent to frequency-dependent selection. Females of the digger wasp *Sphex ichneumoneus* (Sphecidae) nest in underground burrows. They usually dig and provision these by themselves but occasionally a nest is jointly occupied. The two wasps fight whenever they meet and in the end only one of the two females lays an egg in the shared nest. Two models based on the theory of mixed evolutionarily stable strategies were developed and tested on comprehensive field data from two North American populations of these wasps. The first model proposes two strategies called *founding* and *joining*. Founders start burrows alone, but they are more successful when they are joined by a joiner. At equilibrium founders and joiners are equally successful, which amounts to an amicable, sharing relationship. The predictions of this amicable model are decisively rejected by the data. The second model proposes two strategies called *digging* and *entering*. Diggers dig their own burrows but they often have to abandon these burrows because of temporary unsuitability. Enterers move in later, thereby exploiting abandoned burrows as a valuable resource. They do not distinguish an abandoned burrow from one that is still occupied. Therefore sharing of burrows arises as an unfortunate byproduct of selection for entering abandoned burrows, and Model 2 is not an amicable model. Its quantitative predictions are impressively fulfilled in one population, though not in another population. This is one of the only examples yet known of a mixed evolutionarily stable strategy in nature. Yet the word strategy itself can confuse, and this paper tries the experiment of substituting "decision," defined as a moment at which the animal commits future time to a course of action.

1. Introduction

A fashionable idea in modern ethology is that there may coexist among the animals of one population more than one alternative strategy for achieving the same functional end. For instance male crickets *(Gryllus integer)* have been divided into "callers" and "satellites" (Cade, 1978). Callers sing to attract females. Satellites do not sing, but lurk close to a caller and intercept females as they approach him. Some male mammals appear to avoid fighting for harems and concentrate on "sneaking" copulations instead. Gadgil (1972) gave examples and made an important theoretical point which is still too frequently overlooked and which nowadays may be subsumed under the theory of evolutionarily stable strategies—ESS (Maynard Smith & Price, 1973, Maynard Smith, 1974; Maynard Smith & Parker, 1976; for a simple account see Dawkins, 1976). This is that if the two strategies are genuine functional alternatives which coexist in the population for an appreciable time they must on average be equally successful. If they were not, natural selection would soon remove the inferior one from the population. In many of the alleged examples in the literature (see Dawkins, in press) this rigorous condition is probably not met: either, one of the two is in the process of disappearing through natural selection, or they are not truly two alternative strategies at all but one "conditional strategy" such as "if large, fight; if small, sneak." Yet there may still be one or two authentic cases of equally successful alternative strategies maintained in a state of evolutionary stability (Parker, 1978). Here we present clear evidence of such a case.

It is too much to expect that the costs and benefits of the two alternatives should just happen to balance up exactly. The agency most likely to maintain evolutionary stability in the long term is equivalent to the frequency dependent selection long ago invoked by Fisher (1930) in his theories of the sex ratio and of balance in mimic populations. The idea is that when the two strategies exist in the population at a particular equilibrium frequency relative to one another they prosper equally. If the proportion in the population should drift away from equilibrium, selection favours whichever strategy has temporarily drifted down in frequency. So the equilibrium is restored.

It is one of the strengths of ESS theory that it manages to apply the same mathematics to two quite different kinds of equilibrium. One is balanced genetic polymorphism. The other is the possibility that each individual might use two alternative behaviour patterns in random order but with relative frequencies stabilized by natural selection. "Sneak with probability 0.7, fight with probability 0.3" is a single mixed strategy. If it is evolutionarily stable this means that, when more than a critical number of individuals is following it, no other strategy such as "sneak with probability 0.6" or "sneak with probability 0.8" does better. The important point is that evolutionary stability may be achieved by any combination of polymorphism and mixed strategy that leads to the correct equilibrium ratio of behaviour patterns in the population as a whole. A polymorphism of pure fighters and pure sneakers in the correct ratio would be just one end of a continuum, the other end of which would be a uniform mixed strategy.

But this leads us into a terminological difficulty. What are we going to label "a strategy"? We referred to calling in crickets as a strategy, and satellite behaviour as another. But in that case we must use "ESS" to stand for "evolutionarily stable state" (Taylor, in press), because it is not one strategy that is stable but the state of balance between two of them. If, however, we have a mixed strategy with a probabilistic decision rule, ESS may stand for evolutionary stable strategy. But in that case calling and satellite behaviour, fighting and sneaking, or whatever they are, should not be called two "strategies" but two stochastic outcomes of one mixed strategy. In this paper we follow the current convention of using the word strategy for each of two behavioural paths to the same end, although the example we analyze turns out to be closer to a single mixed strategy than to a true polymorphism. But we are uneasily conscious that ESS theorists are soon going to have to standardize their terminology. As a contribution to this, we try the experiment of substituting the less ambiguous term *decision,* to be defined below.

We shall report two behavioural paths to the same end, two "strategies" coexisting in a population of great golden digger wasps *(Sphex ichneumoneus* L.; Hymenoptera, Sphecidae). We demonstrate, using field data, that the two alternatives are equally successful at observed frequencies in a population in New Hampshire, and we develop and test a model of the stabilizing selection which maintains the equilibrium. A more detailed account of the behaviour and a discussion of the model's relevance to the evolution of eusociality will be found in a companion paper (Brockmann & Dawkins, in prep.). The companion paper also gives the statistical justification for many of the facts quoted in the present paper.

2. Biological Background

The normal nesting behaviour of *Sphex ichneumoneus* begins with the female digging a burrow at the bottom of which is a side tunnel ending in an oval brood chamber. She then goes out to fields and hedgerows where she hunts katydids (Orthoptera, Tettigoniidae), paralyses them by stinging, and brings them back to the burrow (Brockmann, 1976). Having accumulated a cache of from one to six katydids, a process which takes from one to ten days, she lays a single egg on the food store, fills in the brood chamber with soil, and begins the cycle again. At this point she may either fill in the whole burrow and dig a new one, or she may dig a new side tunnel and brood chamber higher up in the same main shaft. Meanwhile the larva hatches in its chamber, feeds on the katydids until the food is gone, spins a cocoon in which it remains throughout the winter, pupates the following spring, and emerges as an adult in the early summer. The adults all die at the end of the summer.

Little is known of the males, but the females usually dig their own nests in the same general area as they themselves hatched. In this way "traditional" breeding aggregations develop where several dozen females can be reliably found nesting year after year. The wasps are conventionally regarded as solitary, but occasionally two individuals co-occupy the same nest: while the wasp who dug the burrow is out hunting katydids, a second wasp moves in. This is not a case of "communal" nesting, in which the individuals provision different brood chambers leading off one main burrow shaft (Wilson, 1971; Michener, 1974); our wasps share not only the same burrow but usually the same brood chamber as well (Brockmann & Dawkins, in prep.). They work at provisioning the chamber simultaneously. The two do not often meet because both spend most of their time out hunting. But if they do meet they fight, and the loser usually leaves the burrow for good. The final upshot is that only one of the two wasps ever lays an egg on the jointly provided cache of katydids. It was this appearance of a "winner take all" game that first suggested to us the applicability of ESS theory. The theory of evolutionarily stable strategies was rich in cunning speculation but poor in hard data, and the first author's comprehensive records of individual wasp behaviour and economics seemed capable of filling an important gap.

3. The Data

Brockmann (1976) studied colour-marked individuals in the field. At a site in Exeter, New Hampshire, referred to as NH, all the 30 females in the area were watched continuously through the daylight hours. With a few inevitable breaks of only minutes at a time, the moment by moment actions of all the wasps while at their nests were recorded from 24 July to 22 August 1975, most of their adult life. Similarly comprehensive observations were made for three populations at a site in Dearborn, Michigan, referred to as MI. In 1973, 18 MI wasps were observed, in 1974 12 wasps, and in 1975 eight (with the help of an assistant, T. Manning). In all, this represents over 1500 h of observation, and is a nearly complete record of the histories of 410 burrows and of all the nest-related activities of 68 wasp lifetimes. We know when each one of the wasps in the area dug a burrow, when she left on each hunting trip, when she was joined by a second wasp and which wasp it was who joined her, when she temporarily filled in the burrow for the night, and when she brought in the last katydid, permanently filled in the nest and began a new cycle.

Excavation of a sample of burrows established that a wasp almost never went through the highly characteristic motions of permanently filling a nest unless she had just laid an egg in it. We therefore use permanent filling as an indicator of egg-laying. We also know the approximate size of each katydid caught. Every time a female returned with a prey she left it outside the burrow while she went in for a quick inspection. Before she re-emerged the katydid was swiftly measured and replaced.

We shall use these data to test two models. Model 1 will be rejected, but its development and rejection are instructive and pave the way for Model 2 which is supported by the NH data. The MI data do not support either model, and we postpone consideration of them until the end of the paper. Because Model 1 is simpler it will be used for the explanation of various general points which are equally applicable to both models. We begin with one of these, the use of the decision concept.

4. Decisions in Model 1

Model 1 postulates two alternative behavioural routes to the same functional end—the laying of a properly provisioned egg in a secure underground chamber. The two alternatives are called "founding" and "joining." Joiners do not dig a burrow in the manner described; instead they short-circuit the labour of digging by attempting to move into the burrow of another wasp—a founder—and take it over.

We know that we are not dealing with a polymorphism. Any given individual sometimes founds and sometimes joins. Throughout her life she takes a series of decisions, or choices between founding and joining. A decision is defined as a commitment of a period of future time (Dawkins & Dawkins, 1973). In the present case each decision commits the wasp concerned to a period of association with a particular burrow. When she takes a joining decision she begins her period of association with a burrow by joining an already established foundress. All wasp/burrow associations which do not begin with a joining decision must begin with a founding decision. A founding wasp begins her association with the burrow alone; she may or may not subsequently be joined. A temporary foray into another wasp's nest is not considered to be a true joining decision. To qualify as having committed herself to joining, a wasp has to return to the burrow after leaving it for the first time.

An individual female typically packs a dozen or so decisions into her brief life. ESS theory predicts that the overall frequency of joining decisions in the population is adjusted by natural selection to the level at which a typical joining decision is exactly as successful as a typical founding decision. How are we going to measure "success" in order to test this prediction?

5. Measuring Success Rate

The first thing to note is that, since there is not a polymorphism, we are not interested in measuring the lifetime success of particular individuals. We want to measure the success of a type of decision, averaged over all individuals who make it and across all occasions on which it is made. What would a successful decision be? It would at least have to result in the wasp concerned laying a properly provisioned and securely housed egg. We have no information on subsequent hatching success, and can only record whether or not an egg was laid (strictly whether the wasp showed permanent filling behaviour), and how much food it was laid upon.

This gives us two practical measures of *B*, the benefit resulting from a particular decision. When the burrow concerned has only one chamber, the calculations are straightforward. The "egg score" is simply 1 if the wasp succeeded in laying an egg (i.e. showed permanent filling behaviour) on that occasion, 0 otherwise. The "food score" is also 0 if the wasp failed to lay an egg. If she succeeded, the food score is the volume of food upon which the egg sits (the volume of a katydid is regarded as the cube of its length). But things were not always so simple. A decision is a commitment to one burrow, but this may involve association with a succession of brood chambers within that burrow. In this case the egg score of a decision is the total number of eggs which the wasp concerned laid in the whole burrow, and the food score is the sum of the food scores for the separate chambers in which she laid an egg. These success scores are always credited to the decision which originally initiated her period of association with the burrow.

The rationale for using the food score as well as the plain egg score is this. A wasp who consistently skimped on katydids might run up an impressive egg score but each egg would have low subsequent success because the larva would be inadequately nourished. But just as the egg score probably underestimates the role of food in determining true reproductive success, the simple length-cubed food score probably overestimates it: obviously more means better only up to a point. Fortunately, as far as the predictions of our two models are concerned, the two measures of benefit give almost identical results. We shall quote both sets of results, but for purposes of explaining we shall normally use the egg score because it is simpler.

In any case as the "optimal foraging" theorists have taught us (e.g. Pyke, Pulliam & Charnov, 1977), a vital quantity has so far been left out of the reckoning—time. It takes time to dig a burrow, time to catch a katydid, precious, limited time, for all the wasps die at the end of the six summer weeks of the nesting season. The appropriate measure of success is a rate, B/T, the benefit accrued divided by the time taken to accrue it. The time taken over each burrow is measured from the moment at which the wasp left her previous burrow for the last time. Every minute of her adult life is accounted as spent on one burrow or another. Transit time between one burrow and the next is debited to the second of the two: it is "searching time" or "preparing time" (Larkin & McFarland, 1978). The success rate of a type of decision is the total benefit *B* accrued on all occasions when that type of decision was made divided by the total time, T, spent by wasps on burrows with which they began their association by making that type of decision.

An example may make this clear. At 1330 on 9 August, wasp red/red/yellow (RRY) abandoned her burrow when it was invaded by ants. At this moment we restart our accounting clock, although the decision which is running up the time debt is still in the future. She wandered around the nesting area for the rest of the afternoon, looking furtively into other wasps' nests and occasionally making tentative digging movements. The next morning at 1000 she entered a burrow already occupied by green/green/green (GGG). She returned later that day and is now definitely identified as having made a joining decision. The time "debt" of the joining decision is "back-dated" to the moment at which she left her previous burrow. Both wasps continued to provision the nest regularly, and brought five katydids between them. Then on 13 August at 0800 they both happened to return at the same moment. A fierce fight ensued and GGG ("founder") chased RRY ("joiner") from the nest. RRY never returned to the nest, and her association with it was clocked-off from this moment. She had spent 90.5 h in association with the burrow, and she received no benefit from it since she was driven off before she had a chance to lay an egg. GGG eventually did lay an egg in the joint chamber, and went on to lay another in a new chamber. When she finally left, her association with the burrow had lasted 299 h. She gained a benefit of two eggs. When it comes to computing the food score, GGG is credited with the benefit of all the katydids that her defeated rival caught, as well as her own catches.

We could compute the lifetime benefit and time scores of RRY, GGG, and all the other individual wasps. But this is not the object of the present exercise. RRY joined on this occasion but she sometimes founded; GGG founded, but she sometimes joined. In all, of 136 wasp decisions in NH, 23 were joining decisions and 113 founding decisions. The total number of eggs laid by wasps who, on that occasion, joined, was 8, or 0.348 per decision. The time they spent on joining episodes; whether or not they succeeded in laying an egg, was, in total, 1683.9 h, or 73.2 h per decision. The average B/T score for joining decisions was therefore 8/1683.9 or 0.48 eggs per 100 h.

Notice that we do not calculate the rate B/T separately for each burrow and then average the rates. This would be like computing the average speed of a car on a journey by looking at the speedometer. Our method is equivalent to working out the car's average speed by dividing total distance by total time. The "speedometer method" is appropriate for measuring variance in success rate, and we do use it for purposes of statistical comparison (Brockmann & Dawkins, in prep.).

Our usage of average success rate is based upon some assumptions. The longevity of a wasp is assumed to be unaffected by the strategic decisions she makes. If joining decisions, for instance, shortened the wasp's life expectation, T would not be a suitable denominator for the success rate. "Joining time" would have to be reckoned in a costlier currency than "founding time." A related assumption is that a wasp ends each episode of one type in the same state as she ends each episode of the other: the slate is wiped clean when the clock is restarted for the new nesting episode, and there is no carry-over of influence from one decision to the next. And, indeed, it is true that decisions are not made in runs (Brockmann & Dawkins, in prep.).

6. Outcomes of Model 1

So far, all we have said of Model 1 is that it postulates two types of decision called "found" and "join." Before we can test it we need to express the model in more detail. It is illustrated in Figure 1 . The basic decision is represented by the bifurcation of arrows on the left. In addition to this decision, which is under the wasp's control, we can also distinguish three possible outcomes, which are not necessarily under the wasp's control. If a wasp takes the decision to found, she may be subsequently joined, represented by the box labeled "is joined." Or she may not be joined, in which case the outcome is the one labeled "is alone." The outcome "joins" always results from the decision to join. Which of the two possible outcomes befalls a wasp who has founded her own nest is determined by the relative frequencies of founding and joining in the whole population. If the proportion of decisions per unit time which are founding decisions is q, the proportion which are joining decisions is $1 - q$. Since for every "joins" outcome there must be one "is joined," the proportion of all decisions per unit time which end up in the "is joined" outcome must also be $1 - q$. (Strictly, this statement should be amended. It is theoretically possible for a founding wasp to be joined in the same nest by two different joiners.) Finally, the proportion of all decisions per unit time whose outcome is "is alone" must, by subtraction, be $q - (1 - q) = 2q - 1$.

A wasp who founds has no control over whether she is subsequently joined, but this does not stop us calculating whether she would be better off joined or not. B_1, the average egg score for the "is alone" outcome, is the number of eggs laid, per decision, by all wasps who, on that occasion, made the founding decision and were not subsequently joined. Similarly, B_2 and B_3 are the average egg scores for the "is joined" and the "joins" outcomes, and T_1, T_2 and T_3 are the average times taken, per decision, over nesting episodes for each of the three outcomes. A wasp who decides to join can therefore expect a benefit of B_3 eggs at a cost of T_3 h. For a wasp

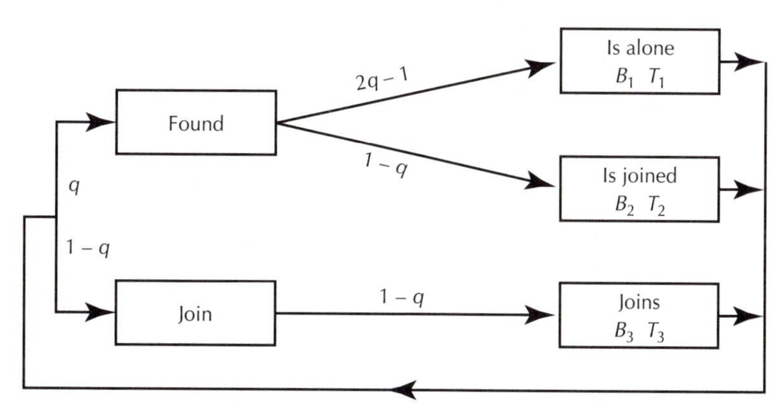

Figure 1. Model 1: There are two alternative kinds of nesting decisions a wasp can make—a proportion q of the decisions are to "found" a nest and a proportion $1 - q$ are to "join" another wasp in her burrow. The "found" decision has two possible outcomes: the wasp either nests alone or she is later joined by another wasp. B_1, B_2 and B_3 are the average benefits associated with each of the outcomes and T_1, T_2 and T_3 are the average times spent in each outcome.

who decides to found, we must make a more complicated calculation. If she happens to be joined she can expect B_2 eggs, but if she is not joined she can expect B_1 eggs. We know the odds of each of these two outcomes: they are $1 - q$ and $2q - 1$ respectively. So the net expectation is that a wasp who founds will lay, on average $(2q-1)B_1 + (1-q)B_2$ eggs per $(2q-1)T_1 + (1-q)T_2$ hours. We now apply the central dogma of this paper: at evolutionary stability the average success rate of founding decisions must equal the average success rate of joining decisions, B_3/T_3. This gives the simple equation

$$\frac{(2q-1)B_1 + (1-q)B_2}{(2q-1)T_1 + (1-q)T_2} = \frac{B_3}{T_3}$$

Solving for q, we obtain the testable prediction

$$q = \frac{T_3(B_1 - B_2) - B_3(T_1 - T_2)}{T_3(2B_1 - B_2) - B_3(2T_1 - T_2)}$$

This equation is the condition for equilibrium, but an equilibrium is not necessarily stable. In order for this one to be stable it is necessary that small perturbations of q away from its equilibrium value should be automatically corrected by natural selection: if founding decisions become more common, joining must receive a higher payoff and vice versa. The condition for stability in Model 1 can be deduced algebraically from the equation above, and is as follows:

$$\frac{B_2}{T_2} > \frac{B_3}{T_3} > \frac{B_1}{T_1}$$

In words, independently of frequencies of occurrence, a decision ending in the "is joined" outcome has to be more successful than a joining decision, and both have to be more successful than a decision ending in the "is alone" outcome. (There is an intuitive rationale for this. If the proportion of joining decisions goes up it is obvious that the number of "is joined" outcomes goes up too. For there to be stabilizing selection, when the rate of joining goes up the average success of founding as a whole must go up; the only way this can happen is if the founding outcome that has automatically increased in frequency, namely the "is joined" outcome, is the generally more prosperous of the two. So "is joined" must be more prosperous than "is alone" in order for the equilibrium to be stable. "Joins" has to be intermediate in payoff between the two outcomes of founding: if it was either better or worse than both of them it could not possibly be equal to their average, and we could never have equilibrium.)

Model 1, therefore, is in spirit an amicable, sharing model. Wasps alone do worse than both partners in a shared nest. This would have to mean that the advantages of sharing—presumably getting help in speedy provisioning, and in keeping fly parasites out of the nest (Brockmann & Dawkins, in prep.)—are so great that they compensate for the fact that only one of the two ends up by laying the egg. Intuitively this seems like a heavy gamble for a wasp. The best thing of all, according to Model 1, is to found a nest and then be joined by a second wasp. Next best is not to found, but to join an incumbent. The worst outcome, according to Model 1, is to be left alone after founding a burrow, for either partner in a sharing menage does better than a lone wasp.

This is all very interesting for the theory of the origins of eusociality, and there could be species to which Model 1 is applicable, but the next section suggests that *Sphex ichneumoneus* is not one of them.

7. Testing Model 1 on the New Hampshire Data

The basic assumption of the model is that the average success rate for founding decisions should equal that for joining decisions. Table 1 shows, on the contrary, that founding decisions are approximately twice as successful. The difference is statistically reliable (Brockmann & Dawkins, in prep.). The inequality conditions for stability are also violated by the NH data. Far from the two sharing outcomes being most successful, the real order of success rates is "is alone" > "is joined" > "joins." As we should expect, the predicted q proportion is also completely wrong.

Model 1, the amicable, sharing model, is decisively rejected. We need a new model, a model in which a wasp does better when alone than when sharing. We go back to the drawing board, and the first thing to do is to look more carefully at the biology of the wasps.

8. Decisions in Model 2

The rationale for Model 1 ignored a vital fact about the wasps. Founders sometimes spontaneously abandon burrows even when no joiner is present. Sometimes there seemed no obvious reason for this. On other occasions the abandoning seemed to be provoked by some catastrophe, invasion by ants, say, or a centipede. The marauder usually left after a while, and the empty burrow was none the worse for it. The original owner had fled, and the burrow was left untenanted and available for easy occupation by another. There was a good surplus of five to ten empty burrows available in the nesting area at any time. Wasps did quite often enter and use burrows which had been abandoned by their original owners.

Perhaps Model 1 postulated the wrong fundamental decision. Maybe there are indeed two alternatives, but they are not found and join, but "dig" and "enter." This is the basis of Model 2. It makes good sense in the light of the actual opportunities available to a wasp. The strategies "founding" and "joining" leap naturally to the mind of a biologist interested in the phylogenetic origins of eusociality. But the decision which actually faces an individual wasp in practice must surely be: shall I enter this hole here, or shall I dig a new one? A wasp who enters a burrow may have no opportunity to even find out whether the original owner is still in occupation until several hours or even days later. Model 2 uses the same 136 wasp decisions as Model 1, but it re-allocates them. In 80 cases the wasp began her association with the burrow by digging it. In 56 cases she began her association with it by entering and adopting it after it had been dug by somebody else. From now on we shall use the word "enter" only in its technical sense as the name of a decision. Once again, we are not talking about a polymorphism. We know that any given individual sometimes dug and sometimes entered, with an overall bias towards digging but not an individually specific bias, and we know that the decisions did not occur in runs (Brockmann & Dawkins, in prep.). When we loosely speak of, for instance, a "digger," we merely mean an individual who on that occasion began her association with the burrow in question by digging it.

Enterers exploit the available surplus of empty burrows. Sometimes they have the misfortune to enter a burrow which is still occupied by its former owner In this case they fall into the "joins" outcome which we formerly

Table 1. The average success rate of found and join decisions and their outcomes in the New Hampshire population of wasps.

Decision	Frequency in pop.	Average success rate B_v/T	Average success rate B_e/T	Outcome	Benefit: B = volume prey provisioned	Benefit: B = number eggs laid	Time T associated with nest	Average success rate B_e/T
Found	0.83(113)	32.42	0.96	is alone	39.85	1.18	116.17	1.01
				is joined	33.98	1.06	142.96	0.74
Join	0.17(23)	13 .68	0.48	joins	10.02	0.35	73.21	0.48

B_v/T is the average rate of accruing benefit based on the volume of prey provisioned per 100 h for all decisions or outcomes of this type.
B_e/T is the average rate of accruing benefit based on the number of eggs laid per 100 h for all decisions or outcomes of this type.

Table 2. The success rates of dig and enter decisions and their outcomes in the New Hampshire population of wasps.

Decision	Frequency in pop.	Average success rate B_v/T	Average success rate B_e/T	Outcome	Benefit: B = volume prey	Benefit: B = number eggs	Time T associated with nest	Average success rate B_e/T
				abandons	1.79†	0.11†	38.97	0.28
Dig	0.59 (80)	33.16	0.96					
				is alone	67.04	1.93	170.48	1.13
				is joined	33.98	1.06	142.96	0.74
Enter	0.41 (56)	26.81	0.84					
				joins	10.02	0.35	73.21	0.48

† There is some benefit associated with the abandons outcome because a few burrows are not abandoned until after one chamber has already been completed.

identified with a strategic decision. But according to Model 2 this is an accident, the worst thing that can happen to an enterer. If she is lucky the burrow she adopts will turn out to have been abandoned by its original owner. Unlike Model 1, then, Model 2 has nothing to do with amicable sharing. Sharing is an unfortunate byproduct of entering. The best outcome for a wasp is to end up alone. She is most likely to achieve this desirable end if she takes the digging decision, but this carries with it the risk that she may have to abandon her burrow after spending time on digging it. If she refrains from digging, and looks for a burrow to enter, she may save time but she runs the risk of finding herself joining an incumbent. If she is fortunate enough to find herself alone, she runs the same risk as a digger of being subsequently joined by a second enterer (Fisher Exact Test $P = 0.46$; Brockmann & Dawkins, in prep.).

Model 2 is more complex than Model 1, and before developing it in detail we should satisfy ourselves that it is going to be worthwhile: is the basic equilibrium assumption upheld? At evolutionary stability Model 2 expects the average success rate for digging decisions to equal that for entering decisions. The total number of eggs laid by wasps who, on that occasion, had begun their association with the burrow by digging it was 82. The total time spent by wasps in association with burrows that they had dug was 8518.7 h. The success rate, B/T for digging decisions is therefore 0.96 eggs per 100 h. For entering decisions the corresponding figures were 57 eggs in 6747.4 h, which gives an average success rate of 0.84 eggs per 100 h. These figures are not identical but they are similar and, together with the comparable figures when benefit is represented by the food score (Table 2), they are encouraging. A one-way analysis of variance of the B/T scores for the separate decisions ("speedometer readings"—see above) suggests no significant difference between the success rates of digging and entering decisions (egg scores, F ratio = 0.0 for 1 and 134 d.f., $P > 0\ 5$; food scores similar, Brockmann & Dawkins, in prep.). It is therefore believable that digging and entering could coexist in the same population without one of them being removed by natural selection.

So, we seem to have an equilibrium. But is it stable; what keeps it so? Qualitatively the outlines of a model of the necessary stabilizing selection seem clear enough. If too many wasps enter, they will tend to saturate available abandoned burrows, and the risk of an entering decision leading to the undesirable outcome of sharing will become too high. If very few wasps enter, the risk of an enterer finding her chosen burrow already occupied is relatively low. About 41% of all decisions in NH are entering decisions. It is easy to imagine this as an equilibrium, maintained by selection in favour of entering whenever the percentage drops below 41% and in favour of digging whenever the entering percentage rises above 41%. We now proceed to develop a more formal model along these lines.

9. Assumptions and Outcomes of Model 2

An important assumption of Model 2—indeed this is one of its essential differences from Model 1—is that an enterer enters empty versus occupied burrows in proportion to their availability. Intuitively this may seem odd, since it implies that a wasp cannot sense the presence of another, and the model surely predicts selection in favour of such discrimination (see below). Yet there is some empirical evidence for it ($\chi^2 = 2.7$ for 1 d.f., $P > 01$; Brockmann and Dawkins, in prep.). It seems less surprising when we reflect that a wasp spends 80%–95% of her time away from her burrow, so the chance of the incumbent being at home at the moment when the new enterer makes her decision is fairly low. Like Model 1, Model 2 assumes that longevity is independent of decision, and that successive decisions are independent of each other. Evidence on the latter point, including a demonstration that the wasps are not deciding in runs, nor using a conditional strategy of the "win-stay-lose-shift" variety, is given in Brockmann & Dawkins (in prep.).

We make two additional assumptions. The first is that the probability that a wasp who has dug a burrow will abandon it is a constant. All this really means is that the likelihood of a given burrow being abandoned is independent of the frequency of digging relative to entering decisions in the population. It seems a reasonable assumption: ants and centipedes are hazards which can assail any burrow at any time. The second assumption is that at equilibrium the rate at which burrows are used up by enterers is equal to the rate at which they are abandoned by diggers. The intuitive rationale for this is as follows. Enterers obviously cannot use up burrows at a faster rate than diggers abandon eventually re-usable burrows. If they did so at a much slower rate, on the other hand, the increasing surplus of re-occupiable burrows would make enterers better off than diggers, and we would not have equilibrium. We assumed that the two rates were actually equal, and further that all abandoned burrow eventually become re-usable. The effect of this is that there is a static market economy of burrows, a fixed stock of empty holes which is being added to at the same rate as it is being drawn upon. Any of these assumptions could be wrong. If so, the predictions of the model will probably also be wrong. A good model is one whose predictions are vulnerable to disproof but are not disproved.

Figure 2 illustrates Model 2 in the same way as Figure 1 did for Model 1. The bifurcation on the left represents the initial strategic decision, in this case between digging and entering. Once again we distinguish decisions which a wasp makes from outcomes which just happen. In this case there are four outcomes. "Abandons" is one possible outcome of a digging decision, and it is assumed never to result from an entering decision. "Joins" is one possible outcome of entering, and it cannot result from digging. The other two outcomes, "is alone" and "is joined," can result from either digging or entering. As already mentioned we assume, with justification, that neither of the two decisions is more likely than the other to lead to any particular one of these two outcomes. This assumption is represented by the merging of the two arrows in the centre of the figure; diggers who do not abandon, and enterers who do not join, merge into a single sub-population of "foundresses." Of these foundresses; a proportion r are subsequently joined, and a proportion $1 - r$ remain alone.

Of all the decisions which occur in a given time, a proportion w lead to the "abandons" outcome, a proportion x are digging decisions that

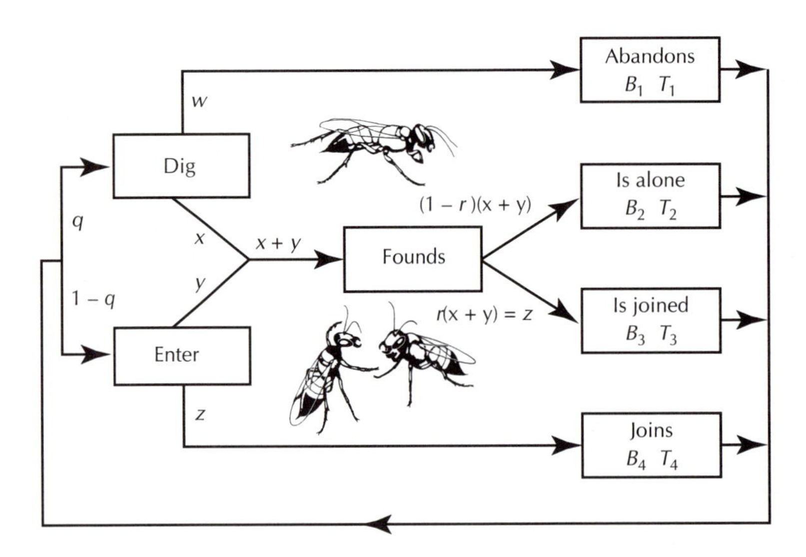

Figure 2. Model 2: There are two alternative kinds of nesting decisions a wasp can make: dig and enter. A digging decision may result in one of three outcomes. The wasp may abandon her burrow, or she may found a nest in which case she may remain alone or she may be joined by another wasp. An entering decision may also result in one of three outcomes. A wasp may join another in her nest or she may found a nest of her own in which case she may end up nesting alone or in company with another wasp. B_1, B_2, B_3 and B_4 are the average benefits associated with each outcome and T_1, T_2, T_3 and T_4 are the average times spent in each outcome.

do not lead to the "abandons" outcome; a proportion z lead to the "joins" outcome, and a proportion y are entering decisions that do not lead to the "joins" outcome, $w + x + y + z = 1$. $w + x = q$, the proportion of all decisions which are digging decisions, $y + z = 1 - q$. $x + y$ is the proportion of all decisions which result in the wasp becoming one of the sub-population called foundresses. $r(x + y)$ is the proportion of all decisions which lead to the outcome "is joined." This must equal z, since for every "is joined" outcome there has to be one "joins" outcome (with the same reservation as mentioned earlier).

As in the case of Model 1, we measure average benefit scores and time scores for each of the four outcomes. T_1 is the time per decision spent on burrows which are eventually abandoned. T_2 is the corresponding time spent on burrows which are occupied alone, and T_3 is the time per decision spent by wasps who, on that occasion "founded" (whether they began by digging or entering) but were later joined. T_4 is the corresponding time spent by wasps who found themselves in the "joins" outcome. B_1, B_2, B_3 and B_4 are the corresponding benefit scores, either egg scores or food scores.

A wasp who decides to dig can expect to lay a number of eggs which is a weighted average of B_1, B_2 and B_3. A wasp who decides to enter can expect a corresponding weighted average of B_2, B_3 and B_4. The times they spend will be homologously weighted averages of the corresponding T scores. The weighting factors are the frequencies with which the various outcomes occur. These are variables in the model, where the B and T scores are parameters. We can read them off Figure 2 as w for the abandoning outcome, z for the joins outcome, $r(x + y)$ for the "is joined" outcome and $(1 - r)(x + y)$ for the "is alone" outcome (where r, it will be remembered, is a short-hand symbol for the proportion of foundresses who are joined, i.e. $z/(x + y)$. The average expectation for a wasp who enters is, therefore $(zB_4 + yrB_3 + y(1 - r)B_2)$ eggs per $(zT_4 + yrT_3 + y(1 - r)T_2)$ h. The corresponding expected success rate for a wasp who takes the digging decision is $(wB_1 + xrB_3 + x(1 - r)B_2)$ eggs per $(wT_1 + xrT_3 + x(1 - r)T_2)$ h.

At evolutionary stability these two success rates must be equal. Therefore we may set up the simple equation

$$\frac{wB_1 + xrB_3 + x(1 - r)B_2}{wT_1 + xrT_3 + x(1 - r)T_2} = \frac{zB_4 + yrB_3 + y(1 - r)B_2}{zT_4 + yrT_3 + y(1 - r)T_2}$$

The four B and four T scores are measured parameters, in terms of which any one of the variables might be expressed. At first sight there appear to be too many unknowns in the equation, namely w, x, y, z and r, but we dispose of all but one as follows. r, as we have seen, is simply a short-hand expression for $z/(x + y)$. By definition, $w = 1 - x - y - z$. Our economic assumption that the rate of recruitment to the pool of available burrows equals the rate at which burrows are used up may be written in the form $y = w$, therefore $y = 1 - x - y - z = (1 - x - z)/2$. Finally, we use our assumption that the probability of a burrow being abandoned is constant. If this constant is K, then $K = w/(w + x) = y/(y + x)$. K is used as a parameter like the B and T scores, and is measured from the NH field data as 0.4625. By means of these supplementary simultaneous equations we can solve our main equilibrium equation, treating any one of the variables as the unknown. We arbitrarily chose to solve for z. z as an algebraic function of the B, T and K parameters is very complicated so we do not give it here, but simply give the predicted value of z from the nine measured parameters. Having obtained the predicted value of z, we can then easily deduce the predicted values of w, x, y and q from it.

The stability condition is also complicated, and so not presented here. It is readily calculated once the rates of benefit of the two decisions are expressed as functions of q; it is only necessary to check that the rate for enterers increases faster than the rate for diggers as q increases. The stability condition is met by the NH data.

10. Testing Model 2 on the New Hampshire Data

Table 3 summarizes the terms and equations of Model 2. Table 4 gives the tests of the predictions. Separate sets of predictions of the four proportions, w, x, y and z are given, one deduced using egg scores for benefit, the other deduced using food scores. Since the four predicted values are proportions which necessarily add up to 1, we really only have three predictions.

The fit of observed to predicted proportions is very close, so close that we should clearly state again which measured parameters were used in making the predictions. In addition to the four B scores and the four T scores we also used the measured value of K, the proportion of dug burrows which were abandoned. K determines that the ratio between the predicted w and x necessarily agrees with the observed ratio. It does not determine the actual magnitudes of w and x.

It is important to stress that this is very much not the kind of model where one fits parameters to observed results so comprehensively that the "predictions" almost cannot help being right. Our predictions really were genuine predictions, and they could very easily have come out wrong. In particular, the fact that we estimated the parameter K did not make the predictions trivial. This can be demonstrated by the following rough randomization test.

Suppose we recalculated our predictions using, not the observed mean value of T_1, but a smaller value, or a larger value. How much difference would it have made to the predictions? Might the fit of predicted to observed z have come out even better? If so, we would not be so impressed by the goodness of fit we actually obtained.

Rather than play around, piecemeal, with artificially distorted values of our eight parameters, we adopted a systematic randomization procedure. For each of the four B (egg scores) and four T parameters we made the computer choose a random number from a normally distributed population having a mean and standard deviation equal to the appropriate observed mean and standard deviation (the distributions were truncated to avoid negative values). We then recalculated our model prediction of z using these randomized parameters instead of the real ones. If the goodness, of fit of the computer prediction to the observed z had been better than the one in Table 4, this would have weakened our confidence in the model's power to explain the real world. In fact this happened in only 9 out of 1000 separate runs of the simulation, less than 1%. On the remaining 991 occasions, the simulated prediction gave a less good fit to the observed value of z than our own real prediction. We conclude that Model 2 is highly sensitive to the data which are fed into it, and that it would be remarkably difficult to obtain, by chance, as good a result as the wasps actually gave.

Table 3. Summary of variables and equations for Model 2.

Definitions

w	Proportion of all decisions which are followed by the digger abandoning her nest
x	Proportion of all decisions which are followed by the digger occupying the nest she dug, i.e. founding
y	Proportion of all decisions which are followed by the entering wasp occupying a nest, i.e. founding
z	Proportion of all decisions which are followed by the entering wasp occupying an already occupied burrow, i.e. where the enterer joins another wasp
r	Proportion of all foundings which are followed by the founder being joined by another wasp

Equations

$w + x + y + z = 1$ (by definition)

$\frac{w}{w + x} = K$ (by assumption)

$z = r(x + y)$ (by assumption)

$w = y$ (by assumption)

$$\frac{wB_1 + x(1 - r)B_2 + xrB_3}{wT_1 + x(1 - r)T_2 + xrT_3} = \frac{zB_4 + y(1 - r)B_2 + yrB_3}{zT_4 + y(1 - r)T_2 + yrT_3} \quad \text{(equilibrium assumption)}$$

Table 4. A comparison between the predicted rates of various outcomes and the observed rates in the New Hampshire population.

	Proportion that	Observed	Predicted B = eggs	Predicted B = food volume
w	dig then abandon	0.272	0.257	0.260
x	dig and do not abandon	0.316	0.298	0.303
v	enter and do not join	0.243	0.257	0.260
z	enter and join	0.169	0.188	0.176

This conclusion is ironically borne out by the total failure of the model to predict the results in the other study area, in Michigan.

11. Testing the Models on the Michigan Data

The Michigan data disprove the predictions of Model 1 as decisively as the New Hampshire data did, but in a different way. Joiners do far better than they did in NH. Indeed they do about as well as founders do (Table 5). But although this might look like an equilibrium, the conditions are not met for it to be stable. The "joins" outcome and "is alone" outcome both have a higher success rate than the "is joined" outcome. Model 1 cannot give a stable equilibrium under these conditions. We must reject the amicable, sharing model here, as in the New Hampshire population, but in this case because joiners are doing so much better than those they join.

But whereas Model 2 successfully accounted for the New Hampshire data, the Michigan data seem irreconcilable with it. Table 6 shows that entering decisions are more

Table 5. The success rates of found and join decisions and their outcomes in the Michigan population of wasps (1973–1975 combined).

Decision	Frequency in pop.	Average success rate B_e/T	Average success rate B_e/T	Outcome	Benefit: B = volume prey	Benefit: B = eggs laid	Time T associated with nest	Average success rate B_e/T
Found	0.95 (259)	48.18	1.69	is alone	32.13	1.12	63.56	1.76
				is joined	35.99	1.42	139.11	1.02
Join	0.05 (15)	53.54	1.86	joins	19.23	0.67	35.92	1.86

Table 6. The success rate of dig and enter decisions and their outcomes in the Michigan population of wasps (1973–1975 combined).

Decision	Frequency in pop.	Average success rate B_e/T	Average success rate B_e/T	Outcome	Benefit: B = volume prey	Benefit: B = eggs laid	Time T associated with nest	Average success rate B_e/T
				abandons	0.89	0.08	16.38	0.50
Dig	0.84 (230)	46.28	1.64					
				is alone	64.10	2.17	110.56	1.96
				is joined	35.99	1.42	139.11	1.02
Enter	0.16 (44)	56.49	1.91					
				joins	19.23	0.67	538.8	1.86

successful than digging decisions. One-way analysis of variance shows that enterers lay more eggs than diggers ($F = 5.85$, for 1 and 272 d.f., $P < 0.05$) and bring more food ($F = 5.17$, $P < 0.05$) and they take no longer over it ($F = 3.0$, $P > 0.05$). Selection, then, should be severely penalizing digging decisions and increasing the frequency of entering decisions.

Is Model 2, perhaps, basically applicable to the Michigan population, but the relative frequency of enterers happens to be below equilibrium at present? If this were so, we should be able to solve the equation of Model 2 to predict how far the swing back toward entering should go: what is the predicted equilibrium frequency? The equation is a quadratic. For the particular values of the nine parameters which were measured at the Michigan site, there are no real roots to the equation: there is no equilibrium frequency at which the two strategies would have equal success rates. Computer simulation shows that, according to the model, enterers would have the advantage over diggers at all frequencies. Diggers should not exist! This is biologically absurd, since there would be no burrows to enter unless the species was an interspecific parasite, which it is not. We cannot, then, salvage Model 2 in Michigan by saying that the population is temporarily away from equilibrium, and is in the process of returning there. We do not understand why the model does not work there, and can only make guesses.

12. What Might We Learn from Model 2's Failure in Michigan?

The failure of a model can be illuminating if it leads us to re-examine our assumptions constructively. Certain of Model 2's assumptions are so basic that to tinker with them would amount to setting up a whole new model. For instance we assumed that the expected benefit and time scores of the four "outcomes" are constant parameters. What would happen if they were frequency-dependent, or density dependent? Is there, perhaps, a Model 3 which replaces "dig" and "enter" by some completely new set of "strategies"? "Join a burrow only if it has at least three katydids in it" should be far more successful than Model 1's "join" and Model 2's "enter." The evidence goes against the wasps having the necessary powers of discrimination (indeed they appear to commit the "Concorde Fallacy" of valuing a nest in proportion to their own prior investment in it, Dawkins & Brockmann, in prep.). But we must not rule out some kind of radical redefinition of the decision options.

Other assumptions of Model 2 seem less fundamental. For instance the "static burrow economy" assumption, $y = w$, might be replaced by $y = uw$, where u is a constant representing the probability that an abandoned burrow can be re-used. In effect the Model 2 which we tested was the special case with $u = 1$. As a matter of fact it is true that u is close to 1 in NH ($u \geq 0.83$, since 0.83 of the abandoned burrows were actually re-used) whereas in MI only 0.27 of the abandoned burrows were re-used. Such a gross violation of an assumption might help to account for Model 2's failure in MI. Out of curiosity, we did try the experiment of plugging the observed minimum value of u into Model 2: the fit to the NH data became even more impressive, but still no stable solution could be found for the MI data. Indeed we showed that no value of u exists such that Model 2 even comes close to making realistic predictions for the MI economic data: the violation of the assumption that $y = w$ is therefore not sufficient to explain the failure of Model 2 in MI. This is an instructive result in itself. Incidentally, the low value of u in MI could just as well be an inevitable consequence of there being fewer enterers than expected, rather than the cause of Model 2's failure.

Setting aside the strong likelihood that Model 2 is fundamentally wrong in MI, we now turn to the second potentially instructive possibility. This is that Model 2 is right but that something is wrong with our data. For instance our assumption that egg-laying success is a good estimator of fitness may be approximately true in NH but false in MI. This could arise if there is a larval or pupal disease in MI which differentially infects shared nests. The important point is that the B scores which we measured may not be the ones which truly apply in MI, and that the overestimate is greatest for shared nests. The latter proviso is needed, since to divide all the B scores by a common mortality constant would leave the predictions unchanged.

Another interesting possibility is that the MI wasps are really adapted to some environment other than the one in which they were studied. If MI conditions have recently changed, the wasps could simply be "out of date." Or there might be continual gene flow into the study area from outlying populations in which digging is more strongly favoured. Incidental observations. and reports from local amateur entomologists, did indeed suggest that the MI aggregation, but not the NH one, was surrounded by outlying lone wasps for whom entering would presumably be bad policy. It is not implausible, then, that the MI study site is subject to a continuous inflow of digger genes which maintain the study site population away from its equilibrium point, thereby inducing continuous selection in favour of entering at the study site itself.

In all this, it should be remembered that the MI population is not behaving as though in accordance with Model 2 but away from the equilibrium point. Something more fundamental is wrong with Model 2 in the Michigan population. The suggestions arising out of this failure should be regarded as stimulation to further work. Meanwhile we return to New Hampshire.

13. What Might We Learn from Model 2's Success in New Hampshire?

If a model's predictions are successful we are entitled to regard this as evidence that its assumptions represent truth about the real world, not overwhelming evidence, but still evidence. We can, therefore, list the assumptions of Model 2, not merely as assumptions but as putative facts about the New Hampshire population of wasps for which we have provided some indirect evidence. They are:

(1) There exist two alternative nesting "strategies" called "dig" and "enter" in the population. This is not a polymorphism in which individuals are consistent specialists. Rather each individual intersperses through her adult life a sequence of choices between the two alternative behaviour patterns. This could be regarded as a simple mixed strategy.

(2) The wasps cannot distinguish an empty abandoned burrow from one that is still occupied by an incumbent. This is a surprising result since the model also predicts selection in favour of such discrimination.

(3) Burrows are re-occupied by enterers at approximately the same rate as they are abandoned by diggers. There is a static market economy of burrows

(4) The two strategies enjoy equal success rate in the population. There is an equilibrium such that selection does not favour either one of them over the other at observed frequencies.

(5) The equilibrium is stable. It is maintained by frequency-dependent advantage whose nature is as follows. Enterers do relatively well when they are rare, because the chance that they will find themselves sharing is relatively low. Enterers do relatively poorly when they are common, because they run a substantial risk of finding themselves joining an incumbent. The equilibrium frequency in the New Hampshire population is about 41% enterers.

Students of Hymenoptera are inevitably awed by the heights of adaptive achievement of the social species, and they tend to see solitary species as the bottom of a kind of entomological "great chain of being." We too began by regarding the nest-sharing of *Sphex ichneumoneus* as a tentative groping toward the brave new world of eusociality. This is why we talked about founding and joining, why we used the word sharing at all. We were prepared to reverse the sign of social benefit, and contemplated regarding joiners as parasites. The one thing we never considered in the early stages was that nest "sharing" might be neither mutualistic nor parasitic but simply *incidental.* This seems obvious now, but it was the decisive failure of Model 1 and the need to think of a radical alternative to the "sharing" assumption which it made, that raised our doubts. And it was the fact that Model 1 had been formulated as a series of precise assumptions that enabled us to question these assumptions systematically and pinpoint the one whose change gave rise to Model 2. "Joining" just does not seem to be an option open to the wasps. It is an incidental consequence of selection in favour of entering and adopting abandoned nests.

As such it is a potential *pre-adaptation* to the evolution of social life. Suppose the parameters of Model 2 were to shift over evolutionary time. Could selection ever begin to favour entering because sharing is an outcome of it, rather than in spite of the fact? If it is a pre-adaptation, it is surely the more promising for being an evolutionarily stable minority habit, rather than a freakish mistake. The relevance of Model 2 to the evolution of social life is the main theme of the companion paper (Brockmann & Dawkins, in prep.).

The great deficiency of our data is the lack of information on post-egg laying success. It would theoretically be possible to watch the young adults emerging from the ground the following year, but it seems impossible to know which of last year's females should get the credit. This is why the first author intends to switch to mud-daubing wasps (*Sceliphron* and *Trypoxylon*). Instead of digging chambers underground they build mud nests on walls like swallows. It will be possible to trap emerging adults of the next generation, and so award posthumous but all important credit to the successful mothers of a summer ago. Otherwise the essential features of their biology are similar to *Sphex ichneumoneus,* with whom they are closely related. In particular it seems likely that, whatever improvements can be made in the numerator of the success score, the denominator will still be the same—time. Time is a currency which an animal spends. Decisions are the moments at which the down payments are made.

Acknowledgments

We would like to thank Richard and Betty Brinckerhoff of Exeter, New Hampshire and Orin Gelderloos and the University of Michigan, Dearborn for their cooperation with this study. The pictures on Figure 2 were drawn by Cheryl Hughes from photographs. John Krebs helpfully criticized the manuscript. HJB was supported by a N.A.T.O. Post-doctoral Research Fellowship.

References

Brockmann, H. J. (1976). Ph.D. Diss. University of Wisconsin, Madison, Wisconsin.

Cade, W. (1978). Nat. Hist. 87:64.

Dawkins, R. (1976). *The Selfish Gene.* Oxford University Press. Oxford.

Dawkins, R. (in press). In: *Sociobiology—Beyond Nature/Nurture.* G. W. Barlow & J. Silverberg eds. Westview Press. Boulder, Colorado.

Dawkins, R. & M. Dawkins. (1973). Behaviour 45:83.

Fisher, R. A. (1930). *The Genetical Theory of Natural Selection.* Oxford University Press. Oxford.

Gadgil, M. (1972). Am. Nat. 106:574.

Larkin, S. & D. J. McFarland. (1978). Anim. Behav. 26:1237.

Maynard Smith, J. (1974). J. theor. Biol. 47:209.

Maynard Smith, J. & O. A. Parker. (1976). Anim. Behav. 24, 159.

Maynard Smith, J. & G. A. Price. (1973). Nature 246:15.

Michner, C. D. (1974). *The Social Behavior of Bees: A Comparative Study.* Harvard University Press. Cambridge, Mass.

Parker, G. A. (1978). In: *Behavioural Ecology.* J. R. Krebs & N. B. Davies, eds. OxFord Blackwell Scientific Publications.

Pyke, G. H., Pulliam, H. R. & E. L. Charnov. (1977). Q. Rev. Biol 52:137.

Taylor, P. (in press). In: *Animal Economics.* M. Dempster & D. J. McFarland, eds. Academic Press. London.

Wilson, E. O. (1971). *The Insect Societies.* Harvard University Press. Cambridge, Mass.

Introduction to
Ronald L. Mumme
Do Helpers Increase Reproductive Success?

by Robert L. Jeanne

Helping behavior in birds poses a conundrum to classical fitness theory: Why should an adult bird opt to be a nonbreeder and help a breeding pair rear their offspring, rather than attempting to breed on its own? This phenomenon is an example of the seemingly altruistic behavior that posed a special difficulty for Darwin as he developed his theory of evolution by natural selection. How can such behavior evolve, if the individuals who show it leave no offspring? The answer appears to be that because helpers tend to help close relatives, they are gaining inclusive fitness. Many of the studies on helping have therefore been designed to test hypotheses about just how the inclusive fitness of a helper is increased. In many cases, however, these studies did not include proper controls for alternative explanations. For example, even if it can be shown that pairs with helpers produce more young than pairs without, the effect could be due to differences in territory quality. Perhaps helpers are more likely to join pairs that have high quality territories. A second problem that remained unaddressed is, even if it can be shown that helping behavior per se increases the number of offspring successfully reared by a pair, it isn't clear that the behavior is the result of a history of natural selection favoring it. The study by Mumme attempts to address these concerns.

The paper is an example of a modern behavioral ecological study in the best tradition of Tinbergen. It is an experimental field study, carefully designed to answer very specific questions left incompletely answered by earlier work. It is a model of how to do a field study: ask specific questions, design the experiments to answer these questions, treat the results statistically, address alternative interpretations of the results, and conclude only what the data allow.

Another point well brought out by this paper is that field studies are subject to year- to-year variation in factors beyond the control of the experimental design. Note that the results in the two years were not the same. Differences in weather, food availability, predator abundance, and so on, all could have caused such a difference. This points up the importance of conducting field experiments over several years as a means of countering the effects of such variables.

Many questions remain unanswered, of course. How do helpers recognize and join their relatives? How does a bird decide whether to be a helper or a breeder? Why does helping occur in some species of birds and not others? Might patterns in the geographic or taxonomic distribution of the behavior suggest hypotheses to test?

Do Helpers Increase Reproductive Success? An Experimental Analysis in the Florida Scrub Jay

Ronald L. Mumme

Original publication in *Behavioral Ecology and Sociobiology* (1992) 31:319–328. Reprinted with permission.

Summary

Although several different hypotheses have been proposed to explain the evolution of helping behavior, most are based on the assumption that helping enhances the reproductive success of recipient breeders. I tested this assumption by removal experiments in the cooperatively breeding Florida scrub jay *(Aphelocoma c. coerulescens)*. This species lives in permanently territorial social units containing a single breeding pair and none to six nonbreeders, which are usually offspring of the breeding pair and which usually act as helpers by feeding the nestlings and fledglings produced by the breeding pair. Although experimental removals of nonbreeders in 1987–1988 had no significant effect on breeder survival, egg production, or hatching success, experimental groups suffered higher rates of predation on nestlings (1987) and lower rates of fledgling survival (both years) than did unmanipulated controls. As a result, experimental groups produced an average of only 0.56 independent juveniles, compared to 1.62 young for controls. Analysis of the factors contributing to nestling and fledgling mortality indicates that helping behavior per se (i.e., the aid that nonbreeders provide to dependent young), not the mere presence of nonbreeders, was responsible for the greater reproductive success observed in control groups. Because survival rates of allofeeders (i.e., those nonbreeders that provisioned dependent young) were virtually identical to those of non-allofeeders, the costs of helping behavior in this species appear to be small. Furthermore, nonbreeders are more likely to provision dependent young within their social unit when those young are closely related. I therefore conclude that nonbreeders increase their indirect fitness by serving as helpers and that helping behavior in the Florida scrub jay is a trait that has current selective utility. It remains debatable, however, whether helping in this species is an adaptation that has been shaped by the process of natural selection.

Introduction

Avian cooperative breeding systems are characterized by the presence of helpers—individuals that show parent-like behavior toward young that are not their own offspring. Because its evolution is difficult to explain through classical notions of individual natural selection, helping behavior has attracted considerable theoretical and empirical study. As is evident from several recent reviews (e.g., Brown 1987; Koenig and Mumme 1990; Stacey and Koenig 1990; Emlen 1991), much of the literature in this field is devoted to developing and testing hypotheses about how individuals might enhance their inclusive fitness by acting as helpers. Many of these hypotheses, based either on direct or indirect selection (Brown 1987), are built upon the assumption that the alloparental behavior performed by helpers increases the reproductive success of recipient breeders.

However, this research paradigm has been subjected to two major criticisms. First, many of the studies purporting to show that helpers enhance reproductive success do not control for the effects of confounding variables. In such cases, it is thus unclear whether increased reproductive output is attributable to alloparental care per se or to differences in correlated traits such as territory quality and breeder quality (Brown 1987; Koenig and Mumme 1990; Emlen 1991). Only two previous studies of cooperatively breeding birds have attempted to address this problem experimentally, and these produced conflicting results. Although Brown et al. (1982) found that experimental removal of helpers significantly reduced reproductive success in the gray-crowned babbler *(Pomatostomus temporalis)*, no such effect was evident in a comparable experiment by Leonard et al. (1989) on the cooperatively breeding moorhen *(Gallinula chloropus)*. Limited experimental work on cooperatively breeding mammals has also produced conflicting results (Solomon 1991). Clearly, additional experimental studies in this area are needed (Brown 1987; Smith 1990; Emlen 1991).

The second major criticism leveled against much of the work on avian helping is that, even in cases where helping behavior has a clear positive effect on helper fitness, its current utility does not necessarily constitute evidence that helping is an evolved adaptation that has been shaped by a history of natural selection. This has been the central issue in a recent debate over the adaptive significance of avian helping (reviewed by Mumme and Koenig 1991).

In this paper I address these criticisms by an experimental analysis of helping behavior in the cooperatively breeding Florida scrub jay *(Aphelocoma c. coerulescens)*. This species lives in permanently territorial family groups that are usually based around a single monogamous breeding pair. Slightly over half of all groups include one to six nonbreeders that are typically offspring of one or both breeders. These nonbreeders are usually 1–4 years of age and include both males and females. Although they do not normally participate in nest building, incubation, or brooding, most nonbreeders act as helpers by feeding the nestlings and/or fledglings produced within their social unit, and by defending them against potential predators. Fledging occurs when the young are 16–20 days old, and even though fledglings begin to forage on their own by day 35, they nonetheless remain highly dependent on adults until they become proficient foragers at about 60 days of age (Woolfenden and Fitzpatrick 1984, 1990; McGowan and Woolfenden 1990).

Previous analyses have shown that Florida scrub jay groups containing nonbreeders (potential helpers) produce significantly more young than do groups without nonbreeders, primarily because of increased hatching success, reduced predation on nestlings, and increased survival of young between fledging and day 60—the age at which young become largely independent of parents and helpers (Woolfenden and Fitzpatrick 1984, 1990; McGowan 1987; McGowan and Woolfenden 1990). However, this result may be confounded by the fact that pairs with potential helpers tend to be experienced breeders residing on high-quality territories. Both of these factors are likely to influence reproductive success independently of any effect of potential helpers. Woolfenden and Fitzpatrick (1984) attempted to control for these confounding variables by performing an additional analysis limited to previously successful pairs that were accompanied by potential helpers in some years but not in others. The results of this more stringent analysis were suggestive, but inconclusive. Although groups containing one or more nonbreeders produced slightly more young than groups without nonbreeders, the difference was not statistically significant (Woolfenden and Fitzpatrick 1984, p. 196). Thus, it is unclear whether the presence of potential helpers truly increases reproductive success in this species. Nor is it clear whether the increase in reproductive output, if it indeed exists, can be attributed to helping per se (the alloparental care that nonbreeders usually provide to dependent young) rather than to incidental benefits associated with group living (Koenig and Mumme 1990).

This study was therefore designed to answer two questions: (1) Does the presence of nonbreeders (potential helpers) have a positive effect of reproductive success independent of the effects of correlated variables such as parental quality and territory quality? (2) If so, is the increase attributable to the alloparental care that nonbreeders provide as helpers?

Methods

Study Area

This study was conducted during three breeding seasons (1987–1989) at Archbold Biological Station in Highlands County, Florida, USA. A color-marked population of Florida scrub jays has been under continuous study at Archbold since 1969 (Woolfenden and Fitzpatrick 1984,1990). However, I focused on a second population of jays adjacent to this primary study site (Schoech et al. 1991; Schaub et al. in press).

Removal Experiments

Removal of potential helpers was performed in 2 years of the 3-year study. In 1987, 20 families containing nonbreeders (potential helpers) were randomly designated as either experimental (n = 8) or control (n = 12) groups. During nest building and early egg laying (mid-March to early April), all 15 nonbreeders from the 8 experimental groups were removed, thereby reducing these groups to a single breeding pair. Nonbreeders were captured either in baited traps (n = 13) or mist nets (n = 1) and maintained in captivity until the end of the breeding season, when they were released on their original territories. (One nonbreeder "removed" itself by dispersing to a nearby territory and thus was not captured.) The mean number of nonbreeders removed from experimental groups was 1.88 (range 1–4). The 12 control groups were not experimentally manipulated, and contained an average of 1.75 nonbreeders (range 1–3).

In 1988, 15 families containing nonbreeders were randomly selected as either experimental (n = 6) or control (n = 9) groups. All 11 nonbreeders from the 6 experimental groups were captured in baited traps and maintained in captivity until the end of the breeding season. Data also were collected on four additional "experimental" groups where natural removal of potential helpers had occurred. These were families consisting of experienced and previously successful pairs whose nonbreeding offspring had disappeared or emigrated prior to the 1988 breeding season. However, one of the ten total experimental groups was excluded from further analysis when the breeding male disappeared immediately after the removal of its potential helper. A second experimental group was excluded when failed breeders from an adjacent territory became helpers at the nest late in the breeding season. Thus, a sample of eight experimental groups and nine unmanipulated control groups were available for analysis in 1988. The mean number of nonbreeders present in experimental groups prior to removal was 1.75 (range 1–3), and 1.89 (range (1–4) for control groups.

General Methods and Data Collection

Basic field methods followed those outlined by Woolfenden and Fitzpatrick (1984). Jays in the study population were censused at monthly or more frequent intervals from March to July during each year of the 3-year study All nests were located, usually during building or before the onset of incubation. Nest contents were usually checked every 2–3 days (see Schaub et al. 1992) until nest failure or fledging. All nestlings were weighed, measured, and color-banded at day 11 (hatching = day 0). Reproductive success was measured in terms of both the number of young surviving to fledging (day 16–20) and the number of young surviving to independence (day 60).

Groups with nestlings or fledglings were observed to determine if nonbreeding group members delivered food to dependent young. Data on the rate at which food was delivered to nestlings were collected during 8-h nest watches conducted at both day 3 and day 10 of the nestling period. Each 8-h nest watch comprised two 4-h segments, the first beginning approximately 30 min after sunrise, and the second ending approximately 30 min before sunset. Methods used in nest watches were similar to those employed by Stallcup and Woolfenden (1978).

Coefficients of relatedness between breeders and nonbreeders were estimated from genealogies of color-banded birds (e g., Woolfenden and Fitzpatrick 1984). Because Florida scrub jays are almost exclusively monogamous (Woolfenden and Fitzpatrick 1990), and recent genetic analysis has found no evidence of extra-pair fertilizations (J. S. Quinn, pers. comm.), genealogical data are appropriate for estimating relatedness in this species.

Statistical analysis was conducted using standard parametric and nonparametric techniques. When appropriate, I used one-tailed procedures to test specific directed hypotheses about the effects of helpers on reproductive success and related aspects of breeding biology and behavior (Siegel 1956).

Terminology

Considerable semantic confusion has surrounded usage of the term "helper" (Brown 1987). In this paper I will follow Brown (1987, p. 300) and define a helper as "an individual that performs parent-like behavior toward young that are not genetically its own offspring." Although most nonbreeding Florida scrub jays meet this definition, others do not. For example, many nonbreeders fail to perform alloparental care simply because no nesting attempt within their social unit advances past the egg stage; such individuals thus never have the opportunity to care for dependent young. In addition, a few nonbreeders may fail to deliver food to nestlings or fledglings even when dependent young are present within their social unit. Such individuals might still be regarded as helpers, however, if they defend nestlings and/or fledglings against potential predators. Because of these ambiguities, in this paper I will use the term "helper" in the generic sense as defined above, and the term "allofeeder" to refer specifically to non-parental individuals that actually deliver food to nestlings or fledglings. Nonbreeding group members, regardless of whether they actually provision dependent young, are collectively referred to as either "nonbreeders" or "potential helpers."

Results

Effects of Removals on Reproductive Success

Experimental removal of nonbreeders had no detectable effect on the survival or basic reproductive characteristics of nesting Florida scrub jays. Experimental groups and unmanipulated control groups did not differ significantly in breeder survival during the nesting season, nesting phenology (first egg date), clutch size, or number of nesting attempts (Table 1). Nor did experimental removal of potential helpers have any significant influence on hatching success, as measured by the percent of nests where at least some eggs hatched, the percent of all eggs that hatched, or hatchability—the hatching success of eggs that survived to the end of incubation period (Table 1).

However, even though removal of nonbreeders had no effect on egg laying or hatching success, experimental groups produced only 0.56 ± 0.73 (mean ± SD) 60-day-old juveniles, compared to 1.62 ± 1.32 juveniles for unmanipulated control groups (Mann-Whitney *U*-test, one-tailed $P = 0.008$). This difference between experimental and control groups persists even if the four "natural" experimental groups from 1988 are excluded. In this more restrictive analysis, experimental groups produced 0.33 ± 0.65 independent offspring ($n = 12$) compared to 1.62 ± 1.32 ($n = 21$) for controls (Mann-Whitney *U*-test, one-tailed $P = 0.004$). A finer-scaled analysis of these data,

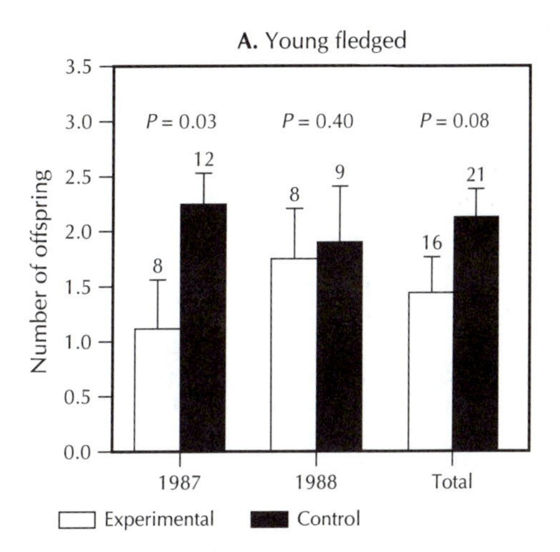

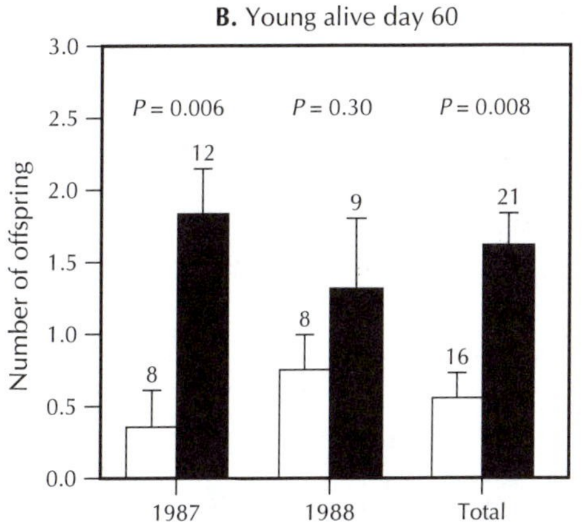

Figure 1A, B. Reproductive success of experimental (potential helpers removed) *(open bars)* and unmanipulated control *(shaded bars)* groups of Florida scrub jays during the 1987 and 1988 breeding seasons. Reproductive success measured as **A,** the number of fledged young and **B,** the number of 60-day-old offspring produced per family. *Means, standard errors,* and *sample sizes* are shown. *P-values* for differences between experimental and control families determined by one-tailed Mann-Whitney *U*-test.

broken down by year and by components of reproductive success, indicates that the difference in offspring production between experimental and control groups can be attributed to

Table 1. Effect of experimental removal of nonbreeding Florida scrub jays on breeder survival, nesting, egg-laying, and hatching success, 1987–1988.

	Experimental groups (nonbreeders removed) $n = 16$ groups	Control groups (nonbreeders present) $n = 21$ groups	*P*
% Breeder survival (n)[a]	96.8 (32)	95.2 (42)	NS
First egg date (n)	1 April ± 8 d (16)	1 April ± 12 d (21)	NS
Clutch size—first nest (n)	3.31 ± 0.87 (16)	3.29 ± 0.46 (21)	NS
Nests per group (n)	1.44 ± 0.73 (16)	1.52 ± 0.60 (21)	NS
% Hatching success—nests (n)	69.6 (23)	78.1 (32)	NS
% Hatching success—eggs (n)	64.8(71)	68.6 (102)	NS
% Hatchability (n)[b]	88.5 (52)	90.8 (76)	NS

Statistical significance of comparisons determined by chi-square contingency analysis (percentage data) or one-tailed Mann-Whitney *U*-test. NS denotes $P > 0.05$.

[a]Breeding season only (March–June)
[b]Eggs surviving to the termination of incubation that hatched

two factors: (1) survival of nestlings and fledging success was lower in experimental groups than in control groups in 1987, but not 1988 (Figure 1, Table 2), and (2) post-fledging survival of young to day 60 was lower in experimental groups than in unmanipulated control groups in both 1987 and 1988 (Figure 2).

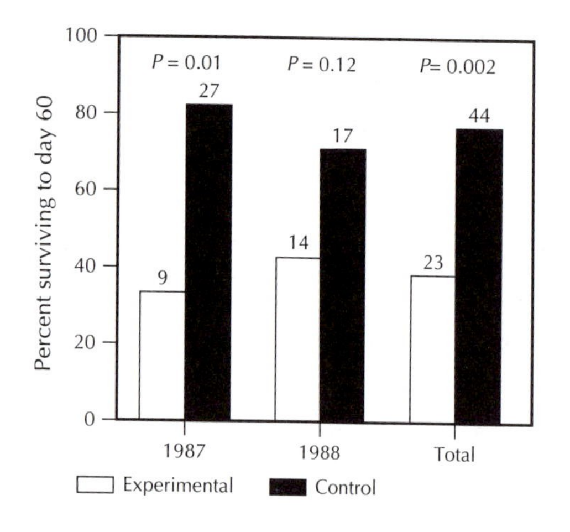

Figure 2. Post-fledging survival of young Florida scrub jays from experimental (potential helpers removed) *(open bars)* and unmanipulated control *(shaded bars)* families, 1987–1988. *Sample sizes* are indicated above the bars. *P-values* for differences between experimental and control groups determined by one-tailed Fisher's exact test.

Survival of Nestlings

The fate of nestlings that hatched from nests of experimental and control groups in 1987–1988 is shown in Table 2. Although hatching success was comparable for both treatment types in 1987, fledging success in experimental groups was significantly reduced. Only 30.0% of young hatched in experimental groups in 1987 ultimately fledged, compared to 62.8% of nestlings from unmanipulated control groups. Starvation and brood reduction, which was recognized by the death or disappearance of underweight or underdeveloped nestlings, was only a minor cause of nestling mortality, and it accounted for fewer than 10% of nestling deaths in both experimental and control groups. Instead, most nestling deaths were attributable to apparent nest predation, the sudden disappearance of healthy nestlings (see Schaub et al. in press). Over 60% of nestlings from 1987 experimental groups were lost as a result of apparent nest predation, compared to 32.6% of nestlings from control groups. In 1988, however, these effects were not evident, and experimental groups did not fledge significantly fewer offspring than did control groups (Table 2).

Table 2. Effect of experimental removal of nonbreeding Florida scrub jays on fate of nestlings, 1987–1988.

	Experimental groups (non breeders removed)	Control groups (non breeders present)	P
1987			
Number of groups	8	12	
Number of nestlings hatched			
Per group(mean ± SD)	3.75 ± 1.83	3.58 ± 1.51	NS
Total	30	43	
Number of nestlings fledging			
Per group (mean ± SD)	1.13 ± 1.25	2.25 ± 1.06	0.03
Total (%)	9 (30.0)	27 (62.8)	
Total number of nestlings lost (%)	21 (70.0)	16 (37.2)	
Starvation/brood reduction (%)	2 (6.7)	2 (4.7)	
Predation (%)	19 (63.3)	14 (32.6)	
1988			
Number of groups	8	9	
Number of nestlings hatched			
Per group (mean ± SD)	2.00 ± 1.51	3.00 ± 0.50	NS
Total	16	27	
Number of nestlings fledging			
Per group (mean ± SD)	1.75 ± 1.28	1.89 ± 1.54	NS
Total (%)	14 (87.5)	17 (63.0)	
Total number of nestlings lost (%)	2 (12.5)	10 (37.0)	
Starvation/brood reduction(%)	1 (6.3)	0(0.0)	
Predation (%)	1 (6.3)	10 (37.0)	

Statistical significance of comparisons determined by one-tailed Mann-Whitney U-test.
NS denotes $P > 0.05$

Survival of Fledglings

Fledged young from experimental groups where nonbreeders had been removed were about half as likely to survive to day 60 as were their counterparts from control groups (Figure 2). This difference in post-fledging survival was statistically significant in 1987 (Fisher's Exact Test one-tailed $P = 0.012$) and in both years combined ($P = 0.002$). In 1988 differences between control and experimental groups fell short of statistical significance ($P = 0.12$). Overall, survival of young between fledging and day 60 was only 39.1% ($n = 23$) for fledglings in experimental groups, compared to 77.3% ($n = 44$) for fledglings in unmanipulated control groups.

Effects of Allofeeders on Provisioning and Growth of Nestlings

The effect of allofeeders (nonbreeders feeding nestlings) on the rate at which food is delivered to 3-day-old and 10-day-old nestling jays is shown in Figure 3. Nestlings are fed more frequently when allofeeders are present than when allofeeders are absent. Although the effect of allofeeders falls short of statistical significance for day-3 nestlings (Mann-Whitney U-test one-tailed $P = 0.06$), by day 10 the effect is highly significant ($P = 0.001$; Figure 3A). On average, each 10-day-old nestling is fed about 2.7 times per hour when allofeeders are present, about 50% more frequently than the rate of 1.8 times per hour for comparable nestlings lacking allofeeders. The provisioning rate of the breeding pair, however, is unaffected by the presence of allofeeders (Figure 3B). Thus, the food that nonbreeding helpers deliver to nestlings is directly responsible for the higher feeding rate seen at nests with allofeeders.

Although potential helpers appear to have no effect on the frequency of nestling starvation and brood reduction (Table 2), the addi-

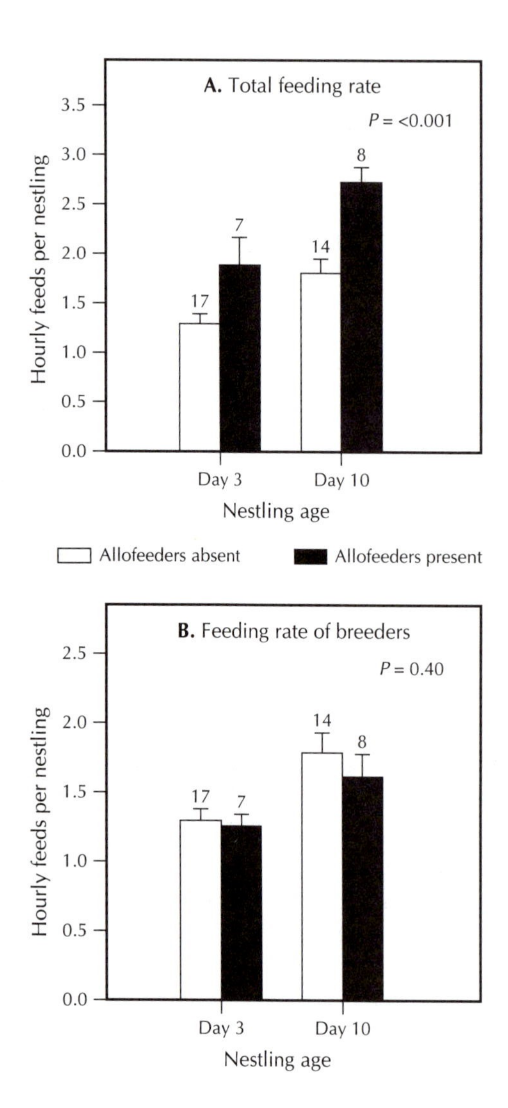

Figure 3A, B. The effect of allofeeders (nonbreeders feeding nestlings) on **A,** the total rate at which food is delivered to 3-day-old and l-day-old nestling Florida scrub jays, and **B,** the rate at which breeders feed nestlings. *Open bars,* allofeeders absent; *shaded bars,* allofeeders present. *Means, standard errors,* and *sample sizes* are shown. *P-value* for differences between groups with and without allofeeders determined by Mann-Whitney *U*-test. Because the presence of helpers was hypothesized to increase total feeding rate, a one-tailed test was used in **A.** Because the presence of helpers could either reduce the parental burden on breeders, or allow breeders to devote more energy to caring for young, a two-tailed test was used in **B.**

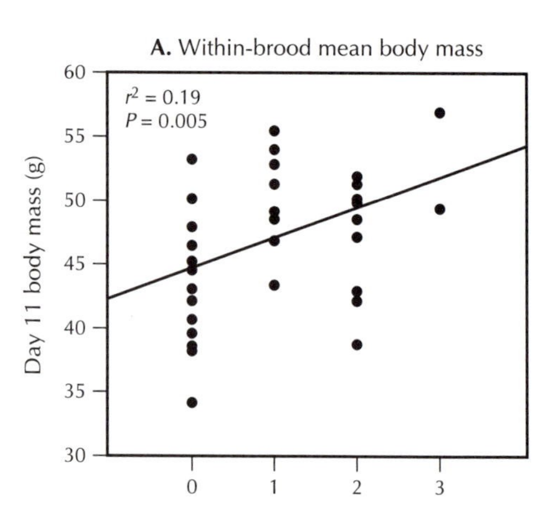

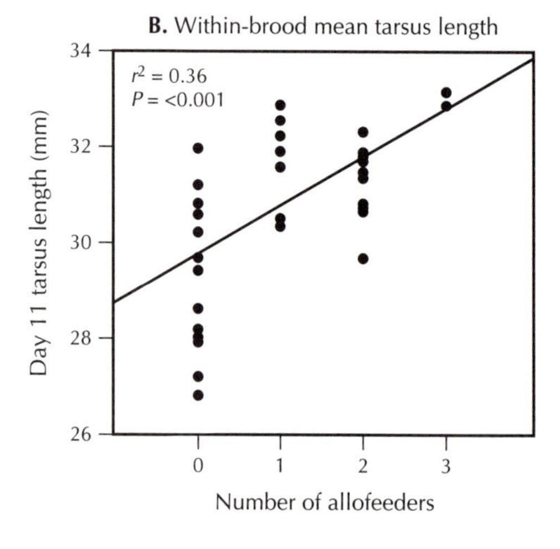

Figure 4A, B. The effect of the number of allofeeders (nonbreeders feeding nestlings) on **A,** the within-brood mean body mass, and **B,** within-brood mean tarsus length of 11-day-old nestling Florida scrub jays. One-nestling broods were excluded from this analysis, and total sample size is 34 broods. Least-squares *regression lines* and one-tailed *P-values* are also shown.

tional food provided by allofeeders resulted in more rapid growth and development of nestlings. The number of allofeeders had a significant positive effect on the within-brood mean weight and within-brood mean tarsus length of 11-day-old nestlings (Figure 4).

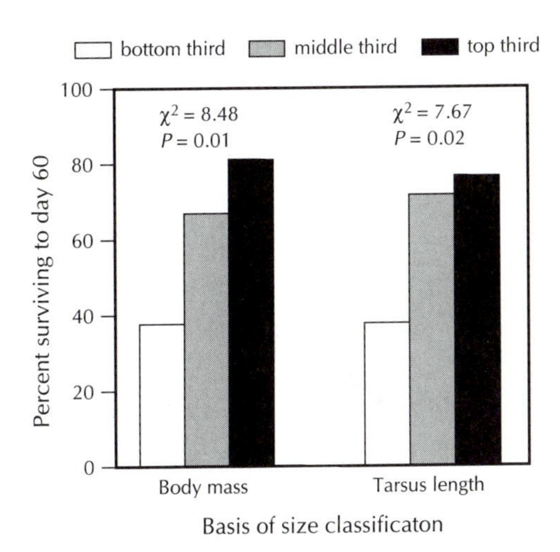

Figure 5. Relationship between post-fledging survival of young Florida scrub jays and their body mass and tarsus length as 11-day-old nestlings. A total of 63 fledglings from experimental and control groups were classified into one of three categories: the smallest third of nestlings *(open bars)* the middle third *(shaded bars)*, and the largest third *(dark bars)* ($n = 21$ for each category).

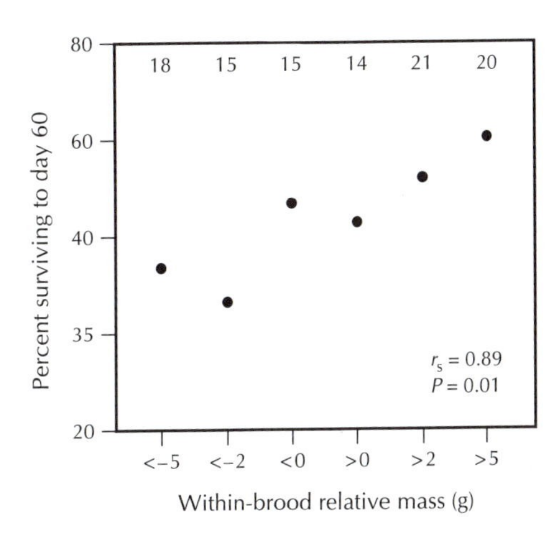

Figure 6. The relationship between post-fledging survival of young Florida scrub jays and their within-brood relative body mass as 11-day-old nestlings. Mass is shown in relation to the mean body mass of their brood-mates at day 11. Only groups where some (but not all) fledged young survived to day 60 were used in this analysis. Spearman's rank *correlation coefficient* one-tailed *P-values,* and *sample sizes* within each category of relative body mass are also shown.

Effects of Nestling Size on Post-fledging Survival

Size as a nestling appears to be related to post-fledging survival in the Florida scrub jay. Data on body mass and tarsus length of 11-day-old nestlings are available for 63 of the 67 fledglings produced by control and experimental groups in 1987–1988. To analyze the effects of nestling size on post-fledging survival, I ranked these 63 fledglings according to their day-11 measurements and divided them into three classes of relative body size corresponding to the bottom, middle, and top third. The results, shown in Figure 5, indicate that fledglings that were larger as day-11 nestlings were significantly more likely to survive to day 60 than were smaller individuals.

It should be noted, however, that the relationship between size as a nestling and post-fledging survival shown in Figure 5 is correlative but not necessarily causal (see Magrath 1991). For example, if families that are above average at provisioning nestlings are also above average at protecting fledged young against predators, a spurious correlation between size as a nestling and post-fledging survival could result.

To control for such confounding influences, I also performed a within-brood analysis of the effect of nestling size on post-fledging survival. A within-brood analysis reduces complications caused by inter-family differences in survival that are unrelated to differences in nestling size. This analysis is based on a larger sample of 103 fledglings from 37 different broods produced during all 3 years of the study (1987–1989). Only broods where two or more young fledged and where some (but not all) of the fledglings survived to day 60 were used in this analysis. Thus, broods containing only a single fledged young, or broods where all fledglings either survived or failed to survive to day 60 were excluded.

For each of the 103 fledglings in this data set I calculated its day-11 body mass relative to the mean mass of its nest-mates. Thus, a 45.0-g nestling with two nest mates that averaged 48.0 g would have a within-brood relative mass of –3.0 g. Individuals were then assigned to six classes of relative body mass (less than –5 g, –5 to –2 g, –2 to 0 g, 0 to 2 g, 2 to 5 g, and greater than 5 g), and the percent of fledglings surviving to day 60 was calculated for each class. The results of this analysis are shown in Figure 6, and a strong positive relationship between within-brood relative mass and post-fledging survival is evident (r_s = 0.89, n = 6, one-tailed P = 0.01).

Relatedness of Potential Helpers and Feeding of Young

When provided with the opportunity to provision dependent young, most nonbreeding Florida scrub jays act as helpers. During 1987–89, 73% (n = 49) of nonbreeding group members exposed to nestlings within their social unit fed those nestlings, and 94% (n = 34) fed fledglings. However, as shown in Figure 7A, the probability of nonbreeders acting as allofeeders increases as relatedness between nonbreeders and dependent young increases; for example, 85% of nonbreeders fed nestlings when the nestlings were full siblings, compared to 60% when the nestlings were half siblings, and 33% in the relatively infrequent cases when the nestlings were unrelated (χ^2 = 8.08, P = 0.02; Figure 7A). Thus, Florida scrub jay helpers usually feed nestlings or fledglings to which they are closely related (Figure 7B).

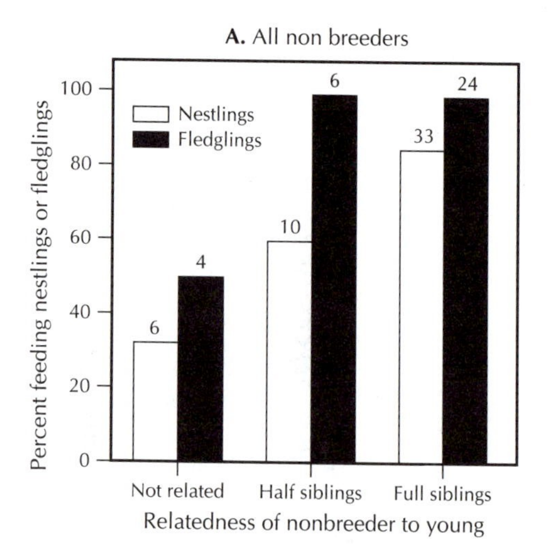

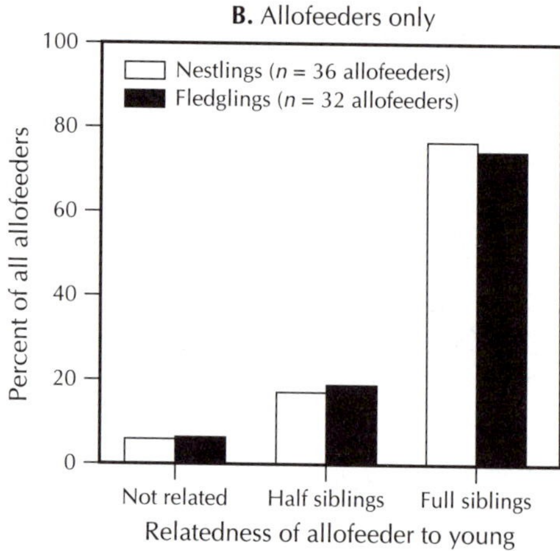

Figure 7. A. Feeding of nestlings (open bars) and fledglings (dark bars) by nonbreeding Florida Scrub jays as a function of relatedness between nonbreeder and young. Sample sizes (number of nonbreeders) for each category are shown. **B.** Relative frequency of allofeeders (nonbreeders feeding nestlings or fledglings) as a function of relatedness between allofeeder and young.

Costs of Helping Behavior

Is helping behavior costly to helpers? I addressed this question by analyzing apparent mortality of "provisioning" and "non-provisioning" nonbreeders during the breeding seasons of 1987–1989. Four (8.9%) of 45 nonbreeders that provisioned nestlings or fledglings disappeared from the study tract (and presumably died) between 1 April and 15 July, the period during which dependent young are normally present on Florida scrub jay territories. Similarly, of the 25 nonbreeders that did not feed dependent young (or simply did not have the opportunity to do so), two (8.0%) disappeared during the same interval. This small difference in apparent mortality between provisioning and non-provisioning nonbreeders is not statistically significant (Fisher exact test, one-tailed $P = 0.64$). Although this analysis should be viewed with caution (e.g., some disappearances may have been a result of long-distance dispersal off of the study area rather than mortality), it nonetheless suggests that the costs of helping behavior in the Florida scrub jay are small and do not adversely affect helper fitness.

Discussion

Removal Experiments and Social Disruption

Experimental manipulations of group composition are generally viewed as a valuable but under-utilized technique in the study of avian cooperative breeding systems (Brown 1987; Smith 1990). Although experimental removals can indeed clarify hypothesized causal relationships and control for the effects of potentially confounding variables, they are not entirely without shortcomings (Koenig and Mumme 1990). Social disruption caused by removing birds from highly integrated cooperative units is one potential problem. For example, was the lower reproductive success of experimental groups observed in this study (Figure 1) caused by the absence of potential helpers in these groups, or was it a by-product of the social disruption caused by the removals?

Although this latter possibility cannot be dismissed entirely, available data do not support such an interpretation. As shown in Table 1, the experimental removal of potential helpers had no detectable effect on the survival, reproductive effort, and hatching success of Florida scrub jays. Thus, no adverse consequences of the removals were detectable until after hatching. If social disruption were a significant problem, we would expect it to have more immediate effects (e.g., delayed onset of breeding or nest failures and abandonments early in incubation). No such effects were evident (Table 1). It is therefore unlikely that the differences in reproductive success between control and experimental groups were caused by social disruption.

Effects on Hatching Success

Because breeding Florida scrub jays typically chase potential helpers away from the nest during egg laying and incubation, nonbreeders almost never begin serving as nest helpers until after eggs have hatched (Stallcup and Woolfenden 1978; Woolfenden and Fitzpatrick 1984). It is thus not surprising that the experimental removals had no detectable effect on hatching success (Table 1). Nonetheless, these findings are contrary to previously published reports suggesting that Florida scrub jays experience higher hatchability and lower rates of egg predation when potential helpers are present (Woolfenden and Fitzpatrick 1984). However, because these earlier analyses did not control for potentially confounding variables, the previously reported relationship between group size and hatching success is probably spurious. For example, pairs with helpers may enjoy greater hatching success simply because such pairs are more likely to have previous breeding experience than pairs lacking helpers. This interpretation is also supported by a recent re-analysis by Schaub et al. (1992). Using a data set more extensive than that available to Woolfenden and Fitzpatrick (1984) and restricted to experienced breeders, Schaub et al. found that the rate of predation on nests with eggs was unrelated to the presence of potential helpers.

Effects on Nestling Survival

Predation on nestlings is the major cause of reproductive failure in the Florida scrub jay (Woolfenden and Fitzpatrick 1984). Diurnal snakes and birds are the most important nest predators, and the great majority of nest predation occurs during daylight hours (Schaub et al. 1992). As shown in Table 2, experimental groups where potential helpers had been removed fledged significantly fewer offspring in 1987 than did unmanipulated control groups, primarily because of higher rates of nestling predation. How, then, could the presence of potential helpers reduce the frequency of such predation?

Helpers reduce the incidence of nest predation in two ways. First, helpers improve "sentinel" behavior (see McGowan and Woolfenden 1989) around nests with nestlings; groups with helpers are much more likely to have at least one vigilant bird near the nest than are groups lacking helpers. Second, helpers improve group mobbing and nest defense (see Francis et al. 1989); groups with helpers mob predators significantly more vigorously than do groups lacking helpers (unpublished data).

But why did the presence of potential helpers have no apparent effect on fledgling production in 1988 (Figure 1, Table 2)? Although the answer to this question is unclear, the same trend was evident in the adjacent study population of Woolfenden and Fitzpatrick; 1988 was one of the few years in the 23-year study during which pairs with potential helpers produced no more fledged young than unassisted pairs (G. E. Woolfenden, pers. comm.). Such year-to-year variation in patterns of reproductive success has been revealed by a number of long-term studies of cooperatively breeding birds (Stacey and Koenig 1990), and serve as a reminder that the results of short-term field experiments should be interpreted cautiously.

Effects on Post-fledging Survival

Several lines of evidence suggest that the higher post-fledging survival observed in families with nonbreeding group members (Figure 2) is at least partially attributable to the additional food that such individuals often provide as helpers. First, nestlings receive more food and grow more rapidly when allofeeders are present than when allofeeders are absent (Figures 3 and 4). Second, the probability of a fledgling surviving to independence is positively correlated with both its absolute size (Figure 5) and its within-brood relative size (Figure 6) as an 11-day-old nestling. Third, food supplementation experiments that I conducted in 1990 indicate that food-supplemented young grew more rapidly as nestlings and had higher post-fledging survival than did young in control groups (unpublished data). These data thus indicate that the relationship between size as a nestling and post-fledging survival is at least partially causal.

The specific reasons why larger fledglings are more likely to survive to independence are unclear. Young jays are relatively immobile, virtually flightless, and highly vulnerable to predators during the week following fledging (McGowan 1987). It may be that well-fed and well-developed young acquire critical motor skills more rapidly than do poorly fed and poorly developed young, thereby reducing the amount of time that fledglings are particularly vulnerable to predators. Well-fed young also may beg less frequently and attract fewer potential predators than do undernourished young. Other possibilities are that large young may be better able to survive brief periods of post-fledging food shortage, or they may be able to adopt foraging behaviors that are less risky than those employed by underdeveloped young (Magrath 1991).

Although increased food delivery and more rapid growth appear to be at least partially responsible for the higher post-fledging survival seen in groups with potential helpers, additional factors also may be at work. For example, by feeding fledged young, helpers may improve sentinel performance around vulnerable fledglings and thereby reduce predation (McGowan and Woolfenden 1990).

Does Helping Have Current Utility?

This study has shown that experimental groups where all nonbreeding group members have been removed produce fewer offspring than do unmanipulated controls, at least in some years. However, is the higher reproductive output of breeders in control groups simply a consequence of their living in social units, or is it a direct result of the aid that nonbreeders usually provide to dependent young as helpers? This is not a trivial question; if we wish to argue that recipients benefit from helping behavior per se, we must be certain that the benefits are derived directly from alloparental care itself and not simply from the presence of additional nonbreeding group members (Koenig and Mumme 1990).

As described above, helpers appear to enhance the reproductive success of recipient breeders by performing anti-predator behavior around nests (including sentinel behavior and mobbing of nest predators), and by feeding dependent young. Thus, the positive effects of nonbreeding helpers on the fitness of recipients are indeed attributable to alloparental care itself and not to either incidental beneficial effects of social living or potentially correlated variables such as territory quality and parental quality.

But how does alloparental care affect the fitness of the helpers themselves? Although nonbreeders almost certainly incur some direct fitness costs when they provision dependent young and protect them from potential predators, these costs have been measured in few previous studies (Reyer 1984; Heinsohn et al. 1990; Rabenold 1990). Data presented here indicate that the alloparental care performed by helpers in the Florida scrub jay has no detectable effect on their survival during the nesting season (see above). Thus, the fitness costs of helping behavior in this species may be small.

In contrast, the fitness benefits of alloparental care are substantial. Because helping is generally directed toward close kin (Figure 7), nonbreeding Florida scrub jays enhance the indirect component of their inclusive fitness by acting as helpers (Brown 1987). A rough estimate of the magnitude of this effect can be calculated from the data—presented in this paper. We must first assume that the costs of helping are negligible (see above), that alloparental care does not significantly influence breeder survival (Table 1), and that the increased reproductive success of groups with potential helpers (Figure 1) is entirely the result of alloparental care (see above). Armed with these basic assumptions, we can calculate that helping behavior increases the production of 60-day-old juveniles by an average of 1.06 young per family (Figure 1). If these 1.06 additional offspring are divided among a mean of 1.81 potential helpers present in each of the 21 control groups (see Methods), an average helper increases reproductive success by 0.59 offspring. Based on the relatedness of 36 nonbreeders known to have provisioned nestlings (Figure 7B), the average coefficient of relatedness between allofeeder and young is 0.43, compared to 0.50 between parents and offspring. Consequently, a Florida scrub jay that acts as a helper increases its inclusive fitness by an average of 0.50 offspring equivalents. Although these calculations should be treated with considerable caution, they nonetheless indicate that alloparental care can have a substantial effect on the indirect component of a helper's inclusive fitness.

Helping behavior in the Florida scrub jay also may affect other aspects of helper fitness. For example, by helping to raise additional young, male nonbreeders can increase their probability of acquiring breeding space through the process of territorial "budding" (Woolfenden and Fitzpatrick 1986). These and other ways in which alloparental care may affect the direct and indirect fitness of helpers are discussed in several recent reviews (Brown 1987; Koenig and Mumme 1990; Emlen 1991).

Is Helping an Adaptation?

The preceding analysis strongly suggests that helping behavior in the Florida scrub jay is a trait with current selective utility. However, current utility alone does not constitute sufficient evidence that helping behavior is an adaptation sensu stricto (e.g., Williams 1966; Gould and Vrba 1982; Baum and Larson 1991) and has been shaped by natural selection acting on heritable genetic variation in the propensity to help (Jamieson 1989, 1991; Mumme 1991; Mumme and Koenig 1991). Nonadaptive alternatives must also be considered.

For example, as shown in Figure 7A, nonbreeding Florida scrub jays are more likely to provision the dependent young within their social unit when those young are closely related. These results, which are similar to those reported in a number of other studies (e.g., Curry 1988; Emlen and Wrege 1988; Clarke 1989, 1990), are consistent with predictions of inclusive fitness theory (Hamilton 1964) and suggest the possibility that alloparental behavior in this species has been adaptively modified by indirect selection acting on helpers.

However, preferential feeding of kin can also be explained by a nonadaptive hypothesis that does not rely on indirect selection. Dominant breeders, who are more tolerant of familiar kin than unfamiliar non-kin, may prevent unrelated nonbreeders from approaching nestlings and/or fledglings, thereby interfering with normal development of the provisioning response (see Jamieson and Craig 1990; Jamieson 1991). Until both adaptive and nonadaptive alternatives can be tested, it thus remains debatable whether helping behavior in the Florida scrub jay is a trait that has been shaped by natural selection.

Acknowledgements

I thank the staff of Archbold Biological Station, particularly John Fitzpatrick, Dave Johnston, and Jim Wolfe, for providing me with access to the Station's outstanding research facilities. I am also grateful to David Dunning, Ron Schaub, Steve Schoech, and Russ Titus for invaluable assistance in the field, and Bob Curry, Maree Elowson, John Fitzpatrick, Jack

Hailman, and Glen Woolfenden for facilitating my research in a variety of ways. Steve Emlen, Paul Sherman, Dave Winkler, and other members of the behavior lunch-bunch at Cornell University offered constructive feedback during the early stages of the study. Helpful criticisms of an early draft of the manuscript were provided by Steve Emlen, Doug Mock, Trish Schwagmeyer, Peter Stacey, Glen Woolfenden, and two anonymous reviewers. Financial support was provided by an NSF postdoctoral fellowship (BSR-8600174) and a grant from the Nongame Wildlife Program of the Florida Game and Freshwater Fish Commission (NG88-043).

References

Baum, D. A. and A. Larson. 1991. Adaptation reviewed: a phylogenetic methodology for studying character macroevolution. Syst Zool 40:1–18.

Brown, J. L. 1987. *Helping and Communal Breeding in Birds: Ecology and Evolution*. Princeton University Press. Princeton.

Brown, J. L, E. R. Brown, S. D. Brown and D. D. Dow. 1982. Helpers: effects of experimental removal on reproductive success. Science 215:421–422.

Clarke, M. F. 1989. The pattern of helping in the bell miner *(Manorina melanophrys)*. Ethology 80:292–306.

Clarke, M. F. 1990. The pattern of helping in the bell miner revisited: a reply to Jamieson and Craig. Ethology 86:250–255.

Curry, R. L. 1988. Influence of kinship on helping behavior in Galapagos mockingbirds. Behav Ecol Sociobiol 22:141–152.

Emlen, S. T. 1991. Evolution of cooperative breeding in birds and mammals. In: *Behavioural Ecology: an Evolutionary Approach*. I. R. Krebs and N. B. Davies, eds. 3rd ed. Blackwell, Oxford. Pp. 301–337.

Emlen, S. T. and P. H. Wrege. 1988. The role of kinship in helping decisions among white-fronted bee-eaters. Behav Ecol Sociobiol 23:305–315.

Francis, A. M., J. P. Hailman and G. E. Woolfenden. 1989. Mobbing by Florida scrub jays: behaviour. sexual asymmetry, role of helpers, and ontogeny. Anim Behav 38:795–816.

Gould, S. J. and E. S. Vrba, 1982. Exaptation—a missing term in the science of form. Paleobiology 8:4–15.

Hamilton, W. D. 1964. The genetical evolution of social behaviour. J Theor Biol 7 1–52.

Heinsohn, R. G., A. Cockburn and R. A. Mulder. 1990. Avian cooperative breeding: old hypotheses and new directions. Trends Ecol Evol 5:403–407.

Jamieson, I. G. 1989. Behavioral heterochrony and the evolution of birds' helping at the nest: an unselected consequence of communal breeding? Am Nat 133:394–406.

Jamieson, I. G. 1991. The unselected hypotheses for the evolution of helping behavior: too much or too little emphasis on natural selection? Am Nat 138:271–282.

Jamieson, I. G. and J. L. Craig. 1990. Reply: evaluating hypotheses on the evolution of helping behaviour in the bell miner, *Manorina melanophrys*. Ethology 85:163–167.

Koenig, W.D. and R. L. Mumme. 1990. Levels of analysis and the functional significance of helping behavior. In: *Interpretation and Explanation in the Study of Animal Behavior*. M. Bekoff and D. Jamieson, eds. Vol 11. Explanation, evolution, and adaptation. Westview, Boulder. Pp. 268–303.

Leonard, M. L., A. G. Horn and S. F. Eden. 1989. Does juvenile helping enhance breeder reproductive success? a removal experiment on moorhens. Behav Ecol Sociobiol 25:357–361.

Magrath, R. D. 1991. Nestling weight and juvenile survival in the blackbird, *Turdus merula*. J Anim Ecol 60:335–351

McGowan, K. J. 1987. Social development in young Florida scrub jays *(Aphelocoma c. coerulescens)*. PhD dissertation, University of South Florida.

McGowan, K. J. and G. E. Woolfenden. 1989. A sentinel system in the Florida scrub jay. Anim Behav 37:1000–1006

McGowan, K. J. and G. E. Woolfenden. 1990. Contributions to fledgling feeding in the Florida scrub jay. J Anim Ecol 59:691–707.

Mumme, R. L. 1991. Helping behaviour in the Florida scrub jay: nonaptation, exaptation, or adaptation? Acta XX Cong Int Ornithol:1317–1324.

Mumme, R. L. and W. D. Koenig. 1991. Explanations for avian helping behavior. Trends Ecol Evol 6:343–344.

Rabenold, K. N. 1990. *Campylorhynchus* wrens: the ecology of delayed dispersal and cooperation in the Venezuelan savanna. In: *Cooperative Breeding in Birds: Long-term Studies of Ecology and Behavior*. P. B. Stacey and W. D. Koenig, eds. Cambridge University Press. Cambridge. Pp. 157–196.

Reyer, H. -U. 1984. Investment and relatedness: a cost/benefit analysis of breeding and helping in the pied kingfisher *(Ceryle rudis)*. Anim Behav 32:116}1178.

Schaub, R., R. L. Mumme and G. E. Woolfenden 1992. Predation on the eggs and nestlings of Florida scrub jays. Auk 109:585–593.

Schoech, S. J., R. L. Mumme and M. C. Moore. 1991. Reproductive endocrinology and mechanisms of breeding inhibition in cooperatively breeding Florida scrub jays *(Aphelocoma c. coerulescens)*. Condor 93:354–364.

Siegel, S. 1956. *Nonparametric statistics for the behavioral sciences*. McGraw-Hill. New York.

Smith, J. N. M. 1990. Summary. In: *Cooperative Breeding in Birds: Long-term Studies of Ecology and Behavior.* P. B. Stacey and W. D. Koenig, eds. Cambridge University Press, Cambridge. Pp. 593–611.

Solomon, N. G. 1991. Current indirect fitness benefits associated with philopatry in juvenile prairie voles. Behav Ecol Sociobiol 29:277–282.

Stacey, P. B. and W. D. Koenig. 1990. *Cooperative Breeding in Birds: Long-term Studies of Ecology and Behavior.* Cambridge University Press. Cambridge.

Stallcup, J. A. and G. E. Woolfenden, 1978. Family status and contribution to breeding by Florida scrub jays. Anim Behav 26:1144–1156.

Williams, G. C. 1966. *Adaptation and Natural Selection.* Princeton University Press. Princeton.

Woolfenden, G. E. and J. W. Fitzpatrick. 1984. *The Florida Scrub Jay: Demography of a Cooperative-breeding Bird.* Princeton University Press. Princeton.

Woolfenden, G. E. and J. W. Fitzpatrick, 1986. Sexual asymmetries in the life history of the Florida scrub jay. In: *Ecological Aspects of Social Evolution: Birds and Mammals.* D. Rubenstein and R. W. Wrangham, eds. Princeton University Press. Princeton. Pp. 87–107.

Woolfenden, G. E. and J. W. Fitzpatrick. 1990. Florida scrub jays: a synopsis after 18 years of study. In: *Cooperative Breeding in Birds: Long-term Studies of Ecology and Behavior.* P. B. Stacey and W. D. Koenig, eds. Cambridge University Press. Cambridge. Pp. 239–266.

Introduction to
Paul Schmid-Hempel, Alejandro Kacelnik, and Alasdair I. Houston
Honeybees Maximize Efficiency by Not Filling Their Crop

by Robert L. Jeanne

"Busy as a bee" is an aphorism that goes back at least to Solomon and Aesop. Indeed, we have all at least casually watched honey bees flit from flower to flower, stopping only long enough to drink the bit of nectar offered by each in return for the opportunity to have its pollen picked up and transported to another flower of the same species. The foragers' behavior seems entirely purposeful: their job is to return energy in the form of nectar to the hive and they seem to do it rapidly and effectively. The seeming 'workaholism' of the honey bee is often held up as a model of industriousness, the implication being that it is a quality that we humans should strive to emulate.

Nectar-feeding animals such as bees, butterflies, and hummingbirds are good choices for studying optimal foraging because the amount of the material that is relevant to the animal's fitness-energy is relatively reliably correlated with its volume and even with the time spent in collecting it. At each visit to a flower, a nectar-foraging bee inserts her proboscis into the pool of a few microliters of nectar produced by the nectaries of the flower. By extending the segments on her abdomen, she draws the liquid up the proboscis and into the crop (the large storage reservoir just ahead of the stomach), much as a syringe draws up liquid when the plunger is retracted. Because the sugar content of nectar produced by a species of flowering plant is reasonably constant, and likewise the rate at which a bee can draw liquid into her crop is relatively constant, we can obtain a reasonable estimate of the amount of energy the bee is collecting by measuring the time she spends at a flower and on a given trip. A bee visits many flowers during the course of a single trip from the hive, returning to unload only, we are inclined to assume, when her crop is full. Early efforts to understand optimal foraging in bees assumed that the bottom line for the bee was to adjust her behavior so as to return energy to the colony at as high a rate as possible. In other words, the 'currency' units bees are using was assumed to be net rate of energy gain.

The paper reprinted here is a good example of an optimality study in which the initial assumptions were wrong. Early studies assumed that the currency honey bees are maximizing is the net rate of return of nectar (energy) to the nest. If this were true, the expectation would be that foragers should continue foraging until they have a full load of nectar. But this is not the case—honey bee foragers often return to the hive with partially-full crops. Schmid-Hempel and his collaborators show that for central-place forager (foragers that return with their food to a nest) taking account of the metabolic costs of carrying a load leads to a better understanding of partial crop loading. They then go on to discuss why the hive will accumulate more resources if foragers adopt this strategy rather than a rate-maximizing one.

The Schmid-Hempel et al. study is an excellent example of how early models to explain how an animal behaves in the context of its environment are modified and refined through repeated alternation between quantitative data collection and mathematical modeling.

Honeybees Maximize Efficiency by Not Filling Their Crop

PAUL SCHMID-HEMPEL, ALEJANDRO KACELNIK, AND ALASDAIR I. HOUSTON

Original publication in *Behavioral Ecology and Sociobiology* (1985), 17:61–66. Reprinted with permission.

Summary

Honeybees often abandon non-depleting food sources with a partially filled crop. This behaviour does not maximize the net rate of energy extraction from the food sources, and thus contradicts predictions of some common models for central place foragers. We show that including the metabolic costs of transport of nectar leads to models that predict partial crop-loading. Furthermore, the observed crop loads of honeybees are less consistent with those predicted by maximization of delivery rate to the hive (net energetic gain/ unit time), than with those predicted by maximization of energetic efficiency (net energetic gain/unit energy expenditure). We argue that maximization of energetic efficiency may be an adaptation to a limited flight-cost budget. This constraint is to be expected because a worker's condition seems to deteriorate as a function of the amount of flight performed.

Introduction

Foraging animals are often bound, temporarily or permanently, to a fixed place, such as a nest or a hive, to which food is carried. Workers of the honeybee *(Apis mellifera)* that collect nectar and carry the load back to the hive in their honeycrops are an example of this kind of central place foraging. It has been suggested that the foraging behaviour of nectar-gathering bees should maximize the net rate of gain of energy (e.g. Pyke 1978; Waddington and Holden 1979; Hodges 1981). For central place foragers, models based on maximization of rate of energy extraction from the food sources (equivalent to energy delivery rate to the central place if all energy costs are ignored) predict that incomplete loads should only be gathered when the animal feeds from a patch where intake rate diminishes with time (Orians and Pearson 1979).

Even in non-depleting patches, however, honeybees do not always fill their crop before returning to the hive (Nuñez 1982). Caged hummingbirds with food ad libitum were also found not to fill their crop before returning to a perch (De Benedictis et al. 1978). De Benedictis et al. showed that meals of limited size resulted in maximization of net energy gain when the cost of transporting the collected nectar was taken into account. Nuñez (1982), on the other hand, suggested that in the highly social honeybee partial crop filling may be an adaptation to the need for exchanging information between workers. Thus, a prolonged absence from the hive would result in a loss of information about alternative food sites and it would pay a worker to return sooner than with a full load. Under certain circumstances, of course, foraging bees may also have to meet other tasks, such as to collect water for cooling the hive (e.g. Lindauer 1954). We do not consider these ideas here; our goal is to investigate how much of the bees' behaviour can be accounted for by purely energetic models of nectar collecting. We present evidence supporting the view that the metabolic costs of carrying a nectar load can help to understand partial crop loading. Our discussion is restricted to only two 'currencies', energetic efficiency and net rate of energy delivery, but more complex currencies could be analyzed (see Kacelnik 1984; Cheverton et al. 1985; Houston, in preparation).

The Model

Consider a worker bee which leaves the hive to forage in a patch of flowers at a given distance from the hive. When the bee arrives at the first flower of the patch (Figure 1), she has already spent a certain amount of energy in flying from the hive to the patch. If a_0 =metabolic rate of the unloaded bee during flight and τ_0 = one-way travel time, then this expenditure is $a_0 \cdot \tau_0$. With each flower visited in the patch, the animal takes up a nectar load, weight w, which is equivalent to an energy of $c \cdot w$, where c = weight-specific energetic value of the nectar. The time to gather the load from each flower is h (handling time). For simplicity, we assume flight velocity to be unaffected by load.

The energy spent flying for time τ from flower to flower increases with load, i.e. with the number of flowers already visited. We assume a linear relationship between metabolic rate during flight and the weight of the nectar load the bee carries. Such a linearity is suggested by findings of Heinrich (1975) for the bumblebee *Bombus edwardsii*, and Beutler (1937) and Heran (1962) for honeybees. An increase in metabolic rate with load while extracting nectar from a flower is probably relatively low because the bee sits on the flower rather than hovering during extraction. We therefore consider this increase only during inter-flower flights. During flower visits, the metabolic rate, a_h, thus remains constant. If the bee visits N flowers before returning to the hive, the total energy expenditure in the patch, C_P, is given by

$$C_p = a_0 \cdot (N-1) \cdot \tau + a \cdot (w + 2w + 3w + \ldots + [N-1]w) \cdot \tau + a_h \cdot N \cdot h$$

where a = linear increment in metabolic rate function of load weight, and w = load increment at each flower. Equation (1a) can be written as

$$C_p = a_0 \cdot (N-1) \cdot \tau + a \cdot \frac{N \cdot (N-1)}{2} \cdot w \cdot \tau + a_h \cdot N \cdot h$$

The return flight to the hive will cost $a_0 + \mathrm{a} \cdot W$ (W= total load = $N \cdot w$), and thus total expenditure during travel is

$$C_T = a_0 \cdot \tau_0 + (a_0 + a \cdot W) \cdot \tau_0$$

Because nectar has to be delivered to recipient bees in the hive, we include the time and energetic expenditure while the bee is in the hive as part the foraging cycle. If hive time is denoted by T_0 and metabolic rate while in the hive by a_τ, then the total expenditure, C, per foraging cycle is

$$C = C_p + C_T + T_0 \cdot a_\tau$$

A bee which visits N flowers in the patch accumulates a gross energy load, G, given by

$$G = N \cdot c \cdot w$$

Finally, the round-trip time for an entire foray excursion, T, is given by

$$T = 2 \cdot \tau_0 + (N-1) \cdot \tau + N \cdot h + T_0$$

By combining Equations (1) to (5) it is possible to find the optimal load size, for both of the 'currencies' used (net rate of energy delivery, and energetic efficiency) expressed in terms of numbers of flowers to be visited (N^*). We solved the problem by numerical iteration on the number of flowers visited per trip (Figure 2).

Methods

Individually marked workers of the honeybee (*Apis mellifera*) were trained to collect food from a feeding site (the patch), 30 m from the hive. The patch consisted of three artificial flowers each containing 1 µl of 50% (weight/weight) sugar solution ("nectar"). 1 µl of 50% sugar solution contains 0.6 mg of sugar, equivalent to 10.05 J (where c= 16.7 J per mg of sugar). The bees typically took a few seconds to empty each flower before taking off to visit one of the other two. Flowers refilled after each visit, so that they could be re-visited without depletion for an indefinite number of times. The distance between the patch and the hive was kept constant, but the inter-flower flight time, τ, within the patch was varied by controlling (with a sliding cover) the interval between leaving one flower and landing on a new one. In the field, such different interflower times would correspond to different flower densities.

Twelve individuals were tested. Each bee experienced a different average inter-flower time τ (range of means: 3.9 s to 49.9 s) during which it spent all the time in flight. Each indi-

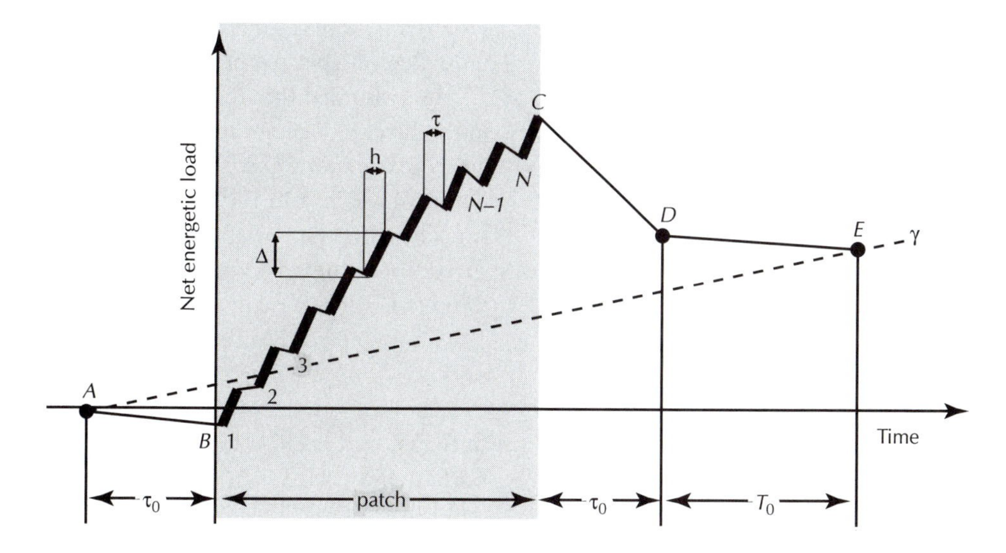

Figure 1. The foraging cycle, shown as the net energetic load carried by a bee as a function of time. The cycle starts at *A* when the bee leaves the hive. After travel time τ_0 the first flower in the patch (*B*) is reached. In the patch *(shaded area)*, the bee visits *N* flowers. Each visit increases net energy balance by an increment Δ, corresponding to the energetic equivalent of nectar reward minus energetic expenditure on flower, and takes time *h (heavy lines)*. For *N* visits, *N*–1 inter-flower flights (duration τ, *thin lines)* are made. Because crop load is increasing, the energetic expenditure on inter- flower flights increases. The bee leaves the patch at *C* to fly back to the hive (*D*). From *D* to *E*, the bee stays in the hive for time T_0, incurring an energetic expenditure at rate a_τ. The slope of γ *(broken line AE)* is the net rate at which energy is delivered to the hive. Leaving the patch after *N* flowers at *C*, rather than earlier or later, gives the steepest slope of γ (cf Figure 2). Hence, *N* is the optimal number (N^*) of flowers to be visited with the rate model. In a graphical representation of the efficiency model, the time axis would be replaced by an axis of energy expenditure. For further definition of parameters see text.

vidual was observed for 8 consecutive trips. The mean loads gathered in the patch before returning to the hive, and the mean of the time measurements (handling, hive time, one-way travel time, and inter-flower flight time) during the last four roundtrips were used for the analysis. Because a bee could visit an unlimited number of nectar-filled flowers, food availability did not diminish with time.

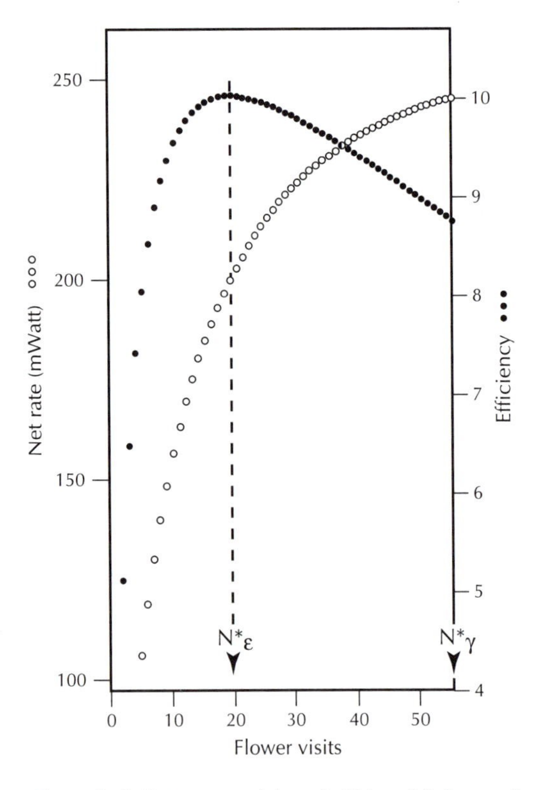

Figure 2. *Ordinate: net rate* (o), and *efficiency* (•) that can be achieved by a bee which leaves the patch to return to the hive after 1, 2, 3...flower visits *(abscissa)*. Net rate and efficiency are scaled such that the maximum has the same ordinate. The average parameter values for all individuals tested were used to generate this example (τ_0 = 22.8 s, h = 12.0 s, T_0 = 235.0 s). Inter-flower flight time is τ = 20 s. When maximizing the net rate, the bee should leave the patch after N^*_γ = 55 flower visits, yielding a net rate of energy delivery to the hive of γ = 245.5 mW. For maximizing efficiency, the bee should leave earlier, after N^*_ϵ = 19 visits (efficiency = 9.93).

Metabolic rates of bees in flight have repeatedly been measured (e. g. Jongbloed and Wiersma 1934; Beutler 1937; Sotavolta 1954; Scholze et al. 1964; Bastian and Esch 1970; Heinrich 1979; Withers 1981; Rothe 1983). We assume that the metabolic rate during flight by an unloaded bee (70 mg) is 82 ml O_2 g h (at average ambient temperatures during the experiment of 30°C, Heinrich 1979) which gives $a_0 = 3.34\ 10^{-2}$ W. This value is close to figures provided by Bastian and Esch (1970), Withers (1981), and Rothe (1983) and takes into account that the bees spent most of their time in a more expensive manoeuvring flight, rather than in straight flight. The linear increase in metabolic rate with load, $a = 5.10^{-5}$ W J^{-1} was estimated from findings of Beutler (1937), Heran (1962), Nuñez (1974), and Heinrich (1975): an increase of approximately 1% of the unloaded rate per mg of additional load.

It is assumed that the metabolic rate while in the hive, a_τ, or handling a flower, a_h, is similar to that experienced by an actively moving bee which is not in flight. Where metabolic rates for this type of activity have been measured, the values given range widely (e.g. Kosmin et al. 1932; Bastian and Esch 1970; Rothe 1983 and references therein). Here, we will assume $a_\tau = a_h = 0.47.10^{-2}$ W. This value amounts to 1/8 of the consumption of the unloaded bee in flight, and is close to figures suggested by Bastian and Esch (1970) for pre-flight activity, and Rothe (1983) for walking bees.

Results

Equations (1) to (5) were used to find the optimal load size for two different models of foraging behaviour: maximizing the *net rate* of energy delivery to the hive = $(G - C)/T$, or maximizing the *efficiency*, i.e. energetic gain per unit of

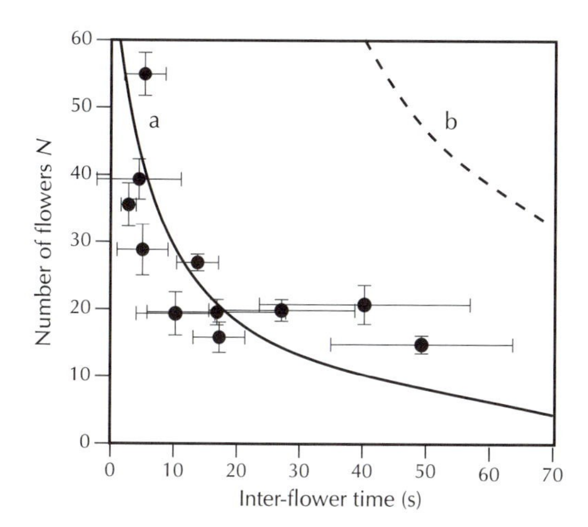

Figure 3. The number of flowers visited, *N (ordinate)*, as a function of inter-flower time τ *(abscissa)*. The numbers predicted by the two models (curves in graph) correspond to the behaviour that maximizes energetic efficiency (net energetic gain/energy expenditure, *solid line (a)*, or maximizes delivery rate to the hive (net energetic gain/time, *broken line (b)*. With both models, partial crop loads are predicted for long inter-flower times. The observations for twelve different individuals are shown (●: mean; bars equal standard deviation). Observations were averaged over four consecutive foraging trips. As in Fig. 2 the predictions *a* and *b* were calculated with the average parameter values for all individuals tested. A full crop load corresponds to approximately 55 flowers.

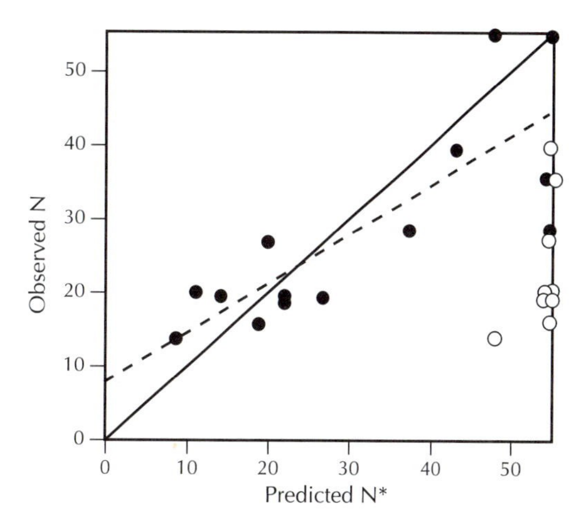

Figure 4. *Ordinate:* the observed number, *N*, of flowers visited in the patch by individual bees before returning to the hive. *Abscissa:* number, N^*, of flowers to be visited, as predicted for the same individual, if either the net rate of energy delivered to the hive is to be maximized (rate model. o), or energetic efficiency is maximized (efficiency model, ●). The predictions are based on foraging parameters (τ, τ_0, h, T_0) as measured for each of the individuals separately. *Each circle* represents the observation as compared to the prediction of the particular model for a given individual. The mean inter-flower times (τ) range from 3.2 s to 49.9 s. The rate model cannot account for the results when observations and predictions are compared pairwise ($t = 8.54$, $P < 0.001$, $n = 12$). For the efficiency model, no significant difference is found (pairwise $t = 0.43$, $P > 0.1$, $n = 12$). The regression relating predictions of the efficiency model to observations is $n = 0.66\ N^* + 8.30$ ($r = 0.831$, $P = 0.008$, n = 12).

energy spent $= (G - C)/C$. The net rate or efficiency that can be achieved in a foraging cycle is, everything else kept constant, a function of the number of flowers visited before returning to the hive (Figure 2). Repeated numerical evaluation of Equations (1) to (5) will thus lead to the number of flowers which results in the highest value for a chosen currency.

Both models predict that for large inter-flower times, the optimal number of flowers to be visited, N^*, is less than maximal crop capacity. The effect is more marked with the efficiency model, whereas the rate model would predict full loads for nearly all of the experimental conditions. As Figure 3 shows, the observed numbers of flowers visited is indeed close to the numbers predicted by the efficiency model. For long inter-flower times, observations and predictions seem to be less consistent than for short inter-flower times.

The predictions in Figure 3 are calculated with the average parameter values for all individuals. Yet, the individuals differed in their foraging parameters (τ_0, h, T_0) due to, for example, somewhat different flight paths to and from the hive. We therefore made a prediction for each bee separately by inserting the appropriate values into the model. The values of N observed and predicted in this manner are shown for both models in Figure 4. As this figure shows, the predictions from the efficiency model do not differ significantly from the observations, but the rate model cannot account for the observations.

Sensitivity to Parameter Values

In order to study the sensitivity of our models to variations in the parameter values, the following modifications were made and their effect on our predictions evaluated numerically. The values of the parameters standing for the metabolic rates of bees (a_0, a, a_h, a_T) were doubled or halved in turn. As Figure 5 clearly shows, under these modifications the efficiency model maintained a better fit than the rate model. The rate model proved to be generally more sensitive to variations in parameter values, as reflected in the wider range of predictions when values were doubled or halved (shaded area around b in Figure 5). If hive time is not included in the foraging cycle, the rate model predicts values of N^* which are closer to observations (Figure 5). Even in this case, however, efficiency yields a significantly better fit than rate when *deviations* (observed minus predicted) are compared pairwise for the two models (Wilcoxon's matched pairs signed rank test, $P = 0.01$, $n = 12$). A proportional decrease in load weight due to sugar consumption in flight was also taken into account, but proved to have little impact on the predictions (Figure 5).

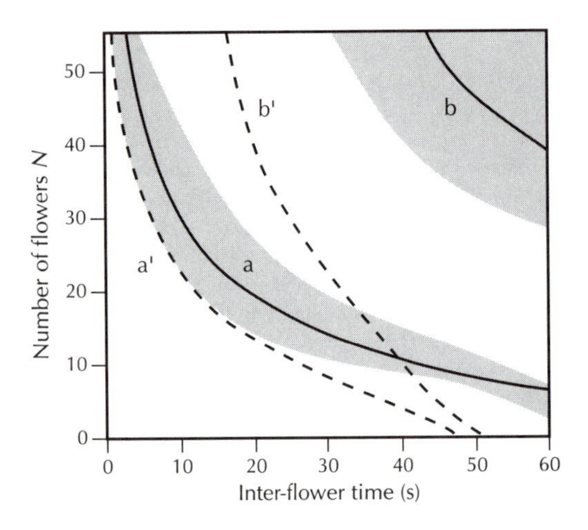

Figure 5. Sensitivity of model predictions to variations in parameter values. *Ordinate, abscissa,* and predictions of the efficiency model *(solid line a)* or rate model *(solid line b)* as in Figure 3. The *shaded areas* around *a* or *b* show the range of prediction (number of flower visits as a function of inter-flower time if the following parameter values are doubled or halved: a_0 a, a_h, a_τ (see text for definitions). If a decrease in load weigh proportional to sugar consumption in flight is taken into account, the predictions are also within the shaded area. The *broken line a'* is predicted if either energetic expenditure while in the hive (a_τ), or hive time (T_0) is omitted from the efficiency model. *Broken line b'* is predicted if hive time is omitted from the rate model.

Discussion

We conclude that the crop-loading of bees can be predicted by assuming that the workers maximize energetic efficiency per foraging trip and incur increasing foraging costs due to the weight of the load. For both models considered here including additional energetic expenditure due to load weight results in smaller rather than larger loads (as in the case of De Benedictis et al. 1978). This finding differs from earlier theoretical claims that the effect of including energy costs should be an increase in load size (Orians and Pearson 1979, see also critique in Kacelnik and Houston 1984).

Whether or not time and energetic expenditure in the hive is included in the foraging cycle has an effect on the quantitative predictions (whereas the qualitative conclusions remain essentially the same in the present case). There is no a priori reason to exclude this period from a complete cycle. However, we would like to point out the importance of the social habit of the honeybee in deciding this question. A worker returning from a foraging trip will typically transfer its load to recipient bees in the hive. If the load is desirable relative to what is delivered by other workers at that time, the forager can unload very quickly and will also convey information about the food source by performing the bee's dance (von Frisch 1965, p. 30 ff; Michener 1974, p. 182 ff). Eventually, the worker will beg for food from others and leave the hive again (von Frisch 1965; Nuñez 1970). Only if a forager is unable to deliver the crop load, or takes a very long time to do so, will she remain in the hive or fly to a different patch (Michener 1974). This would show up in the experiment as the end of a foraging bout, but this was never observed—all bouts were terminated by the experimenter. A worker while in the hive between foraging trips will therefore typically not participate in activities unrelated to the foraging process itself (activities that should be excluded from the cycle or impose constraints on foraging time available, Lucas 1983), such as brood care or cleaning the comb. Rather, an individual adopts the role of a foraging bee late in her life (Lindauer 1952), and there seems to be a relatively sharp distinction between the activities of house bees and field bees in the case of *Apis* (Michener 1974). This temporal division of labour among workers of the honeybee, which is a typical feature of social insect colonies (Oster and Wilson 1978), could thus have profound consequences for how the foraging process should be analyzed theoretically.

The present study is restricted to a situation where workers have to collect food from very close to the hive. In a companion paper (Kacelnik et al., in preparation), we show that the same kind of analysis leads to similar predictions in a situation where food sources are available at different distances from the hive (up to 2000 m). In both cases, partial crop loads are to be expected for short distances and patches with low nectar yield. The question is not merely of theoretical interest, as partial crop-filling is observed under quite realistic conditions (Boch 1956; Nuñez 1982). In contrast, Seeley (1985) suggests that the concentration of nectar brought back from the various food patches within a colony's foraging range is the proximate cue that a bee colony could use to assess the profitability of different food sources, since (the similarly sized) workers usually return to the hive with full loads. This view is not supported by our analysis.

But why should bees, when foraging, maximize energetic efficiency rather than delivery rate to the hive? The rationale underlying the use of models of net rate maximization is the assumption that each calorie spent in foraging is compensated by an extra calorie gained through foraging. This assumption, however, is not necessarily true for animals such as the honeybee. Workers seem to be constrained not by a fixed (life) time available for foraging, but instead by a limited amount of flight performance. As this budget is used, the flight metabolism degenerates and the workers become unable to forage (Neukirch 1982). Therefore, each calorie spent is essentially a non-renewable loss of foraging capacity. If the hive is limited by the number of individuals that can be produced in a given season, as it is likely to be the case, a hive would therefore accumulate more resources if its workers maximize efficiency in foraging rather than if they gathered energy at the greatest instantaneous rate. We therefore suggest that such constraints, set by physiology or population mortality schedules, should be included in the functional analysis of foraging behaviour.

Acknowledgements

We thank Josue Nuñez and John Krebs for ideas and discussion. Paul Emden, Forestry Department Oxford, kindly provided the bees. Tony Price provided technical assistance. Regula Schmid-Hempel made the drawings. The work was supported by a grant from the Swiss N.S.F. (PS), N.E.R.C. to John Krebs (AK), and S.E.R.C. to Nick Davies (AIM)

References

Bastian J. and Esch, H. 1970. The nervous control of the indirect flight muscles of the honey bee. Z Vergl Physiol 67:307–324.

Benedictis, P. A. De, F. B. Gili, F. R. Hainsworth, G. H. Pyke and L. L. Wolf. (1978) Optimal meal size in hummingbirds. Am Nat 112:301–316.

Beutler, R. 1937. Über den Blutzucker der Bienen. Z Vergl Physiol 24:76–115.

Boch, R. 1956. Die Tänze der Bienen bei nahen und fernen Trachtquellen. Z Vergl Physiol 38:136–167.

Cheverton, J., A. Kacelnik and J. R. Krebs.(1985) Optimal foraging: constraints and currencies. In: *Experimental Behavioral Ecology and Sociobiology.* Hölldobler B. and M. Lindauer, eds. Fortschr Zool 31:109–126.

Frisch, K. von. 1965. *Tanzsprache und Orientierung der Bienen.* Springer, Berlin Heidelberg New York

Heinrich, B. 1975. Thermoregulation by bumblebees. II. Energetics of warm-up and free flight. J Comp Physiol 96:155–166.

Heinrich, B. 1979. Keeping a cool head: honeybee thermoregulation. Science 205:1269–1271.

Heran, H. 1962. Wie beeinflußt eine zusätzliche Last die Fluggeschwindigkeit der Honigbiene? Verh Dtsch Zool Gest 962:346–354.

Hodges, C. M. 1981. Optimal foraging in bumblebees: Hunting by expectation. Anim Behav 29:1166–1171.

Jongbloed, J. and C. A. G. Wiersma. 1934. Der Stoffwechsel der Honigbiene während des Fliegens. Z Vergl Physiol 21:519–533.

Kacelnik, A. 1984. Central place foraging in starlings *(Sturnus vulgaris).* I. patch residence time. J Anim Ecol 53:283–300.

Kacelnik, A. and A. I. Houston. 1984. Some effects of energy costs on foraging strategies. Anim Behav 32:609–614.

Kosmin, N. P., W. W. Alpaton and M. A. Resnitschenko. 1932. Zur Kenntnis des Gaswechsels und des Energieverbrauchs der Biene in Beziehung zu deren Aktivität. Z Vergl Physiol 17:408–422.

Lindauer, M. 1952. Ein Beitrag zur Frage der Arbeitsteilung im Bienenstaat. Z Vergl Physiol 34:299–345.

Lindauer, M. 1954. Temperaturregulierung und Wasserhaushalt im Bienenstaat. Z Vergl Physiol 36:391–432.

Lucas, J. R. 1983. The role of foraging constraints and variable prey encounter in optimal diet choice. Am Nat 122:191–209.

Michener, C. D. 1974. *The Social Behavior of the Bees.* Belknap, Harvard University. Cambridge, Mass.

Neukirch, A. 1982. Dependence of the life span of the honeybee *(Apis mellifera)* upon flight performance and energy consumption. J Comp Physiol 146:35–40.

Nuñez, J. 1970. The relationship between sugar flow and foraging and recruiting behavior of honey bees *(Apis mellifera* L.). Anim Behav 18:527–538.

Nuñez, J. 1974. Metabolism and activity of the worker. Proc 24th Int Beekeep Congr, Buenos Aires. Pp 298–299.

Nuñez, J. 1982. Honeybee foraging strategies at a food source in relation to its distance from the hive and the rate of sugar flow. J Apic Res 21:139–150.

Orians, G. H. and Pearson, N. E. 1979. On the theory of central place foraging. In: *Analysis of Ecological Systems.* Horn, D. J., G. R. Stairs and R. D. Mitchell, eds. Ohio University Press.

Oster, G. F. and E. O. Wilson. 1978. *Caste and Ecology in the Social Insects.* Monographs in population biology, vol 12. Princeton University Press. Princeton, NJ.

Pyke, G. H. 1978. Optimal foraging in bumblebees and coevolution with their plants. Oecologia (Berl) 36:281–293.

Rothe, U. 1983. Stoffwechselphysiologische Untersuchungen an ruhenden, laufenden und fliegenden Honigbienen (*Apis mellifera carnica*). PhD thesis, Universitat Saarbrucken.

Scholze, E., H. Pichler and H. Heran. 1964. Zur Entfernungsschätzung der Bienen nach dem Kraftaufwand. Naturwissenschaften 3:69–70.

Seeley, T. D. 1985. The information-center strategy of the honeybee. In: *Experimental Behavioral Ecology and Sociobiology.* Hölldobler, B. and M. Lindauer, eds. Fortschr Zool 31:75–90.

Sotavolta, O. 1954. On the fuel consumption of the honeybee *(Apis mellifica* L.) in flight experiments. Ann Zool Soc Vanamo 16:1–27.

Waddington, K. D. and L. R. Holden. 1979. Optimal foraging: On flower selection by bees. Am Nat 114:179–196.

Withers, P. 1981. The effect of ambient air pressure on oxygen consumption of resting and hovering honeybees. J C Physiol 141:433–437.

Readings for

POPULATION ECOLOGY

The Waxing and Waning of Populations

Anthony R. Ives

- **Joel E. Cohen**
  Population Growth and Earth's Human Carrying Capacity
- **Joseph H. Connell**
  The Influence of Interspecific Competition and Other Factors on the Distribution of the Barnacle Chthamalus stellatus
- **Charles J. Krebs, Stan Boutin, Rudy Boonstra, et al.**
  Impact of Food and Predation on the Snowshoe Hare Cycle

Introduction to
Joel E. Cohen
Population Growth and Earth's Human Carrying Capacity

by Anthony R. Ives

We are currently in the midst of one of the most dramatic events in the earth's history: the increase in abundance of a single species that now dominates all biomes of the world. One could argue that the increase in the human population is "unnatural." Certainly, the rate of increase and the geographical extent of the human population is impressive. Even more impressive are the diverse ways in which humans affect the environment and other species; we mine the earth for minerals and energy, cut trees for energy and construction, produce an astounding variety of chemicals and release them into the air and water, change the dominant vegetation types from wild to cultivated plants, and intentionally or unintentionally redistribute other species throughout the world. Nonetheless, the human population is still natural in the sense that human population growth is governed by the same general principles of population regulation as all other species—reproduction and/or death rates must ultimately depend on population density, and when they are equal the human population size will level off.

Joel Cohen (1995) addresses the problem of how high the human population will rise before it levels off. There have been many different estimates of the earth's human carrying capacity. Remarkably, the average prediction for the human carrying capacity has remained roughly constant since the first estimates in the 17th century. However, the differences among estimates have increased through history, suggesting that the accumulation of knowledge has provided no greater certainty about the human carrying capacity.

Cohen's main point is that the human carrying capacity is not a fixed number, but instead is a function of the human population size. This is because humans change the world, so with the birth of each additional human, the carrying capacity may increase. To illustrate this, consider the Sindh state in southern Pakistan along the Indus River. Naturally, the Sindh is a dry and seemingly inhospitable place. Nonetheless, it supports a population of millions, primarily due to the largest single irrigation project in the world that takes water from the Indus River. Thus, the carrying capacity of the Sindh has increased with the construction of the irrigation system, and today the Sindh can support far more people than it could 50 years ago. Now, however, salination and water-logging of agricultural land are increasing problems, and the carrying capacity of the Sindh may well be on a decline.

Population Growth and Earth's Human Carrying Capacity

JOEL E. COHEN

Earth's capacity to support people is determined both by natural constraints and by human choices concerning economics, environment, culture (including values and politics), and demography. Human carrying capacity is therefore dynamic and uncertain. Human choice is not captured by ecological notions of carrying capacity that are appropriate for nonhuman populations. Simple mathematical models of the relation between human population growth and human carrying capacity can account for faster-than-exponential population growth followed by a slowing population growth rate, as observed in recent human history.

Scientific uncertainty about whether and how Earth will support its projected human population has led to public controversy: will humankind live amid scarcity or abundance or a mixture of both (1, 2)? This article surveys the past, the present, and some possible futures of the global human population; compares plausible United Nations population projections with numerical estimates of how many people Earth can support; presents simplified models of the interaction of human population size and human carrying capacity; and identifies some issues for the future.

The Past and Some Possible Futures

Over the last 2000 years, the annual rate of increase of global population grew about 50-fold from an average of 0.04% per year between A.D. 1 and 1650 to its all-time peak of 2.1% per year around 1965 to 1970 (3). The growth rate has since declined haltingly to about 1.6% per year (4) (Figure 1). Human influence on the planet has in creased faster than the human population. For example, while the human population more than quadrupled from 1860 to 1991, human use of inanimate energy increased from 10^9 (1 billion) megawatt hours/year (MW hours/year) to 93 billion MW hours/year (Figure 2). For many people, human action is linked to an unprecedented litany of environmental problems (5), some of which affect human-well-being directly. As more humans contact the viruses and other pathogens of previously remote forests and grasslands, dense urban populations and global travel increase opportunities for infections to spread (6): The wild beasts of this century and the next are microbial, not carnivorous.

Along with human population, the inequality in the distribution of global income has grown in recent decades (7). In 1992, 15% of people in the world's richest countries enjoyed 79% of the world's income (8). In every continent, in giant city systems, people

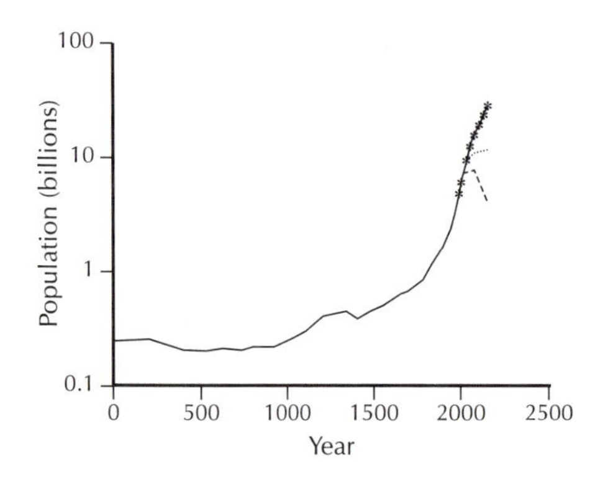

Figure 1. Recent world population history A.D. 1 to 1990 (solid line) (53) and 1992 population projections of the UN (11) from 1990 to 2150; high (solid line with asterisks); medium (dotted line); and low (dashed line). Population growth was faster than exponential from about 1400 to 1970.

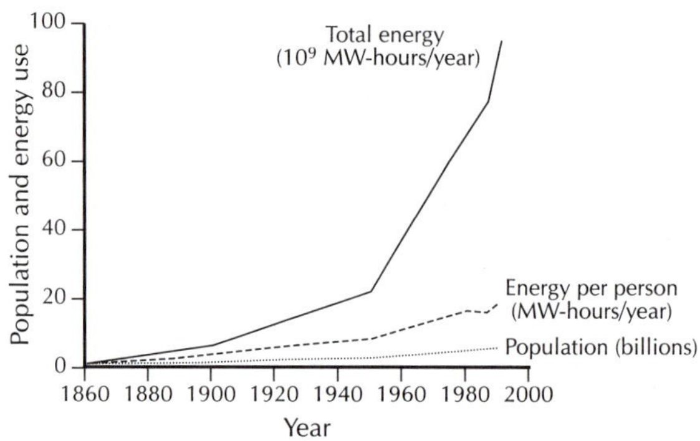

Figure 2. Inanimate energy use from all sources from 1860 to 1991: aggregate (solid line) (54) and per person (dashed line). Global population size is indicated by the dotted line.

increasingly come into direct contact with others who vary in culture, language, religion, values, ethnicity, and socially defined race and who share the same space for social, political, and economic activities (9). The resulting frictions are evident in all parts of the world.

Today, the world has about 5.7 billion people. The population would double in 43 years if it continued to grow at its present rate of 1.6% per year, though that is not likely. The population of less developed regions is growing at 1.9% per year, while that of more developed regions grows at 0.3 to 0.4% per year (10). The future of the human population, like the futures of its economies, environments, and cultures, is highly unpredictable. The United Nations (UN) regularly publishes projections that range from high to low (Figure 1). A high projection published in 1992 assumed that the worldwide average number of children born to a woman during her lifetime at current birthrates (the total fertility rate, or TFR) would fall to 2.5 children per woman in the 21st century; in this scenario population would grow to 12.5 million by 2050 (11). The UN's 1992 low projection assumed that the worldwide average TFR would fall to 1.7 children per woman; in this case, the population would peak at 7.8 billion in 2050 before beginning to decline.

There is much more uncertainty about the demographic future than such projections suggest (12). At the high end, the TFR in less developed countries today, excluding China, is about 4.2 children per woman; that region includes 3.25 billion people. Unless fertility in the less developed countries falls substantially, global fertility could exceed that assumed in the UN's high projection. At the low end, the average woman in Italy and Germany has about 1.3 children, and in Spain, 1.2. Fertility could fall well below that assumed in the UN's low projection.

Can Earth support the people projected for 2050? If so, at what levels of living? In 1679, Antoni van Leeuwenhoek (1632–1723) estimated that the maximum number of people Earth can support is 13.4 billion (13). Many more estimates of how many people Earth

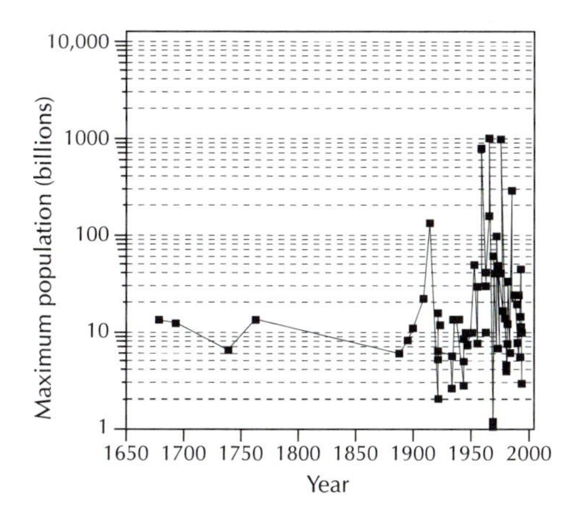

Figure 3. Estimates of how many people Earth can support, by the date at which the estimate was made. When an author gave a range of estimates or indicated only an upper bound, the highest number stated is plotted here (55).

could support followed (14) (Figure 3). The estimates have varied from <1 billion to >1000 billion. Estimates published in 1994 alone ranged from <3 billion to 44 billion (15). Since 1679, there has been no clear increasing or decreasing trend in the estimated upper bounds. The scatter among the estimates increased with the passage of time. This growing divergence is the opposite of the progressive convergence that would ideally occur when a constant of nature is measured. Such estimates deserve the same profound skepticism as population projections. They depend sensitively on assumptions about future natural constraints and human choices.

Many authors gave both a low estimate and a high estimate. Considering only the highest number given when an author stated a range, and including all single or point estimates, the median of 65 upper bounds on human population was 12 billion. If the lowest number given is used when an author stated a range of estimates, and all point estimates are included otherwise, the median of 65 estimated bounds on human population was 7.7 billion. This range of low to high medians, 7.7 to 12 billion, is very close to the range of low and high UN projections for 2050: 7.8 to 12.5 billion. A historical survey of estimated limits is no proof that limits lie in this range. It is merely a warning that the human population is entering a zone where limits on the human carrying capacity of Earth have been anticipated and may be encountered.

Methods of Estimating Human Carrying Capacity

Calculations of estimates of Earth's maximum supportable human population use one of six methods, apart from those that are categorical assertions without data. First, several geographers divided Earth's land into regions, assumed a maximum supportable population density in each region, multiplied each assumed maximal population density by the area of the corresponding region, and summed over all regions to get a maximum supportable population of Earth. The assumed maximum regional population densities were treated as static and were not selected by an objective procedure. Second, some analysts fitted mathematical curves to historical population sizes and extrapolated them into the future (16). As the causal factors responsible for changes in birth rates and death rates were, and are, not well understood, there has been little scientific basis for the selection of the fitted curves.

Third, many studies focused on a single assumed constraint on population size, without checking whether some other factors might intervene before the assumed constraint comes into play. The single factor most often selected as a likely constraint was food (17). In 1925, the

German geographer Albrecht Penck stated a simple formula that has been widely used (18):

$$\text{Population that can be fed} = \frac{\text{food supply}}{\text{individual food requirement}} \quad (1)$$

This apparently objective formula can lead to extremely different estimates of maximum supportable population because it depends on estimates of the food supply and of individual requirements. The food supply depends on areas to be planted and watered, choice of cultivars, yields, losses to pests and waste, cultural definitions of what constitutes acceptable food, and random fluctuations of weather. Individual requirements depend on the calories and protein consumed directly as well as on nutrients used as animal fodder (19). Besides food, other factors proposed as sole constraints of human numbers include energy, biologically accessible nitrogen, phosphorus, fresh water, light, soil, space, diseases, waste disposal, nonfuel minerals, forests, biological diversity, and climatic change.

Fourth, several authors reduced multiple requirements to the amount of some single factor. For example, in 1978 Eyre reduced requirements for food, paper, timber, and other forest products to the area of land required to grow them (20). Other factors that cannot be reduced to an area of land, such as water or energy, are sometimes recognized indirectly as constraints on the extent or productivity of cultivable land. The authors who combined different constraints into a single resource assumed that their chosen resource intervened as a constraint before any other factor.

Fifth, several authors treated population size as constrained by multiple independent factors. For example, Westing in 1981 estimated the constraints on population imposed independently by total land area, cultivated land area, forest land area, cereals, and wood (21). Constraints from multiple independent resources are easily combined formally. For example, if one assumes, in addition to a food constraint, a water constraint

$$\text{Population that can be watered} = \frac{\text{water supply}}{\text{individual water requirement}} \quad (2)$$

and if both constraints (1) and (2) must be satisfied independently, then

$$\text{Population that can be fed and watered} = \text{minimum of} \left\{ \frac{\text{food supply}}{\text{individual food requirement}} \quad \frac{\text{water supply}}{\text{individual water requirement}} \right\} \quad (3)$$

This formula is an example of the law of the minimum proposed by the German agricultural chemist Justus Freiherr von Liebig (1803–1873) (22). Liebig's law of the minimum asserts that under steady-state conditions, the population size of a species is constrained by whatever resource is in shortest supply (23). This law has serious limitations when it is used to estimate the carrying capacity of any population. If different components of a population have heterogeneous requirements, aggregated estimates of carrying capacity based on a single formula will not be accurate; different portions of the global human population are likely to have heterogeneous requirements. In addition, Liebig's law does not apply when limiting factors fluctuate, because different factors may be constraining at different times—an average over time may be misleading. Liebig's law assumes that the carrying capacity is strictly proportional to the limiting factor (within the range where that factor is limiting); strictly linear responses are not generally observed (24). Liebig's law assumes no interactions among the inputs; independence among limiting factors is not generally observed. (For example, Equation 3 neglects the possibility that changes in the water supply may affect the food supply through irrigation.) Liebig's law also assumes that adaptive responses will not alter requirements or resources during the time span of interest. But economic history (including the inventions of agriculture and industry) and biological history (including the rise of mutant infections and the evolution of resistance to pesticides and drugs) are full of such adaptive responses.

Sixth and finally, several authors have treated population size as constrained by multiple interdependent factors and have described this interdependence in system models. System models are large sets of difference equations (deterministic or stochastic), which are usually solved numerically on a computer. System models of human population and other variables have often embodied relations and assumptions that were neither mechanistically derived nor quantitatively tested (25).

The first five methods are deterministic and static. They make no allowances for changes in exogenous or endogenous variables or in functional relations among variables. Although a probabilistic measure of human carrying capacity has been developed for local populations in the Amazon (26), no probabilistic approach to global human population carrying capacity has been developed. Yet, stochastic variability affects local and global human populations through weather, epidemics, accidents, crop diseases and pests, volcanic eruptions, the El Niño Southern Oscillation in the Pacific Ocean, genetic variability in viruses and other microbes, and international financial and political arrangements. Stochastic models of human carrying capacity would make it possible to address questions that deterministic models cannot, including (conditional on all assumptions that go into any measure of human carrying capacity) what level of population could be maintained 95 years in 100 in spite of anticipated variability (27).

Some ecologists and others claim that the ecological concept of carrying capacity provides special insight into the question of how many people Earth can support. In basic and applied ecology, the carrying capacity of nonhuman species has been defined in at least nine different ways, none of which is adequate for humans (28). Human carrying capacity depends both on natural constraints which are not fully understood, and on individual and collective choices concerning the average level and distribution of material well-being, technology, political institutions, economic arrangements, family structure, migration and other demographic arrangements, physical, chemical, and biological environments, variability and risk, the time horizon, and values, tastes, and fashions. How many people Earth can support depends in part on how many will wear cotton and how many polyester; on how many will eat meat and how many bean sprouts, on how many will want parks and how many will want parking lots. These choices will change in time and so will the number of people Earth can support.

Some have urged that individual nations or regions estimate their human carrying capacity separately (29). Although specific resources such as mineral deposits can be defined region by region the knowledge, energy, and technology required to exploit local resources often depend on other regions; the positive and negative effects of resource development commonly cross national borders. Human carrying capacity cannot be defined for a nation independently of other regions if that nation trades with others and shares the global resources of the atmosphere, oceans, climate, and biodiversity.

Mathematical Cartoons

If a current global human carrying capacity could be defined as a statistical indicator, there would be no reason to expect that indicator to be static. In 1798, Thomas Robert Malthus (1766–1834) described a dynamic relation between human population size and human carrying capacity: "The happiness of a country does not depend, absolutely, upon its poverty or its riches, upon its youth or its age, upon its being thinly or fully inhabited, but upon the rapidity with which it is increasing, upon the degree in which the yearly increase of food approaches to the yearly increase of an unrestricted population" (30). Malthus opposed the optimism of the Marquis de Condorcet (1743–1794), who saw the human mind as capable of removing all obstacles to human progress. Malthus predicted wrongly that the population growth rate would always promptly win a race against the rate of growth of food. Malthus has been wrong for nearly two centuries because he did not foresee how much people can expand the human carrying capacity of Earth, including but not limited to food production. To examine whether Malthus will continue to be wrong, economists, demographers, and system analysts have constructed models in which population growth drives technological change, which permits further population growth (31).

I describe here idealized mathematical models for the race between the human population and human carrying capacity (32). Suppose that it is possible to define a current human carrying capacity $K(t)$ as a numerical quantity measured in numbers of individuals. Suppose also that $P(t)$ is the total number of individuals in the population at time t and that

$$\frac{dP(t)}{dt} = rP(t)\,[K(t)-P(t)] \qquad (4)$$

The constant $r > 0$ is called the Malthusian parameter (33). I will call Equation 4, the equation of Malthus. It is the same as the logistic equation except that the constant K in the logistic equation is replaced by variable carrying capacity $K(t)$ here.

To describe changes in the carrying capacity $K(t)$, let us recognize, in the phrase of former U. S. President George H. Bush Jr., that "every human being represents hands to work, and not just another mouth to feed" (34). Additional people clear rocks from fields, build irrigation canals, discover ore deposits and antibiotics, and invent steam engines; they also clear-cut primary forests, contribute to the erosion of topsoil, and manufacture chlorofluorocarbons and plutonium. Additional people may increase savings or dilute and deplete capital; they may increase or decrease the human carrying capacity.

Suppose that the rate of change of carrying capacity is directly proportional to the rate of change in population size. Call Equation 5, the equation of Condorcet:

$$\frac{dK(t)}{dt} = c\,\frac{dP(t)}{dt} \qquad (5)$$

The Condorcet parameter c can be negative, zero, or positive.

In this model, population size changes in one of three distinct ways: faster than exponentially, exponentially, and logistically (35). When $c > 1$, each additional person increases the human carrying capacity enough for her own wants plus something extra. Then $K(t) - P(t)$ increases with time t, population growth accelerates faster than exponentially, and finally, after some finite period of time, $P(t)$ explodes to infinity. When $c = 1$, each additional person adds to carrying capacity just as much as he consumes. Thus, $K(t) - P(t) = K(0) - P(0)$ for any t and $P(t)$ grows exponentially. When $c < 1$, $P(t)$ grows logistically, even though $K(t)$ will change if $c \neq 0$. The population growth rate falls smoothly toward zero. When $c < 1$, the net effect on population size of changes in $K(t)$ is equivalent to having a "virtual" constant carrying capacity K'. The virtual K' equals the initial carrying capacity $K(0)$ if and only if $c = 0$, when changes in $P(t)$ do not alter $K(t)$. $K' > K(0)$ if $0 < c < 1$: in this case, each additional person increases the carrying capacity, but not by as much as the person consumes. When $c < 0$, population growth diminishes $K(t)$, as in situations of congestion, pollution, and overgrazing, and $K' < K(0)$. The Malthus-Condorcet model integrates the exponential growth model of Euler in the 18th century, the logistic growth model of Verhulst in the 19th century, and the doomsday (faster-than-exponential) growth model of von Foerster et al. in the 20th (36).

The discrete-time equations of Malthus and Condorcet replace the derivatives dP/dt and dK/dt by the corresponding finite differences $[P(t + \Delta t) - P(t)]/\Delta t$ and $[K(t + \Delta t) - K(t)]/\Delta t$. This model can display exponential ($c = 1$) and faster-than-exponential ($c > 1$) growth as well as all the dynamic behaviors of the discrete-time logistic equation (logistic growth, overshoot and damped oscillations, and periodic oscillations with various periods, chaotic behavior, and overshoot and collapse) (37). Overshoots become possible in discrete time because population and carrying capacity respond to current conditions with a time lag.

If an additional person can increase human carrying capacity by an amount that depends on the resources available to make her hands productive, and if these resources must be shared among more people as the population increases, then the constant c should be replaced by a variable $c(t)$ that declines as population size increases. Suppose, for example, that there is a constant $L > 0$ such that $c(t) = L/P(t)$. The assumption that $c(t) = L/P(t)$ is positive, no matter how big $P(t)$ is, models the dilution of resources, but not their depletion or degradation. Replacing c by $L/P(t)$ gives the Condorcet-Mill equation (6), which I name after the British philosopher John Stuart Mill (1806–1873), who foresaw a stationary population as both inevitable and desirable (38); L is the Mill parameter.

$$\frac{dK(t)}{dt} = \frac{L}{P(t)} \frac{dP(t)}{dt} \qquad (6)$$

Assume further that $c(0) = L/P(0) > 1$. Then the population initially grows faster than exponentially. As $P(t)$ increases past L, $c(t)$ passes through 1 and the population experiences a brief instant of exponential growth. Then $c(t)$ falls below 1 and the population size thereafter grows sigmoidally. The overall trajectory looks sigmoidal on a logarithmic scale of population (Figure 4). Population size rises to approach a unique stationary level, which is independent of r. The bigger $K(0)$ and L are,

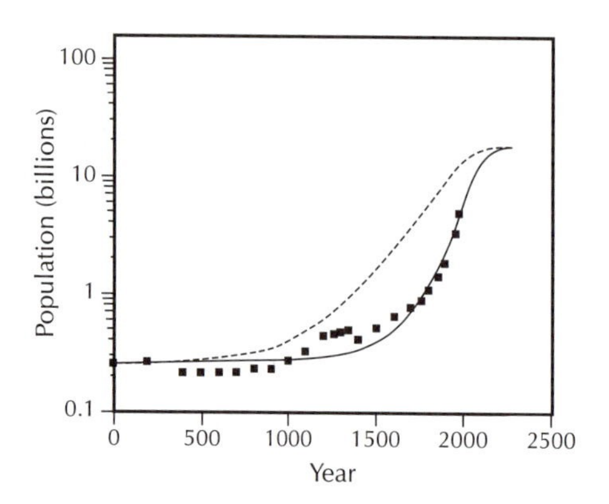

Figure 4. Numerical illustration of the equations of Malthus and Condorcet-Mill: human carrying capacity $K(t)$ is shown by the dashed line and model population size $P(t)$ by the solid line; for comparison, estimated actual human population (solid rectangles) is shown. Equations: $P(t+\Delta t) - P(t) = rP(t)[K(t) - P(t)]\Delta t$, $K(t+\Delta t) - K(t) = Lr[K(t) - P(t)]\Delta t$. Initial conditions and parameters: $\Delta t = 20$ years, $P(0) = 0.252$, $K(0) = 0.252789$, $r = 0.0014829$, and $L = 3.7$. $P(0)$, $K(0)$, and L are measured in billions (10^9).

the bigger the stationary level is, other things being equal.

Figure 4 shows a trajectory of human carrying capacity $K(t)$ above and population size $P(t)$ below according to the Malthus-Condorcet-Mill model; $P(t)$ is compared with the estimated human population history over the past 2000 years (39). Values of $P(t)$ beyond $t = 1995$ are intended only to illustrate the qualitative behavior of the model, not to predict future human population; nothing guarantees that the actual human population will reach or remain at the high plateau shown. For example, the model neglects the possibilities that people could increasingly choose to divide the available material resources among fewer offspring, trading numbers for wealth, and that pollution or exogenous climatic changes could diminish human carrying capacity.

Up to about $t = 1910$, population sizes (theoretical and actual) are convex on the logarithmic scale; after roughly $t = 1970$, they are concave. The human carrying capacity $K(t)$, initially only slightly above $P(t)$, began to exceed $P(t)$ substantially at times corresponding to the 9th and 10th centuries and experienced nearly exponential growth (linear increase on the logarithmic scale shown) from the 11th to the mid-20th century. According to the model, the acceleration of population growth in the 17th century was preceded by a long period of increasing human carrying capacity (40).

These models illuminate Earth's human carrying capacity. First, the statement that "every human being represents hands to work, and not just another mouth to feed" does not specify the cultural, environmental, and economic resources available to make additional hands productive and therefore does not specify by how much the additional hands can increase (or decrease) human carrying capacity. Yet, the quantitative relation between an increment in population and an increment in carrying capacity is crucial to the future trajectory of both the population and the carrying capacity. Second, the historical record of faster-than-exponential population growth, accompanied by an immense improvement in average well-being, is logically consistent with many alternative futures, including a continued expansion of population and carrying capacity, or a sigmoidal tapering off of the growth in population size and carrying capacity, or oscillations (damped or periodic), or chaotic fluctuations, or overshoot and collapse. Third, to believe that no ceiling to population size or carrying capacity is imminent entails believing that nothing in the near future will stop people from increasing Earth's ability to satisfy their wants by more than, or at least as much as, they consume. The models focus attention on, and provide a framework in which to interpret, quantitative empirical studies of the relation between rapid population growth and changing human carrying capacity.

Issues for the Future

Three valuable approaches have been advocated to ease future trade offs among population, economic well-being, environmental quality, and cultural values. Each of these approaches is probably necessary, but is not sufficient by itself, to alleviate the economic, environmental, and cultural problems described above. First, the "bigger pie" school says: develop more technology (41). Second, the "fewer forks" school says: slow or stop population growth (42). In September 1994 at the UN population conference in Cairo, several approaches to slowing population growth by lowering fertility were advocated and disputed. They included promoting modern contraceptives; promoting economic development; improving the survival of infants and children; improving the status of women; educating men; and various combinations of these. Unfortunately, there appears to be no believable information to show which approach will lower a country's fertility rate the most, now or a decade from now, per dollar spent. In some developing countries such as Indonesia, family planning programs interact with educational, cultural, and economic improvements to lower fertility by more than the sum of their inferred separate effects (43). Some unanswered questions are how soon will global fertility fall, by what means, and at whose expense.

Third, the "better manners" school says: improve the terms under which people interact (for example, by defining property rights to open-access resources; by removing economic irrationalities; and by improving governance) (44). When individuals use the environment as a source or a sink and when they have additional children, their actions have consequences for others. Economists call "externalities" the consequences that fall on people who are not directly involved in a particular action. That individuals neglect negative externalities when they use the environment has been called "the tragedy of the commons" (45); that individuals neglect negative externalities when they have children has been called "the second tragedy of the commons" (46). The balance of positive and negative externalities in private decisions about fertility and use of the environment depends on circumstances. The balance is most fiercely debated when persuasive scientific evidence is least available. Whatever the balance, the neglect by individuals of the negative externalities of childbearing biases fertility upward compared to the level of aggregate fertility that those same individuals would be likely to choose if they could act in concert or if there were a market in the externalities of childbearing. Voluntary social action could change the incentives to which individuals respond in their choices concerning childbearing and use of the environment.

References and Notes

1. Brown, L R. and H. Kane. 1994. *Full House: Reassessing the Earth's Population Carrying Capacity.* Norton, New York.
2. Duchin, F. and G. Lange. 1994. *The Future of the Environment: Ecological Economics and Technological Change.* Oxford Univ. Press, New York.
Myers, N. and J. L. Simon. 1994. *Scarcity or Abundance? A Debate on the Environment.* Norton, New York .
3. Livi-Bacci, M. A. 1992. *Concise History of World Population.* Carl Ipsen, translator. Blackwell, Cambridge, MA. Estimates of global population size at A.D. 1 vary from 133 million [E. S. Deevey Jr., Sci. Am. 203:195] to 330 million (47).
4. Horiuchi, S. 1992, Science 257:761.

5. Demeny, P. 1991. In: *Resources, Environment and Population: Present Knowledge, Future Options.* Davis, K. and M. S. Bernstam, Eds. Oxford Univ. Press, New York. Pp. 408–421. Gives a grim list of such environmental problems: "loss of topsoil, desertification, reforestation, toxic poisoning of drinking water, Oceanic pollution, shrinking wetlands, overgrazing, species loss, shortage of firewood, exhaustion of oil reserves and of various mineral resources, siltation in rivers and estuaries. encroachment of human habitat on arable land, dropping water tables, erosion of the ozone layer, loss of wilderness areas, global warming, rising sea levels, nuclear wastes, acid rain" (p. 416).

6. Lederberg, J. 1988. J. Am. Med. Assoc. 260:684.
Morse, S. S. ed. 1993. *Emerging Viruses.* Oxford Univ. Press, New York.
Anderson, R. M. and R. M. May. 1991. *Infectious Diseases of Humans: Dynamics and Control.* Chapter 23. Oxford Univ. Press, Oxford.

7. In 1960, the richest countries with 20% of world population earned 70.2% of global income, while the poorest countries with 20% of world population earned 2.3% of global income. Thus, the ratio of income per person between the top fifth and the bottom fifth was 31:1 in 1960. In 1970, that ratio was 32:1; in 1980, 45:1; in 1991, 61:1. In constant 1989 U.S. dollars, the absolute gap between the top fifth and the bottom fifth rose from $1864 in 1960 to $15,149 in 1989. United Nations Development Programme, Human Development Report 1992. Oxford Univ. Press, New York, p. 36; Human Development Report 1994, p. 63].

8. Demeny, P. 1994. Population and Development, International Conference on Population and Development 1994. International Union for the Scientific Study of Population, Lege, Belgium. These numbers are based on World Bank estimates.

9. From 1950 to 1995, the world's urban population increased more than 3.5-fold from 0.74 billion to 2.6 billion and grew from 29% to 45% of the total population [United Nations, World Urbanization Prospects: The 1992 Revision, publication ST/ESA/SER.A/136. United Nations, New York, 1993, pp. 74–75 and 82–83.

10. World Population Data Sheet. Population Reference Bureau, Washington, DC, 1994; World Population 1994. United Nations, New York.

11. United Nations, Department of International Economic and Social Affairs. 1992. *Long-Range World Population Projections: Two Centuries of Population Growth 1950–2150.* United Nations, New York. Publication ST/ESA/SER.A/125.

12. Systematic retrospective analyses of past population projections indicate that more confidence has been attached to projections than was justified by their eventual predictive accuracy. Stoto, M. A. 1983. J. Am. Stat. Assoc. 78:13. Keyfitz, N. 1982. *Population Change and Social Policy.* Chapter 13. Abt Books, Cambridge, MA.

13. van Leeuwenhoek, A. *Collected Letters* (Swets and Zeitlinger, Amsterdam, 1948). Letter 43, 25 April 1679, vol. 3, pp. 4–35. Leeuwenhoek multiplied his estimate of the population of Holland (1 million people) by his estimate of the ratio of Earth's inhabited land area to Holland's area (13,385).

14. Cohen, J. E. *How Many People Can the Earth Support?* Norton, New York. In press.

15. V. Smil [Popul. Dev. Rev. 20, 255 (1994)] estimated 10 to 11 billion; D. Pimentel, R. Harman, M. Pacenza, J. Pecarsky, and M. Pimentel [Popul. Environ. 15, 347 (May 1994)] estimated <3 billion; P. E. Waggoner (48) estimated at least 10 billion; Brown and Kane (1) estimated that a projected world grain harvest of 2.1 billion tons in 2030 could feed 2.5 billion people at the U. S. consumption level of 800 kg/year per person or just over 10 billion people at the Indian level of consumption of 200 kg/year per person; and the Wetenschappelijke Raad voor het Regeringsbeleid [(Scientific Council for the Dutch Government), Duurzeme riskos: een blijvend gegeven (Sustainable risks: An endunng given) (Sdu Uitgeverq Plantijnstraat, Den Haag, Netherlands, 1994), p. 9] estimated 11 to 44 billion people, depending on the scenario.

16. For example, R. Pearl and L. J. Reed [in: *Studies in Human Biology*, R. Pearl, Ed. (Williams and Wilkins, Baltimore, MD, 1924), chap. 25, p. 632] fitted a logistic curve to past world population sizes and confidently estimated a maximum world population of 2 billion. The world's population passed 2 billion around 1930. Undeterred, R. Pearl and S. Gould [Hum. Biol. 8:399 (1936)] again used the logistic curve to project 2.645 billion people as an ultimate limit to be nearly approached by the end of the 21st century. That population size was surpassed before 1955. On a logarithmic scale of population, the logistic curve is concave while the observed trajectory of global population size was convex until about 1970. The failures of Pearl's logistic projections and the usefulness of A. J. Lotka's theory of population growth and age-compositor [Theorie analytique des associations biologiques, II Analyse demographique avec application particuliere a l'espece humaine (Hemmann, Pans, 1939)] led demographers to abandon studying the absolute size of populations in favor of studying population structure and change. Since World War II, estimates of Earth's human carrying capacity have been published almost exclusively by non-demographers. Demography, like economics, still lacks a working theory of scale. In another example of curve-fitting, A. L. Austin and J. W. Brewer (49) modified the logistic curve to allow for faster-than-exponential growth followed by leveling off; they fitted their curve to past global population sizes and predicted an asymptote around 50 billon people.

17. Brown, H. 1954. *The Challenge of Man's Future: An Inquiry Concerning the Condition of Man During the Years That Lie Ahead.* Viking, New York.

Brown, H., J. Bonner, J. Weir., 1954. *The Next Hundred Years: Man's Natural and Technological Resources.* Viking, New York.

Clark, C. 1958. Nature 181:1235. Reprinted in *Readings in Human Population Ecology,* 1971. W. Y. Davis, Ed. Prentice-Hall, Englewood Cliffs, NJ. Pp. 101–106.

Cepede, M., F. Houtart, L. Grond. 1964. *Population and Food.* Sheed & Ward, New York.

Schmin, W. R. 1985. Ann. MY. Acad. Sci. 118:645.

Lieth, H. 1973. Hum. Ecol.1:303.

Blaxter, K. 1986. *People, Food and Resources.* Cambridge Univ. Press, Cambridge.

Buringh, P., H. D. J. van Heemst, G. J. Staring. 1975. *Computation of the Absolute Maximum Food Production of the World.* Agricultural Press, Wageningen, Netherlands.

Buringh, P. and H. D. J. van Heemst. 1977. *An Estimation of World Food Production Based on Labour-Oriented Agricuture.* Agricultural Press, Wageningen, Netherlands.

Higgins, G. M., A. H. Kassam, L. Naiken, G. Fischer and M. M. Shah. 1983. "Potential population supporting capacities of lands in the developing world" technical report of project INT/75/P13; "Land resources for populations of the future;" FPA/INT/513. Food and Agricultural Organization of the United Nations, Rome.

Chen, R. S. et al. 1990. Eds. *The Hunger Report: 1990* HR-90-1, Alan Shawn Feinstein World Hunger Program, Brown University, Providence, RI.

Millman, S. R. et al. 1991. *The Hunger Report: Update 1991.* HR-91-1, Alan Shawn Feinstein World Hunger Program, Brown University, Rhode Island, April 1991. See also (1), p. 31. It is remarkable that food continues to be viewed as a limiting constraint on population size even though, globally, the countries with the lowest fertility and the lowest population growth rates are among those where food is most abundant [J. Mayer, Daedalus 93,830 (1964)].

18. Penck, A. 1925. Z. Geopolitik 2:330. Sitzungsberichte der Preußischen Akademie der Wissenschaften 22:242 (1924). The formula was used in 1917 but not stated explicitly by G. H. Knibbs [*The Mathematical Theory of Population, of Its Character and Fluctuations, and of the Factors Which Influence Them,* appendix A. Vol. 1. 1917. Census of the Commonwealth of Australia, Minister of State for Home and Territories, Melbourne. McCarron, Bird & Co., Melbourne. P. 455.

19. In 1972, domestic animals were fed 41% of all grain consumed; in 1992, 37%. See (50), p. 296; (1), p. 67.

20. Eyre, S. R. 1978. *The Real Wealth of Nations.* St. Martin's, New York.

21. Westing, A. H. 1981. Environ. Conserv. 8:177.

22. von Liebig, J. F. 1855. *Principles of Agricultural Chemistry.* Wiley, New York. German edition: *Die Grundsatze der Agiculturchemie.* Vieweg und Sohn, Braunscheweig, 1855. Also see DeAngelis, D. L. 1992. *Dynamics of Nutrient Cycling and Food Webs.* Chapman and Hall, London. Pp.38–45, 228.

23. Lebig's law extends to any number of independent constraints. When population on the left side of the formula is replaced by production, the formula is known in economic theory as the Walras-Leontie-Harrod-Domar production function.

24. Smil, V. (1991, Popul. Dev. Rev. 17:569. Pp. 586 and 597) reported that in the 1980s, nitrogen applied in the Zhejiang and Shandong provinces of China increased rice yields by amounts that were only 50 to 80% as large as the increments from an additional kilogram per hectare applied in the 1960s.

25. Examples of system models are J. W. Forrester, *World Dynamics* (Wright-Allen Press, Cambridge, MA, 1971); D. Meadows, D. L. Meadows, J. Randers, W. W. Behrens III *The Limits to Growth* (Signet, New American Library, New York, 1972, ed. 2, 1974); M. Mesarovic and E. Pestel, *Mankind at the Turning Point* (Dutton and Reader's Digest Press, New York, 1974). See also (51); J. Gever, R. Kaufmann, D. Skole, C. Vorosmarty, *Beyond Oil: The Threat to Food and Fuel in the Coming Decades* (Ballinger, Cambridge, MA, 1986); A. J. Gilbert and L. C. Braat, Eds., *Modeling for Population and Sustainable Development* (Routledge, London, 1991). See also (52) and W. Lutz, Ed., *Population-Development-Environment: Understanding Their Interactions in Mauritius* (Springer-Verlag, Heidelberg, 1994). Critiques of system models are C. Kaysen, Foreign Aff. 50:660 (1972) H. S. D. Cole, C. Freeman, M. Jahoda, K. L. R. Pavitt Eds., *Model of Doom: A Critique of The Limits to Growth, With a Reply by the Authors of The Limits to Growth* (Universe Books, New York, 1973); W. D. Nordhaus, Econ. J. 83:1156 (1973); E. van de Walle. Science 189:1077 (1975), D. Beninsh, *On Systems Analysis* (MIT Press, Cambridge, MA, 1976); and P. R. Ehrlich, A. H. Ehrlich, J. P. Holdren, *Ecoscience: Population, Resources, Environment* (Freeman, San Francisco,1977), pp. 730–733.

26. Fearnside, P. M. 1986. *Human Carrying Capacity of the Brazilian Rainforest.* Columbia Univ. Press, New York.

27. In many regions, the average amount of fresh water available annually is more than twice the amount of water that can be counted on 95 years in 100 [P. P. Rogers, H R. Repetto, Ed., The Global Possible: Resources, Development and the New Century (Yale Univ. Press, New Haven, 1985), p.294].

28. B. Zaba and I. Scoones. 1994. In: *Environment and Population Change.* Zaba, B. and J. Clarke, Eds. Ordina Editions, Liege. Pp. 197–219.

Pulliam,H. R. and N. M. Haddad. 1994. Bull. Ecol. Soc. Amer. 75:141. See also (14).

29. Paragraph 5.23 of Agenda 21, final document of the UN Conference on Environment and Development, Rio de Janeiro, June 1992 (Rio Earth Summit).

30. Malthus, T. R. 1960. An Essay on the Principle of Population [1798, repent, with part of ed. 7. In: *On Population,* Chap. VII. Himmelfarb, G. Ed. Modem Library, New York.

P. 51]. Demeny, P. (1989, In: *Population and Resources in Western Intellectual Traditions,* M. S. Teitelbaum and J. M. Winter, Eds. (Population Council, New York. P. 232) generalized Malthus's view to incorporate all aspects of economic output, not just food: "Posed in the simplest terms, the economics of population reduces to a race between two rates of growth: that of population and that of economic output."

31. Pryor, F. L. and S. B. Maurer, 1982. J. Dev. Econ. 10:325. Lee, R. D. 1986. In: *The State of Population Theory: Forward from Malthus.* Coleman, D. and R. Schofield, Eds. Basil Blackwell, Oxford. Pp. 96–130.
Math. Popul. Studies 1,265 (1988) Population 47:1533 (1992); Explor. Econ. Hist. 30,1 (1993)
Kremer,M. 1993. Q. J. Econ. 108:681. See also (49, 51, 52).
32. These models assume no migration and ignore the population's age composition, geographical distribution, and the distribution of well-being. The models also omit any changes in human carrying capacity that depend on the past history of human carrying capacity and population see, rather than on their present magnitudes. The models also ignore stochastic fluctuations in environmental and human factors. When these models are extended to describe two or more; regional populations that interact through migration as well as by influencing one another's carrying capacity (for example, through transboundary air pollution), additional modes of behavior appear.
33. The dimensions of r are $[T^{-1}P^{-1}]$, where T is time.
34. Carrying Capacity Network, Focus: Carrying Capacity Selections 1:57 (1992).
35. At $t = 0$, assume that $K(0) > P(0) > 0$ and $r > 0$. The form of $P(t)$ depends on c.

Case i: $c < 1$. Because $K(0) > P(0) > cP(O)$,

$$P(t) = \frac{1}{\left(\frac{1}{P(0)} + \frac{c-1}{K(0) - cP(0)}\right) e^{-r[K(0) - cP(0)]t} - \frac{c-1}{K(0) - cP(0)}} \tag{7}$$

The solution $P(t)$ is a logistic curve with the "virtual" Malthusian parameter $r' = r(1 - c) > 0$ and the constant "virtual" carrying capacity $K' = [K(0) - cP(0)]/(1 - c)$. [The special case $c = 0$ reproduces the logistic equation with $r' = r$ and $K' = K(0)$.] $K(t)$ is logistic in shape after possible shifting by $K(0) - cP(0)$ and rescaling by c. $K' > K(0)$, $K' = K(0)$, or $K' < K(0)$ according to whether $1 > c > 0$, $c = 0$, or $0 > c$.

Case ii: $c = 1$. Then $P(t)$ grows exponentially with "virtual" Malthusian parameter $r[K(0) - P(0)]$.

Case iii: $c > 1$. For some finite time T (where $0 < T <= \infty$), $\lim_{t \uparrow T} P(t) = \infty$ When $K(0) - cP(0) \neq 0$, Equation 7 describes $P(t)$ and the population size becomes infinite when t rises to

$$T = \frac{1}{r[K(0) - cP(0)]} \times \log\left\{\left[\frac{1}{P(0)} + \frac{c-1}{K(0) - cP(0)}\right] \Big/ \left[\frac{c-1}{K(0) - cP(0)}\right]\right\} \tag{8}$$

When $K(0) - cP(0) = 0$, then $P(t) = P(0)/[1 - r(c - 1)P(0)t]$, which is the solution of the "doomsday equation" of von Foerster et al. The population size becomes infinite at time $T = 1/[r(c - 1)P(0)]$.

36. von Foerster, H., P. M. Mora and L. W. Amiot. 1960. Science 132:1291.
Robertson, J. S., V. P. Bond and E. P. Cronkite. 1961. Science 133:936.
Coale, A. J. Science 133:1931
von Foerster, H,. P. M. Mora, L. W. Amiot. 1962. Science 136:173.
37. Myrberg, P. J. and J. Math. 1962. Pures Appl. 41:339.
May, R. M. 1974. Science 186:645.
May, R. M. 1975 J. Theor Biol. 51:511.
May, R. M. and G. F. Oster. 1976. Am. Nat. 110:573.
May, R. M. 1986, Ecology 67:1115.
Hastings, A., C. L. Hom, S. Ellner, P. Turchrn and H. C. J. Godfray. 1993. Annul Rev. Ecol. Syst. 24:1.
Cohen, J. E. and I. Barradas. In preparation.
These behaviors include all those D. M. Meadows, D. L. Meadows, and J. Flanders [(52), pp.108–9] attributed to the system models called World3 and World3/91 in *Beyond the Limits* (52), and more.
38. Mill, J. S. (1848) 1965. *Principles of Political Economy with Some of Their Applications to Social Philosophy.* V. W. Bladen and J. M. Robson, Eds. Univ. of Toronto Press, Toronto.
39. I chose $P(0) = 0.252$ billion, the estimated population in A.D. 1, and $L = 3.7$ billion, the estimated population size in 1970, then varied the remaining parameters r and $K(0)$ numerically to obtain a close fit by using "fmins" of MATLAB version 4.1, based on the Nelder-Meade simplex search; see MATLAB Reference Guide (Mathworks, Natick, MA, August 1992), pp.208–10. The criterion of goodness of fit was

$$\sum(\text{observed population} - \text{calculated population})^2 / (\text{observed population})$$

summed over the dates of observation shown in the figure.
40. Europe grew technologically and economically for a millennium before the Industrial Revolution; England developed economically from the 12th century onward. In the 13th century, English forests were cleared, swamps drained, and new lands exploited for cultivation; yields

improved as a result of liming, plowing straw ash into the field, and planting new varieties of seeds. At the same time, the mining and smelting of tin, lead, and iron; the manufacture of pottery; and the production of salt and wool all increased. Additional surges of economic development occurred in the 16th century [C. M. Cipolia, *Before the Industrial Revolution: European Society and Economy 1000–1700* (Norton, New York, ed. 3, 1994), pp.137–159; R. G. Wilkinson, *Poverty and Progress: An Ecological Perspective on Economic Development* (Praeger, New York, 1973), D. L. Hardesty, *Ecological Anthropology* (Wiley, New York, 1977), pp. 209–10].
41.Ausubel, J. H. 1993. The Sciences 33:14. See also (48.
42. Bongaarts, J. 1994. Science 263:771.
43. Gertler P. J. and J. W. Molyneaux. 1994. Demogr. 31:33.
44. Repetto, R., ed. 1985. *The Global Possible: Resources, Development, and the New Century.* Yale Univ. Press, New Haven. Pearce, D. W. and J. J. 1993. *Warford, World Without End: Economics, Environment, and Sustainable Development.* Oxford Univ. Press, New York.
45. Hardin, G. 1968. Science 162:1243.
46. Lee, R. D. 1991. In: *Resources, Environment, and Population: Present Knowledge, Future Options.* Davis, K. and M. S. Bernstam, eds. Oxford Univ, Press, New York. Pp. 315–332. Earlier discussions of the negative externalities of childbearing include H. F. Dorn, Science 135,283 (1962); P. E. Sorenson, in: *Population, Environment and People,* N. Hinrichs, Ed. (McGraw-Hill, New York, 1971), pp. 113–121; and P. Dasgupta and R, J Willis, in: *Population Growth and Economic Development: Issues and Evidence,* D. G. Johnson and R. D. Lee, Eds., Univ. of Wisconsin Press, Madison, 1987, pp: 631–702.
47. Durand, J. 1977. Popul. Dev. Rev. 3:253.
48. Waggoner, P. E. 1994. "How much land can ten billion people spare for nature?" Task Force Report 121, Council for Agricultural Science and Technology, Ames, IA.
49. Austin, A. L. and J. W. Brewer. 1971. Technol. Forecasting Soc. Change 3:23.
50. World Resources Institute. 1994. *World Resources 1994–95.* Oxford Univ. Press.
51. House, P. W. and E. R. Williams. 1975. *The Carrying Capacity of a Nation: Growth and the Quality of Life.* Lexington Books, Lexington, MA.
52. Meadows, D. H., D. L. Meadows, J. Randers. 1992. *Beyond the Limits: Global Collapse or a Sustainable Future.* Earthscan Publications, London.
53. McEvedy, C. and R. Jones. 1978. Atlas of World Population History (Viking Penguin, New York.
Biraben, J.-N. 1979. Population 34:13. See also (11, 47).
54. Cipolla, C. M. 1974. *Economic History of World Population.* Penguin, Harmondsworth, England, ed. 6. Pp. 56–59. See also (50), pp. 332–334.
55. Data are from (14). The estimate by J. H. Fremlin [New Sci. 24, 285 (1964)] would be off the scale and is omitted.
56. J. E. C. acknowledges with thanks NSF grant BSR9207293 and the hospitality of Mr. and Mrs. W. T. Golden. J. H. Ausubel, G. M. Feeney, R. B. Gailagher, S. Horiuchi, and R. M. May reviewed and improved previous drafts.

Introduction to
Joseph H. Connell
The Influence of Interspecific Competition and Other Factors on the Distribution of the Barnacle Chthamalus stellatus

by Anthony R. Ives

For many people, science and experimentation are synonymous. Nonetheless, many ecological studies do not contain experiments but instead consist of detailed observations. For example, to construct a food web of who eats whom in a community, detailed observations can be employed to determine the diets of each species. Careful comparisons among different species or different ecosystems may also be used to give insight into ecological processes. The power of careful observations and comparisons is illustrated by Charles Darwin; *The Origin of Species* relies almost entirely on observations and comparisons to support the idea of evolution through natural selection.

But experimentation can be a powerful tool in ecological studies. The main advantage of an experimental approach is that the experimenter can control how things differ. This control can reveal the causes of ecological patterns, rather than just the correlations observed in ecological patterns.

The study by Joe Connell (1961) is one of the classic early experimental studies in ecology. Connell observed that the barnacle *Chthamalus stellatus* occurred on the rocky coast of Scotland only in the highest areas of the intertidal zone, even though juveniles settled everywhere above the mid-tide mark. There are three possible causes of this pattern: (1) some environmental factor that varies with height above the mid-tide mark, such as wave action, (2) predation by the snail *Thais lapillus*, and (3) competition with the barnacle *Balanus balanoides*. To test among these, Connell attached small rocks with *C. stellatus* juveniles at differing heights on the shore. He covered some of the rocks with cages that excluded the predator *T. lapillus*, and on one side of each rock he removed all of the competing *B. balanoides*. By thus varying all three possible factors simultaneously, Connell showed that competition with *B. balanoides* had the greatest impact on the survival of *C. stellatus* in the lower intertidal areas, while predation by *T. lapillus* had a smaller yet noticeable effect. Although abiotic factors did not seem to play a role in setting the lower limit of the *C. stellatus* distribution, desiccation set the upper limit. Thus, all three factors (environment, predation, and competition) played some role in explaining the distribution of *C. stellatus*.

The power of this experimental approach is that it separates the effects of factors by manipulating each independently. The difficulty with trying to understand the distribution of *C. stellatus* without an experiment is that the environment, predation pressure from *T. lapillus*, and competition with *B. balanoides* all change together along the gradient from the mid-tide mark to the top of the intertidal. By manipulating all three factors independently, Connell was able to identify the individual effects of each. Furthermore, interactions between factors were apparent. In lower areas where predation by *T. lapillus* was greater on both *C. stellatus* and its competitor, *B. balanoides*, the competitive effect of *B. balanoides* on *C. stellatus* decreased simply because there were fewer *B. balanoides*.

The Influence of Interspecific Competition and Other Factors on the Distribution of the Barnacle *Chthamalus stellatus*

Joseph H. Connell

Original publication in *Ecology* (1961), 42(4):710–723. Reprinted with permission.

Introduction

Most of the evidence for the occurrence of interspecific competition in animals has been gained from laboratory populations. Because of the small amount of direct evidence for its occurrence in nature, competition has sometimes been assigned a minor role in determining the composition of animal communities.

Indirect evidence exists, however, which suggests that competition may sometimes be responsible for the distribution of animals in nature. The range of distribution of a species may be decreased in the presence of another species with similar requirements (Beauchamp and Ullyott 1932, Endean, Kenny and Stephenson 1956). Uniform distribution in space is usually attributed to intraspecies competition (Holme 1950, Clark and Evans 1954). When animals with similar requirements, such as 2 or more closely related species, are found coexisting in the same area, careful analysis usually indicates that they are not actually competing with each other (Lack 1954, Mac Arthur 1958).

In the course of an investigation of the animals of an intertidal rocky shore I noticed that the adults of 2 species of barnacles occupied 2 separate horizontal zones with a small area of overlap, whereas the young of the species from the upper zone were found in much of the lower zone. The upper species, *Chthamalus stellatus* (Poll) thus settled but did not survive in the lower zone. It seemed probable that this species was eliminated by the lower one, *Balanus balanoides* (L), in a struggle for a common requisite which was in short supply. In the rocky intertidal region, space for attachment and growth is often extremely limited. This paper is an account of some observations and experiments designed to test the hypothesis that the absence in the lower zone of adults of *Chthamalus* was due to interspecific competition with *Balanus* for space. Other factors which may have influenced the distribution were also studied. The study was made at Millport, Isle of Cumbrae, Scotland.

I would like to thank Prof. C. M. Yonge and the staff of the Marine Station, Millport, for their help, discussions and encouragement during the course of this work. Thanks are due to the following for their critical reading of the manuscript: C. S. Elton, P. W. Frank, G. Hardin, N. G. Hairston. E. Orias, T. Park and his students, and my wife.

Distribution of the Species of Barnacles

The upper species, *Chthamalus stellatus*, has its center of distribution in the Mediterranean; it reaches its northern limit in the Shetland

Islands, north of Scotland. At Millport, adults of this species occur between the levels of mean high water of neap and spring tides (M.H.W.N. and M.H.W.S.: see Figure 5 and Table I). In southwest England and Ireland, adult *Chthamalus* occur at moderate population densities throughout the intertidal zone, more abundantly when *Balanus balanoides* is sparse or absent (Southward and Crisp 1954, 1956). At Millport the larvae settle from the plankton onto the shore mainly in September and October; some additional settlement may occur until December. The settlement is most abundant between M.H.W.S. and mean tide level (M.T.L.) in patches of rock surface left bare as a result of the mortality of *Balanus*, limpets, and other sedentary organisms. Few of the *Chthamalus* that settle below M.H.W.N. survive, so that adults are found only occasionally at these levels.

Balanus balanoides is a boreal-arctic species, reaching its southern limit in northern Spain. At Millport it occupies almost the entire intertidal region, from mean low water of spring tides (M.L.W.S.) up to the region between M.H.W.N. and M.H.W.S. Above M.H.W.N. it occurs intermingled with *Chthamalus* for a short distance. *Balanus* settles on the shore in April and May, often in very dense concentrations (see Table IV).

The main purpose of this study was to determine the cause of death of those *Chthamalus* that settled below M.H.W.N. A study which was being carried on at this time had revealed that physical conditions, competition for space, and predation by the snail *Thais lapillus* L. were among the most important causes of mortality of *Balanus balanoides*. Therefore, the observations and experiments in the present study were designed to detect the effects of these factors on the survival of *Chthamalus*.

Methods

Intertidal barnacles are very nearly ideal for the study of survival under natural conditions. Their sessile habit allows direct observation of the survival of individuals in a group whose positions have been mapped. Their small size and dense concentrations on rocks exposed at intervals make experimentation feasible. In addition, they may be handled and transplanted without injury on pieces of rock, since their opercular plates remain closed when exposed to air.

The experimental area was located on the Isle of Cumbrae in the Firth of Clyde, Scotland. Farland Point, where the study was made, comprises the southeast tip of the island; it is exposed to moderate wave action. The shore rock consists mainly of old red sandstone, arranged in a series of ridges, from 2 to 6 ft high, oriented at right angles to the shoreline. A more detailed description is given by Connell (1961). The other barnacle species present were *Balanus crenatus* Brug and *Verruca stroemia* (O. F. Muller), both found in small numbers only at and below M.L.W.S.

To measure the survival of *Chthamalus*, the positions of all individuals in a patch were mapped. All barnacles which were empty or missing at the next examination of this patch must have died in the interval, since emigration is impossible. The mapping was done by placing thin glass plates (lantern slide cover glasses, 10.7 x 8.2 cm, area 87.7 cm^2) over a patch of barnacles and marking the position of each *Chthamalus* on it with glass-marking ink. The positions of the corners of the plate were marked by drilling small holes in the rock. Observations made in subsequent censuses were noted on a paper copy of the glass map.

The study areas were chosen by searching for patches of *Chthamalus* below M.H.W.N. in a stretch of shore about 50 ft long. When 8 patches had been found, no more were looked for. The only basis for rejection of an area in this search was that it contained fewer than 50 *Chthamalus* in an area of about $^1/_{10}$ m^2. Each numbered area consisted of one or more glass maps located in the $^1/_{10}$ m^2. They were mapped in March and April, 1954, before the main settlement of *Balanus* began in late April.

Very few *Chthamalus* were found to have settled below mid-tide level. Therefore pieces of rock bearing *Chthamalus* were removed from levels above M.H.W.N. and transplanted to and below M.T.L. A hole was drilled through each piece: it was then fastened to the rock by a stainless steel screw driven into a plastic screw anchor fitted into a hole drilled into the rock. A hole $^1/_4$" in diameter and 1" deep was found to be satisfactory. The screw could be removed and replaced repeatedly and only one stone was lost in the entire period.

For censusing, the stones were removed during a low tide period, brought to the laboratory for examination, and returned before the tide rose again. The locations and arrange-

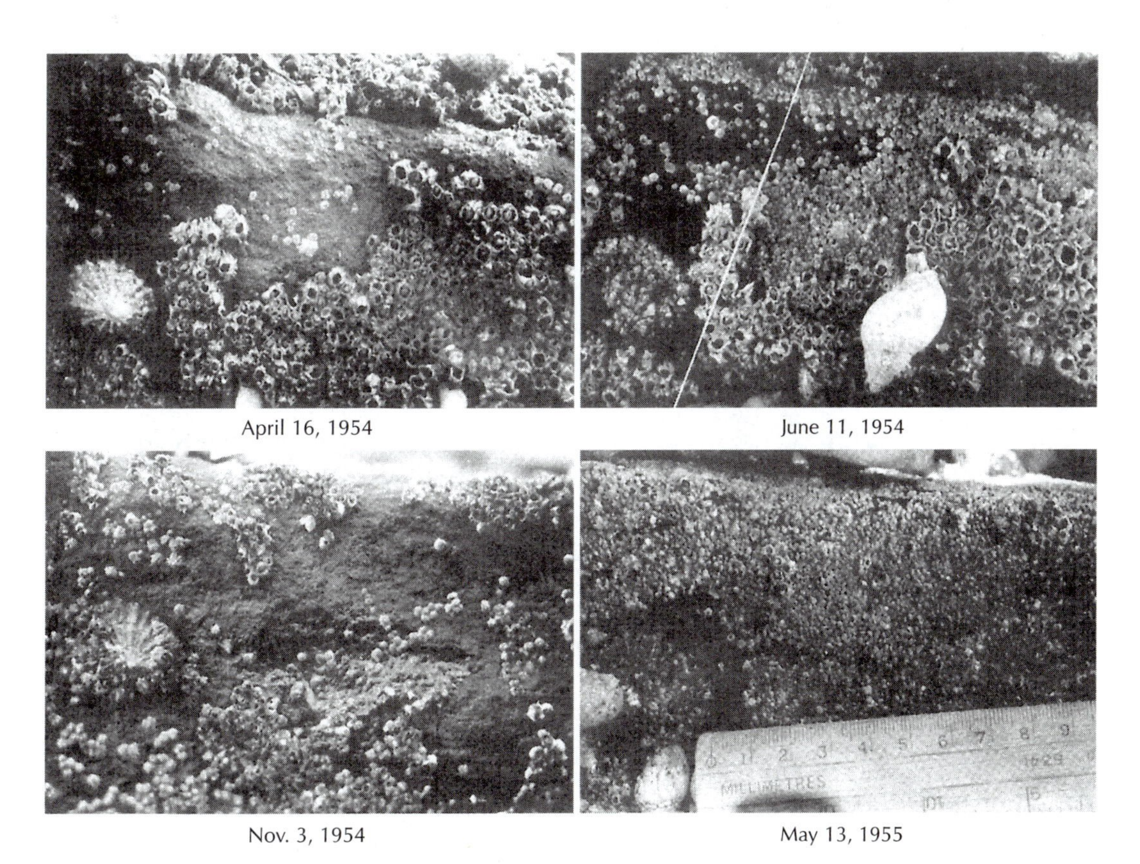

Figure 1. [Four views of] Area 7b. In the first photograph the large barnacles are *Balanus,* the small ones scattered in the bare patch, *Chthamalus.* The white line on the second photograph divides the undisturbed portion (right) from the portion from which *Balanus* were removed (left). A limpet, *Patella vulgata,* occurs on the left, and predatory snails, *Thais lapillus,* are visible.

Table I. Description of experimental areas*

Area no.	Height in ft from M.T.L.	% of time sub-merged	Population Density: no./cm² in June, 1964 — *Chthamalus*, autumn 1953 settlement: Undisturbed portion	*Chthamalus*, autumn 1953 settlement: Portion without *Balanus*	All barnacles, undisturbed portion	Remarks
M.H.W.S.	+4.9	4	–	–	–	–
1	+4.2	9	2.2	–	19.2	Vertical, partly protected
2	+3.5	16	6.2	4.2	–	Vertical, wave beaten
M.H.W.N.	+3.1	21	–	–	–	–
3a	+2.2	30	0.6	0.6	30.9	Horizontal, wave beaten
3b	+2.2	30	0.5	0.7	29.2	Horizontal, wave beaten
4	+1.4	38	1.9	0.6	–	30° to vertical, partly protected
5	+1.4	38	2.4	1.2	–	30° to vertical, partly protected
6	+1.0	42	1.1	1.9	38.2	Horizontal, top of a boulder, partly protected
7a	+0.7	44	1.3	2.0	49.3	Vertical, protected
7b	+0.7	44	2.3	2.0	51.7	Vertical, protected
11a	0.0	50	1.0	0.6	32.0	Vertical, protected
11b	0.0	50	0.2	0.3		Vertical, protected
12a	0.0	100	1.2	1.2	18.8	Horizontal, immersed in tide pool
12b	0.0	100	0.8	0.9		Horizontal, immersed in tide pool
13a	−1.0	58	4.9	4.1	29.5	Vertical, wave beaten
13b	−1.0	58	3.1	2.4	–	Vertical, wave beaten
14a	−2.5	71	0.7	1.1	–	45° angle, wave beaten
14b	−2.5	71	1.0	1.0	–	45° angle, wave beaten
M.L.W.N.	−3.0	77	–	–	–	–
M.L.W.S.	−5.1	96	–	–	–	–
15	+1.0	42	32.0	–	–	*Chthamalus* of autumn, 1954 settlement; densities of Oct., 1954.
7b	+0.7	44	6.0	3.7	–	

*The letter "a" following an area number indicates that this area was enclosed by a cage; "b" refers to a closely adjacent area which was not enclosed. All faced either east or south except 7a and 7b, which faced north.

ments of each area are given in Table I; the transplanted stones are represented by areas 11 to 15.

The effect of competition for space on the survival of *Chthamalus* was studied in the following manner: After the settlement of *Balanus* had stopped in early June, having reached densities of 49/cm^2 on the experimental areas (Table I) a census of the surviving *Chthamalus* was made on each area (see Figure 1). Each map was then divided so that about half of the number of *Chthamalus* were in each portion.

One portion was chosen (by flipping a coin), and those *Balanus* which were touching or immediately surrounding each *Chthamalus* were carefully removed with a needle: the other portion was left untouched. In this way it was possible to measure the effect on the survival of *Chthamalus* both of intraspecific competition alone and of competition with *Balanus.* It was not possible to have the numbers or population densities of *Chthamalus* exactly equal on the 2 portions of each area. This was due to the fact that, since *Chthamalus* often occurred in groups, the *Balanus* had to be removed from around all the members of a group to ensure that no crowding by *Balanus* occurred. The densities of *Chthamalus* were very low, however, so that the slight differences in density between the 2 portions of each area can probably be disregarded; intraspecific crowding was very seldom observed. Censuses of the *Chthamalus* were made at intervals of 4–6 weeks during the next year; notes were made at each census of factors such as crowding, undercutting or smothering which had taken place since the last examination. When necessary, *Balanus* which had grown until they threatened to touch the *Chthamalus* were removed in later examinations.

To study the effects of different degrees of immersion, the areas were located throughout the tidal range, either in situ or on transplanted stones, as shown in Table I. Area 1 had been under observation for 1½ years previously. The effects of different degrees of wave shock could not be studied adequately in such a small area of shore but such differences as existed are listed in Table I.

The effects of the predatory snail, *Thais lapillus,* (synonymous with *Nucella* or *Purpura,* Clench 1947), were studied as follows: Cages of stainless steel wire netting, 8 meshes per inch, were attached over some of the areas. This mesh has an open area of 60% and previous work (Cornell 1961) had shown that it did not inhibit growth or survival of the barnacles. The cages were about 4 x 6 inches, the roof was about an inch above the barnacles and the sides were fitted to the irregularities of the rock. They were held in place in the same manner as the transplanted stones. The transplanted stones were attached in pairs, one of each pair being enclosed in a cage (Table I).

These cages were effective in excluding all but the smallest *Thais.* Occasionally small *Thais,* ½ to 1 cm in length, entered the cages through gaps at the line of juncture of netting and rock surface. In the concurrent study of *Balanus* (Connell 1961), small *Thais* were estimated to have occurred inside the cages about 3% of the time.

All the areas and stones were established before the settlement of *Balanus* began in late April, 1954. Thus the *Chthamalus* which had settled naturally on the shore were then of the 1953 year class and all about 7 months old. Some *Chthamalus* which settled in the autumn of 1954 were followed until the study was ended in June, 1955. In addition, some adults which, judging from their large size and the great erosion of their shells, must have settled in 1952 or earlier, were present on the transplanted stones. Thus records were made of at least 3 year-classes of *Chthamalus.*

Results

The Effects of Physical Factors

In Figures 2 and 3, the dashed line indicates the survival of *Chthamalus* growing without contact with *Balanus.* The suffix "a" indicates that the area was protected from *Thais* by a cage. In the absence of *Balanus* and *Thais,* and protected by the cages from damage by water-borne

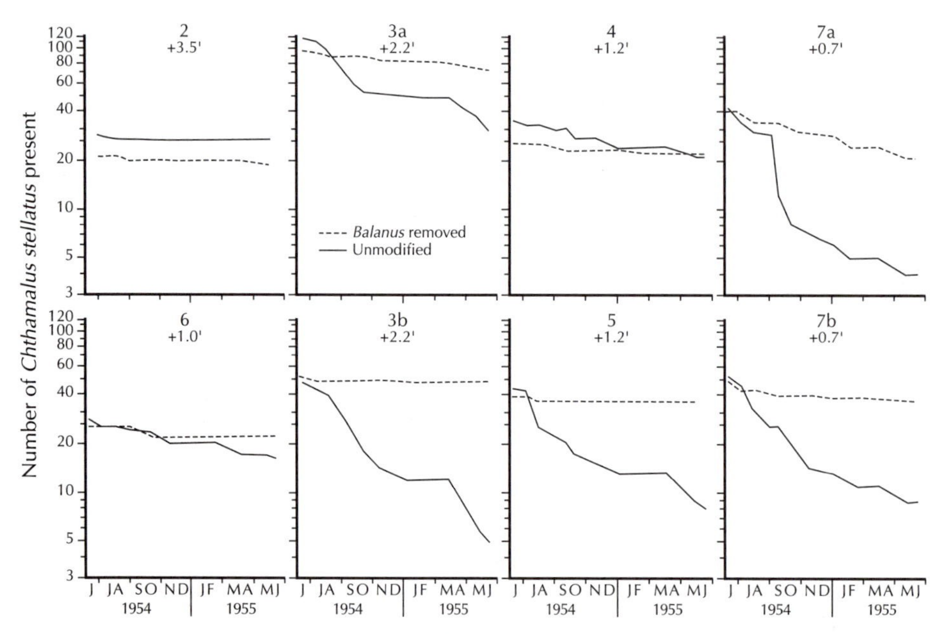

Figure 2. Surviorship curves of *Chthamalus stellatus* which had settled naturally on the shore in the autumn of 1953. Areas designated "a" were protected from predation by cages. In each area the survival of *Chthamalus* growing without contact with *Balanus* compared to that in the undisturbed area. For each area the vertical distance in feet from M.T.L. is shown.

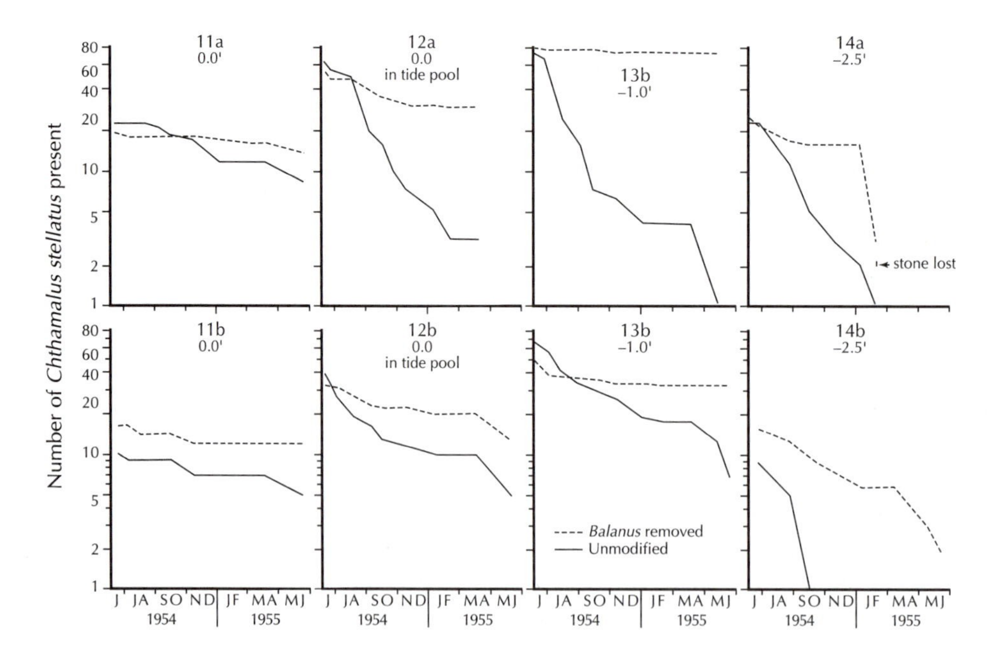

Figure 3. Survivorship curves of *Chthamalus stellatus* on stones transplanted from high levels. These had settled in the autumn of 1953; the arrangement is the same as that of Figure 2.

objects, the survival of *Chthamalus* was good at all levels. For those which had settled normally on the shore (Figure 2), the poorest survival was on the lowest area, 7a. On the transplanted stones (Figure 3, area 12), constant immersion in a tide pool resulted in the poorest survival. The reasons for the trend toward slightly greater mortality as the degree of immersion increased are unknown. The amount of attached algae on the stones in the tide pool was much greater than on the other areas. This may have reduced the flow of water and food or have interfered directly with feeding movements. Another possible indirect effect of increased immersion is the increase in predation by the snail, *Thais lapillus,* at lower levels.

Chthamalus is tolerant of a much greater degree of immersion than it normally encounters. This is shown by the survival for a year on area 12 in a tide pool, together with the findings of Fischer (1928) and Barnes (1956a), who found that *Chthamalus* withstood submersion for 12 and 22 months. respectively. Its absence below M.T.L. can probably be ascribed either to a lack of initial settlement or to poor survival of newly settled larvae. Lewis and Powell (1960) have suggested that the survival of *Chthamalus* may be favored by increased light or warmth during emersion in its early life on the shore. These conditions would tend to occur higher on the shore in Scotland than in southern England.

The effects of wave action on the survival of *Chthamalus* are difficult to assess. Like the degree of immersion, the effects of wave action may act indirectly. The areas 7 and 12, where relatively poor survival was found, were also the areas of least wave action. Although *Chthamalus* is usually abundant on wave-beaten areas and absent from sheltered bays in Scotland, Lewis and Powell (1960) have shown that in certain sheltered bays it may be very abundant. Hatton (1938) found that in northern France. settlement and growth rates were greater in wave-beaten areas at M.T.L., but, at M.H.W.N., greater in sheltered areas.

At the upper shore margins of distribution *Chthamalus* evidently can exist higher than *Balanus* mainly as a result of its greater tolerance to heat and/or desiccation. The evidence for this was gained during the spring of 1955. Records from a tide and wave guage operating at this time about one-half mile north of the study area showed that a period of neap tides had coincided with an unusual period of warm calm weather in April so that for several days no water, not even waves, reached the level of Area 1. In the period between the censuses of February and May, *Balanus* aged one year suffered a mortality of 92%, those 2 years and older, 51%. Over the same period the mortality of *Chthamalus* aged 7 months was 62%, those 1 1/2 years and older, 2%. Records of the survival of *Balanus* at several levels below this showed that only those *Balanus* in the top quarter of the intertidal region suffered high mortality during this time (Cornell 1961).

Competition for Space

At each census, notes were made for individual barnacles of any crowding which had occurred since the last census. Thus when one barnacle started to grow up over another, this fact was noted and at the next census 4–6 weeks later the progress of this process was noted. In this way a detailed description was built up of these gradually occurring events.

Intraspecific competition leading to mortality in *Chthamalus* was a rare event. For areas 2 to 7, on the portions from which *Balanus* had been removed, 167 deaths were recorded in a year. Of these, only 6 could be ascribed to crowding between individuals of *Chthamalus.*

Table II. The causes of mortality of *Chthamalus stellatus* of the 1953 year group on the undisturbed portions of each area.

Area no.	Height in ft from M.T.L.	No. at start	No. of deaths in the next year	Percentage of Deaths Resulting from: Smothering by *Balanus*	Undercutting by *Balanus*	Other crowding by *Balanus*	Unknown causes
2	+3.5	28	1	0	0	0	100
3a	+2.2	111	81	61	6	10	93
3b	+2.2	47	42	57	5	2	36
4	+1.4	34	14	21	14	0	65
5	+1.4	43	35	11	11	3	75
6	+1.0	27	11	9	0	0	91
7a	+0.7	42	38	21	16	53	10
7b	+0.7	51	42	24	10	10	56
11a	0.0	21	13	64	8	0	38
11b	0.0	10	5	40	0	0	60
12a	0.0	60	57	19	33	7	41
12b	0.0	39	34	9	18	3	70
13a	−1.0	71	70	19	24	3	54
13b	−1.0	69	62	18	8	3	71
14a	−2.6	22	21	24	42	10	24
14b	−2.6	9	9	0	0	0	100
Total, 2–7	–	383	264	37	9	16	38
Total, 11–14	–	301	271	19	21	4	56

On the undisturbed portions no such crowding was observed. This accords with Hatton's (1938) observation that he never saw crowding between individuals of *Chthamalus* as contrasted to its frequent occurrence between individuals of *Balanus*.

Interspecific competition between *Balanus* and *Chthamalus* was, on the other hand, a most important cause of death of *Chthamalus*. This is shown both by the direct observations of the process of crowding at each census and by the differences between the survival curves of *Chthamalus* with and without *Balanus*. From the periodic observations it was noted that after the first month on the undisturbed portions of areas 3 to 7 about 10% of the *Chthamalus* were being covered as *Balanus* grew over them; about 3% were being undercut and lifted by growing *Balanus*; a few had died without crowding. By the end of the 2nd month about 20% of the *Chthamalus* were either wholly or partly covered by *Balanus*; about 4% had been undercut; others were surrounded by tall *Balanus*. These processes continued at a lower rate in the autumn and almost ceased during the later winter. In the spring *Balanus* resumed growth and more crowding was observed.

In Table II, these observations are summarized for the undistributed portions of all the areas. Above M.T.L., the *Balanus* tended to overgrow the *Chthamalus*, whereas at the lower levels, undercutting was more common. This same trend was evident within each group of areas, undercutting being more prevalent on area 7 than on area 3, for example. The faster growth of *Balanus* at lower levels (Hatton 1938, Barnes and Powell 1953) may have resulted in more undercutting. When *Chthamalus* was completely covered by *Balanus* it was recorded as dead; even though death may not have occurred immediately, the buried barnacle was obviously not a functioning member of the population.

In Table II under the term "other crowding" have been placed all instances where *Chthamalus* were crushed laterally between 2 or more *Balanus*, or where *Chthamalus* disappeared in an interval during which a dense population of *Balanus* grew rapidly. For example; in area 7a the *Balanus*, which were at the high population density of 48 per cm^2, had no room to expand except upward and the barnacles very quickly grew into the form of tall cylinders or cones with the diameter of the opercular opening greater than that of the base. It was obvious that extreme crowding occurred under these circumstances, but the exact cause of the mortality of the *Chthamalus* caught in this crush was difficult to ascertain.

In comparing the survival curves of Figures 2 and 3 within each area it is evident that *Chthamalus* kept free of *Balanus* survived better than those in the adjacent undisturbed areas on all but areas 2 and 14a. Area 2 was in the zone where adults of *Balanus* and *Chthamalus* were normally mixed; at this high level *Balanus* evidently has no influence on the survival of *Chthamalus*. On Stone 14a, the survival of *Chthamalus* without *Balanus* was much better until January, when a starfish, *Asterias rubens* L., entered the cage and ate the barnacles.

Much variation occurred on the other 14 areas. When the *Chthamalus* growing without contact with *Balanus* are compared with those on the adjacent undisturbed portion of the area, the survival was very much better on 10 areas and moderately better on 4. In all areas, some *Chthamalus* in the undisturbed portions escaped severe crowding. Sometimes no *Balanus* happened to settle close to a *Chthamalus*, or sometimes those which did died soon after settlement. In some instances, *Chthamalus* which were being undercut by *Balanus* attached themselves to the *Balanus* and so survived. Some *Chthamalus* were partly covered by *Balanus* but still survived. It seems probable that in the 4 areas, nos. 4, 6, 11a. and 11b, where *Chthamalus* survived well in the presence of *Balanus*, a higher proportion of the *Chthamalus* escaped death in one of these ways.

The fate of very young *Chthamalus* which settled in the autumn of 1954 was followed in detail in 2 instances, on stone 15 and area 7b. The *Chthamalus* on stone 15 had settled in an irregular space surrounded by large *Balanus*. Most of the mortality occurred around the edges of the space as the *Balanus* undercut and lifted the small *Chthamalus* nearby. The following is a tabulation of all the deaths of young *Chthamalus* between Sept. 30, 1954 and Feb. 14, 1955 on Stone 15, with the associated situations:

Lifted by *Balanus*	29
Crushed by *Balanus*	4
Smothered by *Balanus* and *Chthamalus*	2
Crushed between *Balanus* and *Chthamalus*	1
Lifted by *Chthamalus*	1
Crushed between two other *Chthamalus*	1
Unknown	3

This list shows that crowding of newly settled *Chthamalus* by older *Balanus* in the autumn mainly takes the form of undercutting, rather than of smothering as was the case in the spring. The reason for this difference is probably that the *Chthamalus* are more firmly attached in the spring so that the fast growing young *Balanus* grow up over them when they make contact. In the autumn the reverse is the case, the *Balanus* being firmly attached, the *Chthamalus* weakly so.

Although the settlement of *Chthamalus* on Stone 15 in the autumn of 1954 was very dense, $32/cm^2$ so that most of them were touching another, only 2 of the 41 deaths were caused by intraspecific crowding among the *Chthamalus*. This is in accord with the findings from the 1953 settlement of *Chthamalus*.

TABLE III. Comparison of the mortality rates of young and older *Chthamalus stellatus* on transplanted stones.

Stone No	Shore level	Treatment	Number of *Chthamalus* present in June, 1954: 1953 year group	Number of *Chthamalus* present in June, 1954: 1952 or older year group	% mortality over one year (or for 6 months for 14a) of *Chthamalus*: 1953 year group	% mortality over one year (or for 6 months for 14a) of *Chthamalus*: 1952 or older year group
13b	1.0 ft below MTL	*Balanus* removed	51	3	35	0
		Undisturbed	69	16	90	31
12a	MTL, in a tide pool, caged	*Balanus* removed	50	41	44	37
		Undisturbed	60	31	95	71
14a	2.5 ft below MTL, caged	*Balanus* removed	25	45	40	36
		Undisturbed	22	8	86	75

The mortality rates for the young *Chthamalus* on area 7b showed seasonal variations. Between October 10, 1954 and May 15, 1955 the relative mortality rate per day x 100 was 0.14 on the undisturbed area and 0.13 where *Balanus* had been removed. Over the next month, the rate increased to 1.49 on the undisturbed area and 0.22 where *Balanus* was absent. Thus the increase in mortality of young *Chthamalus* in late spring was also associated with the presence of *Balanus*.

Some of the stones transplanted from high to low levels in the spring of 1954 bore adult *Chthamalus*. On 3 stones, records were kept of the survival of these adults, which had settled in the autumn of 1952 or in previous years and were at least 20 months old at the start of the experiment. Their mortality is shown in Table III; it was always much greater when *Balanus* was not removed. On 2 of the 3 stones this mortality rate was almost as high as that of the younger group. These results suggest that any *Chthamalus* that managed to survive the competition for space with *Balanus* during the first year would probably be eliminated in the 2nd year.

Censuses of *Balanus* were not made on the experimental areas. However, on many other areas in the same stretch of shore the survival of *Balanus* was being studied during the same period (Connell 1961). In Table IV some mortality rates measured in that study are listed; the *Balanus* were members of the 1954 settlement at population densities and shore levels similar to those of the present study. The mortality rates of *Balanus* were about the same as those of *Chthamalus* in similar situations except at the highest level, area 1, where *Balanus* suffered much greater mortality than *Chthamalus*. Much of this mortality was caused by intraspecific crowding at all levels below area 1.

In the observations made at each census it appeared that *Balanus* was growing faster than *Chthamalus*. Measurements of growth rates of the 2 species were made from photographs of the areas taken in June and November, 1954. Barnacles growing free of contact with each other were measured; the results are given in Table V.

Table IV. Comparison of annual mortality rates of *Chthamalus stellatus* and *Balanus balanoides**

Area no.	Height in ft from M.T.L.	Population density: no./cm^2 June, 1954	% mortality in the next year
	Chthamalus stellatus autumn 1953 settlement		
1	+4.2	21	17
3a	+2.2	31	72
3b	+2.2	29	89
6	+1.0	38	41
7a	+0.7	49	90
7b	+0.7	52	82
11a	0.0	32	62
13a	−1.0	29	99
12a	(tide pool)	19	95
	Balanus balanoides spring 1954 settlement		
1 (top)	+4.2	21	99
1:Middle Cage 1	+2.1	85	92
1:Middle Cage 2	+2.1	25	77
1:Low Cage 1	+1.5	26	88
Stone 1	−0.9	26	86
Stone 2	−0.9	68	94

*Population density includes both species. The mortality rates of *Chthamalus* refer to those on the undisturbed portions of each area. The data and area designations for *Balanus* were taken from Connell (1961); the present area 1 is the same as that designated 1 (top) in that paper.

The growth rate of *Balanus* was greater than that of *Chthamalus* in the experimental areas; this agrees with the findings of Hatton (1938) on the shore in France and of Barnes (1956a) for continual submergence on a raft at Millport.

After a year of crowding, the average population densities of *Balanus* and *Chthamalus* remained in the same relative proportion as they had been at the start, since the mortality rates were about the same. However, because of its faster growth, *Balanus* occupied a relatively greater area and, presumably, possessed a greater biomass relative to that of *Chthamalus* after a year.

The faster growth of *Balanus* probably accounts for the manner in which *Chthamalus* were crowded by *Balanus*. It also accounts for the sinuosity of the survival curves of *Chthamalus* growing in contact with *Balanus*. The mortality rate of these *Chthamalus*, as indicated by the slope of the curves in Figures 2 and 3, was greatest in summer, decreased in winter and increased again in spring. The survival curves of *Chthamalus* growing without contact with *Balanus* do not show these seasonal variations which, therefore, cannot be the result of the direct action of physical factors such as temperature, wave action or rain.

Seasonal variations in growth rate of *Balanus* correspond to these changes in mortality rate of *Chthamalus*. In Figure 4 the growth of *Balanus* throughout the year as studied on an intertidal panel at Millport by Barnes and Powell (1953), is compared to the survival of

Table V. Growth rates of *Chthamalus stellatus* and *Balanus balanoides*. Measurements were made of uncrowded individuals on photographs of areas 3a, 3b and 7b. Those of *Chthamalus* were made on the same individuals on both dates; of *Balanus* representative samples were chosen.

	Chthamalus		Balanus	
	No. measured	Average size, mm	No. measured	Average size, mm
June 11, 1954	25	2.49	39	1.87
November 3,1954	25	4.24	27	4.83
Average size in the interval	3.36		3.35	
Absolute growth rate per day x 100	1.21		2.04	

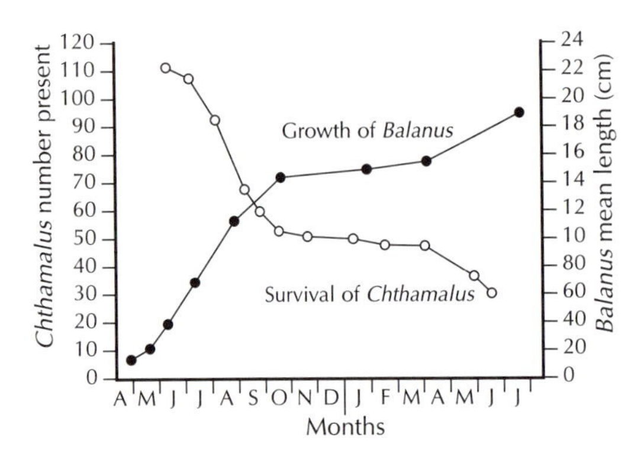

Figure 4. A comparison of the seasonal changes in the growth of *Balanus balanoides* and in the survival of *Chthamalus stellatus* being crowded by *Balanus*. The growth of *Balanus* was that of panel 3, Barnes and Powell (1953), just above M.T.L. on Keppel Pier, Millport, during 1951–52. The *Chthamalus* were on area 3a of the present study, one-half mile south of Keppell Pier, during 1954–55.

Chthamalus at about the same intertidal level in the present study. The increased mortality of *Chthamalus* was found to occur in the same seasons as the increases in the growth rate of *Balanus*. The correlation was tested using the Spearman rank correlation coefficient. The absolute increase in diameter of *Balanus* in each month, read from the curve of growth, was compared to the percentage mortality of *Chthamalus* in the same month. For the 13 months in which data for *Chthamalus* was available, the correlation was highly significant, $P = .01$.

From all these observations it appears that the poor survival of *Chthamalus* below M.H.W.N. is a result mainly of crowding by dense populations of faster growing *Balanus*.

At the end of the experiment in June, 1955, the surviving *Chthamalus* were collected from 5 of the areas. As shown in Table VI, the average size was greater in the *Chthamalus* which had grown free of contact with *Balanus*; in every case the difference was significant ($P < .01$, Mann-Whitney U test Siegel 1956). The survivors on the undisturbed areas were often misshapen in some cases as a result of being lifted onto the side of an undercutting *Balanus*. Thus the smaller size of these barnacles may have been due to disturbances in the normal pattern of growth while they were being crowded.

These *Chthamalus* were examined for the presence of developing larvae in their mantle cavities. As shown in Table VI, in every area the proportion of the uncrowded *Chthamalus* with larvae was equal to or more often slightly greater than on the crowded areas. The reason for this may be related to the smaller size of the crowded *Chthamalus*. It is not due to separation, since *Chthamalus* can self-fertilize (Barnes and Crisp 1956). Moore (1935) and Barnes (1953) have shown that the number of larvae in an individual of *Balanus balanoides* increases with increase in volume of the parent. Comparison of the cube of diameter, which is proportional to the volume, of *Chthamalus* with and without *Balanus* shows that the volume may be decreased to $\frac{1}{4}$ normal when crowding occurs. Assuming that the relation between larval numbers and volume in *Chthamalus* is similar to that of *Balanus*, a decrease in both frequency of occurrence and abundance of larvae in *Chthamalus* results from competition with *Balanus*. Thus the process described this paper satisfies both aspects of interspecific competition as defined by Elton and Miller (1954): "in which one species affects the population of another by a process of interference i.e., by reducing the reproductive efficiency or increasing the mortality of its competitor."

The Effect of Predation by Thais

Cages which excluded *Thais* had been attached 6 areas (indicated by the letter "a" following number of the area). Area 14 was not included in the following analysis since many

TABLE VI. The effect of crowding on the size and presence of larvae in *Chthamalus stellatus,* collected in June, 1955.

Area	Treatment	Level, feet above M.T.L.	Number of *Chthamalus*	Diameter in mm: Average	Diameter in mm: Range	% of individuals which had larvae in mantle cavity
3a	Undisturbed	2.2	18	3.5	2.7–4.6	61
3a	*Balanus* removed	2.2	50	4.1	3.0–5.5	65
4	Undisturbed	1.4	16	2.3	1.8–3.2	81
4	*Balanus* removed	1.4	37	3.7	2.5–5.1	100
5	Undisturbed	1.4	7	3.3	2.8–3.7	70
5	*Balanus* removed	1.4	13	4.0	3.5–4.5	100
6	Undisturbed	1.0	13	2.8	2.1–3.9	100
6	*Balanus* moved	1.0	14	4.1	3.0–5.2	100
7a & b	Undisturbed	0.7	10	3.5	2.7–4.5	70
7a & b	*Balanus* removed	0.7	23	4.3	3.0–6.3	81

starfish were observed feeding on the barnacles at this level; one entered the cage in January, 1955, and ate most of the barnacles.

Thais were common in this locality, feeding on barnacles and mussels, and reaching average population densities of 200/m^2 below M.T.L. (Connell 1961). The mortality rates for *Chthamalus* in cages and on adjacent areas outside cages indicated by the letter "b" after the number are shown on Table VII.

If the mortality rates of *Chthamalus* growing without contact with *Balanus* are compared in and out of the cages, it can be seen that at the upper levels mortality is greater inside the cages, at lower levels greater outside. Densities of *Thais* tend to be greater at and below M.T.L. so that this trend in the mortality rates of *Chthamalus* may be ascribed to an increase in predation by *Thais* at lower levels. Mortality of *Chthamalus* in the absence of *Balanus* was appreciably greater outside than inside the cage only on area 13. In the other 4 areas it seems evident that few *Chthamalus* were being eaten by *Thais.* In a concurrent study of the behavior of *Thais* in feeding on *Balanus balanoides,* it was found that *Thais* selected the larger individuals as prey (Cornell 1961). Since *Balanus* after a few month's growth was usually larger than *Chthamalus,* it might be expected that *Thais* would feed on *Balanus* in preference to

TABLE VII. The effect of predation by *Thais lapillus* on the annual mortality rate of *Chthamalus stellatus* in the experimental areas.*

Area	Height in ft. from M.T.L.	% mortality of *Chthamalus* over a year (The initial numbers are given in parentheses) — a: Protected from predation by a cage: With *Balanus*	a: Without *Balanus*	a: Difference	b: Unprotected, open to predation: With *Balanus*	b: Without *Balanus*	b: Difference
Area 3	+2.2	73 (112)	25 (96)	48	89 (47)	6 (50)	83
Area 7	+0.7	90 (42)	47 (40)	43	82 (51)	23 (47)	59
Area 11	0	62 (21)	28 (18)	34	50 (10)	25 (16)	25
Area 12	0†	100 (60)	53 (50)	17	87 (39)	59 (32)	28
Area 13	–1.0	98 (72)	9 (77)	89	90 (69)	35 (51)	55

*The records for 12a extend over only 10 months; for purposes of comparison the mortality rate for 12a has been multiplied by 1.2.

†Tide pool.

Chthamalus. In a later study (unpublished) made at Santa Barbara, California, *Thais emarginata* Deshayes were enclosed in cages on the shore with mixed populations of *Balanus glandula* Darwin and *Chthamalus fissus* Darwin. These species were each of the same size range as the corresponding species at Millport. It was found that *Thais emarginata* fed on *Balanus glandula* in preference to *Chthamalus fissus*.

As has been indicated, much of the mortality of *Chthamalus* growing naturally intermingled with *Balanus* was a result of direct crowding by *Balanus*. It therefore seemed reasonable to take the difference between the mortality rates of *Chthamalus* with and without *Balanus* as an index of the degree of competition between the species. This difference was calculated for each area and is included in Table VII. If these differences are compared between each pair of adjacent areas in and out of a cage, it appears that the difference, and therefore the degree of competition, was greater outside the cages at the upper shore levels and less outside the cages at the lower levels. Thus as predation increased at lower levels, the degree of competition decreased. This result would have been expected if *Thais* had fed upon *Balanus* in preference to *Chthamalus*. The general effect of predation by *Thais* seems to have been to lessen the interspecific competition below M.T.L.

Discussion

"Although animal communities appear qualitatively to be constructed as if competition were regulating their structure, even in the best studied cases there are nearly always difficulties and unexplored possibilities" (Hutchinson 1957).

In the present study direct observations at intervals showed that competition was occurring under natural conditions. In addition, the evidence is strong that the observed competition with *Balanus* the principal factor determining the local distribution of *Chthamalus*. *Chthamalus* thrived at lower levels when it was not growing in contact with *Balanus*.

However, there remain unexplored possibilities. The elimination of *Chthamalus* requires a dense population of *Balanus*, yet the settlement of *Balanus* varied from year to year. At Millport, the settlement density of *Balanus balanoides* was measured for 9 years between 1944 and 1958 (Barnes 1956b, Connell 1961). Settlement was light in 2 years, 1946 and 1958. In the 3 seasons of *Balanus* settlement studied in detail, 1953–55, there was a vast oversupply of larvae ready for settlement. It thus seems probable that most of the *Chthamalus* which survived in a year of poor settlement of *Balanus* would be killed in competition with a normal settlement the following year. A succession of years with poor settlements of *Balanus* is a possible, but improbable occurrence at Millport, judging from the past record. A very light settlement is probably the result of a chance combination of unfavorable weather circumstances during the planktonic period (Barnes 1956b). Also, after a light settlement, survival on the shore is improved, owing principally to the reduction in intraspecific crowding (Connell 1961); this would tend to favor a normal settlement the following year, since barnacles are stimulated to settle by the presence of members of their own species already attached on the surface (Knight-Jones 1953).

The fate of those *Chthamalus* which had survived a year on the undisturbed areas is not known since the experiment ended at that time. It is probable, however, that most of them would have been eliminated within 6 mouths; the mortality rate had increased in the spring

(Figures 2 and 3), and these survivors were often misshapen and smaller than those which had not been crowded (Table VI). Adults on the transplanted stones had suffered high mortality in the previous year (Table III).

Another difficulty was that *Chthamalus* was rarely found to have settled below mid-tide level at Millport. The reasons for this are unknown; it survived well if transplanted below this level, in the absence of *Balanus*. In other areas of the British Isles (in southwest England and Ireland, for example) it occurs below mid-tide level.

The possibility that *Chthamalus* might affect *Balanus* deleteriously remains to be considered. It is unlikely that *Chthamalus* could cause much mortality of *Balanus* by direct crowding; its growth is much slower, and crowding between individuals of *Chthamalus* seldom resulted in death. A dense population of *Chthamalus* might deprive larvae of *Balanus* of space for settlement. Also, *Chthamalus* might feed on the planktonic larvae of *Balanus*; however, this would occur in March and April when both the sea water temperature and rate of cirral activity (presumably correlated with feeding activity), would be near their minima (Southward 1955).

The indication from the caging experiments that predation decreased interspecific competition suggests that the action of such additional factors tends to reduce the intensity of such interactions in natural conditions. An additional suggestion in this regard may be made concerning parasitism. Crisp (1960) found that the growth rate of *Balanus balanoides* was decreased if individuals were infected with the isopod parasite *Hemioniscus balani* (Spence Bate). In Britain this parasite has not been reported from *Chthamalus stellatus*. Thus if this parasite were present, both the growth rate of *Balanus*, and its ability to eliminate *Chthamalus* would be decreased, with a corresponding lessening of the degree of competition between the species.

The Causes of Zonation

The evidence presented in this paper indicates that the lower limit of the intertidal zone of *Chthamalus stellatus* at Millport was determined by interspecific competition for space with *Balanus balanoides*. *Balanus*, by virtue of its greater population density and faster growth, eliminated most of the *Chthamalus* by directing crowding. At the upper limits of the zones of these species no interaction was observed. *Chthamalus* evidently can exist higher on the shore than *Balanus* mainly as a result of its greater tolerance to heat and/or desiccation.

The upper limits of most intertidal animals are probably determined by physical factors such as these. Since growth rates usually decrease with increasing height on the shore, it would be less likely that a sessile species occupying a higher zone could, by competition for space, prevent a lower one from extending upwards. Likewise, there has been, as far as the author is aware, no study made which shows that predation by land species determines the upper limit of an intertidal animal. In one of the most thorough of such studies, Drinnan (1957) indicated that intense predation by birds accounted for an annual mortality of 22% of cockles (*Cardium edule* L.) in sand flats where their total mortality was 74% per year.

In regard to the lower limits of an animal's zone, it is evident that physical factors may act directly to determine this boundary. For example, some active amphipods from the upper levels of sandy beaches die if kept submerged. However, evidence is accumulating that the lower limits of distribution of intertidal animals are determined mainly by biotic factors.

Connell (1961) found that the shorter length of life of *Balanus balanoides* at low shore levels could be accounted for by selective predation by *Thais lapillus* and increased intraspecific competition for space. The results of the experiments in the present study confirm the suggestions of other authors that lower limits may be due to interspecific competition for space. Knox (1954) suggested that competition determined the distribution of 2 species of barnacles in New Zealand. Endean, Kenny and Stephenson (1956) gave indirect evidence that competition with a colonial polychaete worm, *(Galeolaria)* may have determined the lower limit of a barnacle *(Tetraclita)* in Queensland, Australia. In turn the lower limit of Galeolaria appeared to be determined by competition with a tunicate, *Pyura,* or with dense algal mats.

With regard to the 2 species of barnacles in the present paper, some interesting observations have been made concerning changes in their abundance in Britain. Moore (1936) found that in southwest England in 1934, *Chthamalus stellatus* was most dense at M.H.W.N., decreasing in numbers toward M.T.L. while *Balanus balanoides* increased in numbers below M.H.W.N. At the same localities in 1951, Southward and Crisp (1954) found that *Balanus* had almost disappeared and that *Chthamalus* had increased both above and below M.H.W.N. *Chthamalus* had not reached the former densities of *Balanus* except at one locality, Brixham. After 1951, *Balanus* began to return in numbers, although by 1954 it had not reached the densities of 1934; *Chthamalus* had declined, but again not to its former densities (Southward and Crisp 1956).

Since *Chthamalus* increased in abundance at the lower levels vacated by *Balanus,* it may previously have been excluded by competition with *Balanus.* The growth rate of *Balanus* is greater than *Chthamalus* both north and south (Hatton 1938) of this location, so that *Balanus* would be likely to win in competition with *Chthamalus.* However, changes in other environmental factors such as temperature may have influenced the abundance of these species in a reciprocal manner. In its return to southwest England after 1951, the maximum density of settlement of *Balanus* was 12 per cm^2; competition of the degree observed at Millport would not be expected to occur at this density. At a higher population density, *Balanus* in southern England would probably eliminate *Chthamalus* at low shore levels in the same manner as it did at Millport.

In Loch Sween, on the Argyll Peninsula, Scotland, Lewis and Powell (1960) have described an unusual pattern of zonation of *Chthamalus stellatus.* On the outer coast of the Argyll Peninsula *Chthamalus* has a distribution similar to that at Millport. In the more sheltered waters of Loch Sween, however, *Chthamalus* occurs from above M.H.W.S. to about M.T.L., judging the distribution by its relationship to other organisms. *Balanus balanoides* is scarce above M.T.L. in Loch Sween, so that there appears to be no possibility of competition with *Chthamalus,* such as that occurring at Millport, between the levels of M.T.L. and M.H.W.N.

In Figure 5 an attempt has been made to summarize the distribution of adults and newly settled larvae in relation to the main factors which appear to determine this distribution. For *Balanus* the estimates were based on the findings of a previous study (Cornell 1961); intraspecific competition was severe at the lower levels during the first year, after which predation increased in importance. With *Chthamalus,* it appears that avoidance of settlement or early mortality of those larvae which settled at levels below M.T.L., and elimination

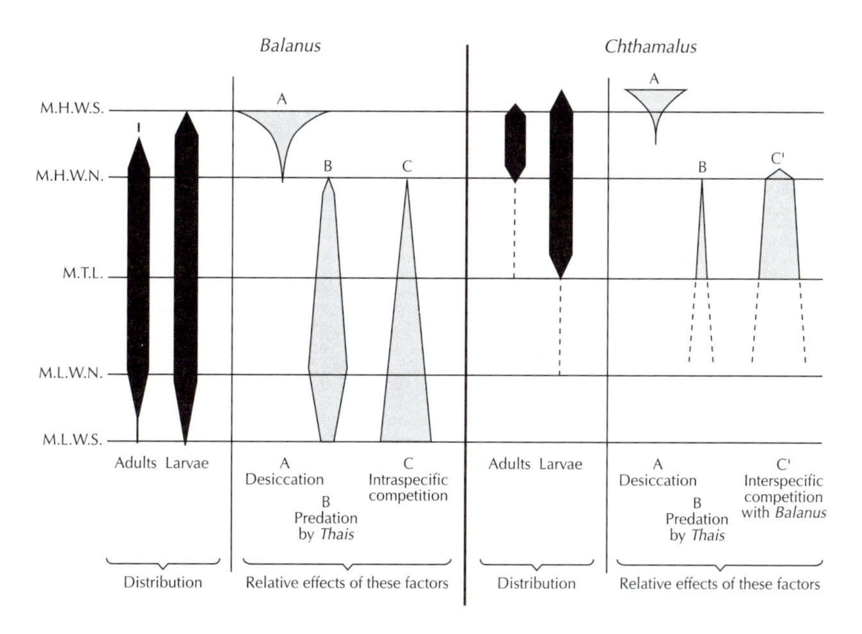

Figure 5. The intertidal distribution of adults and newly settled larvae of *Balanus balanoides* and *Chthamalus stellatus* at Millport, with a diagrammatic representation of the relative effects of the principal limiting factors.

by competition with *Balanus* of those which settled between M.T.L. and M.H.W.N., were the principal causes for the absence of adults below M.H.W.N. at Millport. This distribution appears to be typical for much of western Scotland.

Summary

Adults of *Chthamalus stellatus* occur in the marine intertidal in a zone above that of another barnacle, *Balanus balanoides*. Young *Chthamalus* settle in the *Balanus* zone but evidently seldom survive, since few adults are found there.

The survival of *Chthamalus* which had settled at various levels in the *Balanus* zone was followed for a year by successive censuses of mapped individuals. Some *Chthamalus* were kept free of contact with *Balanus*. These survived very well at all intertidal levels, indicating that increased time of submergence was not the factor responsible for elimination of *Chthamalus* at low shore levels. Comparison of the survival of unprotected populations with others, protected by enclosure in cages from predation by the snail, *Thais lapillus* showed that *Thais* was not greatly affecting the survival of *Chthamalus*.

Comparison of the survival of undisturbed populations of *Chthamalus* with those kept free of contact with *Balanus* indicated that *Balanus* could cause great mortality of *Chthamalus*. *Balanus* settled in greater population densities and grew faster than *Chthamalus*. Direct observations at each census showed that *Balanus* smothered, undercut, or crushed the *Chthamalus*; the greatest mortality of *Chthamalus* occurred during the seasons of most rapid growth of *Balanus*. Even older *Chthamalus* transplanted to low levels were killed by *Balanus* in this way. Predation by *Thais* tended to decrease the severity of this interspecific competition.

Survivors of *Chthamalus* after a year of crowding by *Balanus* were smaller than uncrowded ones. Since smaller barnacles produce fewer offspring, competition tended to reduce reproductive efficiency in addition to increasing mortality.

Mortality as a result of intraspecies competition for space between individuals of *Chthamalus* was only rarely observed.

The evidence of this and other studies indicates that the lower limit of distribution of intertidal organisms is mainly determined by the action of biotic factors such as competition for space or predation. The upper limit is probably more often set by physical factors.

References

Barnes, H. 1953. Size variations in the cyprids of some common barnacles. J. Mar. Biol. Ass. U. K. 32:297–304.

Barnes, H. 1956a. The growth rate of *Chthamalus stellatus* (Poli). J. Mar. Biol. Ass. U K. 35:355–361.

Barnes, H. 1956b. *Balanus balanoides* (L.) in the Firth of Clyde: The development and annual variation of the larval population, and the causative factors. J. Anim. Ecol. 25:72–84.

Barnes, H. and H. T. Powell. 1953. The growth of *Balanus balanoides* (L.) and *B. crenatus* Brug under varying conditions of submersion. J. Mar. Biol. Ass. U. K. 32:107–128.

Barnes, H. and D. J. Crisp. 1956. Evidence of self-fertilization in certain species of barnacles. J. Mar. Biol. Ass. U. K. 35:631–639.

Beauchamp, R. S. A. and P. Ullyott. 1932. Competitive relationships between certain species of freshwater Triclads. J. Ecol. 20:200–208.

Clark, P. J. and F. C. Evans. 1954. Distance to nearest neighbor as a measure of spatial relationships in in populations. Ecology 35:445–453.

Clench, W. J. 1947. The genera *Purpura* and *Thais* in the western Atlantic. Johnsonia 2, No. 23:61–92.

Connell, J. H. 1961. The effects of competition, predation by *Thais lapillus*, and other factors on natural populations of the barnacle, *Balanus balanoides*. Ecol. Mon. 31:61–104.

Crisp, D. J. 1960. Factors influencing growth-rate in *Balanus balanoides*. J. Anim. Ecol. 29:95–116.

Drinnan, R. E. 1957. The winter feeding of the oyster catcher *(Haematopus ostralegus)* on the edible cockle *(Cardium edule)*. J. Anim. Ecol. 26:441–469.

Elton, Charles and R. S. Miller. 1954. The ecological survey of animal communities: with a practical scheme of classifying habitats by structural characters. J. Ecol. 42:460–496.

Endean, R., R. Kenny and W. Stephenson. 1956. The ecology and distribution of intertidal organisms on the rocky shores of the Queensland mainland. Aust. J. Mar. Freshw. Res. 7:88–146.

Fischer, E. 1928. Sur la distribution geographique de quelques organismes de rockier, le long des cotes de la Manche. Trav. Lab. Mus. Hist. Nat. St. Servan 2:1–16.

Hatton, H. 1938. Essais de bionomie explicative sur quelques especes intercotidales d'algues et d'animaux. Ann. Inst. Oceanogr. Monaco 17:241–348.

Holme, N. A. 1950. Population-dispersion in *Tellina tenuis* Da Costa. J. Mar. Biol. Ass. U. K. 29:267–280.

Hutchinson, G. E. 1957. Concluding remarks. Cold Spring Harbor Symposium on Quant. Biol. 22:415–427.

Knight-Jones, E. W. 1953. Laboratory experiments on gregariousness during setting in *Balanus balanoides* and other barnacles. J. Exp. Biol. 30:584–598.

Knox, G. A. 1954. The intertidal flora and fauna of the Chatham Islands. Nature Lond. 174:871–873.

Lack, D. 1954. *The Natural Regulation of Animal Numbers*. Oxford, Clarendon Press.

Lewis, J. R. and H. T. Powell. 1960. Aspects of the intertidal ecology of rocky shores in Argyll, Scotland. I. General description of the area. II. The distribution of *Chthamalus stellatus* and *Balanus balanoides* in Kintyre. Trans. Roy. Soc. Edin. 64:45–100.

MacArthur, R. H. 1958. Population ecology of some warblers of northeastern coniferous forests. Ecology 39:599–619.

Moore, H. B. 1935. The biology of *Balanus balanoides*. III. The soft parts. J. Mar. Biol. Ass. U. K. 20:263–277.

Moore, H. B. 1936. The biology of *Balanus balanoides*. V. Distribution in the Plymouth area. J. Mar. Biol. Ass. U. K. 20:701–716.

Siegel, S. 1956. *Nonparametric statistics*. New York, McGraw Hill.

Southward, A. J. 1955. On the behavior of barnacles. I. The relation of cirral and other activities to temperature. J. Mar. Biol. Ass. U. K. 34:403–422.

Southward, A. J., and D. J. Crisp. 1954. Recent changes in the distribution of the intertidal barnacles *Chthamalus stellatus* Poli and *Balanus balanoides* L. in the British Isles. J. Anim. Ecol. 23:163–177.

Southward, A. J.,1956. Fluctuations in the distribution and abundance of intertidal barnacles. J. Mar. Biol. Ass. U. K. 35:211–229.

Introduction to

Charles J. Krebs, Stan Boutin, Rudy Boonstra, A. R. E. Sinclair, J. N. M. Smith, Mark R. T. Dale, K. Martin and R. Turkington

Impact of Food and Predation on the Snowshoe Hare Cycle

by Anthony R. Ives

Cyclic population dynamics are thought to be a key feature of predator-prey dynamics. Simple mathematical models of predator-prey interactions show cycles, as do simple laboratory experiments. Therefore, an obvious hypothesis explaining cyclic population dynamics in nature is that the cycles are caused by predator-prey interactions. Nonetheless, demonstrating that natural population cycles are caused by predator-prey interactions has proved to be remarkably difficult.

Charles Krebs and colleagues (1995) report on one of the latest experiments aimed at understanding snowshoe hare cycles. Currently there are two explanations for the cycles. The first involves predator-prey interactions in which plants are the prey and hares the predator. High hare densities depress not only the quantity of food plants but also the quality, as plants respond by producing chemicals that protect them from hare feeding. Evidence to support this explanation is that at high hare densities, the condition of hares (measured by fat reserves) decreases. The second explanation involves the interaction between hares and predators. This explanation is supported by the fact that predation is the main source of hare mortality, in many locations causing the death of about 90% of hares. A variety of predators attack snowshoe hares, including hawks, coyotes, and lynx, with the relative importance of different predators varying across the hare's geographical range. Some predators, such as lynx, show cyclic dynamics with cycles lagging behind those of the hares.

Krebs and colleagues performed a large-scale experiment to manipulate food and predation in 1 km^2 plots in the boreal forests of western Canada. Food quality and quantity were manipulated by adding commercial rabbit chow, and predation was manipulated by surrounding two plots with an electric fence. The timing of the manipulations was selected to coincide with the peak and decline of the regular hare cycle in unmanipulated (control) plots. During the natural period of decline from peak densities, both supplementing food and reducing predation lead to higher hare densities. This implicates both food and predation in driving the snowshoe hare cycle. Furthermore, the effects of food and predation are likely linked. The direct effect of food on mortality was small, since starvation only accounted for 9% of the mortality in the control plots. Nonetheless, food limitation could increase predation, either by forcing hares to search for food more extensively and thereby increasing their exposure to predators, or by reducing the health of hares and making them less able to escape predators.

This experiment was clearly large and difficult, and the results link food availability and predation to declines in snowshoe hare abundance. However, additional information is needed to demonstrate conclusively that snowshoe hare cycles are driven by a combination of predation and food availability.

Impact of Food and Predation on the Snowshoe Hare Cycle

CHARLES J. KREBS, STAN BOUTIN, RUDY BOONSTRA, A. R. E. SINCLAIR, J. N. M. SMITH, MARK R. T. DALE, K. MARTIN, R. TURKINGTON

Snowshoe hare populations in the boreal forests of North America go through 10-year cycles. Supplemental food and mammalian predator abundance were manipulated in a factorial design on 1-square-kilometer areas for 8 years in the Yukon. Two blocks of forest were fertilized to test for nutrient effects. Predator exclosure doubled and food addition tripled hare density during the cyclic peak and decline. Predator exclosure combined with food addition increased density 11-fold. Added nutrients increased plant growth but not hare density. Food and predation together had a more than additive effect, which suggests that a three-trophic-level interaction generates hare cycles.

The 10-year cycle of snowshoe hare populations and those of their predators is one of the dominant perturbations of the boreal forests of North America. Predation and food shortage have been postulated as the major factors causing these fluctuations (1). Because in all cyclic populations many factors will change in a manner correlated with population density, necessary conditions can be recognized only by experimental manipulations (2). From 1976 to 1984, we manipulated food supplies of snowshoe hares *(Lepus americanus)* in the southern Yukon and showed that the cyclic decline could not be prevented by either artificial or natural food addition (3). Single-factor manipulations have been criticized in field ecology because they may miss important interactions between factors (4). For the past 8 years, we have carried out large-scale experiments on nutrients, supplemental food, and predation in the Yukon to untangle the causes of the hare cycle and the consequences the hare cycle has for the vertebrate community. By crossing a predator reduction manipulation with food addition we estimated interaction effects caused by the failure of factors to combine additively.

We chose 1-km^2 blocks of undisturbed boreal forest near Kluane Lake, Yukon, as our experimental units (5). The boreal forest in this region is dominated by white spruce *(Picea glauca)* and was not disturbed by logging, fire, or extensive fur trapping during our studies. We used a factorial design to untangle the effects of food and predation on hares. Three areas were used as controls (6). Two experimental areas were provided with ad lib supplemental food year-round. We excluded mammalian predators by building one electric fence in the summer of 1987. In the summer of 1988, we built a second electric fence to use for the combined predator reduction-food addition treatment (7). Since January 1989, the electric fences have worked effectively to prevent mammalian predators from entering the two areas. The fences are per-

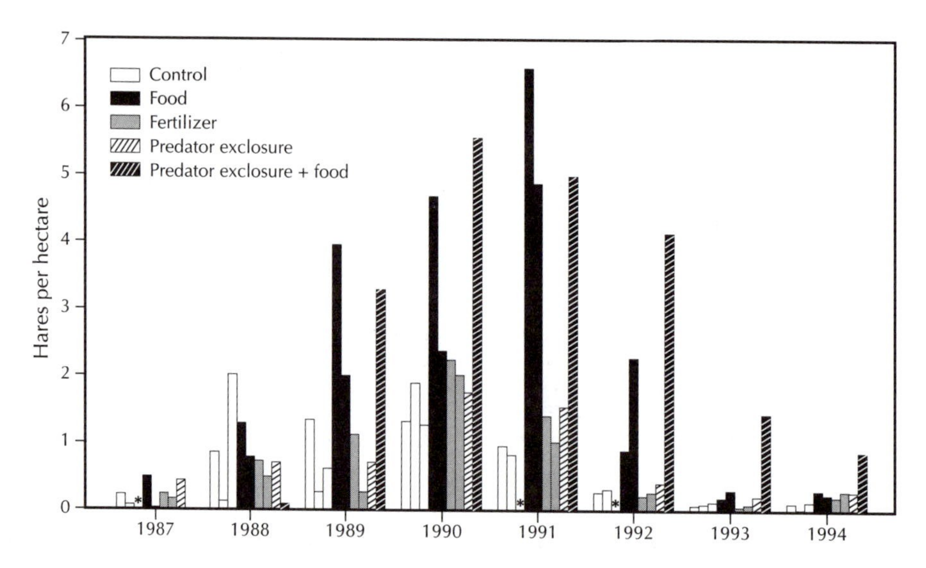

Figure 1. Spring densities of snowshoe hares in three control and six treatment areas, Kluane Lake, Yukon, 1987 to 1994. Densities were estimated from mark-recapture live trapping for 4 to 5 days in late March and early April each year with the use of the jackknife estimator in the program Capture (10). Asterisk indicates one of the three control areas that was not trapped in 1987, 1991, and 1992.

meable to snowshoe hares. Beginning in 1987, we added nitrogen-potassium-phosphorus (NPK) fertilizer to two blocks of forest to increase plant growth (8). We chose to manipulate a few large areas rather than many small areas because of the failure of most field experiments to address large-scale issues (9). We captured, marked, and released snowshoe hares every March and October and estimated densities with the robust design (10).

Snowshoe hares in the control areas increased from a low in the mid-1980s to a peak in 1989 and 1990 (Figure 1). The increase phase from 1986 to 1988 showed considerable variation among the three control populations, but from the peak phase onward all the controls were similar in their year-to-year dynamics. The cyclic decline began in autumn and winter 1992 and continued until the spring of 1993 when hares had reached low numbers of approximately one hare per 15 ha. Population increase in snowshoe hares is stopped both by increased mortality and by reduced reproductive output (11). This previously described syndrome of demographic changes was consistent over the cycle we observed. Juvenile mortality increased while the population was still in the increase phase of the cycle (Figure 2), whereas adult losses did not become severe until the decline phase. The decline phase in 1991 and 1992 was characterized by poor survival of both juveniles and adults and by reduced reproductive output by females through restriction or elimination of their second and third summer litters (11).

The impact of our experimental treatments can be measured in several ways. We concentrate here on changes in the population density of hares in the treated areas and on the survival rates of radio-collared hares.

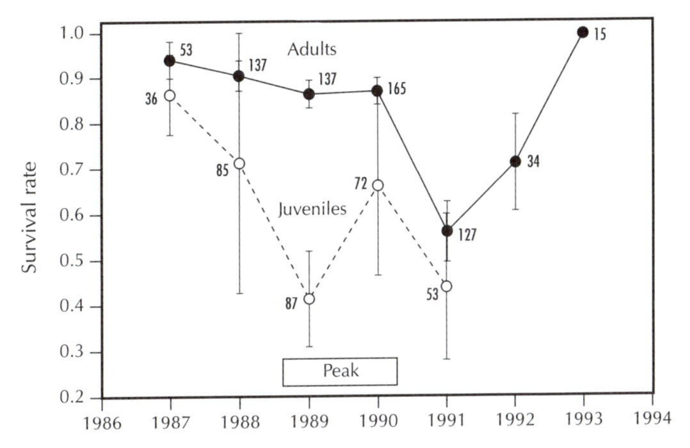

Figure 2. Survival rates of adult (●) and juvenile (○) snowshoe hares in the most intensively studied control area (control area 1). Juvenile survival is already deteriorating in the late phase (1988) and is low in the peak and decline phases of the cycle. We could not estimate juvenile survival in 1992 because few of the 13 juveniles caught were ever recaptured. We presume juvenile survival was poor in 1992. Survival rates (per 28 days) were estimated from mark-recapture data with the use of the Jolly-Seber Model B (10). Juvenile survival refers to trappable juveniles only, which are those more than 6 to 8 weeks old. Sample sizes of hares are given next to data points.

Density effects can be most simply expressed as ratios of the density in the treated area to the density in the control areas. We estimated these each spring and autumn for all treatments (Figure 3). Effects were small during the increase phase in 1987 and 1988 because the treatments were just being established. All treatments were effective by spring 1989. The food addition effect was always positive and produced densities ranging from 1.5- to 6-fold over control levels during the peak and decline phases (Figure 3B). The predator exclosure effect was negligible in the peak phase from 1989 to 1990 but became pronounced in the late decline and low phases, producing densities ranging from 1.4- to 6-fold over control levels (Figure 3A). The largest effect was shown by predator exclosure and food treatment combined, particularly in the late decline phase when densities exceeded control levels by 36-fold (Figure 3C). Averaged over both the peak and the decline phase, predator exclosure approximately doubled the density of hares, food addition approximately tripled density, and the combined treatment increased density 11- fold.

In contrast to the strong effects shown by manipulation of predation and food supply, the addition of nutrients had virtually no effect on snowshoe hare numbers. In spite of increased growth of herbs, grasses, shrubs, and trees (8), the fertilized plots contained virtually the same number of hares as did the control plots (Figure 1). Fertilized vegetation in the boreal forest cannot duplicate either the quantity or quality of the artificial food that we added in our experiments, and for this reason fertilization is a relatively ineffective method of food addition for hares.

Survival rates can be estimated from mark-recapture methods or from radiotelemetry (12). Treatments had little impact on survival rates during the peak phase of the hare cycle. Monthly adult survival rates were greater than

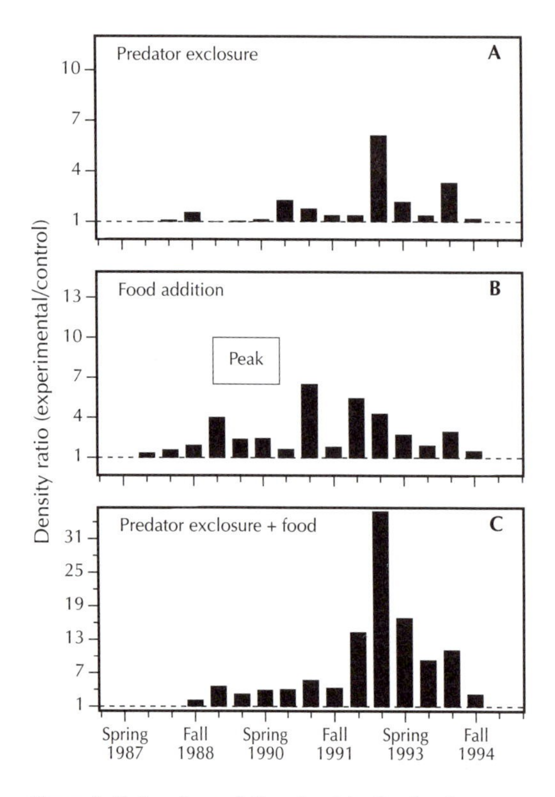

Figure 3. Ratio of population densities for the three treatments to average control population densities at the same time. If there is no treatment effect, we expect a ratio of 1.0. During the peak and decline phases, the predator exclosure **(A)** doubled density on average, food addition **(B)** tripled density, and the combined treatment **(C)** increased density 11-fold.

90% in the peak phase, leaving little room for improvement.

The major effects of the treatments on survival were visible in the decline phase (Figure 4). The probability of a hare living for 1 year in the control areas during the decline was 0.7%. In fertilized areas, this probability was 1.9%, which is slightly but not significantly higher than in the control areas. This probability improved to 3.7% in the food addition grids and to 9.5% in the predator exclosure areas. The best chances of survival occurred in areas treated with a combination of mammalian predator reduction and food, where the probability of survival was 20.8% for 1 year during the decline. The effects of food and predation on survival during the decline phase were nearly additive and showed no sign of an interaction. The addition of food by itself was not sufficient to prevent large losses to predators, and the rapid population collapse in the food areas from 1991 to 1992 (Figure 1) was due to heavy predation.

The numbers of both avian and mammalian predators follow the hare cycle, but with a 1- to 2-year time lag (1). Virtually all snowshoe hares in our study area die from predator attack in the immediate sense. From 1989 to 1993, we found that 83% of the deaths of all radio-collared hares were due to predation and only 9% were attributed to starvation (13). We presume that hares suffering from food stress will be more susceptible to death from predation.

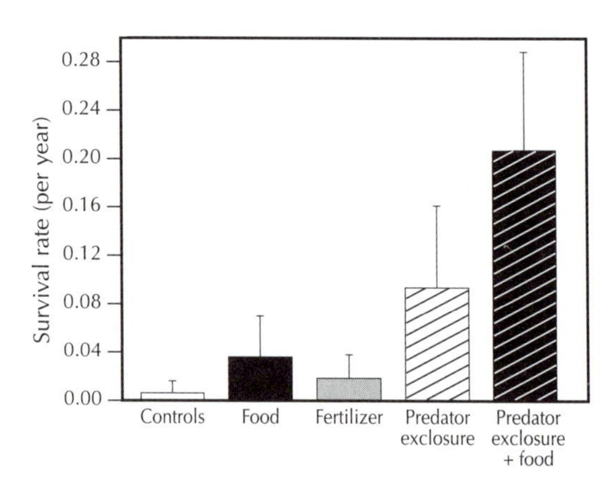

Figure 4. Annual survival rates for snowshoe hares with radio collars during the decline phase from autumn 1990 to autumn 1992. Ninety-five percent confidence limits are shown. Sample sizes for the estimates are (in order): 278, 206, 197, 246, and 262 hares. Radio collars were placed only on fully grown animals and thus are used to measure adult mortality rates.

Table 1. Total production of live young by female snowshoe hares over the summer breeding season during the late increase (1988), peak (1989 and 1990), and decline (1991 and 1992) phases of the hare cycle. Litter sizes at birth were not available for all treatments in all years. We assume that total production in the predator exclosure plus food treatment area would have been the same as in the food areas in the peak years of 1989 and 1990. Numbers in parentheses indicate number of females sampled.

Year	Control areas [no. of young ± SE (n)]	Food areas [no. of young ± SE (n)]	Predator exclosure + food area [no. of young ± SE (n)]
1988	16.4 ± 0.44 (10)	—	—
1989	13.7 ± 0.39 (21)	14.1 ± 0.43 (36)	—
1990	13.7 ± 0.40 (33)	15.1 ± 0.43 (36)	—
1991	7.8 ± 0.37 (18)	—	16.3 ± 0.73 (15)
1992	3.3 ± 0.25 (4)	—	17.1 ± 0.31 (50)

Because we used live trapping as our primary technique of study, we have less data on the reproductive output of snowshoe hares in relation to these three treatments. During the peak phase, the food treatment areas have the same reproductive output as do the control areas (14). Table 1 gives the total production of live young hares at birth for a female over the summer breeding period for the control areas, the food areas, and the predator exclosure plus food grid. The collapse of reproduction in the control areas was prevented in the combined treatment area. Because we do not know the reproductive schedule for the food or the predator exclosure treatment areas during the decline phase, we cannot assess the separate contributions of food and predation to the reproductive output of females.

There are three possible explanations of these differences in reproductive output. Reproductive changes may be driven by food limitations in the decline period. Alternatively, hares may respond to predation risk in the decline phase by altering their habitat use so that they cannot achieve adequate nutrition (15). In both of these cases, this reproductive curtailment is due to food shortage, but in the first case it is absolute food shortage and in the second case it is relative food shortage caused as an indirect effect of predation. The third possibility is that these reproductive effects are a direct result of stress and the physiological derangement associated with stress (16). We cannot yet determine which of these explanations is correct. Absolute winter food shortage does not necessarily occur during the peak or decline phase (3), and the weight of evidence is against the first explanation. Behavioral evidence suggests that there may be a relative shortage of food (15).

If both food and predation are together sufficient to explain population cycles in snowshoe hares, why were we not able to prevent the decline entirely in the combination treatment area? Hare densities fell from about seven per hectare in 1989 to about one per hectare in 1994 in the combination treatment area, even though hare density in this area remained at or above normal peak densities for 7 years from 1988 to 1995. The combination treatment delayed the decline but did not prevent it. There are three possible reasons for this and they illustrate one difficulty of large scale experiments. First, hares could move freely into and out of the predator exclosures. Individuals were often killed by predators when they moved out, but others could emigrate into nearly unoccupied landscape outside the fence in the low years of 1992 to 1994 (17). Second, we could not prevent raptor predation inside the predator exclosure. Our monofilament treatment covered only a small fraction of the predator exclosure and was ineffective in preventing raptors such as goshawks from invading the area. Because raptors and owls cause about 40% of the predation mortality in our hare populations, the effect of the exclosures was to reduce total predation losses, not eliminate them. Great horned owls and goshawks continued to kill hares inside the predator exclosures during the decline and low phases. Third, another factor in addition to food and predation may be sufficient to cause the decline.

These results support the view that population cycles in snowshoe hares in the boreal forest are a result of the interaction between food supplies and predation. They do not support either the plant-herbivore model or the predator- prey model for cycles but suggest that hare cycles result from a three-trophic-level interaction (18). Our experimental results are consistent with the general ideas of Keith (19) and Wolff (20) that both food and predation play a role in generating hare cycles, but they do not support Keith's sequential two-factor model that states that food shortage effects are followed by predation effects in causing cyclic declines. Further work will be required to determine if the nutritional effects on hares are an indirect effect of predation that is explicable in terms of the hares' behavioral responses to predation risk. Our studies have provided little data on the causes of the low phase of the hare cycle, which can persist for 3 to 4 years. Food supplies recover quickly after the peak has passed, and predator numbers collapse during the hare decline. Whether the direct or indirect effects of predation can also explain the low phase remains an open question.

References and Notes

1. Keith, L. B. 1963. *Wildlife's Ten Year Cycle.* U. of Wisconsin Press, Madison, WI. • Keith, L. B., J. R.Cary, O. J. Rongstad, M. C. Brittingham. 1984. Wildl. Monogr.90:1 • L. B. Keith. 1981. *Proceedings of the World Lagomorph Conference, Guelph, Ontario, 12 to 16 August 1979.* Myers, K. and C. D. MacInnes, Eds. Univ. of Guelph, Ontario, Canada. Pp. 395–440 • Vaughan, M. R. and L. B. Keith. 1981. J. WildL Manage. 45:354 • Finerty, J. P. 1980. *The Population Ecology of Cycles in Small Mammals.* Yale Univ. Press, New Haven, CT.

2. Krebs, C. J. 1988. Oikos 52:143. • Chitty, D. 1960. Can. J. ZooL 38:99.

3. Krebs, C. J., B. S. Gilbert. S. Boutin, A. R. E. Sinclair and J. N. M. Smith. 1986. J. Anim. Ecol. 55:963. • Sinclair, A. R. E., C. J. Krebs, J. N. M. Smith, S. Boutin. 1988. J. Anim. Ecol. 57:787. • Smith, J. N. M., C. J. Krebs, A. R. E. Sinclair, R. Boonstra. 1985. J. Anim. Ecol. p. 269. • Krebs, C. J., S. Boutin, B. S. Gilbert, Oecologia 70:194.

4. Desy, E. A. and G. O. Batzli. 1989. Ecology 70:411. • Batzli, G. O. 1992. In: *Wildlife 2001: Populations.* McCullough, D. R and R. H. Barrett, Eds. Elsevier, London. Pp. 831–850.

5. Blocks were spaced at least 1 km apart. Within each block, we surveyed checkerboard grids of 20-by-20 points with 30.5-m spacing and used these grids for snowshoe hare live trapping. Two experimental areas were provided with supplemental food (commercial rabbit chow, 16% protein) year round. In the summer of 1987, we built one electric fence around 1 km^2 to exclude mammalian predators, and over the following year we covered 10 ha with monofilament to reduce avian predation. The monofilament was never effective in preventing avian predation inside the electric fences, and consequently we did not rely on it as a part of the treatment. In the summer of 1988, we built a second electric fence around 1 km^2 to use for the combined predator reduction food addition treatment. We modified the design of the electric fences in 1988 to make them more effective, and since then they have worked effectively to prevent mammalian predators from entering the area. The fences are permeable to snowshoe hares. We could not replicate either the predator reduction or the predator reduction-food addition treatment because of maintenance costs and the difficulty of maintaining electric fences in the Yukon winter with –45°C temperatures. The fences had to be checked every day during

winter. From 1976 to 1985, we trapped hares in six areas within the main study region and found that their population trajectories were very similar (3). We thus have no reason to suspect strong area effects on the unreplicated predator reduction plots.
6. We used three control areas but were not able to trap hares in all of them every year. We have more detailed data on hares from control area 1. The three control areas had quite different histories during the increase phase from 1986 to 1988. Control area 3 reached its greatest hare density in 1988 and remained at a plateau until 1990. Control area 2 reached its peak density in 1990, and control area 1 reached its peak in 1989. By the late peak in 1990 and during the decline phase, the control areas were much more similar to each other in hare densities.
7. The electric fence was 10-stranded, 2.2 m in height, and carried 8600 V. Snow tracking of mammalian predators meeting the fence illustrated its effectiveness. We excluded mammalian predators virtually continuously from January 1989 onward. Our attempts to use monofilament fishing line as a deterrent to birds of prey was largely ineffective because ice formation and snow accumulation on the lines in winter caused them to break or collapse to the ground. We used monofilament on 10 ha of the predator exclosure but did not attempt to use it on the combination treatment area. The predator exclosures thus were mammalian predator exclosures and were still subject to avian predation.
8. We fertilized two 1-km^2 blocks of forest with commercial fertilizer. In May 1987, we used ammonium nitrate at 25 g/m^2. In May 1988, we switched to NPK fertilizer and used 17.5 9 of N/m^2, 5 g of P/m^2, and 2.5 g of K/m^2. In 1989, we used half this amount, and in the years 1990 to 1994 we used the full amount as in 1988, The fertilizer was spread aerially and we did ground checks to make sure it was uniformly spread. We do not present the data here to show the plant growth responses, but all elements of the flora responded dramatically to the added nutrients (C. J. Krebs et al., unpublished data).
9. Tilman,D. 1989. In: *Long-Term Studies in Ecology*. Likens,G. E., Ed. Springer-Verlag, New York. Pp.136–157. • Franklin, J. F. 1989. In: *Long-Term Studies in Ecology*. Likens,G. E., Ed. Springer-Verlag, New York. Pp. 3–19.
10. Pollock, K. H., J. D. Nichols, C. Brownie and J. E. Hines. 1990. Wildl Monogr. 107:1. • Pollock, K. H. 1982. J. Wildl.Manage. 46:752. • Confidence limits were typically ± 15% of the population estimates. Snow tracking in winter confirmed density estimates.
11. Keith, L. B. 1990. Curr. Mammal. 2:119.
12. Boutin, S. and C. J. Krebs. 1986. J. Wildl. Manage. 50:592. • Pollock, K. H., S. R. Winterstein, C. M. Bunck and P. D. Curtis. 1989 J. Wildl. Manage. 53:7.
13. Hik, D. 1995. Wldl. Res. 22:115. • Boutin, S. et al. Unpublished data.
14. O'Donoghue, M. and C. J. Krebs. 1992. J. Anim. Eco 61:631. • Cary, J. R. and L. B. Keith. 1979. Can. J. Zool. 57:375.
15. Hik, D. Unpublished data. • Boonstra, R. and G. R. Singleton. 1993. Gen. Comp. Endocrinol. 91:126.
16. Boonstra, R. 1994. Evol. Ecol. 8:196. • Hik, D., G. R. Singleton. In preparation.
17. In snowshoe hares, dispersal occurs main in the juvenile stage; adult hares are mostly sedentary [Boutin, S., B. S. Gilbert, C. J. Krebs, A. R. E. Sinclair, J. N. M. Smith, 1985. Can. J. Zool. 63:106]. Artificial food addition causes immigration of adult hares [Boutin, S. 1984. Oecologia 62:393] and population changes in food addition areas are more affected by movements than are changes at other treatment or control sites.
18. Oksanen, L. 1990 In: *Perspectives on Plant Consumption*. Tilman, D. and J. Grace, Eds. Academic Press, New York. Pp. 445–474 • Fretwell, S. D. 1987. Oikos 50:291. • Carpenter, S. R., J. F. Kitchell and J. R. Hodgson. 1985. Bioscience 35:634.
19. Keith, L. B. 1983. Oikos 40:385.
20. Wolff, J. O. 1980. Ecol. Monogr. 50:111.
21. We thank the Natural Sciences and Engineering Research Council of Canada for supporting this research program through a Collaborative Special Project grant; the Arctic Institute of North America, University of Calgary, for the use of the Kluane Lake research station; and all the graduate students, technicians, and undergraduates who have assisted in this project over the past 9 years, in particular V. Nams, S. Schweiger, and M. O'Donoghue.

Readings for

COMMUNITY ECOLOGY

The Issue At The Center

T. F. H. Allen

- **Timothy F. H. Allen, Robert V. O'Neill and Thomas W. Hoekstra**
  Interlevel Relations in Ecological Research and Management
- **Garry Peterson, Craig R. Allen, and C. S. Holling**
  Ecological Resilience, Biodiversity, and Scale
- **Eric D. Schneider and James J. Kay**
  Life as a Manifestation of the Second Law of Thermodynamics

Introduction to

T. F. H. Allen, R. V. O'Neill and T. W. Hoekstra

Interlevel Relations in Ecological Research and Management: Some Working Principles From Hierarchy Theory

by Timothy F. H. Allen

While ecologists have always been aware of questions of scale, this paper presses the issue until changes of scale lead to qualitative differences instead of mere quantitative change. More is not just more, it becomes different. This paper is the first succinct modern statement of hierarchy theory in ecology. It is the first to lay out the difference between the extent of a universe of discourse and the grain of the finest distinction made in a set of observations. The nickname given to this paper by its early users is "The eyes paper," because of the eyes looking inwards and outwards as the diagrams systematically lay out the consequences of moving across barriers. The strength of the paper is its systematic treatment of scale and level, and the translation of those notions into practical terms for ecologists.

The paper starts by recognizing surfaces as they define ecological entities. The role of the observer is crucial here. First there is a choice in recognizing a structure and another in selecting a spatiotemporal scale of observation. A change in scale causes aggregation or disaggregation in the ecological items observed, but again the observer chooses the criteria for aggregating to find upper level entities. Emergent properties arise from changing the criteria for explanation by disaggregation from the aggregation criteria that define the whole. This paper is a map so one does not become lost in moves up and down, between, and across levels in observing and defining complex ecological systems.

Interlevel Relations in Ecological Research and Management: Some Working Principles From Hierarchy Theory

T. F. H. ALLEN, R. V. O'NEILL AND T. W. HOEKSTRA

This report clarifies the role of the observer and what he observes as an aid to ecological research and natural resource management. It describes what is involved in observing complex natural systems, the role of the observer in moving through surfaces of natural systems, and finally summarizes, from these, the important principles in linking levels of natural systems that emerge.

Original publication in USDA Forest Service General Technical Report RM–110, July 1984. Reprinted with permission.

Introduction

Ecologists study complex systems which span many levels of organization (e.g., organisms, populations, and communities). As a result, ecologists face the difficult problem of linking levels of organization. Differences in criteria for organization and dynamic behavior make simple aggregation of lower levels insufficient to explain higher levels. What is needed is a set of working principles that allow the ecologist to keep track of terms and data between levels.

Although criteria for organization change between levels, there are consistent patterns that order interlevel relationships and perception of those relationships. For example observations made with coarser grain resolution will necessarily detect high levels of organization as long as observational criteria are held constant. Furthermore, this is true independent of which criteria are actually used for observation be they physiological, floristic, or hydrological. Also, sets of observation extending over larger areas and/or longer time will similarly be able to detect higher levels of organization. Thus, grain and extent are crucial to ordering perception of levels. Perception of higher or lower levels involves the passage of the ecological observer through what are functionally ecological surfaces. Surfaces are crucial to defining levels.

Ecologists often are faced with coupling the level that contains readily perceived objects (i.e., organisms) to other levels where there are valid but less tangible entities (e.g., ecosystems). Moving to lower levels sometimes demands observation of entities (e.g., microbes) and processes (e.g., stomatal opening) much smaller than ordinary human perception. Moving to higher levels of organization, the entities are also hard to perceive and several observations must be conceptually linked before the structure emerges.

Higher levels of organization are particularly troublesome. Ecologists have only recently possessed computational power commensurate with physically large systems. Inexperience with complex systems is sometimes brought home with unkind force, when dust bowls arrive in real time or when computer simulations produce absurdities. Nevertheless, society recognizes pressing problems precisely in these large scale systems, and it demands, quite fairly, that ecologists contribute to solutions. For example, the 1974 Forest and Rangeland Renewable Resources Planning Act requires that each decade the USDA Forest Service not only report on the status and trend of forest, wildlife, range, recreation, and fisheries resources at a national level, but also that it translate effects of local management practices into national level programs to support national needs. It seems to be ecology's responsibility to link small scale systems production estimates to production in large scale complex systems, but how to start?

Prior to the American effort in the International Biological Program, it was believed that ecological systems could be almost perfectly simulated, given enough electronic memory and sufficiently fine-grained data. Few would maintain that today. Apparently, there is more to complex systems than lots of little bits of information. Part of that "something more" may be found in the hierarchical organization that structures complex systems (Allen and Starr 1982).

This report examines general ways to study complex systems, and develops general principles pertinent to such problems. These principles are derived from a detailed analysis of what an observer of a complex system will experience, and the precautions that must be taken in linking levels of organization. While the principles have much generality, the focus is on complexity in systems of particular concern to ecologists.

Observing Complexity

Role of Surfaces

Two distinctions are needed to characterize a hierarchical system. First, there is the distinction between structural entities at a given level of organization (e.g., the distinction between two trees). Second, there is the distinction between successive levels (e.g., between trees and the forest). The two distinctions are related. On the level at which trees are distinguished, the interesting behavior involves how one tree interacts with another. At the "forest" level distinctions between trees are lost, but boundaries, such as forest edges, are recognized.

Hierarchies are, in the most general terms, partially ordered sets (Sugihara 1983) where there is an asymmetry of relationship between elements. Several criteria are particularly helpful for defining asymmetry in ecological hierarchies. These criteria define higher levels as (1) containing (in nested systems), (2) constraining, (3) the context of, (4) behaving at a lower frequency than, and (5) exhibiting less bond strength than, lower levels (Webster 1979).

The definition of any structure may be seen in terms of these criteria. A complex structure is made of smaller structures and is a component of a higher level of the hierarchy (Figure 1). The entity to be defined is contained by its outer surface (criteria 1–3). Surfaces in space are those places around which the strength of interaction is most variable. Inside a surface there is a collection of parts with strong interactions (criterion 5) and rapid exchange (criterion 4) (Simon 1962, 1973). Thus, surfaces define separate entities and are responsible for the characterization of discrete levels of organization (Allen and Starr 1982).

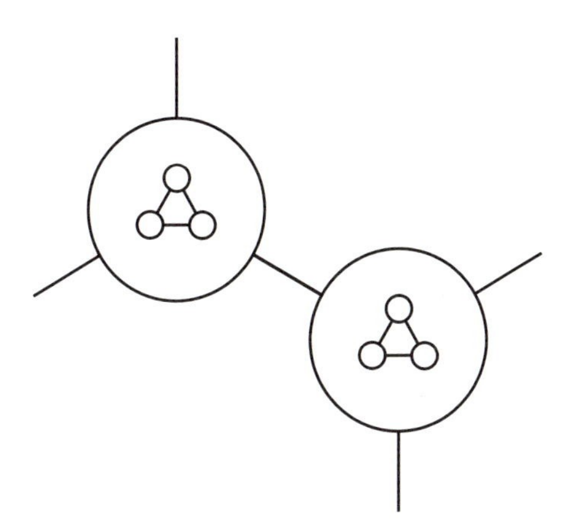

Figure 1. A hypothetical system with various parts visible under different protocols for observation. The full system is composed of two entities, each with three parts. The complete system is not visible within any single observation set.

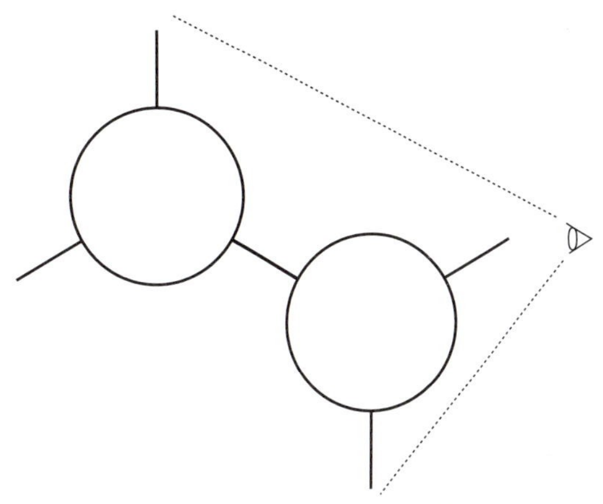

Figure 3. If the observer moves far enough away from the surface, the other whole is identifiable as a separate entity, responsible for part of the environmental influence. (The eye indicates the position from which the system is observed.)

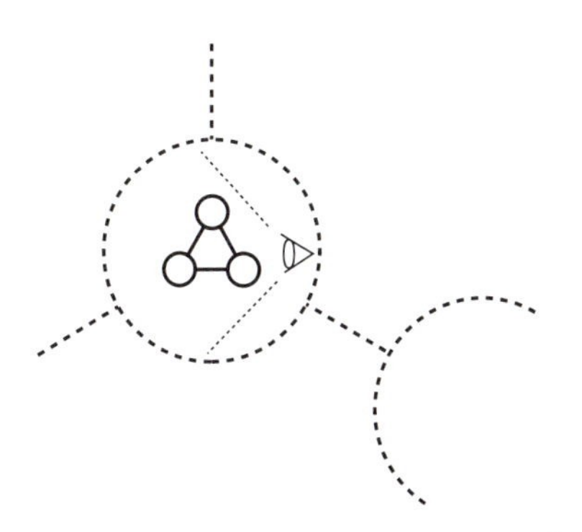

Figure 2. Inside the surface, looking inward, is the only position from which the parts and their interconnections can be seen without distortion. (The eye indicates the position from which the system is observed.)

Now introduce an observer to see what effects surfaces have upon what is observed. Consider exchange across a surface that is weak and sluggish (Platt 1969). An observer inside the surface detects high frequency behavior associated with rapid exchanges among components (Figure 2). An observer outside sees slower behavior characteristic of interactions among higher level entities (Figure 3). Thus, the surface is defined by a change of frequency, sandwiched between high frequency behavior inside and low frequency exchanges outside. The relationship of the surface to the observer is crucial to determining what is observed.

Relax the requirement that surfaces be situated in certain places in space, and the essential features remain; configurations that define observable entities can still be detected. For example, the Krebs Cycle can be distinguished as an entity, even though its enzymes, substrates and products all interdigitate with a mixture of other organic molecules. Similarly, cycling pathways may be detected in ecological nutrient data. The pathway is adequately defined by frequency characteristics even though the cycle cannot be mapped on the ground. It is the difference in reaction rates within, as opposed to without, that defines the entity (Levins 1973).

A critical characteristic of surfaces is their power of integration (Platt 1969; Simon 1962, 1973). Thus, for example, the surface of a community may be recognized by the way component biota are integrated into an interacting unit. The fact that this surface does not map easily onto a particular part of three-dimensional experiential space is, for present purposes, superfluous. This approach to surfaces provides a tool for dealing with levels where there are intangibles above and below the level of commonplace human experience. Discussion of surfaces, henceforth, will include these intangible surfaces as a normal case.

Observation Sets

Because of the importance of surfaces in distinguishing entities and levels of organization, researchers and managers must be explicit about how they look at an ecological system and detect the surfaces. The critical concept here is the observation set. An observation set defines how the investigator decides to look at a system. Two decisions are required. First, the investigator chooses the objects of study. He picks the level on which he will focus; this, in turn, determines the structural entities he will discriminate (Rosen 1977). In other words, the investigator must decide if he will look at cells, organisms, guilds, or communities. Second, the investigator must decide on the phenomena of interest. He chooses those changes in the objects of study which will be considered as interesting (Pattee 1978). Thus, although he has decided to look at populations as his object of study, he must still decide whether he is interested in, say, the phenomenon of growth (i.e., changes in numbers through time) or migration (i.e., changes in numbers over space).

The observation set begins with a data set, but goes beyond just a collection of raw measurements. The observation set is the means to observe in a scientific manner. It is coupled to a procedure for collecting data, such as tree densities in 0.1-ha plots or fluorometric readings of plankton chlorophyll. More than this, scientific observation involves methods of analysis such as Fourier transformation or ordination of binary transformed data. Observation also involves criteria for identifying significance (statistical or otherwise) in the results of analysis. Only when the raw data are analyzed and the significance of the analysis is recognized, is the critical role of observation complete. The observation set depends upon the scientist's paradigm (Kuhn 1962) and is deeply influenced by the way he considers his object of study.

A pH meter reading alone does not constitute an observation set because observation is not complete without fixing (1) the procedure for taking the readings, (2) the interval between the readings, (3) the degree to which the readings are to be integrated, e.g., by averaging over time and space, (4) the extent in time and space of the entire universe of observation, and (5) the criteria for significant change in the readings. In an observation set the concern is not just with system configurations frozen in a set of encounters with the meter, but also with the dynamics inferred from differentials between individual measurements.

Grain and Extent

Observation sets distinguish entities and levels by virtue of the surfaces detected. Two particularly relevant aspects of observation sets at this point are grain and extent, both of which position the observation set with respect to surfaces.

Grain determines the fineness of the distinctions that can be made in an observation set. Sampling more often or employing an analytical procedure that preserves fine distinctions, both make the observation set more fine-grained. Lower level entities can only be seen in a fine-grained observation set which preserves the fine distinctions needed to discriminate between small things. Thus, in a time series, sampling intervals must be kept short to detect high frequency behavior. Grain determines the lower limits of observation and fixes the finest possible level of resolution in an observation set.

In contrast, extent determines the largest distinctions (i.e., the largest surfaces) that can be seen. If the characteristic behavior of a relevant large entity takes longer to occur than the period of the entire sampling regime, then the behavior can not be seen. Sampling for one summer season does not allow a plankton ecologist to address phenomena associated with the annual cycle. The difference between successive summers can only be studied if the observation set is extended to cover two or more summers. Sampling for only one summer fixes the extent so as to deny access to annual phenomena no matter what the mode of analysis.

The detection limits imposed by grain upon the smallness and, by extent, upon the grandness of phenomena are absolute. It is not possible to go beyond these limits once the observation set is fixed. Analysis and interpretation may impose further limitations even if the sampling aspects of grain are small and the extent is large. Understanding involves rejection of the full set of ways to look at a sampled system in favor of a powerful subset that allows relevant prediction. It is precisely because of limits of human comprehension that analyses are performed explicitly to remove fine-grained distinctions so that large scaled phenomena may emerge in interpretation. The preliminary sampling aspects of grain and extent in observation sets give the limits of what it is possible to see; within these limits, transformation and analysis further confine what is seen to something commensurate with human comprehension. Furthermore, even with modern computers and a defined system like the game chess, computational limits are reached remarkably quickly (Pattee 1973). In systems where the rules are unknown, such as the interrelationships of the leaves on just one tree, computational limits close in immediately (Weinberg 1975).

Looking In and Out

Now consider the difference between looking in at entities as opposed to looking out at the environment. Looking inward, the observer sees the discreteness of entities (Figures 2, 3, and 4). Looking outward, the observer focuses his attention on the background rather than on any discrete object (Figures 5 and 6). Looking out directs attention at the environment, not the discrete entities.

Koestler (1967) emphasized the part-whole duality of entities in a hierarchical system. He viewed entities as windows of interconnection between their parts and the rest of the universe. Note how, in that conception, the rest of the universe is undifferentiated; it corresponds to our notion of environment. Looking out at the environment, in the present scheme, endeavors to fit the entity of interest into the larger whole. The entity is conceived as a part. By contrast, looking in toward its outer surface, the entity is seen as a whole.

Consider looking in, with an observation set of small extent so that only a single entity can be differentiated (Figure 4). Everything else is considered to be part of the environment. Environment is explicitly not differentiated into entities. It is that "everything else" which can be defined only in terms of its dynamical influence on the structures of definition. Now, increase the extent so that a second entity is detected and seen to be interacting

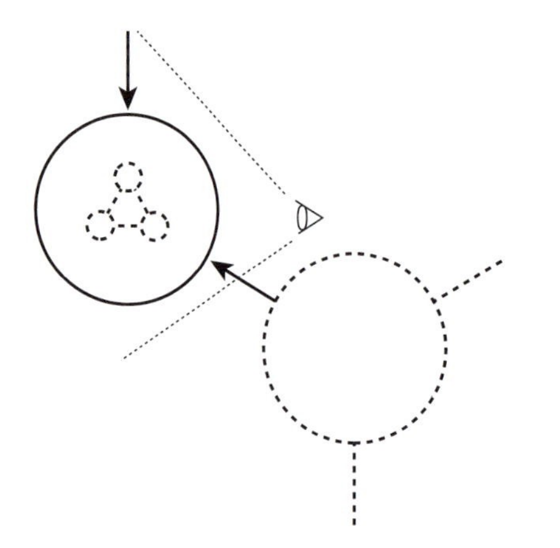

Figure 4. Seen from outside, the parts are obscured by the intervening surface and the other entity is manifested only as an environmental influence of undefined origin. (The eye indicates the position from which the system is observed.)

with the first (Figure 3). Now a part of the original set of "environmental influences" can be seen to be interactions between the two entities. If structure is recognized in the environment, then one is no longer looking at environment, but instead, at a pair consisting of the original entity and the new entity which emerged from the environment. The new structure emerged from the dynamical background as the extent of the observation set is increased. Looking outward focuses on the fact that the original entity, and now the pair of them, is part of a greater whole, as yet structurally undefined, and subject to influences from that context.

To match the part/whole duality of the entities in a hierarchical system, there is a corresponding duality for dynamics. Dynamics link parts to make the whole and deliver the influence of the whole to the parts. Entities sandwiched between the dynamical influences of their environment are in one direction, and the dynamical interaction of their parts are in the other direction.

Structures thus possess two sets of behaviors or processes. One set is directed upward to the next level and pertains to the "part" mode of the structural part/whole duality. The other behavior is directed downward, and is the dynamic that pertains to the "whole" mode of the structural part/whole duality. This downward directed dynamic gives the structure as a whole control over its parts. It is not always immediately apparent to which set (i.e., upward or downward directed) any given behavior belongs. That only emerges when grain and extent are manipulated for several criteria so that dynamics leading to parts can be discriminated from behavior leading to higher levels. It is possible for a given behavior to have aspects that lead upwards and other aspects that lead down, as will be demonstrated later.

Moving Through Surfaces

When a surface is viewed from the outside, the separateness of the parts is ordinarily lost (Figure 4 compared to Figure 2). Conversely, when a surface is viewed from the inside, the separate identities of other higher level entities are lost (Figure 2). Distinctions that can be made on one side of the surface cannot be made on the other, because the signal is integrated and filtered out by the surface. The separate parts are merged into the integrated whole (Figure 4) and the other entities are merely part of an undifferentiated environmental influence (Figure 5). In both cases, structure on the other side cannot be seen through the surface (Platt 1969). When an observer moves far enough through a surface, the distinctions on the other side become clear. Either formerly obscure parts can be detected (Figures 2, 4, and 5) or separate entities emerge from the environment (Figures 2 to 6).

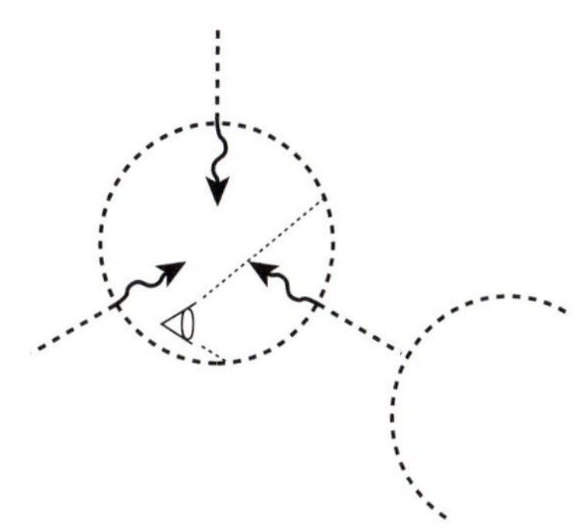

Figure 5. From inside the surface, looking outward, the observer cannot see the other entity because of the intervening surface. All he can see are the wavy lines which denote environmental influences modified by the surface. (The eye indicates the position from which the system is observed.)

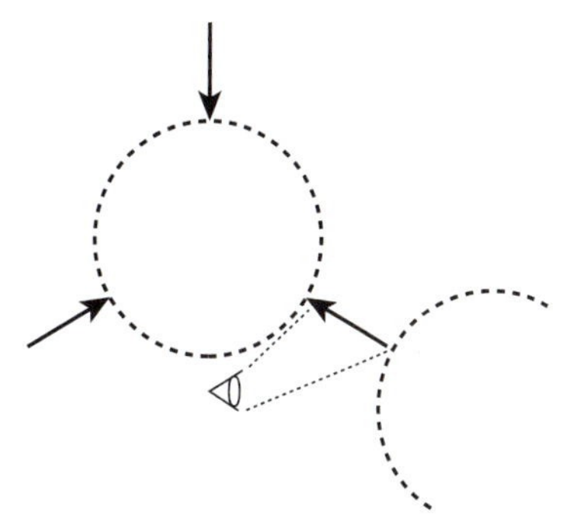

Figure 6. When the observer looks outward from a surface, he can identity unmodified environmental influences, but he does not have sufficient scope of vision to see the other entity. (The eye indicates the position from which the system is observed.)

The way appearances and relationships change as one moves through a surface can be clarified by an example. Species association (Williams and Lambert 1959) is a relative matter, relative particularly to the size of the universe in which the comparison is made (i. e., the extent of the observation set). Consider a two-by-two contingency table for tree species association within a forest. Such a table might well show a negative association between species occurrences, indicating distinct distributions of the species within the forest. The association is negative, because most samples contain either one species or the other, and relatively few contain both.

Now increase the extent of the observation set with additional samples taken in adjacent grasslands. Because neither species appears in these new samples, the number of mutual occurrences does not change. Previously, expectations of mutual occurrence were high, because both species were common in the data set: common species should occur together commonly. As the observation set is extended, the proportion of samples containing either species becomes smaller. As the species become less common, there should be fewer mutual occurrences, and the observed values move closer to expectations. As a result, the negative correlation disappears. Move out through a surface and the distinction between the species distributions (inferred from the negative correlation) is obscured (Figure 2 compared to Figure 4).

Further expansion of the observation set to include more and more grassland samples makes both species rare, and their mutual occurrences now appear higher than expected. That is, the two species are now very rare and generally would not be expected to occur together at all. However, they still do occur together: now more frequently than one would expect if each were randomly distributed across the entire sampling universe. At this point they become positively associated and are seen as sharing the status of "forest species." Now, positive associations appear among forest species, and negative associations separate the forest species as a group from the grassland species as a group (Beals 1973).

Moving out through the surface of the forest altered what can be observed. Within the forest (Figure 2) one could discriminate between species distributions. Pioneer trees were distinguished from climax trees by negative associations (i.e., if one is present, the other is likely to be absent). As the observation set is expanded, the distinction between species (Figure 4) is lost. Pioneer and climax trees are both forest species, and that common identity begins to emerge. Further expansion of the observation set (Figure 3) makes it impossible to discriminate among tree species, which are now seen together as part of an entity distinguished from a second entity (i.e., grassland species) by negative associations.

Moving Out through a Surface

Movement out through any surface, such as the forest, is achieved through a new observation set which is greater in extent. Usually, it is also more coarse grained. Coarse grain erects an opaque surface, because it does not allow perception of the smaller entities on the other side of the surface (Figure 4). Movement in through a surface may be achieved by the converse operations, using an observation set which has smaller extent and is also finer grained. The observation set now has a narrower extent, and the observer is trapped within the surface of any entity larger than the extent of the observation set (Figure 2).

Because of the change in grain, surfaces are ordinarily opaque. The only exception is when a new observation set involves only a change in extent. When a higher level emerges by virtue of a simple increase in extent, the smaller entities still may be detectable and are not necessarily obscured by the surface. It is this special case that allows one to tunnel from one level to the next, and to draw direct perceptual links between levels.

Transparent surfaces are crucial for linking levels, but are limited as a way to look at the higher level entity. When a surface is transparent, the new entity is simply an aggregate. It is being viewed by the same criteria which make the parts interesting objects of study. New properties which might be interesting at the higher level, sometimes called emergent properties, will not ordinarily be seen in this observation set. Higher levels of organization are important precisely because they have properties which are not immediately relevant to the parts. If organisms seen as aggregates of organs and tissues did not also display homeostasis and reproduction, there would be little reason to designate them as special objects of study. It is only when new properties are seen that the differentiation of the higher level entity advances understanding and enriches experience. These properties do not appear unless the criteria for observing the higher level entity differ from those used to study the parts. Thus, the observation sets appropriate for different levels ordinarily differ in criteria. Although there is a special case where surfaces appear transparent, the normal change in observation criteria that occurs as the observer changes his level ensures that the surfaces of most entities are opaque.

Returning Through the Surface

Once the interesting properties of the higher level entity are identified, it is often profitable to change grain and extent and move back within the surface. However, now the criteria for observation have changed. For example, "forest" might be identified as a higher level entity in an extensive observation set on tree species. Now the interest is in the forest as an entity that fixes carbon. At this point the focus is on a new property of the forest. The motivation for moving back across the surface of the forest is to identify relationships or interactions (which may amount to mechanisms) operating between the parts, which explain the new property. The surface of the larger entity is transparent under this new criterion, since only extent manipulated and primary production by individual trees within the forest is observed. However, the new observation set on the lower level entities is based on new criteria and is not identical to first observation of this level. In the first observation set, trees were grouped into species categories. Now, when the observer re-enters the forest, entities may be grouped according to whether they are in the canopy or the understory, irrespective of species. New entities, herbaceous plants which were not considered in the original observation set, will also be included.

It is satisfying to the observer if the parts and their interactions in the new observation set are identical to those observed in the original analysis of the trees. In this case he feels confident in the linkage established between the levels. However, in the face of emergent properties, there is the possibility that the parts and processes observed by moving back inside the surface will be somewhat different.

If the same parts and processes cannot be observed as one moves back within the surface, the linkage between levels (i.e., between the new properties of the higher level and the old parts in the original observation set) remains inferential rather that perceptual, and the surface is opaque. The point of significance here is that there is no reason to expect criteria for observing one level to apply to observing another. The only time a match can be hoped for is when only grain and extent are changed. When criteria of observation change (i.e., when new properties are identified at the higher level) there is no a priori reason to believe that the new properties can be reduced to interactions among the old parts. New parts and new processes may be expected. Just because large scale entities remain observable under new criteria (e.g., forest defined by carbon fixation rate as opposed to species composition), there is no need to expect a concomitant mapping (Rosen 1977) of the parts. The large scale entity may be robust (i.e., preserved under different observation criteria), even though the parts needed to explain behavior under the new criteria may change. The result is that trying to link properties of interest at one level to properties of interest at a different level is a difficult process.

Inferring the Whole From the Dynamics of the Parts

Despite the difficulties, ecologists must draw inferences across levels. Therefore, one must analyze what is involved in such inferences. With increase in the extent of an observation set, a surface is reached. At this point, a set of interacting entities have formed a sort of closure, so that they are interacting among themselves with high frequency and show only lower frequency interactions with the undifferentiated environment. The higher level entity has now emerged (Figure 7).

By comparing observation sets on either side of this surface, one can now infer that the entities observed inside the surface indeed are the interacting parts which make up the new whole and can be observed from outside the surface (Figure 7). It is precisely the rapid interactions among the parts which result in the structural integrity of this new whole. The new structure is defined by the dynamics of the parts.

Note here that multiple observation sets are required to draw inferences about linkages between levels. Observation sets that differ in

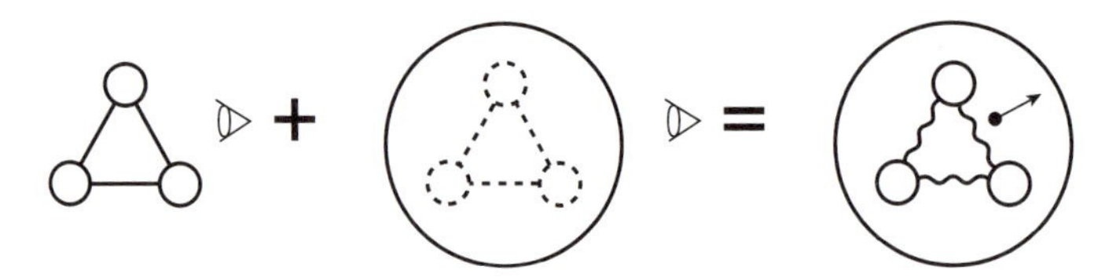

Figure 7. Drawing inferences about the relationship between parts and wholes. The first observation set sees the collection of parts and their interactions but cannot see the larger whole because of narrow extent. The second observation set uses the same criteria for distinction, but with coarser grain and larger extent. This second set sees the whole but, at best, only a ghost of the parts. Together these two observation sets allow inferences to be drawn on how relationships among the parts lead to the whole. The dot-arrow symbol points to the whole, but the dot is within, because the whole is conceived as being derived from separate parts. The wavy lines denote inferred rather than perceived relationships.

criteria or in both grain and extent may only be linked by inference, because each is a separate mapping of the world. The closest one can come perceptually to link between levels is a change in extent, with every effort made to keep the grain and criteria for distinction the same. If criteria change, one cannot be sure of correspondence.

Inferring Environmental Filtering

Moving through a surface while focusing attention outward (i.e., on the environment) leads to a completely new situation (Figures 2 to 5). Now inferences are sought about the manner in which the whole modifies the experience of its parts by filtering of environmental influences. Because the focus is not on entities, the required change in observation set can seldom, if ever, be made by a simple change in extent. The situation will ordinarily involve at least two observation sets and an inference drawn from the comparison (Figure 8).

There is a significant difference in the environment viewed from inside (Figure 5) and outside (Figure 3) of the surface. Outside, an undifferentiated background is seen as influence, manifested in the behavior of the whole through its correlation with environmental measurements. Environmental influence on the parts cannot be perceived, because the parts themselves cannot be seen (Figure 3). Inside, one can see environmental signal modified by passage through the surface. Comparing the observation sets one can infer that the whole modifies environmental influences before they operate directly on the parts (Figure 8).

An example here might help clarify the point. Consider diurnal temperature fluctuations and a forest. Outside the forest, temperature fluctuations are quite large. Inside, the biomass of the trees moderates these fluctuations. Forest trees experience cooler temperatures during the day and warmer temperatures at night than a single tree standing alone in a field. The forest tree experiences only the modified influence, and it is beside the point that the night temperature outside the forest is low. Thus, it is by drawing an inference from the observation of the temperature inside and outside the forest that it can be stated that the environmental influence is modified by the forest. Only then can the observer conclude that the temperature difference results from the mod-

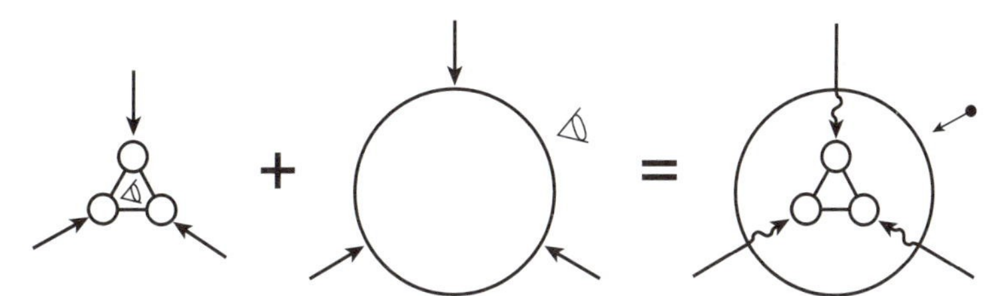

Figure 8. Drawing inferences about environment based on two observation sets. The first observation set sees only the parts and their immediate environmental influences. The second observation set considers the whole as itself a part of some undifferentiated higher level. Together, these two observation sites allow inferences to be drawn about how the whole ameliorates the environmental experience of the parts. The dot-arrow symbol points to the whole and the dot is outside since the whole is conceived as a modifier of the environmental influence coming from the outside. The wavy lines denote inferred, not perceived, relationships.

erating influence of the forest, which filters the temperature signal from the environment. In general, in order to draw conclusions about the role of the whole in modifying the environment or constraining the parts, a procedure like this must be followed.

Drawing Inferences Versus Changing Extent

Ecologists often draw inferences using observation sets based on different criteria. There is an important difference between crossing a surface by manipulation of grain and extent, and crossing by change of criteria. When grain and extent are changed to give a different pattern but the criteria for distinction are held constant, then an orderly change has taken place. The parts and whole are linked in the change that took the observer through a surface. When a change occurs by the imposition of different criteria, the observation sets represent different mappings of the world. It may be that there is coherence across the change, and indeed the change does reflect a part/whole relationship, but this is not necessarily the case.

It should be apparent that errors could easily be made whenever the observer unwittingly changes the criteria for significance. A case in point could involve drawing inferences between species composition of a forest and nutrient cycling. Reduction of grain and extent in an observation set on the forest may show the system parts not as species but as functional entities. These entities may be formed by the interaction of soil, water, organic matter, parts of some plants (root hairs), all of some fungi (mycorrhizae), and communities of organisms such as microbes and nematodes that migrate in and out and are not always included in the functional part. The community with its species will not reappear. As a result, inferences drawn between species composition and the components of nutrient cycling must be tenuous, at best.

Often such inferences can be made more explicit by comparing observation sets that differ only in grain and extent. A case in point arises in comparing ordination analyses based on different criteria. Focusing on different phenomena reveals different aspects of the community. Each aspect refers to organismal relationships based on some linking factor such as disturbance, soil moisture, or intensity of competition. Different factors are based on different criteria of observation, and give constellations of species relationships that reflect different scales in time and space.

As an example, it is possible to interpret communities as determined by either competition or environment. By focusing on competition over a narrow range of mesic conditions, environment becomes merely the context within which competition occurs. On a larger geographic scale which includes environmentally stressed sites, one can view environment as the direct cause of species distributions (Allen and Starr 1982). The difference between the two emphases can be seen, for example, in the way that the tolerant, poor competitors (i.e., losers in one observation set) occupy stressed sites successfully (i.e., winners in the other observation set). Notice how it is difficult to hold these two concepts of community at the same time; for environments, stress is the frame in one, whereas competition is the frame in the other. Furthermore, competition is explicitly irrelevant in the stress frame. Putting it another way, a view that focuses on one set of phenomena cannot give a perfect account of a system if the interest is in some different set of phenomena.

To look at a specific case, Loucks (1962) performed ordinations of forests in New Brunswick by two different criteria. One was a Bray and Curtis (1957) ordination of stand species composition. The other was based on scalars of environmental gradients. This is a clear case of change of criteria. Because of this, Loucks was forced into an extended and deeply inferential argument to relate the results of the two ordinations.

The difference between the ordination results was great in that the moisture gradient curved around in the vegetation analysis. Black spruce and cedar were positioned together in the vegetation ordination at the ends of the distorted moisture gradient. In the environmental ordination, the moisture gradient must necessarily be straight, and black spruce and cedar are bimodally distributed, appearing at either end in very moist or very dry sites.

While Loucks was forced to explain these changes indirectly, he recognized that a series of observation sets of increasing extent would be a more direct approach:

> "The data for black spruce suggest that if the samples in the lowest [stressed] range of the Nutrient scalar were considered alone, black spruce is not bimodal. It becomes so only when considered with preponderance of other communities that develop at higher levels of the Nutrient scalar; these nutrient levels occur only on intermediate moisture positions."
>
> (Loucks 1962)

If he had begun with the low nutrient sites and expanded the extent to include more and more mesic sites, he would have seen black spruce and cedar change in an orderly fashion. Both low nutrient status and wet/dry extremes of the moisture gradient represent environmental stress. At the extremes, abiotic stress allows survival of the tolerant in a circumstance where competition is unimportant. Black spruce and cedar would have become rare as the number of mesic sites increased, because neither of them do well under mesic conditions where competition with other species is important. Then the fine-grained concept (i.e., structured by competition) and the coarse-grained concept (i.e., structured by environment) could have been directly related through a simple change of grain and extent.

Correct and Incorrect Linkages

At this point ecologists can begin to identify the types of error they encounter when endeavoring to link levels of organization. Some rules and warning are already part of ecological practice, having emerged from the exercise of common sense. Even so, there is something to be said for a scheme that derives from general principles instead of common practice. The former would be a general prescriptive scheme, the latter merely local and descriptive.

Linking Levels Through Phenomena

Any scientific investigation starts by designating the entities one chooses to discriminate. The entities are structural and arise from the criteria for recognizing entities in an observation set. For example, one can choose to study individual organisms. Then he observes that the individuals demonstrate behavior (i.e., they can change through time). Once the entities are fixed, the observer has no choice as to whether or not his measuring device detects change. In this sense, the behavior of the individuals is forced on the observer.

Although the dynamics of the entities are forced on the observer, he must still make a second, observer-dependent choice concerning which specific dynamics he will consider to be of importance. The observer must enter the process again and decide the particular changes of state he will select as significant. The changes he selects constitute the "phenomena" of interest. Thus phenomena are dynamical behavior but are quite distinct from the raw dynamics that are imposed on the observer when he infers differentials for the measured states of the structure of interest. The raw dynamics are objective and imposed on the observer, while the dynamic which is a phenomenon is the special subjectively chosen subset which represents only the dynamics asserted to be interesting.

Observation, therefore, depends on two distinct choices separated by dynamics. The steps involved in setting up any observation set are (1) the choice of the level of interest and the entities or structures at the level, (2) the development of a means to observe changes in these entities, and (3) the choice of which changes will be considered significant (Rosen 1977). The observer decides about both entities and phenomena even though the dynamics are imposed on him.

Phenomena form crucial links between levels. Consider observations made on individual organisms through time (Figure 9). Among other things, the observer notices that eventually each of the individuals ceases to function. He decides this change of state is interesting, and "death" is established as the phenomenon (Figure 9). By increasing the extent of our observation set to include the death of many individuals it is possible to calculate a death rate.

The dynamic now links the observations to a higher structural level since death rate is a behavioral dynamic of the population (Figure 9). The individual and the population are linked through their dynamics by having a phenomenon, death, in common. Whenever a phenomenon is significant at two levels, it can be used to link the levels. The link is made perceptually since it only involves changing the extent of the data set on dynamics. This is another example of the fact that a direct or perceptual linkage between levels can only be made by a simple change in grain and extent while maintaining the same criterion of distinction.

An analogous path can be followed to connect the organism to lower levels (Figure 9). Once again observations will be made on the individuals, but now special attention will be given to change in size for each organism.

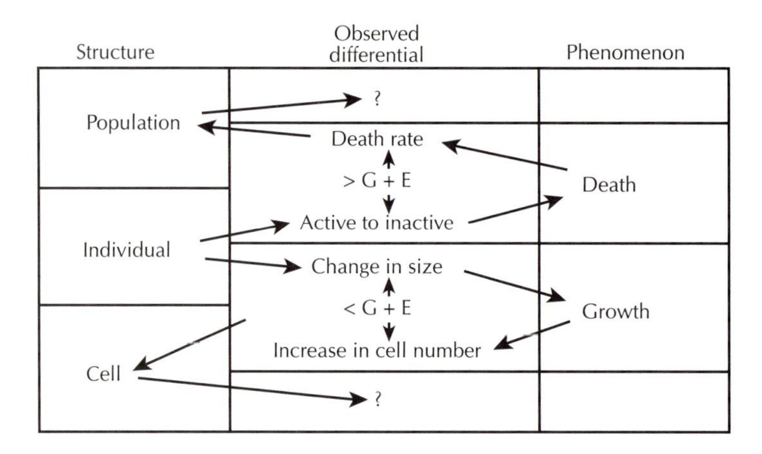

Figure 9. Entities at structural levels above and below the individual are linked by first observing a differential (i.e., a change in state), then denoting an appropriate phenomenon, and finally observing again with modified grain and extent to observe a new behavior that links the phenomenon to the other level.

Thus, "growth" is selected as the phenomenon. If the extent of the observation set is maintained, but a much finer grain is selected, changes in the number of cells might be seen (Figure 9). This leads to a lower structural level, because change in number is a dynamic definable on the entity, "cell." Thus, once again levels have been linked by a change in grain and extent using a phenomenon that they have in common. Note, however, that we could choose the phenomenon "growth," see trees grow, expand extent to see many trees growing, and then infer growth as an internal dynamic of a larger structure, namely forest.

Of course, it is possible to explain the phenomenon of growth in other ways. One might decide to examine caloric inputs, arriving at growth as the net difference. This approach is certainly legitimate, but, because of the change in criteria, it is unlikely that cells would be discovered as a lower level entity. The change in criterion would destroy the ability to link these levels considered above.

This approach for linking levels may seem pedantic and overdrawn. However, the necessity for the distinctions can only be appreciated when we see the errors which can be made when the linkage is made on less rigorous grounds.

Error in Linking Levels

The level to which the link is made may change depending on the phenomenon chosen. In the end, different levels may be linked than those the observer had in mind because the phenomenon chosen had different consequences than those he intended. The starting point for this example is the population, and the dynamic involves foraging and eating. If the behavior, eating, is seen to be associated with the phenomenon, pollination, then an expanded conservation set could lead to an upper level entity called "guild" (Figure 10). Alternatively, given the same structure and behavior, if the phenomenon is recognized as primary consumption, then an increase in

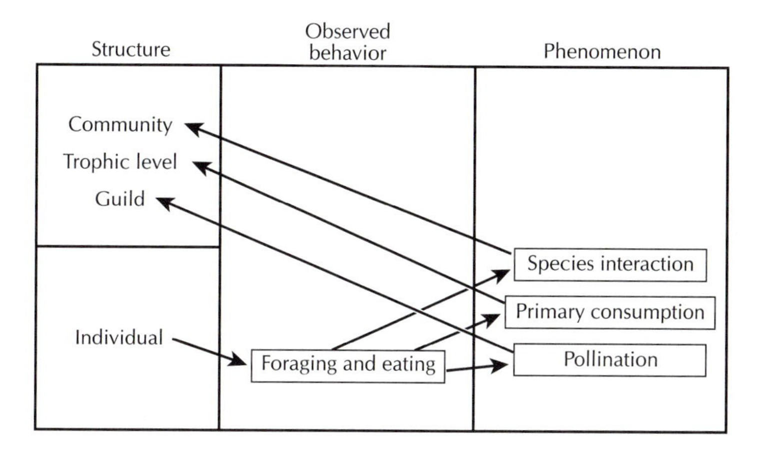

Figure 10. The structures at the individual level are observed to change state and so exhibit behavior. Three different phenomena can be chosen on the same observation set and each links the individual level with a different upper level structure.

extent of observation set would lead to "trophic level." As a further alternative, recognition of the behavior as an interspecific relationship might lead one up the "community" (Figure 10). Thus, the linkages depend on the phenomenon, not just the raw observed changes in state.

Similarly, the levels which the ecologist links may change, if he changes the dynamics which he observes. Consider Figure 9 once again; but now observe whether or not individuals mate with each other. The relevant phenomenon might now be constancy of mating among individuals or a similar type. Now a change in extent would not lead to the population or community, but to the taxonomic species as the next higher level.

These examples illustrate a problem commonly found in linking levels in ecology. Given the definition of level and phenomenon, it is clear from Figure 10 that there are no absolute levels of organization, independent of the observer. There is no Platonic Chain of Being. Given a prior choice of structure defining the starting level, it is the choice of phenomenon that determines what is the next level in the hierarchy, whether it is the next level up or down. The phenomenon is chosen by the observer, not imposed by the external world; and it is the phenomena that both explain the workings of higher levels and describe the role of lower levels. Thus, the analysis presented here reveals that any attempt to find the true ecological hierarchy (e.g., MacMahon et al. 1981) is founded on an inadequate epistemological base and must of necessity fail. Intermediate or alternative levels always can be located by a change of phenomenon. When one moves up and down from the starting point of choice, the other levels are not fixed by what other investigators, with other observation sets, might choose as their starting points. Thus, the first problem with this analysis is the widespread practice of choosing the levels in the hierarchy a priori. The higher and lower levels must be revealed to the observer by a change in grain and extent with the entities needed to explain the mechanics (lower) and function (higher) of the chosen phenomenon.

This problem often arises in linking levels which are widely separated. Allen and Starr (1982) note that it becomes much harder to connect levels directly when there are intervening levels which are relevant to the chosen phenomenon but ignored. A case in point would involve explaining environmental control of phytoplankton communities by recourse to cell physiology and uptake rates of nutrients. Allen et al. (1977) suggest that the intervening levels of species, guilds, and strategies (Allen and Koonce 1973) attenuate and confound the linkage. Difficulty arises because the levels of observation are too far apart and articulation is lost. The levels of organization were imposed instead of being allowed to emerge from observations.

A similar problem arises in attempts to link ecosystems to communities or populations (Figure 11). Once again, the levels are prescribed, and the linkage is assumed. It is true that ecosystems have populations in them, but this "observation" actually involves a number of observation sets, based on very different criteria. In this report, it has been demonstrated that such a specification of the complex problem is inadequate, and depends upon insufficient evidence to establish the linkage. As a result, there is an epistemological problem with defining the ecosystems as composed of plant and animal populations and their environment.

To link ecosystems and populations, it is necessary to establish common phenomena and show that the observation sets that lead upward to the ecosystem and downward to the population involve a change of grain and extent but not change in criteria. In fact, it is possible to make this connection for some phenomena but not for others (Figure 11). Consider, for example, primary production. Here, a simple change of grain and extent will reveal that plant populations, and nothing else, are involved in this dynamic. However, other ecosystem phenomena, such as nutrient retention, do not provide adequate linkages. It was previously identified how a relevant component here, the rhizosphere, is a strange mixture of organism wholes and parts. The problem is that the dynamic does not result only from interactions among biotic populations. Nutrient retention may be in part biotic (tree boles), it may be abiotic (soil organic matter), or it may be due to a complex interaction. In this case, the ecosystem dynamic will not emerge from any observation set on the populations and their interactions. There is no simple increase in grain and extent that takes the population observations to nutrient retention dynamics of the ecosystem. The links only occur when an ecosystem phenomenon is given account by a single population or an aggregate of populations.

Another common error in linking ecological levels results from ambiguity in the words used to delineate a phenomenon. For example, the word "competition" implies a link between individuals and populations. However, the meaning to the word is very level-dependent. As Harper (1967) points out, individual competition is commonly between the mature of one species and the juveniles of another. Therefore, there is much confounding of neighborhood and population competition. Although the treatment of competition as something very local permeates much of J. L. Harper's earlier work, "Neighborhood competition" did not even arise as a specific technical

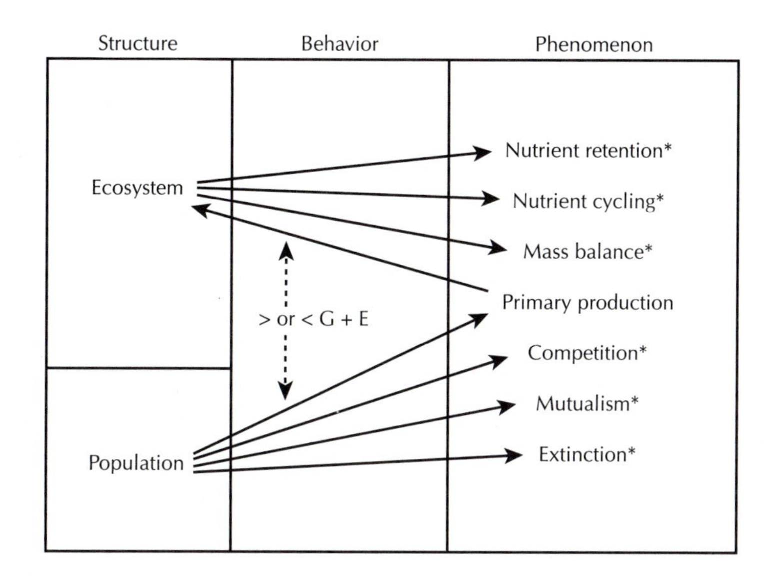

Figure 11. Ecosystems can be linked to population, but only if a phenomenon can be identified that is common to both, such as primary production. The asterisks denote phenomena that are not common to both levels and cannot serve to link the levels.

term for individual to individual competition until recently (Mack and Harper 1977). The taxonomy of the individuals in neighborhood competition is usually unhelpful in explaining the observation since size and maturity are the determining factors. At the population level, "competition" refers to interactions among populations. Thus, if the observation set used to detect neighborhood competition does not contain information on the species, it cannot be changed by grain and extent to detect the competition which is of interest at the population level. If the observation set that detects population competition does not (as most do not) contain size hierarchy information, then a narrowing of grain and extent will not lead to neighborhood competition. The relationship between the two types of competition is by metaphor or, at best, analogy. The mistake is to assume that they are homologous.

Neighborhood competition viewed as resource capture does lead to populations seen in terms of resource base, because behavior of populations with respect to resource base is a simple extension of neighborhood resource capture, that extension being achieved through manipulation of only grain and extent. It is a population behavior that may be seen as directed downward. Population competition, however, involves behavior that is most readily seen as linking populations to make higher, not lower, hierarchical levels.

Similar ambiguities arise with the word "stability" used to link ecosystems and populations. At the ecosystem level, stability may refer to the constancy of a function, such as a primary production. At the population or community level, stability may refer to the fact that all of the populations are still there following a disturbance. However, it is quite possible the primary production has remained constant following a disturbance precisely because of species replacement resulting from resource competition. One population has disappeared and another has expanded to take over its function. Thus, the system can be considered simultaneously as stable and unstable depending on the level of interest. As a result, the phenomenon "stability" is a poor candidate for linking the levels. Simply because the same word is used for phenomena on two levels does not insure adequate linkage. One must be certain that the criteria have not changed even though the word has remained the same.

Summary: General Principles for Linking Levels

The following summarizes these analyses in a set of general principles.

1. The level at which an inquiry begins is arbitrary. The observer chooses the entities on which he will focus and they are differentiated by observer-defined criteria. Effective criteria emphasize surfaces that (1) coincide with significant changes in rates of interaction, and (2) can be detected in different observation sets (i.e., are robust under transformation, as forest edges that coincide for change in both temperature and biomass).
2. A scientific description of the level chosen for study requires explicit recognition of part/whole duality (Koestler 1967). The entity of interest will consist of parts (i.e., lower level entities) which it constrains and contains. At the same time, the entity of interest must be seen as part of a greater whole, although this may be described in undifferentiated terms as its environment. To describe the entity as both a whole composed of parts and as a part of a greater whole will ordinarily require drawing inferences based on different observation sets; and the inferences will only be validly grounded on a perceptual basis if the criteria for significance have not changed between the observation sets. To move upward in the hierarchy will require an observation set with sufficient extent that it includes a number of entities and their interactions. To move downward requires an observation set of sufficiently fine grain that the individual parts and their interactions can be observed.
3. Once the observer begins to make observations on the entities, he can detect changes which correspond to their behavior unique to that level. A full description of dynamics, like structure, is also dual, and both delivers constraints downward to parts, as well as showing how the entity responds upwards to the greater whole of which it is a part. To understand both the upward and downward aspects of dynamics will ordinarily require inferences based on multiple observation sets.
4. Following observation of dynamics, a further observer-dependent choice must be made. This arbitrary decision involves the aspect of dynamics which will be given special significance and called the "phenomenon" of interest. The choice of phenomena will then determine the higher and lower-level entities with which we can establish linkages.
5. Dynamic linkages between levels are established through phenomena. If the phenomenon in combination with a change of grain and/or extent can be established

as interesting at both higher and lower structural levels, and if it can be related to each of these levels through observation sets on dynamics that differ only in extent or grain, then the linkage is justified. Care must be taken that the word used to describe the phenomenon has the same meaning on both levels.

6. Linkages between levels using only a change in grain and extent are crucial for established unequivocal or at least justified assertions of relationship. Only one phenomenon should be involved with only one set of criteria for observation. The relationship is simple because upper levels here are only aggregates of lower levels.

7. The simplicity of relationships so defined severely limits what insights can be derived. Fortunately levels which emerge from such simple relationships may persist when criteria for observing them are changed. Once they are found, emerging levels may be robust under transformation. Reverse manipulations of grain and extent using these new criteria do not usually lead back to the original level because a new phenomenon associated with the new criteria has become involved. Explanations of the new phenomenon do not return to the original level of concern but to a new level defined by the new criterion for observation.

8. From the above, the relationship between levels can be explained by alternative criteria (e.g., the relationship of a forest to alternative lower levels is complex: First, move from one lower level to a forest using one criterion, increasing the grain and extent; next, shift the criteria for defining a forest; finally, reduce to the alternative lower level by a reduction in grain and extent of the second criteria).

9. Attempts to link prescribed levels are difficult because the linking phenomena are undefined, and it may not be possible to find them under any observation set.

Literature Cited

Allen, T. F. H. and J. F. Koonce. 1973. Multivariate approaches to algal stratagems and tactics in systems analysis of phytoplankton. Ecology 54:1234–1246.

Allen, T. F. H., Steven M. Bartell, Joseph F. Koonce. 1977. Multiple stable configurations in ordination of phytoplankton community change rates. Ecology 58:1076–1084.

Allen, T. F. H. and Thomas B. Starr. 1982. *Hierarchy: Perspectives for Ecological complexity.* The University of Chicago Press, Chicago.

Beals, Edward W. 1973. Ordination: Mathematical elegance and ecological naivete. Journal of Ecology 61:23–36.

Bray, J. Roger and J. T. Curtis. 1957. An ordination of the upland forest communities of southern Wisconsin. Ecol. Monogr. 27:325–349.

Harper, J. L. 1967. A Darwinian approach to plant ecology. Journal of Ecology 55:246–270.

Koestler, Arthur 1967. *The Ghost in the Machine.* Macmillan, New York.

Kuhn, T. S. 1962. *The Structure of Scientific Revolutions.* The University of Chicago Press, Chicago.

Levins, Richard 1974. Discussion paper: The qualitative analysis of partially specified systems Annals of the New York Academy of Sciences 231:123–138.

Loucks, O. L. 1962. Ordinating forest communities by means of environmental scalars and phytosociological indices. Ecol. Monogr. 32:137–166.

Mack, Richard N., and John L. Harper. 1977. Interference in dune annuals: Spatial pattern and neighborhood effects. Journal of Ecology 65:345–363.

McMahon, James A., David J. Schimpf, Douglas C. Anderson, Kimberly G. Smith, Robert L. Bayn, Jr. 1981. An organism-centered approach to some community and ecosystem concepts. Journal of Theoretical Biology 88:287–307.

Miller, J. R. 1978. *Living Systems.* McGraw-Hill, New York.

Pattee, Howard H. 1973. *Hierarchy theory: The Challenge of Complex Systems.* Braziller, New York. 156 pp.

Pattee, H. H. 1978. The complementarity principle in biological and social structures. Journal of Social and Biological Structures 1:191–200.

Platt, J. 1969. Theorems on boundaries in hierarchical systems. In: *Hierarchical Structures.* L. L. Whyte, A. G. Wilson, and D. Wilson, eds. American Elsevier, New York. Pp. 201–214.

Rosen, Robert 1977. Observation and biological system. Bulletin of Mathematical Biology 39:663–678.

Simon, Herbert A. 1962. The architecture of complexity. Proceedings of the American Philosophical Society 106:467–482.

Simon, Herbert A. 1973. The organization of complex systems. In: *Hierarchy theory: The Challenge of Complex Systems.* Braziller, New York. Pp. 1–28.

Sugihara, George 1983. Peeling apart nature. Nature 304:94.

Webster, Jackson R. 1979. Hierarchical organization of ecosystems. In: *Theoretical System Ecological.* Efraim Halfon, editor. Academic Press, New York. Pp. 119–129.

Weinberg, Gerald M. 1975. *An Introduction to General Systems Thinking.* Wiley, New York.

Williams, W. T. and J. M. Lambert. 1959. Multivariate methods in plant ecology. I. Association analysis in plant communities. Journal of Ecology 47:83–101.

Introduction to
Garry Peterson, Craig R. Allen, and C. S. Holling
Ecological Resilience, Biodiversity, and Scale

by Timothy F. H. Allen

One would not be off the mark too far to say that diversity was one of the weakest intellectual constructs in ecology. Although an easy measurement to make, diversity languished for a concept to make it nontrivial. Until this paper, the absence of scalar notions in the literature made diversity a nonstarter; and while most ecologists claim to know that scale matters, most considerations of scale have been simplistic. Peterson, Allen and Holling, by contrast, are very sophisticated about scale. By showing explicitly what was missing in the literature, offering a systematic treatment of the various traditional models of diversity and laying out their assumptions, the authors how such assumptions are so gross as to make the issue of diversity uninteresting.

Communities are not easily seen as things in a place, because the different species that make up the community occupy the landscape at different scales. Accordingly, diversity with its lumping of species as all somehow equivalent has critical problems until one identifies the critical scales of operation in the system. Diversity is linked with stability in the literature with limited success. For systems to remain stable, the components cannot be linked together too tightly. Consider the stability of the prey-predator relationships for two different systems: lions versus ladybugs. Each predator has a scale of operation where it is too efficient and destabilizes the resource (a single kill for lions, and populations of insects exterminated on a whole plant by ladybugs). On the other hand, lions and ladybugs both also have a level of inefficient exploitation that leaves enough slack for long term stability (low predation levels for lions at the population level, and ladybug ineptitude at finding new populations on new plants). The paper by Peterson et al. lays out how to put scale differences into diversity, and so shows how diversity can be related to stability. This work is a prime example of how a muddled literature suddenly can be made clear by introducing the scale issue in a sophisticated way.

Ecological Resilience, Biodiversity, and Scale

GARRY PETERSON, CRAIG R. ALLEN, AND C. S. HOLLING

Original publication in *Ecosystems* (1998), 1:6–18. Reprinted with permission.

Abstract

We describe existing models of the relationship between species diversity and ecological function, and propose a conceptual model that relates species richness, ecological resilience, and scale. We suggest that species interact with scale-dependent sets of ecological structures and processes that determine functional opportunities. We propose that ecological resilience is generated by diverse, but overlapping, function within a scale and by apparently redundant species that operate at different scales, thereby reinforcing function across scales. The distribution of functional diversity within and across scales enables regeneration and renewal to occur following ecological disruption over a wide range of scales.

Introduction

One of the central questions in ecology is how biological diversity relates to ecological function. This question has become increasingly relevant as anthropogenic transformation of the earth has intensified. The distribution and abundance of species have been radically transformed as massive land-use changes have eliminated endemic species (Turner and others 1993), and the expansion of global transportation networks has spread other species (McNeely and others 1995). This biotic reorganization is co-occurring with a variety of other global changes, including climate change, alteration of nutrient cycles, and chemical contamination of the biosphere. Maintaining the ecological services that support humanity, and other life, during this extensive and rapid ecological reorganization requires understanding how ecological interactions among species produce resilient ecosystems.

Species perform diverse ecological functions. A species may regulate biogeochemical cycles (Vitousek 1990; Zimov and others 1995), alter disturbance regimes (Dublin and others 1990; D'Antonio and Vitousek 1992), or modify the physical environment (Jones and others 1994; Naiman and others 1994). Other species regulate ecological processes indirectly, through trophic interactions such as predation or parasitism (Kitchell and Carpenter 1993; Prins and Van der Jeud 1993), or functional interactions such as pollination (Fleming and Sosa 1994) and seed dispersal (Brown and Heske 1990). The variety of functions that a species can perform is limited, and consequently ecologists frequently have proposed that an increase in species richness also increases functional diversity, producing an increase in ecological stability (Tilman and others 1996).

The idea that species richness produces ecological stability was originally proposed by Darwin (1859), reiterated by MacArthur (1955), and modeled by May (1973). Recently, Tilman and colleagues (Tilman 1996; Tilman and others 1996) experimentally demonstrated that in small systems, over ecologically brief periods, increased species richness increases the efficiency and stability of some ecosystem functions, but decreases population stability. Despite the demonstrated link between species richness and ecological stability over small scales, the nature of this connection remains uncertain.

Models of Ecological Organization

Many competing models attempt to describe how an increase in species richness increases stability. Following previous authors, we divide these models into four classes: "species richness-diversity" (MacArthur 1955), "idiosyncratic" (Lawson 1994), "rivet" (Ehrlich and Ehrlich 1981), and "drivers and passengers" (Walker 1992). These models all explicitly or implicitly assume that a species has ecological function, and that the function of a species can be represented as occupying an area of multidimensional ecological function space (Grinnell 1917; Hutchinson 1957; Sugihara 1980). For illustrative purposes, we compress multidimensional functional space into one dimension in which breadth represents the variety of a species' ecological function (Clark 1954). For example, a species such as a beaver, that strongly influences tree populations, hydrology, and nutrient cycles, has a broad function, whereas a fig wasp that pollinates a single species of fig would have a narrow function. We represent the intensity of a species' ecological function by height. For example, a "keystone species" (Paine 1969; Power and others 1996) has a stronger influence than a "passenger" species (Walker 1992).

We emphasize the differences between these models before discussing their similarities. We then present our model of "cross-scale resilience," which incorporates scale into an expanded model of the relationship between diversity and ecological function.

Species Diversity

Darwin (1859) proposed that an area is more ecologically stable if it is occupied by a large number of species than if it is occupied by a small number. This idea was formalized by MacArthur (1955), who proposed that the addition of species to an ecosystem increases the number of ecological functions present, and that this increase stabilizes an ecosystem (Figure 1).

Although many experimental studies have demonstrated that increasing the number of species increases the stability of ecosystem function (Schindler 1990; Naeem and others 1994; Frost and others 1995; Holling and others 1995; Ewel and Bigelow 1996; Tilman 1996), apparently no investigations of the relationship between species richness and stability have indicated that additional species continue to increase stability at a constant rate, indicating that the species-diversity model is excessively simplistic. Consequently, we focus our attention upon models that propose more complex relationships between species richness and ecological stability.

Idiosyncratic

A competing model of the relationship between species and ecological function proposes that strong ecological interactions among species result in an ecosystem that is extremely variable, and contingent on the particular nature of interspecific interactions (Lawson 1994). This model proposes that the degree of stability in a community depends idiosyncratically upon which species are present (Figure 2).

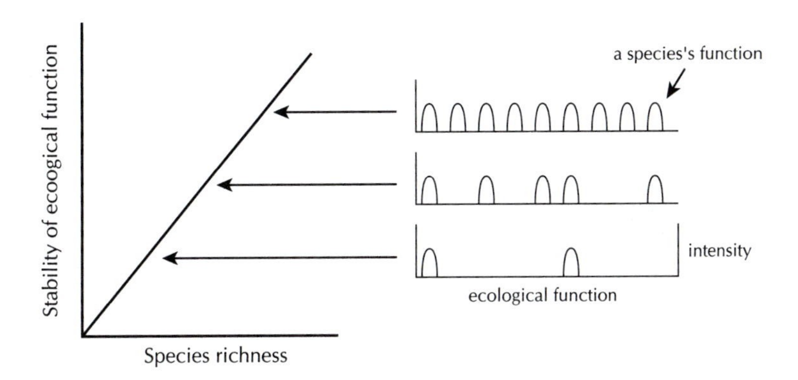

Figure 1. A representation of the Darwin/MacArthur model: increasing species richness increases the stability of ecological function. This model, and the other models we discuss, implicitly represents species ecological function as occupying a portion of a multidimensional ecological function space that is analogous to niche space (MacArthur 1955). As species accumulate, they fill this space. The width and height dimensions of the inset diagrams represent the breadth and intensity of a species' ecological function. This model assumes that function space is relatively empty and therefore species can be continually added to a community without saturating it. It also assumes that the strength and breadth of ecological functions do not vary among species.

For example, fire ants have had great impacts on ecosystems of the southeastern United States (Porter and Savignano 1990; Allen and others 1995), but have a much different role in the Pantanal of Brazil and Paraguay (Orr and others 1995). Such situations suggest that ecosystem function is contingent on the ecological history of a region and the evolutionary history of interacting species. However, ecosystems are not only products of historical contingency, ecosystem ecology has demonstrated that many ecosystems are similarly organized.

Many ecosystem studies have revealed that despite dissimilar species compositions, ecosystems can have striking ecological similarities. For example, lake studies have demonstrated that similar ecological function can be maintained over a wide mix of species and population densities (Schindler 1990; Frost and others 1995). Mediterranean climate ecosystems provide a good example of functional convergence. The world's five Mediterranean climate regions, despite geographic and evolutionary isolation that has produced radically different floras and faunas, are extremely similar in ecological structure and function (Di Castri and Mooney 1973; Kalin Arroyo and others 1995). This convergence suggests that species are organized into functional groups, and that these groups are determined by regional ecological processes. Both the "rivet" (Ehrlich and Ehrlich 1981) and "drivers and passengers" (Walker 1992) models of functional diversity assume that some sort of functional redundancy exists, but they differ in the importance they assign to functional groups.

Rivets

Empirical evidence suggests that the effect of species removal from or addition to an ecosystem varies. Ehrlich and Ehrlich's (1981) rivet hypothesis, which is similar to Frost and colleagues' (1995) model of compensating complementarily, likens the ecological func-

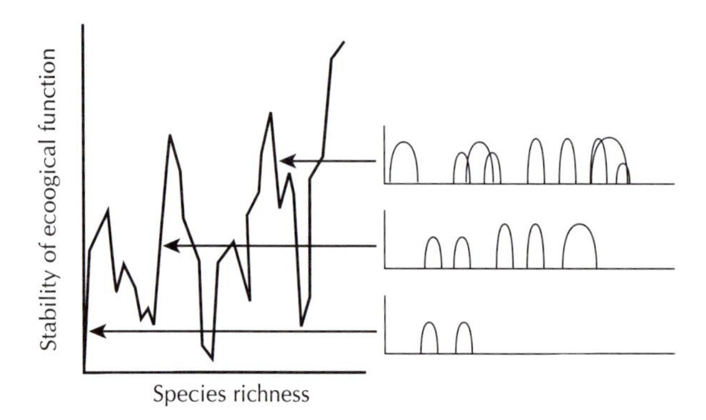

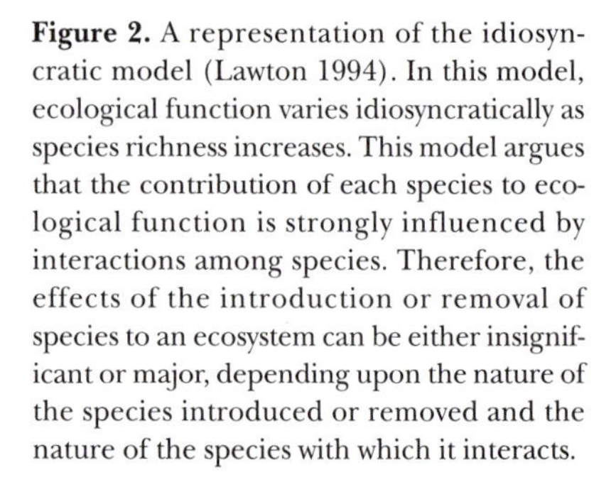

Figure 2. A representation of the idiosyncratic model (Lawton 1994). In this model, ecological function varies idiosyncratically as species richness increases. This model argues that the contribution of each species to ecological function is strongly influenced by interactions among species. Therefore, the effects of the introduction or removal of species to an ecosystem can be either insignificant or major, depending upon the nature of the species introduced or removed and the nature of the species with which it interacts.

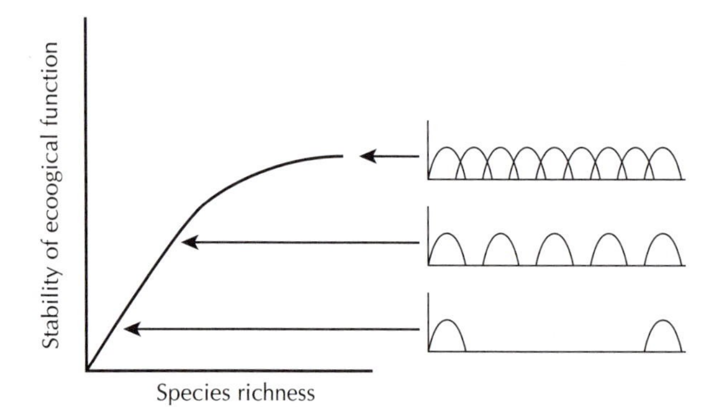

Figure 3. The "rivet" model of ecological function (Ehrlich and Ehrlich 1981) presumes that ecological function space is relatively small. Therefore, as species are added to an ecosystem, their functions begin to overlap or complement one another. This overlap allows ecological function to persist despite the loss of a limited number of species, since species with similar functions can compensate for the elimination or decline of other species. However, the increase of stability gained by adding new species decreases as species richness increases and functional space becomes increasingly crowded.

tion of species to the rivets that attach a wing to a plane. Several rivets can be lost before the wing falls off. This model proposes that the ecological functions of different species overlap, so that even if a species is removed, ecological function may persist because of the compensation of other species with similar functions (Figure 3).

In the rivet model, an ecological function will not disappear until all the species performing that function are removed from an ecosystem. Overlap of ecological function enables an ecosystem to persist. Compensation masks ecosystem degradation, because while a degraded system may function similarly to an intact system, the loss of redundancy decreases the system's ability to withstand disturbance or further species removal.

Drivers and Passengers

Walker's "drivers and passengers" hypothesis accepts the notion of species complementarily and extends it by proposing that ecological function resides in "driver" species or in functional groups of such species (Walker 1992, 1995). It is similar to Holling's (1992) "extended keystone hypothesis." Walker defines a driver as a species that has a strong ecological function. Such species significantly structure the ecosystems in which they and passenger species exist. Passenger species are those that have minor ecological impact. Driver species can take many forms. They may be "ecological engineers" (Jones and others 1994), such as beavers (Neiman and others 1994), or gopher tortoises (Diemer 1986), which physically structure their environments. Or drivers may be "keystone species" (Paine 1969), such as sea otters (Estes and Duggins 1995) or asynchronously fruiting trees (Terborgh 1986), that have strong interactions with other species (Power and others 1996). Walker (1995) proposes that since most ecological function resides in the strong influence of driver species, it is their presence or absence that determines the stability of an ecosystem's ecological function (Figure 4).

Model Synthesis

Whereas the "rivet" hypothesis assumes that ecological function is evenly partitioned among species, Walker's model assumes there are large differences between drivers that have strong ecological function and passengers that have weak ecological function (Figure 4). Both hypotheses recognize that different types of ecological functionality are required to produce ecological stability, and that as additional species are added to an ecosystem the increasing redundancy of function decreases the rate at which ecological stability increases. The existence of some type of ecological redundancy is supported by experiments conducted in Minnesota grasslands, tropical rainforests, artificial mesocosms, and lakes (Schindler 1990; Naeem and others 1994; Ewel and Bigelow 1996; Tilman and others 1996).

Tilman, for example, demonstrated that more diverse plots (4 x 4 m) have greater plant cover and more efficiently utilize nitrogen (Tilman 1996). Tilman and colleagues demonstrated that ecological function was more stable in diverse communities despite, or perhaps because of, large fluctuations in populations of species (Tilman and others 1996). These results echo those of Frank and McNaughton (1991), who demonstrated that more diverse natural grass communities recovered faster than less diverse communities following drought.

In a series of experiments, Ewel and coworkers constructed a set of tropical ecosystems with different levels of species richness and compared their functioning to adjacent rainforest. They demonstrated that relatively few species, if drawn from different functional groups, can duplicate the ecological flows of a diverse rainforest (Ewel and others 1991). Herbivory per leaf area was lower and less variable in species-rich plots (Brown and Ewel 1987).

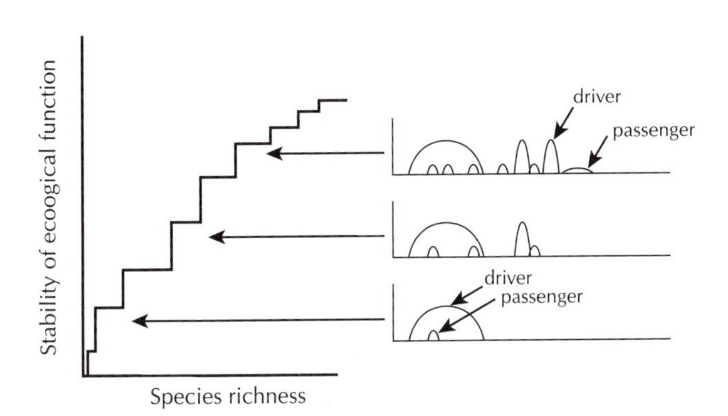

Figure 4. Walker's "drivers and passengers" model of redundant ecological function (1992, 1995) proposes that ecological function is unevenly distributed among species. Drivers have a large ecological impact, while passengers have a minimal impact. The addition of drivers increases the stability of the system, while passengers have little or no effect.

They also demonstrated that a variety of ecosystem variables, such as soil organic matter, increase rapidly as one adds different functional types to a plot (Ewel and Bigelow 1996), and that simple agroecosystems function quite similarly to much more species-rich rainforests, at least in areas of about ⅓ ha (80 x 40 m) for 5 years (Ewel and others 1991).

Naeem and coworkers (1994) assembled replicate artificial ecosystems at a number of levels of species richness. They demonstrated that carbon dioxide consumption, vegetative cover, and productivity increased with species richness. These increases were greater between 9 and 15 species than between 15 and 31 species, providing support for the hypothesis that an increase in species richness increases ecological redundancy. Water and nutrient retention did not vary with species richness.

Frost and coworkers (1995) demonstrated that ecological function is preserved if population declines of zooplankton species are compensated for by population increases in other species with similar ecological functions. Their results suggest that lakes with fewer species in a functional group would exhibit decreased ability to compensate for population declines in other species. Similarly, Schindler (1990) observed that the largest changes in ecological processes and food-web organization occurred when species that were the only remaining member of a functional group were eliminated.

These studies demonstrate that the stability of many, but not all, ecological processes increases with species richness. They also suggest that the ecological stability is generated more by a diversity of functional groups than by species richness. These results suggest a possible synthesis of the various models relating stability to species richness.

The model that best describes an ecosystem appears to depend upon the variety of functional roles that are occupied in that system, and the evenness of the distribution of ecological function among species. An ecosystem consisting of species that each perform different ecological functions will be less redundant than an ecosystem consisting of the same number of species that each perform a wide variety of ecological functions. Similarly, if there is little difference between the ecological impact of different species, there is little point in differentiating driver and passenger species; they can all be considered rivets. We propose that these models of how species richness influences the stability of ecological function can be collapsed into a simple model that can produce specific versions of these models by varying the degree of functional overlap and the degree of variation in ecological function among species (Figure 5).

The experimental results just discussed suggest ecosystems possess considerable functional redundancy. Indeed, it is difficult to envision how ecosystems without redundancy could continue to persist in the face of disturbance. We assume that since no species are identical, redundancy does not reside in groups of species, but rather it emerges from the interactions of species. Therefore, it is not possible to substitute species for one another; rather, there are many possible combinations and organizations of species that can produce similar ecological functions. Redundancy quickly emerged in the experimental ecosystems, but these experiments were all conducted over relatively small areas and short time periods. Ewel and his coworkers (1991) conducted the longest and largest experimental manipulations of diversity, but even 5 years and ⅓ ha are small in comparison to the spatial and temporal dynamics of

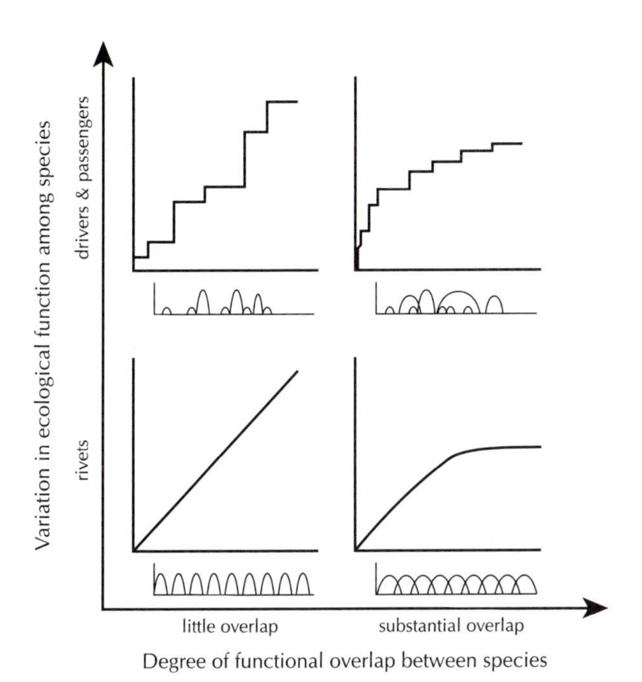

Figure 5. The relationship between stability and species richness varies with the degree of overlap that exists among the ecological function of different species and the amount of variation in the ecological impact of species ecological function. Overlap in ecological function leads to ecological redundancy. If the ecological impact of different species is similar they are "rivets," whereas if some species have relatively large ecological impact they are "drivers" and others are "passengers."

an ecosystem, or even the life span and home range of a medium-sized mammal.

Understanding of stability and ecological function developed at small scales can not be easily extended to larger scales, since the type and effect of ecological structures and processes vary with scale. At different scales, different sets of mutually reinforcing ecological processes leave their imprint on spatial, temporal and morphological patterns. Change may cause an ecosystem, at a particular scale, to reorganize suddenly around a set of alternative mutually reinforcing processes. For example, Hughes (1994) described an epidemic that caused a 99% decline in the population of an algae-eating fish in a Jamaican near-shore coral community. The loss of these herbivores caused the community to shift from being dominated by corals to being dominated by fleshy macroalgae. Similar reorganizations are demonstrated in paleo-ecological (Carpenter and Leavitt 1991), historical (Pries and Jeud 1993), and long-term ecological research (Hughes 1994).

Resilience

Assessing the stability of ecosystems that can reorganize requires more than a single metric. One common measure, what we term engineering resilience (Holling 1996), is the rate at which a system returns to a single steady or cyclic state following a perturbation. Engineering resilience assumes that behavior of a system remains within the stable domain that contains this steady state. When a system can reorganize (that is, shift from one stability domain to another), a more relevant measure of ecosystem dynamics is ecological resilience (Holling 1973). Ecological resilience is a measure of the amount of change or disruption that is required to transform a system from being maintained by one set of mutually reinforcing processes and structures to a different set of processes and structures. Note that this use of resilience is different from its use by others [for example, Pimm (1984)], who define resilience as what we term engineering resilience (Holling 1996).

The difference between ecological and engineering resilience can be illustrated by modeling an ecological "state" as the position of a ball on a landscape. Gravity pulls the ball downward, and therefore pits in the surface of the landscape are stable states. The deeper a pit, the more stable it is, because increasingly strong disturbances are required to move an ecological state away from the bottom of the pit. The steepness of the sides of a stability pit corresponds to the strength of negative feedback processes maintaining an ecosystem near its stable state, and consequently engineering resilience increases with the slope of the sides of a pit (Figure 6).

Ecological resilience assumes that an ecosystem can exist in alternative self-organized or "stable" states. It measures the change required to move the ecosystem from being organized around one set of mutually reinforcing structures and processes to another. Using the landscape metaphor, whereas engineering resilience is a local measure of slope of the stability landscape, ecological resilience is a measure of regional topography. The ecological resilience of a state corresponds to the width of its stability pit. This corresponds to the degree to which the system would have to be altered before it begins to reorganize around another set of processes (Figure 7).

Ecological and engineering resilience reflect different properties. Ecological resilience concentrates on the ability of a set of mutually reinforcing structures and processes to persist. It allows ecologists or managers to focus upon transitions between definable states, defined by sets of organizing processes and structures, and the likelihood of such occurrence. Engineering resilience, on the other hand, concentrates on conditions near a steady state where transient measurements of rate of return are made following small disturbances. Engineering resilience focuses upon small portions of a system's stability landscape, whereas ecological resilience focuses upon its contours. Engineering resilience does not help assess either the response of a system to large perturbations or when gradual changes in a system's stability landscape may cause the system to move from one stability domain to another. For these reasons we concentrate on ecological resilience.

Scale

Ecosystems are resilient when ecological interactions reinforce one another and dampen disruptions. Such situations may arise due to compensation when a species with an ecological function similar to another species increases in abundance as the other declines (Holling 1996), or as one species reduces the impact of a disruption on other species. However, different species operate at different temporal and spatial scales, as is clearly demonstrated by the scaling relationships that relate body size to ecological function (Peters 1983).

We define a scale as a range of spatial and temporal frequencies. This range of frequencies is defined by resolution below which faster and smaller frequencies are noise, and the extent above which slower and larger frequencies are background. Species that operate at the

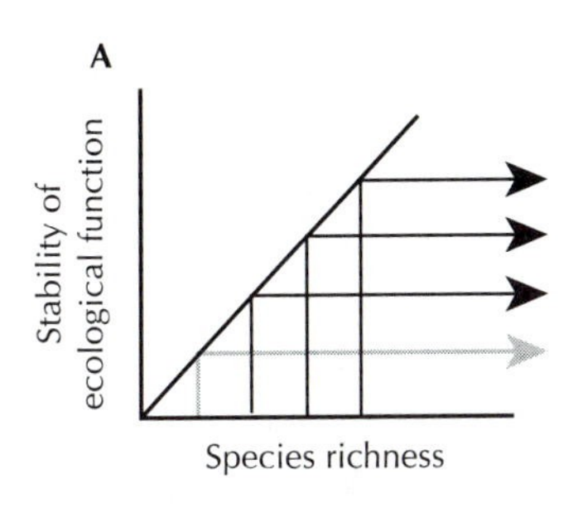

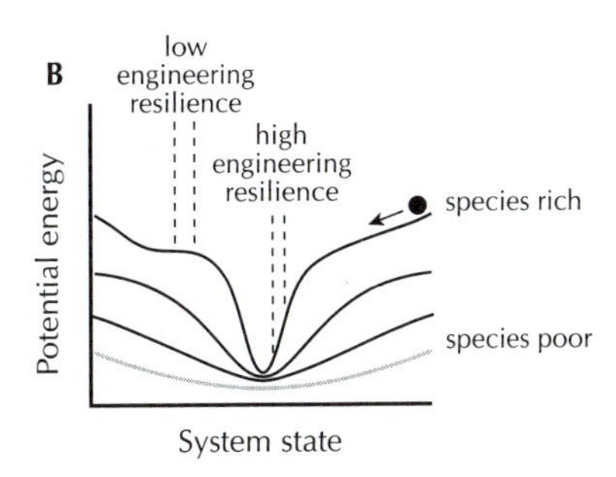

Figure 6. The relationship between stability and species richness can be represented by a set of stability landscapes. The dynamics of a system are expressed by a landscape, and its "state" is represented by a ball that is pulled into pits. Different landscape topographies may exist at different levels of species richness. In this model, the stability of a state increases with the depth of a pit. Zones of the stability surface that have low slopes have less engineering resilience than do areas that have steep slopes.

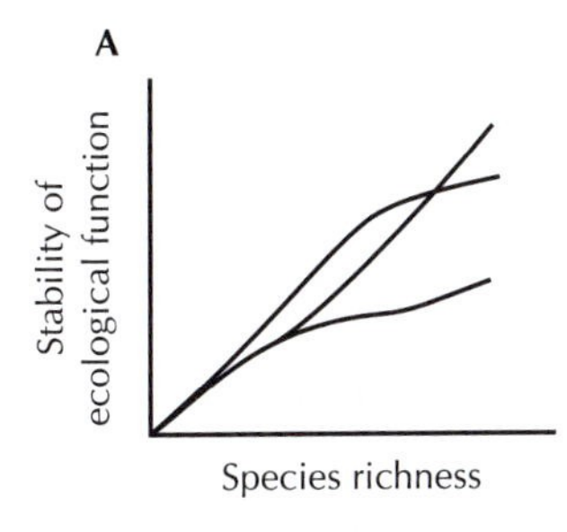

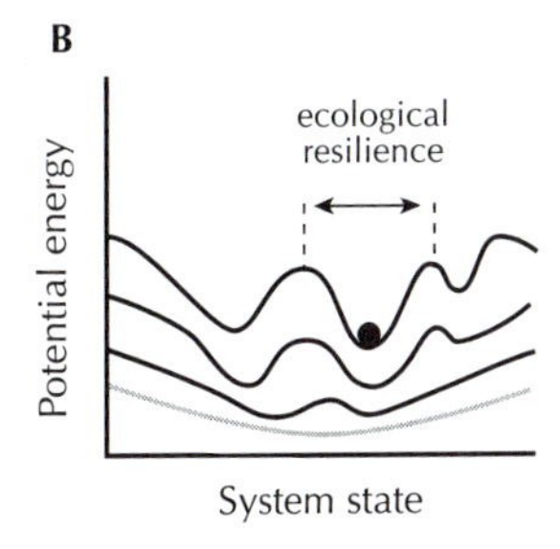

Figure 7. A system may be locally stable in a number of different states. Disturbance that moves the system across the landscape and slow systemic changes that alter the shape of the landscape both drive the movement of a system between states. The stability of a state is a local measure. It is determined by the slope of the landscape at its present position. The resilience of a state is a large-scale measure, as it corresponds to the width of the pit the system is currently within.

same scale interact strongly with one another, but the organization and context of these interactions are determined by the cross-scale organization of an ecosystem. Consequently, understanding interactions among species requires understanding how species interact within and across scales.

Many disturbance processes provide an ecological connection across scales. Contagious disturbance processes such as fire, disease, and insect outbreaks have the ability to propagate themselves across a landscape, which allows small-scale changes to drive larger-scale changes. For example, the lightning ignition of a single tree can produce a fire that spreads across thousands of square kilometers. Such disturbances are not external to ecological organization, but rather form integral parts of ecological organization (Holling 1986). Disturbance dynamics affect and are affected by species and their ecological functions (D'Antonio and Vitousek 1992). Consequently, the processes regulating contagious disturbances are as much determinants of ecological resilience as are more local interactions among species.

Current models of the relationship between species richness and stability implicitly model species and their ecological functions at the same scale; however, ecological systems are not scale invariant. A growing body of empirical evidence, theory, and models suggests that ecological structure and dynamics are primarily regulated by a small set of plant, animal, and abiotic processes (Carpenter and Leavitt 1991; Levin 1992; Holling and others 1995). Processes operate at characteristic periodicities and spatial scales (Holling 1992). Small and fast scales are dominated by biophysical processes that control plant physiology and morphology. At the larger and slower scale of patch dynamics, interspecific plant competition for nutrients, light, and water influences local species composition and regeneration. At a still larger scale of stands in a forest, mesoscale processes of fire, storm, insect outbreak, and large mammal herbivory determine structure and successional dynamics from tens of meters to kilometers, and from years to decades. At the largest landscape scales, climate, geomorphological, and biogeographical processes alter ecological structure and dynamics across hundreds of kilometers and over millennia (Figure 8). These processes produce patterns and are in turn reinforced by those patterns; that is, they are self-organized (Kauffman 1993).

Ecological processes produce a scale-specific template of ecological structures that are available to species (Morse and others 1985; Krummel and others 1987; O'Neill and others 1991). Ecological structure and patterns vary across landscapes and across scales. Many species may inhabit a given area, but if they live at different scales they will experience that area quite differently. For example, a wetland may be inhabited by both a mouse and a moose, but these species perceive and experience the wetland differently. A mouse may spend its entire life within a patch of land smaller than a hectare, while a moose may move among wetlands over more than a thousand hectares (Figure 8). This scale separation reduces the strength of interactions between mice and moose relative to interactions among animals that operate at similar scales (Allen and Hoekstra 1992). In the next section, we propose a conceptual model that relates species richness, ecological resilience, and scale.

Species, Scale, and Ecological Function

Species can be divided into functional groups based upon their ecological roles (Clark 1954; Komer 1996). Species can be also be divided into groups based upon the specific scales that they exploit. The ecological scales at which species operate often strongly correspond with average species body mass, making this measure a useful proxy variable for determining the scales of an animal's perception and influence (Holling 1992). We propose that the resilience of ecological processes, and therefore of the ecosystems they maintain, depends upon the distribution of functional groups within and across scales.

We hypothesize that if species in a functional group operate at different scales, they provide mutual reinforcement that contributes to the resilience of a function, while at the same time minimizing competition among species within the functional group (Figure 9). This cross-scale resilience complements a within-scale resilience produced by overlap of ecological function among species of different functional groups that operate at the same scales. Competition among members of a multitaxa functional group may be minimized if

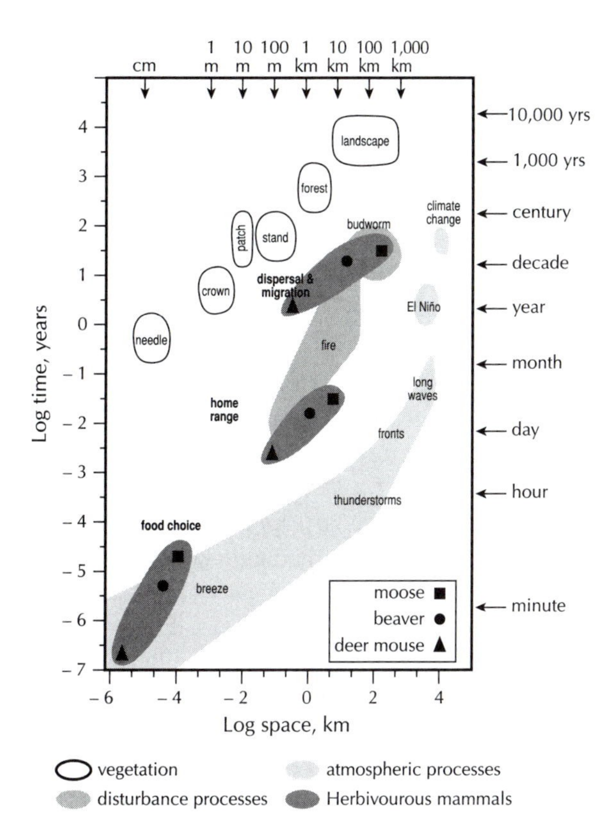

Figure 8. Time and space scales of the boreal forest (Holling 1986) and their relationship to some of the processes that structure the forest. These processes include insect outbreaks, fire, atmospheric processes. and the rapid carbon dioxide increase in modern times (Clark 1985). Contagious mesoscale disturbance processes provide a linkage between macroscale atmospheric processes and microscale landscape processes. Scales at which deer mouse, beaver, and moose choose food items, occupy a home range, and disperse to locate suitable home ranges very with their body size (Holling 1992; Macdonald 1985; Nowak and Paradiso 1983).

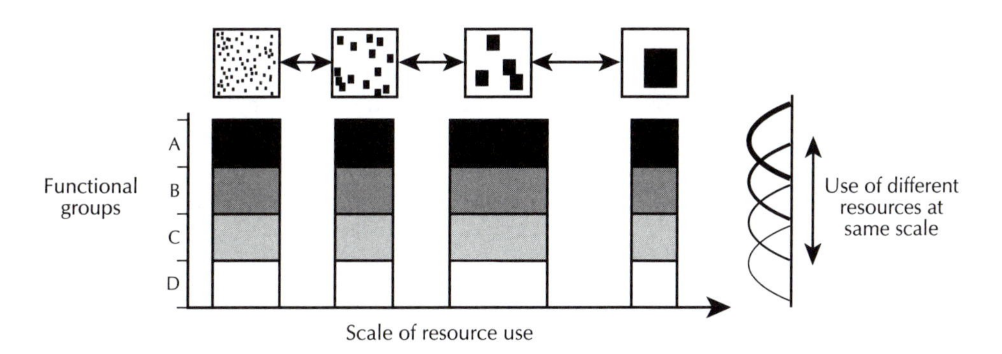

Figure 9. Our hypothesized relationship between the scale of species interactions and their membership in a functional group. Different species use resources at different spatial and temporal scales. Members of a functional group use similar resources, but species that operate at larger scales require those resources to be more aggregated in space than do species that operate at smaller scales. Within scales, the presence of different functional groups provides robust ecological functioning, whereas replication of function across scales reinforces ecological function. The combination of a diversity of ecological function at specific scales and the replication of function across a diversity of scales produces resilient ecological function.

group members that use similar resources exploit different ecological scales. Ecological resilience does not derive from redundancy in the traditional engineering sense; rather, it derives from overlapping function within scales and reinforcement of function across scales.

We illustrate these two features of resilience by summarizing the effects of two functional groups on ecosystem dynamics and diversity in two different systems. The first example summarizes the results of field and modeling investigations of the role of avian predators in the dynamics of spruce/fir forests of eastern North America. The second summarizes field and modeling studies of the role of mammalian seed dispersers in the tropical forests of East Africa.

Avian Predation of Insect Defoliators

The combination of within-scale and cross-scale resilience enables an ecological function such as predation of keystone defoliators to be maintained despite sudden variations in resource availability or environmental conditions. It is well known that if a particular insect becomes more common, species that would not normally exploit it may switch to using it (Murdoch 1969). This occurs as the increasing relative abundance of a resource makes its utilization less costly. We argue that as resources become increasingly aggregated they become available to larger animals that are unable to exploit dispersed resources efficiently. This mechanism introduces strong negative feedback regulation of resource abundance over a wide range of resource densities.

A well-studied example of such a situation is found in the forests of New Brunswick, Canada, where outbreaks of a defoliating insect, spruce budworm *(Choristoneura fumiferana)*, periodically kill large areas of mature boreal fir forest. The initiation of these outbreaks is controlled by the interactions between the slowly changing volume of a growing forest susceptible to budworm, the more quickly changing densities and feeding responses of budworm's avian predators, and rapidly changing weather conditions (Morris 1963; Clark and Holling 1979).

Avian predation on budworm regulates the timing of budworm outbreaks by having its largest influence when budworm densities are low and forests stands are young. At least 31 species of birds prey upon budworm (Holling 1988). These bird species can be divided into five distinct body-mass classes or body-mass lumps, separated by gaps in their body-mass distributions (Holling 1992). The existence of budworm predators in these different body-size classes makes the influence of predation robust over a broad range of budworm densities. This robustness emerges not because the predators exhibit redundant functional forms of predation, but rather because the scales at which predators are effective overlap, spreading their impact over a wide range of densities and spatial aggregations of budworms.

The predatory effectiveness of a bird is largely determined by its body size. The amount of food that a bird can consume—its functional response (Holling 1959)—is a function of its body size, and a bird's search rate is greatly influenced by the scale at which it searches. Kinglets *(Regulus* sp.), chickadees *(Paws* sp.) and warblers (Emberizidae), small birds with an average body mass of about 10 g, concentrate on recognizing prey at the scale of needles or tufts of needles. Medium-sized birds focus their foraging upon branches, while larger birds such as evening grosbeaks *(Coccothraustes vespertinus,* 45 g) react to stand-level concentrations of food such as irruptions of seeds during good mast years or stand-level budworm outbreaks. The movement of birds over a landscape also is scaled to its body size. Larger birds forage over wider areas than do smaller birds. Consequently, both the body mass of birds attracted to budworm and the distance from which they are attracted will increase as the size of local aggregations of budworm increase. A diversity of foraging strategies within and across scales thus provides a strong and highly resilient predation on budworm populations (Holling 1988), particularly at low densities of budworm within stands of young trees (< 30 years old).

Members of functional groups maintain and therefore determine the resilience of ecosystems by spreading their influence over a range of scales. When a functional group consists of species that operate at different scales, that group provides cross-scale functional reinforcement that greatly increases the resilience of its function. This interpretation of the partitioning of ecological function suggests that what is often defined as redundancy, is not. The apparent redundancy of similar function replicated at different scales adds resilience to an ecosystem: because disturbances are limited to specific scales, functions that operate at other scales are able to persist. The production of resilience by cross-scale functional diversity can be illustrated in a model of seed dispersal.

Mammalian Seed Dispersal in an African Tropical Forest

In Uganda's Kibale National Park, seed dispersers vary in size from small mice that range over areas of less than a hectare, to chimpanzees that range over tens of square kilometers. In a simple model of seed dispersal, when the area disturbed annually and the total amount of dispersal are held constant. the population growth rate of mammal-dispersed trees is determined by the distance over which its seeds are dispersed and the size of disturbance. A diverse set of dispersers, functioning at different scales, enables the tree population to persist despite disturbance. If, however, large, long-distance seed dispersers are absent, the tree population declines, especially when large disturbances occur (Figure 10). Mammal-dispersed trees are more aggregated when dispersal is only by small mammals that move the seeds small distances. When disturbance sizes are large, this limited dispersal is unable to maintain populations of mamma-dispersed trees (G. Peterson and C. A. Chapman, unpublished data).

Due to cross-scale functional reinforcement, and the nonlinear fashion in which ecosystem behavior can suddenly flip from one set of mutually reinforcing structures and processes to another, the gradual loss of species in a functional group may initially have little apparent effect, but their loss would nevertheless reduce ecological resilience. This decrease in resilience would be recognized only at specific spatial and temporal scales, and even then may be compensated for within or across scales. However, the ecosystem would become increasingly vulnerable to perturbations that previously could have been absorbed without changes in function or structure.

An indirect consequence of species loss is that it limits the potential number of ways a system can reorganize. Especially troubling is the possibility that the loss of large species, such as moose (Pastor and others 1993) or elephants (Dublin and others 1990), that generate mesoscale ecological structure may also eliminate forms of ecological organization.

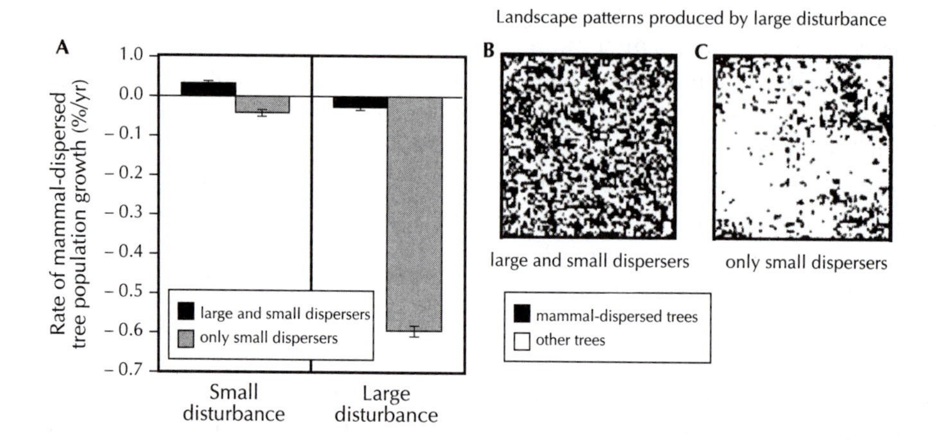

Figure 10. Results from a simple model of forest dynamics and seed dispersal by mammalian frugivores in Kibale National Park, Uganda. **(A)** Forest disturbance size interacts with the disperser community to determine the success of mammal-dispersed trees. When both large and small seed dispersers are present, the mammal-dispersed trees are resilient to both small and large disturbance events. When large dispersers are absent, mammal-dispersed trees slowly decline after small disturbances, but rapidly decline after large disturbances. Large differences in landscape pattern can be seen after 200 years, when the forest is subjected to large disturbances, between **(B)** a forest containing both large and small seed dispersers and **(C)** a forest with only small seed dispersers. The model demonstrates that seed dispersal at a diversity of scales is more resilient to disturbance than is seed dispersal over small scales. The model assumes lottery colonization of disturbed sites (Hubbell 1979) by either mammal-dispersed or other tree species (Chapman and Chapman 1996). Total mammal seed dispersals assumed to be constant. Dispersal range was estimated for large mammals (1010 m for *Cercocebus albigena,* and 1930 m for *Pan trogolyptes)* and small mammals (355 m for *Cercopithecus mitus,* 245 m for *Cercopithecus ascanius,* and 30 m for various Rodentia) (C. A. Chapman, unpublished data). The disturbance rate was held constant at 1.5%/year in the model, with only the spatial scale of disturbance varying between the small (0.04 ha) and large (10.24 ha) disturbance regimes.

This may have occurred during the Pleistocene extinctions of megaherbivores (Owen-Smith 1989; Flannery 1994; Zimov and others 1995). These losses appear to be particularly difficult to reverse even with large-scale ecological engineering projects (Flannery 1994).

Potential tests of Cross-scale Resilience

Our model expands theory relating biodiversity to ecological resilience by incorporating scale. The scaling relationships we propose can be tested through the analysis of empirical data, simulation, and field experimentation.

The proposition that ecological function is distributed across scales can be tested by analyzing the distribution of ecological function of an ecosystems species, and determining whether species belonging to the same guild or functional group are dispersed across scales as we predict. The proposition that competition within a scale drives the dispersion of guilds across scales can be tested by determining whether species are more evenly morphologically dispersed within a scale than across scales.

Our model of cross-scale resilience can be tested by creating simulations that use various assemblages of species, divided by function and scale, to assess the resilience of a system to a fluctuating environment. We advocate two approaches, one focusing on the role of scale in function, and the other focusing on the plausibility of our model of ecological organization. The first approach is the one followed in the model of Kibale Forest that was just described. An ecological function that is performed by a number of species at different scales can be modeled, and then this model may be perturbed by disrupting function and species composition to analyze ecological resilience. Our idea that ecological resilience derives from cross-scale functional redundancy resulting from strong within-scale interactions can be tested by simulating an evolving community of organisms that compete for a set of resources. Allowing the resource preference and scale of the organisms to evolve allows one to evaluate our hypothesis that competitive interactions could lead to the distribution of similar function across scales and functional diversity within scales.

Finally, field experiments can be designed to test the response of species to resource availability at different scales. We hypothesize that limited, nonaggregated resources will be used by species that live at small scales (for example, small birds such as warblers), whereas if resources are aggregated they will be used by larger species. We predict that resource utilization by animals is determined by the density of resources at their foraging scale. Since density is a scale-dependent measure, as resources are increasingly aggregated we expect that they will be used by larger animals.

These tests will provide partial evaluation of our model. To test our theory more fully, and better understand ecological resilience in general, requires long-term and extensive experiments that manipulate species composition and ecological structure at different scales.

Conclusions

We argue that ecosystems are usefully considered not as fixed objects in space, but as interacting, self-organized sets of processes and structures that vary across scales. Our approach integrates existing models of the relationship between species and ecological function, and extends these models to incorporate scale. Ecological organization at a specific scale is determined by interactions between species and processes operating within that scale. Competitive interactions are strongest among species that have similar functions and operate at similar scales. These interactions encourage functional diversity within a scale, and the distribution of ecological function across scales, enhancing cross-scale resilience. We suggest that it is possible to identify critical scales of landscape change that may be altered by species extinctions or introductions, or alternatively to identify which species may be affected by changes in landscape structure. Ultimately, we argue that understanding interactions between the scaling of species and scaling of ecological processes should be a central goal of ecology.

Our model of cross-scale resilience has several consequences for ecological policy. The history of resource exploitation and development reveals that ecological crisis and surprises often emerge from unexpected cross-scale interactions (Holling 1986; Regier and Baskerville 1986; Gunderson and others 1995). Management of natural resources often produces high short-term yields and, either purposefully or unintentionally, creates ecosystems that are less variable and diverse over space and time. Management channels ecological productivity into a reduced number of ecological functions and eliminates ecological functions at many scales. This simplification reduces cross-scale resilience, leaving systems increasingly vulnerable to biophysical, economic, or social events that otherwise could have been absorbed—disease, weather anomalies, or market fluctuations. In Jamaica, for example, off-shore fishing reduced the diversity of herbivorous fish species, leading to the replacement of coral reefs by macroalgae (Hughes 1994). Similarly, in New Brunswick, forestry eliminated landscape and age-class diversity, leading to a long period of chronic spruce budworm infestation (Regier and Baskerville 1986). In both of these cases, management reduced the resilience of these ecosystems, leaving the existing people and biota vulnerable to abrupt ecological reorganization. To avoid repeating the ecological management disasters of the past, it is necessary that ecologists understand how the scale-dependent organization of ecosystems and functional reinforcement across scales combine to produce ecological resilience.

We propose that ecological resilience is generated by diverse, but overlapping, function within a scale and by apparently redundant species that operate at different scales. The distribution of functional diversity within and across scales allows regeneration and renewal to occur following ecological disruption over a wide range of scales. The consequences of species loss may not be immediately visible, but species loss decreases ecological resilience to disturbance or disruption. It produces ecosystems that are more vulnerable to ecological collapse and reduces the variety of possible alternative ecological organizations. Ecological resilience must be understood if humanity is to anticipate and cope with the ecological crises and surprises that accelerating global change will bring.

Acknowledgments

We appreciate the support of a NASA/EOS Interdisciplinary Scientific Investigations of the Earth Observing Systems grant (NAG 2524), a NASA Terrestrial Ecology grant (NAG 3698), and a NASA Earth System Science Fellowship to G.P. Our manuscript was improved by comments from S. Bigelow, C. Chapman, K. Sieving, F. Putz, T. Allen, T. Frost, and S. Carpenter.

References

Allen, C. R., R. S. Lutz and S. Demaraiss. 1995. Red imported fire ant impacts on Northern Bobwhite populations. Ecol Appl 5:632–638.

Allen, T. F. H. and T. W. Hoekstra. 1992. *Toward a Unified Ecology*. New York. Columbia University.

Brown, B. J. and J. J. Ewel. 1987. Herbivory in complex and simple tropical successional ecosystems. Ecology 68:108–116.

Brown, J. H. and E. J. Heske. 1990. Control of a desert-grassland by a keystone rodent guild. Science 250:1705–1707.

Carpenter, S. R. and P. R. Leavitt. 1991. Temporal variation in paleolimnological record arising from a trophic cascade. Ecology 72:277–285.

Chapman, C. A and L. J. Chapman. 1996. Frugivory and the fate of dispersed and non-dispersed seeds of 6 African tree species. J Trop Ecol 12:491–504.

Clark, G. L. 1954. *Elements of Ecology*. New York: John Wiley.

Clark, W. C. 1985. Scales of climate impacts. Climatic Change 7:5–27.

Clark, W. C. and C. S. Holling. 1979. Process models. equilibrium structures, and population dynamics: on the formulation and testing of realistic theory in ecology. Popul Ecol 25:29–52.

D'Antonio, C. M and P. M. Vitousek. 1992. Biological invasions by exotic grasses, the grass/fire cycle and global change. Annu Rev Ecol Syst 23:63–87.

Darwin, C. 1859. *On the Origin of Species by means of natural selection or the preservation of favoured races in the struggle for life* [reprinted 1964]. Cambridge (MA): Harvard University.

Di Castri, F. and H. A. Mooney. 1973. *Mediterranean Type Ecosystems: Origins and Structure*. New York: Springer-Verlag.

Diemer, J. E. 1986. The ecology and management of the gopher tortoise in the southeastern United States. Herpetologica 42:125–133.
Dublin, H. T, A. R. E. Sinclair and J. McGlade. 1990. Elephants and fire as causes of multiple stable states in the Serengeti-Mara woodlands. J Anim Ecol 59:1147–1164.
Ehrlich, P. R and A. H. Ehrlich. 1981. *Extinction: the Causes and Consequences of the Disappearance of Species.* New York: Random House.
Estes, J. A. and D. O. Duggins.1995. Sea otters and kelp forests in Alaska: generality and variation in a community ecological paradigm. Ecol Monogr 65:75–100.
Ewel, J. J. and S. W. Bigelow. 1996. Plant life-forms and tropical ecosystem functioning. In: *Biodiversity and Ecosystem Processes in Tropical Forests.* Orians, G. H., R. Dirzo and J. J. Cushman, eds. Heidelberg: Springer-Verlag. Pp. 101–126.
Ewel, J. J., M. J. Mazzarino and C. W. Berrish. 1991. Tropical soil fertility changes under monocultures and successional communities of different structure. Ecol Appl 1 :289–302.
Flannery, T. 1994. *The Future Eaters: an Ecological History of the Australasian Lands and People.* New York: George Braziller.
Fleming, T. H. and V. J. Sosa. 1994. Effects of nectarivorous and frugivorous mammals on reproductive success of plants. J Mammal 75:845–851.
Frank D. A. and S. J. McNaughton. 1991. Stability increases with diversity in plant communities: empirical evidence from the 1988 Yellowstone drought. Oikos 62:360–362.
Frost, T. M., S. R. Carpenter, A. R. Ives and T. K. Kratz. 1995. Species compensation and complementarily in ecosystem function. In: *Linking Species and Ecosystems.* Jones, C. G. and J. H. Lawton, eds. New York: Chapman and Hall. Pp. 220–239.
Grinnell, J. 1917. The niche relations of the California thrashers. Auk 34:427–433.
Gunderson, L., C. Holling and S. Light. 1995. *Barriers and Bridges to the Renewal of Ecosystems and Institutions.* New York: Columbia University.
Holling, C. S. 1959. The components of predation as revealed by a study of small mammal predation of the European pine sawfly. Can Entomol 9 :293–320.
Holling, C. S. 1973. Resilience and stability of ecological systems. Annu Rev Ecol Syst 4:1–23.
Holling, C. 5. 1986. The resilience of ecosystems: local surprise and global change. In: *Sustainable Development of the Biosphere.* Clark, W. C and R. E. Munn. eds. Cambridge (UK): Cambridge University. Pp. 292–317.
Holling, C. S. 1988. Temperate forest insect outbreaks, tropical deforestation and migratory birds. Mem Entomol Soc Can 146:21–32.
Holling, C. S.1992. Cross-scale morphology, geometry and dynamics of ecosystems. Ecol Monogr 62:447–502.
Holling, C. S. 1996. Engineering resilience versus ecological resilience. In: *Engineering within Ecological Constraints.* Schulze P, ed. Washington (DC): National Academy. Pp. 31–44.
Holling, C. S., D. W. Schindler, B. W. Walker and J. Roughgarden. 1995. Biodiversity in the functioning of ecosystems: an ecological synthesis. In: *Biodiversity Loss: Economic and Ecological Issues.* Perrings, C., K. G. Maler, C. Folke, C. S. Holling and B. O. Jansson, eds New York: Cambridge University. Pp. 44–83.
Hubbell, S. R. 1979. Tree dispersion. abundance, and diversity in a tropical dry forest. Science 203:1299–1309.
Hughes, T. R 1994. Catastrophes, phase shifts, and large-scale degradation of a Caribbean coral reef. Science 265:1547–1551.
Hutchinson, G. E. 1957. Concluding remarks. Cold Spring Harbor Symp Quant Biol 22:415–427.
Jones, C. G., J. H. Lawton and M. Shachak.1994. Organisms as ecosystem engineers. Oikos 69:373–386.
Kalin Arroyo, M. T., P. H. Zedler and M. D. Fox. 1995. *Ecology and biogeography of Mediterranean Ecosystems in Chile, California, and Australia.* New York: Springer-Verlag.
Kauffman, S. A. 1993. *Origins of Order: self-organization and selection in evolution.* Oxford: Oxford University.
Kitchell, J. E. and S. R. Carpenter.1993. Synthesis and new directions. In: *The Trophic Cascade in Lakes.* Carpenter S. R. and J. F. Kitchell, eds. New York: Cambridge University. Pp. 332–350.
Korner, C. 1996. Scaling from species to vegetation: the usefulness of functional groups. In: *Biodiversity and Ecosystem Function.* Schulze, E. D. and H. A. Mooney, eds. New York: Springer- Verlag. Pp. 117–140.
Krummel, J. R, R. H. Gardner, G. Sugihara, R. V. O'Neill and P. R. Coleman. 1987. Landscape patterns in a disturbed environment. Oikos 48:321–324.
Lawton, J. H. 1994. What do species do in ecosystems? Oikos 71:367–374.
Levin, S. A. 1992. The problem of pattern and scale in ecology. Ecology 73:1943–1967.
MacArthur, R. H. 1955. Fluctuations of animal populations and a measure of community stability. Ecology 36:533–536.
Macdonald, D. 1985. The encyclopedia of mammals. New York: Facts On File.
May, R. M. 1973. Stability and complexity in model ecosystems. Princeton (NJ): Princeton University.
McNeely, J. A., M. Gadgil, C. Leveque, C. Padoch and K. Redford. 1995. Human influences on biodiversity. In: *Global Biodiversity Assessment.* Heywood, V, ed Cambridge (UK): Cambridge University. Pp. 711–822.
Morris, R. F. 1963. The dynamics of epidemic spruce budworm populations. Mem Entomol Soc Can 31:1–322.
Morse, D. R., J. H. Lawton and M. M. Dodson. 1985. Fractal dimension of vegetation and the distribution of arthropod body lengths. Nature 314:731–733.
Murdoch, W. W. 1969. Switching in general predators:

experiments on predator specificity and stability of prey populations. Ecol Monogr 39:335–354.

Naeem S., L. J.Thompson, S. P. Lawler, J. H. Lawton and R. M. Woodfin. 1994. Declining biodiversity can alter the performance of ecosystems. Nature 368:734–737.

Naiman, R. J., G. Pinay, C. A. Johnston and J. Pastor. 1994. Beaver influences on the long-term biogeochemical characteristics of boreal forest drainage networks. Ecology 75:905–921.

Nowak, R. M and J. L. Paradiso. 1983. Walker's mammals of the world. Baltimore: John Hopkins University.

O'Neill, R., S. J. Turner, V. I. Cullinam , D. P. Coffin, T. Cook. W. Conley, J. Brunt, J. M. Thomas, M. R. Conley and J. Gosz. 1991. Multiple landscape scales: an intersite comparison. Landscape Ecol 5:137–144.

Orr, M. R., S. H. Seike, W. W. Benson and L. E. Gilbert. 1995. Flies suppress fire ants. Nature 373:292–293.

Owen-Smith, N. 1989. Megafaunal extinctions: the conservation message from 11,000 years B.C. Conserv Biol 3:405–411.

Paine, R. T. 1969. A note on trophic complexity and community stability. Am Nat 103:91–93.

Pastor, J., B. Dewey, R. J. Naiman, P. E. McInnes and Y. Cohen. 1993. Moose browsing and soil fertility in the boreal forests of Isle Royale National Park. Ecology 74:467–480.

Peters, R. H. 1983. *The Ecological Implications of Body Size.* Cambridge (UK): Cambridge University.

Pimm, S. L. 1984. The complexity and stability of ecosystems. Nature 307:321–326.

Porter, S.D. and D. A. Savignano. 1990. Invasion of polygyne fire ants decimates native ants and disrupts arthropod community. Ecology 71:2095–2116.

Power, M. E., D. Tilman, J. A. Estes, B. A. Menge, W. J. Bond, L. S. Mills, G. Daily, J. C. Castilla, J. Lubchenco and R. Paine. 1996. Challenges in the quest for keystones. BioScience 46:609–620.

Prins, H. H. T. and H. P. Van der Jeud. 1993. Herbivore population crashes and woodland structure in East Africa. J Ecol 81:305–314.

Regier, H.A. and G. L. Baskerville . 1986. Sustainable redevelopment of regional ecosystems degraded by exploitive development. In: *Sustainable Development of the Biosphere.* Munn, W. C and R. E. Munn, eds. Cambridge (MA): Cambridge University. Pp. 75–101.

Schindler, D. W. 1990. Experimental perturbations of whole lakes as tests of hypotheses concerning ecosystem structure and function. Oikos 57:25–41.

Sugihara, G.1980. Minimal community structure: an explanation for species abundance patterns. Am Nat 116:770–787.

Terborgh, J.1986. Keystone plant resources in the tropical forest. In: *Conservation Biology: the science of scarcity and diversity.* Soule, M. E., ed. Sunderland (UK): Sinauer. Pp. 330–344.

Tilman, D.1996. Biodiversity: population versus ecosystem stability. Ecology 77:350–363.

Tilman, D., D. Wedin and J. Knops. 1996. Productivity and sustainability influenced by biodiversity in grasslands ecosystems. Nature 379:718–720.

Turner, B. L., W. C. Clark, R. W. Kates, J. E. Richards, J. T. Mathews and W. B. Meyer. 1993. *The Earth as Transformed by Human Action.* New York: Cambridge University

Vitousek, P. M. 1990. Biological invasions and ecosystem processes: towards an integration of population biology and ecosystem studies. Oikos 57:7–13.

Walker, B. 1992. Biological diversity and ecological redundancy. Conserv Biol 6:18–23.

Walker, B. 1995. Conserving biological diversity through ecosystem resilience. Conserv Biol 9:747–752.

Zimov, S.A., V. I. Chuprynin, A. P. Oreshko, I. F. S. Chapin, J. E. Reynolds and M. C. Chapin. 1995. Steppe-tundra transition: a herbivore-driven biome shift at the end of the Pleistocene. Am Nat 146:765–794.

Introduction to

Eric D. Schneider and James J. Kay

Life as a Manifestation of the Second Law of Thermodynamics

by Timothy F. H. Allen

The following paper by Schneider and Kay shows how landscapes can be tied to the health of ecosystems that occur on them. It indicates how the complicated patterns in communities can be applied to the processes and function of the whole material ecological system. In a most remarkable treatment of system energetics, it shows exactly what is required for organization to emerge, be it elaborations of physical form of vegetation, regulation of the entire planet, or for patterns of human land use. Furthermore, the ideas in Schneider and Kay are indeed cutting edge, and matter for issues of sustainability and the modern technology we have available to monitor it.

Life as a Manifestation of the Second Law of Thermodynamics

E. D. Schneider and J. J. Kay

Reprinted with permission from *Mathematical and Computer Modelling,* Vol. 19, pp, 25–48, E. D. Schneider and J. J. Kay, Life as a manifestation of the second law of thermodynamics, 1994, with permission from Elsevier Science.

Abstract

We examine the thermodynamic evolution of various evolving systems, from primitive physical systems to complex living systems, and conclude that they involve similar processes which are phenomenological manifestations of the second law of thermodynamics. We take the reformulated second law of thermodynamics of Hatsopoulos and Keenan and Kestin and extend it to nonequilibrium regions, where nonequilibrium is described in terms of gradients maintaining systems at some distance away from equilibrium.

The reformulated second law suggests that as systems are moved away from equilibrium they will take advantage of all available means to resist externally applied gradients. When highly ordered complex systems emerge, they develop and grow at the expense of increasing the disorder at higher levels in the system's hierarchy. We note that this behaviour appears universally in physical and chemical systems. We present a paradigm which provides for a thermodynamically consistent explanation of why there is life, including the origin of Life, biological growth, the development of ecosystems, and patterns of biological evolution observed in the fossil record.

We illustrate the use of this paradigm through a discussion of ecosystem development. We argue that as ecosystems grow and develop, they should increase their total dissipation, develop more complex structures with more energy flow, increase their cycling activity, develop greater diversity and generate more hierarchical levels, all to abet energy degradation. Species which survive in ecosystems are those that funnel energy into their own production and reproduction and contribute to autocatalytic processes which increase the total dissipation of the ecosystem. In short, ecosystems develop in ways which systematically increase their ability to degrade the incoming solar energy. We believe that our thermodynamic paradigm makes it possible for the study of ecosystems to be developed from a descriptive science to predictive science founded on the most basic principle of physics.

Introduction

In 1943, E. Schrödinger [1] wrote his small book *What is Life?* in which he attempted to draw together the fundamental processes of biology and the sciences of physics and chemistry. He noted that life was comprised of two fundamental processes; one "order from order" and the other "order from disorder." He observed that the gene with its soon-to-be discovered DNA, controlled a process that generated order from order in a species, that is, the progeny inherited the traits of the parent. Schrödinger recognized that this process was controlled by an aperiodic crystal, with unusual stability and coding capabilities. Over a decade later, these processes were uncovered by Watson and Crick. Their work provided biology with a framework that allowed for some of the most important findings of the last thirty years.

However, Schrödinger's equally important and less understood observation was his "order from disorder" premise. This was an effort to link biology with the fundamental theorems of thermodynamics. He noted that at first glance, living systems seem to defy the second law of thermodynamics as it insists that, within closed systems, entropy should be maximized and disorder should reign. Living systems, however, are the antithesis of such disorder. They display marvelous levels of order created from disorder. For instance, plants are highly ordered structures, which are synthesized from disordered atoms and molecules found in atmospheric gases and soils.

Schrödinger solved this dilemma by turning to nonequilibrium thermodynamics, that is, he recognized that living systems exist in a world of energy and material fluxes. An organism stays alive in its highly organized state by taking energy from outside itself, from a larger encompassing system, and processing it to produce, within itself, a lower entropy, more organized state. Schrödinger recognized that life is a far-from-equilibrium system that maintains its local level of organization at the expense of the larger global entropy budget. He proposed that to study living systems from a nonequilibrium perspective would reconcile biological self-organization and thermodynamics. Furthermore, he expected that such a study would yield new principles of physics.

This paper takes on the task proposed by Schrödinger and expands on his thermodynamic view of life. We explain that the second law of thermodynamics is not an impediment to the understanding of life but rather is necessary for a complete description of living processes. We further expand thermodynamics into the causality of the living process and assert that the second law is a necessary but not sufficient cause for life itself. In short, our re-examination of thermodynamics shows that the second law underlies and determines the direction of many of the processes observed in the development of living systems. This work harmonizes physics and biology at the macro level and shows that biology is not an exception to physics; we have simply misunderstood the rules of physics.

Central to our discussion is a fresh look at thermodynamics. Since the time of Boltzmann and Gibbs there have been major advances in thermodynamics especially by Carathéodory, Hatsopoulos and Keenan, Kestin, Jaynes, and Tribus. We take the restated laws of thermodynamics of Hatsopoulos and Keenan and Kestin and extend them so that in nonequilibrium regions processes and systems can be described in terms of gradients maintaining systems away from equilibrium. In this context, the second law mandates that as systems are moved away from equilibrium they will take advantage of all means available to them to resist externally applied gradients. Our expansion of the second law immediately applies to complex systems in nonequilibrium settings unlike classical statements which are restricted to equilibrium or near equilibrium conditions. Away from equilibrium, highly ordered stable complex systems can emerge, develop, and grow at the expense of more disorder at higher levels in the system's hierarchy.

We will demonstrate the utility of these restatements of the second law by considering one of the classic examples of dissipative structures, Bénard Cells. We argue that this paradigm can be applied to physical and chemical systems, and that it allows for a thermodynamically consistent explanation of the development of far-from-equilibrium complex systems including life.

As a case study, we focus on the applications of these thermodynamic principles to the science of ecology. We view ecosystems as open thermodynamic systems with a large gradient impressed on them by the sun. The thermodynamic imperative of the restated second law is that these systems will strive to reduce this gradient by all physical and chemical processes available to them. Thus, ecosystems will develop structures and functions selected to most effectively dissipate the gradients imposed on them while allowing for the continued existence of the ecosystem. We examine one ecosystem closely and using analyses of carbon flows in stressed and unstressed conditions, we show that the unstressed ecosystem has structural and functional-attributes that lead to more effective degradation of the energy entrained within the ecosystem. Patterns of ecosystem growth, cycling, trophic structure and efficiencies are explained by this paradigm.

A rigorous test of our hypothesis is the measurement of reradiated temperatures from terrestrial ecosystems. We argue that more mature ecosystems should degrade incoming solar radiation into lower quality exergy, that is, have lower reradiated temperatures. We then provide data to show that not only are more mature ecosystems better degraders of energy (cooler) but that airborne infrared thermal measurements of terrestrial ecosystems may offer a major breakthrough in providing measures of ecosystem health or integrity.

Classical Thermodynamics

Because the basic tenets of this paper are built on the principles of modern thermodynamics, we start this paper with a brief discussion of thermodynamics. We ask the reader who is particularly interested in ecology to bear with us through this discussion, because an understanding of these aspects of thermodynamics will make much of our discussion of ecology self-evident. For the reader who has mastered thermodynamics, we believe that our approach to the theoretical issues of nonequilibrium thermodynamics is original and permits a more satisfactory discussion of observed far-from-equilibrium phenomena.

Comparatively speaking, thermodynamics is a young science but has been shown to apply to all work and energy systems including the classic temperature-volume-pressure systems, chemical kinetic systems, electromagnetic and quantum systems. The development of classical thermodynamics was initiated by Carnot in 1824 through his attempts to understand steam engines. He is responsible for the notion of mechanical work, cycles, reversible processes, and early statements of the first and second law. Clausius in the period 1840 to 1860 refined Carnot's work, formalizing the first and second law and the notion of entropy.

The first law arose from efforts to understand the relation between heat and work. Most simply stated, the first law says that energy cannot be created or destroyed and that despite the transformations that energy is constantly undergoing in nature (i.e., from heat to work, chemical potential to light), the total energy within a closed or isolated system remains unchanged. It must be remembered that although the total quantity of energy in a closed system will remain unchanged, the quality of the energy in the system (i.e., the free energy or the exergy content) may change.

The second law requires that if there are any physical or chemical processes underway in a system, then the overall quality of the energy in that system will degrade. The second law of thermodynamics arose from Carnot's experiments with steam engines and his recognition that it was impossible to convert all the heat in such a system completely to work. His formal statement of the second law may be stated as: It is impossible for any system to undergo a process in which it absorbs heat from a reservoir at a single temperature and converts it completely into mechanical work, while ending at the same state in which it began. The second law notes that work may be dissipated into heat, whereas heat may not be converted entirely into work, thus proving the existence of irreversibility in nature. (This was a novel, paradigm shattering suggestion for its time.)

The second law can also be stated in terms of the quantitative measure of irreversibility, entropy. Clausius discovered that for any cyclic process[1]:

$$\int \frac{dQ}{T} \leq 0$$

where equality holds only for reversible processes (for example, a Carnot cycle). Clausius, borrowing from the Greek for transformation, called the quantity $\int -\frac{dQ}{T}$ which measures the irreversibility of a process, the "entropy" of the process. The second law can be stated as: any real process can only proceed in a direction which results in an entropy increase. This universal increase in entropy draws the "arrow of time" [2] in nature and represents the extent to which nature becomes more disordered or random.

All natural processes can be viewed in light of the second law and in *all* cases this one-sided aspect of nature is observed. Heat always flows spontaneously from a hotter reservoir to a colder reservoir until there is no longer a temperature difference or gradient; gas will always flow from high pressure to low pressure until there is no longer a pressure difference or gradient. If one mixes hot and cold water, the mixture comes to a uniform temperature. The resulting luke warm water will not spontaneously unmix itself into hot and cold portions. Boltzmann would have restated the above sentence as: it is highly improbable that water will spontaneously separate into hot and cold portions, but it is not impossible.

Boltzmann recast thermodynamics in terms of energy microstates of matter. In this context, entropy reflects the number of different ways microstates can be combined to give a particular macrostate. The larger the number of microstates for a given macrostate, the larger the entropy. Consider a ten compartment box with 10,000 marbles in one of the ten compartments and the rest of the compartments being empty. If doors are opened between the compartments and the box is subjected to a pattern of random shaking one would expect, over time, to see a distribution of about 1,000 marbles per compartment, the distribution which has the largest number of possible microstates. This randomization of the marbles to the equiprobable distribution corresponds to the macrostate with the maximum entropy for the closed system. If one continued the shaking it would be highly improbable but not impossible for all the marbles to reseparate themselves into the low entropy configuration with 10,000 marbles in one compartment. The same logic is applied by Boltzmann to explain the macroscopic phenomena of thermodynamics in terms of microstates of matter. Systems will tend to the macrostate which has the largest number of corresponding accessible microstates.

The Extended Laws of Thermodynamics

In 1908, thermodynamics was moved a step forward by the work of Carathéodory [3] when he developed a proof that showed that the law of "entropy increase" is not the general statement of the second law. The more encompassing statement of the second law of thermodynamics is that: "In the neighbourhood of any given state of any closed system, there exist states which are inaccessible from it along any adiabatic path, reversible or irreversible." This statement of the second law, unlike earlier ones, does not depend on the concepts of entropy or temperature and applies equally well in the positive and negative temperature regimes.

More recently, Hatsopoulos and Keenan [4] and Kestin [5] have put forward a principle which subsumes the other 1st and 2nd laws: "When an isolated system performs a process after the removal of a series of internal constraints, it will reach a unique state of equilibrium: this state of equilibrium is independent of the order in which the constraints are removed." (This is called the Law of Stable Equilibrium by Hatsopoulos and Keenan and the Unified Principle of Thermodynamics by Kestin.)

The importance of this statement is that, unlike all the earlier statements which show that all real processes are irreversible, it dictates a direction and an end state for all real processes. All previous formulations of the second law tell us what systems cannot do. This statement tells us what systems will do. An example of this phenomena are two flasks, connected with a closed stopcock. One flask holds 10,000 molecules of a gas. Upon removing the constraint (opening the stopcock) the system will come to its equilibrium state of 5,000 molecules in each flask, with no gradient between the flasks. These principles hold for closed isolated systems. However, a more interesting class of phenomena belong to systems that are open to energy and/or material flows and reside at stable states some distance from equilibrium.

These much more complex thermodynamic systems are the ones investigated by Prigogine and his collaborators [6,7] These systems are open and are moved away from equilibrium by the fluxes of material and energy across their boundary. These systems maintain their form or structure by continuous dissipation of energy and, thus, are known as dissipative structures. Prigogine showed that nonequilibrium systems, through their exchange of matter and/or energy with the outside world, can maintain themselves for a period of time away from thermodynamic equilibrium in a locally reduced entropy steady-state. This is done at the cost of increasing the entropy of the larger "global" system in which the dissipative structure is imbedded; thus, following the mandate of the second law that overall entropy must increase. Nonliving organized systems like convection cells, tornados and lasers) and living systems (from cells to ecosystems) are dependent on outside energy fluxes to maintain their organization in a locally reduced entropy state.

Prigogine's description of dissipative structures is formally limited to the neighbourhood of equilibrium. This is because his analysis depends on a linear expansion of the entropy function about equilibrium. This is a severe restriction on the application of his theory and in particular precludes its formal application to living systems.

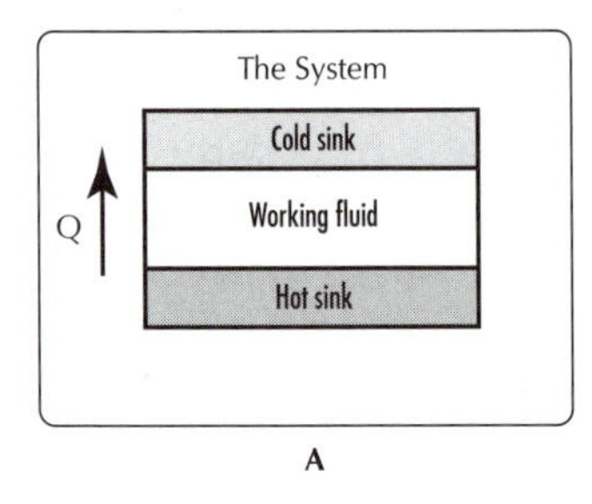

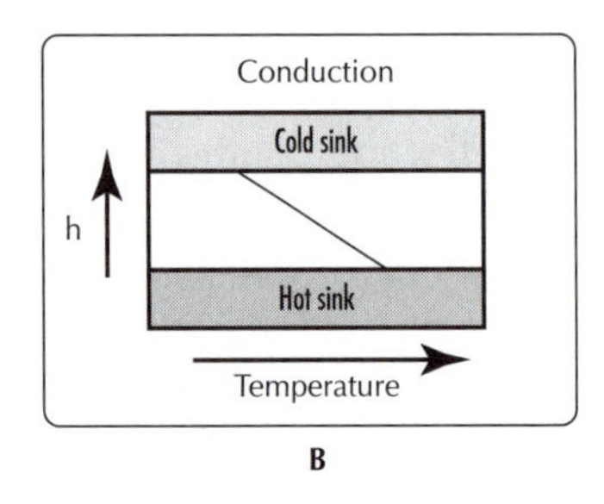

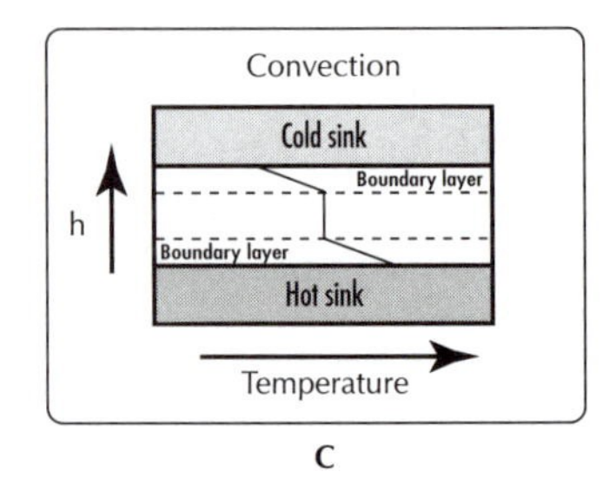

Figure 1. Bénard Cells. Schematics of the Bénard cell apparatus (in **A**, Q = heat transfer) and temperature profiles (lines inside apparatus in **B** and **C**, h = height) in the working fluid before and after the transition to convection. The fluid is heated from below and the top of the apparatus acts as a cold sink.

Initially, all dissipation through the fluid occurs via conduction and molecule to molecule interaction. When the gradient reaches a critical level (Rayleigh number 1760) the transition to highly organized convection occurs.

As the gradient is increased, it becomes harder and harder (more work is required) to maintain the higher rate of dissipation (see Figure 2). The further the system is moved away from its equilibrium state the more exergy is destroyed, the system produces more entropy, and more work is required to maintain it in its nonequilibrium state.

Due to the convective overturn, most of the working fluid in the container becomes vertically isothermal (with little gradient) and only the boundary layers on the edge of the system carry the gradient. As a gradient is increased the boundary layers become thinner and more dissipation occurs.

To deal with the thermodynamics of nonequilibrium systems, we propose the following corollary that follows from the proof by Kestin of the Unified Principle of Thermodynamics. His proof shows that a system's equilibrium state is stable in the Lyapunov sense. Implicit in this conclusion is that a system will resist being removed from the equilibrium state. The degree to which a system has been moved from equilibrium is measured by the gradients imposed on the system.

> The thermodynamic principle which governs the behaviour of systems is that, as they are moved away from equilibrium they will utilize all avenues available to counter the applied gradients. As the applied gradients increase, so does the system's ability to oppose further movement from equilibrium.

(In the discussion that follows, we shall refer to this statement as the "restated second law." The pre-Carathéodory statements (i.e., entropy will increase) will be referred to as the classical second law).

A simple example of this phenomena is the Bénard cell. The experimental apparatus for studying the Bénard cell consists of a highly instrumented insulated container enclosing a fluid. The bottom of the container is a heat source and the top is a cold reservoir. (See Figure 1.) When the fluid is heated from below it resists the applied gradients (ΔT) by dissipating heat through conduction. As the gradient is increased, the fluid develops convection cells. These convection cells increase the rate of dissipation. (These convection cells are called Bénard cells [8].)

Figure 2C shows a plot of the gradient (Ra, the Rayleigh number, which is proportional to the gradient ΔT) against the available work which is expended in maintaining the gradient. The "dynamics of the system are such that it becomes more and more difficult to move the system away from equilibrium (that is, propor-

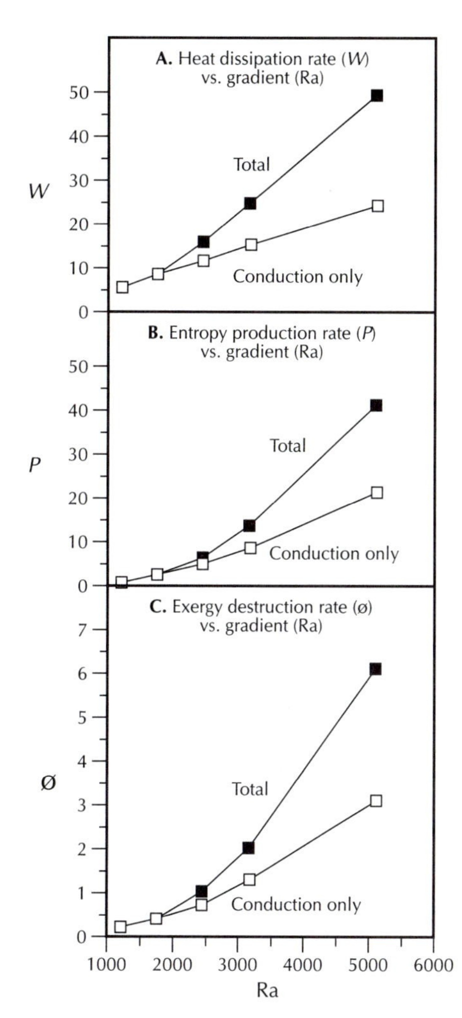

Figure 2. These graphs are calculations by us of entropy production and exergy destruction during heat transfer across a fluid. The data used is from the classic experiments of Silveston [10]. A fluid is heated from below. The temperature difference between the top and bottom of the fluid is increased and the heat transfer from the bottom to the top is recorded. At a Rayleigh number (a dimensionless measure of gradient) of about 1760, Bénard Cells appear, that is, convection starts. As the Rayleigh number (temperature difference) increases, convection becomes the dominant mode of heat transfer.

The particular analysis is for Silicon 350 at a depth of 6.98 mm with a surface area of 0.0305 m^2. The first graph is the heat transfer rate (kcal/hour) across the fluid as a function of Rayleigh number (temperature difference). The bottom curve is the heat transfer rate without Bénard Cells and the top curve is the heat transfer rate with the formation of Bénard Cells. The point is that the emergence of the ordered structure (Bénard Cells) dissipates more energy.

The second graph is of entropy production rate (kcal/hour/K) versus Rayleigh number (the gradient). Again, we see the emergence of the ordered structure results in more entropy production. The third graph shows the available work[2] (exergy) (kcal/hour) which must be provided from an external source in order to maintain the gradient. As the gradient increases a greater amount of work must be done to incrementally increase the gradient. It becomes more difficult to maintain the gradient as the system becomes more organized.

tionally more available work must be expended for each incremental increase in gradient as the system gets further from equilibrium, i.e., ΔT increases).

In chemical systems, LeChatelier's principle is another example of the restated second law. Fermi, in his 1936 lectures on thermodynamics, noted that the effect of a change in external conditions on the equilibrium of a chemical reaction is prescribed by the LeChatelier's principle. "If the external conditions of a thermodynamic system are altered, the equilibrium of the system will tend to move in such a direction as to oppose the change in the external conditions" [9]. Fermi noted that if a chemical reaction were exothermal (i.e., A + B = C + D + heat), an increase in temperature will shift the chemical equilibrium to the left hand side. Since the reaction from left to right is exothermal, the displacement of the equilibrium towards the left results in the absorption of heat and opposes the rise in temperature. Similarly, a change in pressure (at a constant temperature) results in a shift in the chemical equilibrium of reactions which opposes the pressure change.

Thermodynamic systems exhibiting temperature, pressure, and chemical equilibrium resist movement away from these equilibrium states. When moved away from their local equilibrium state, they shift their state in a way which opposes the applied gradients and moves the system back towards its local equilibrium attractor. The stronger the applied gradient, the greater the effect of the equilibrium attractor on the system.

The reason that our restatement of the second law is a significant step forward for thermodynamics is that it tell us how systems will behave as they are moved away from equilibrium. Implicit in this is that this principle is applicable to nonequilibrium systems, something which is not true for classical formulations of the second law.

In particular, our "restated second law" avoids the problems associated with state variables such as entropy which are only defined for equilibrium. Our restatement of the second law sidesteps the problems of defining entropy and entropy production in nonequilibrium systems, an issue that has plagued nonequilibrium thermodynamics for years. By focusing on gradient destruction, we avoid completely the problems encountered by Prigogine [14], and more recently Swenson [15], who use extremum principles based on the concept of entropy to describe self-organizing systems. Nonequilibrium systems can be described by their forces and requisite flows using the well-developed methods of network thermodynamics [16–18].

Dissipative Structures as Gradient Dissipators

In this section, we examine the behaviour of dissipative structures in light of the restated second law. Prigogine and his colleagues have shown that dissipative structures self-organize through fluctuations, small instabilities which lead to irreversible bifurcations and new stable system states. Thus, the future states of such systems are not deterministic. Dissipative structures are stable over a finite range of conditions and are sensitive to fluxes and flows from outside the system. Glansdorff and Prigogine [19] have shown that these thermodynamic relationships are best represented by coupled nonlinear relationships, i.e., autocatalytic positive feedback cycles, many of which lead to stable macroscopic structures which exist away from the equilibrium state. Convection cells, hurricanes, autocatalytic chemical reactions, and living systems are all examples of far-from-equilibrium dissipative structures which exhibit coherent behavior.

The transition in a heated fluid between conduction and convection is a striking example of emergent coherent organization in response to an external energy input. A thorough analysis of these simple physical systems has allowed us to develop some general thermodynamic principles applicable to the development of complex systems as they emerge at some distance away from equilibrium. Bénard, Rayleigh, (see [8]), Silveston [10] and Brown [20] conducted carefully designed experiments to study this transition.

The lower surface of the experimental apparatus is heated and the upper surface is kept at a cooler temperature. (See Figure 1). Hence, a temperature gradient is induced across the fluid. The initial heat flow through the system is controlled by conduction. Energy transfer is by molecule to molecule interaction. When the heat flux reaches a critical value of the temperature gradient, the system becomes unstable and the molecular action of the fluid becomes coherent and convective overturning emerges. These convective structures result in highly structured coherent hexagonal surface patterns (Bénard Cells) in the fluids. This coherent kinetic structuring increases the rate of heat transfer and gradient destruction in the system.

This transition between noncoherent, molecule to molecule, heat transfer to coherent

Table 1

Definitions:

Nusselt Number:	$Nu = Q/(k\Delta T/h)$	
Rayleigh Number:	$Ra = g\alpha\Delta T/h^3/\kappa\nu$	

where Q = heat flow, T = temperature, g = gravity, κ = coefficient of thermometric conductivity, k = coefficient of heat conduction, h = depth, α = coefficient of volume expansion, ν = coefficient of kinematic viscosity.

structure results in excess of 10^{23} molecules acting together in an organized manner. This seemingly improbable occurrence is the direct result of the applied temperature gradient and is the system's response to attempts to move it away from equilibrium.

Recently, we studied the Bénard Cell phenomena in detail, using the original data sets collected by Silveston [10] and Brown [20] (which they most graciously provided us with). We believe that these analyses are significant, in that we have calculated for the first time the entropy production, exergy drop and available work destruction, resulting from these organizing events. (See Figure 2.) Our analysis clearly shows that as the gradient increases, it becomes harder to increase the gradient. Initially, the temperature gradient in the apparatus is being accommodated solely by random conductive activity. When the gradient is raised, a combination of factors including surface tension effects and gravitational fluid instability, converts the system to a mixed conductive-convective heat transfer system. The transition to coherent behaviour occurs at Ra = 1760 (See Table 1 and Figure 1).

Figure 2 consists of plots of Silveston's original data using silicon as the working fluid with a separation between the hot and cold sinks of 6.98 mm. The curve labeled "Conduction" is the dissipation which would occur without the emergent coherent behaviour. The difference between the curve labeled "Total," and the one labeled "Conduction," is the increase in dissipation due to the emergence of the dissipative structure. As is shown in Figure 2, with the onset of convection there is a dramatic increase in the heat transfer rate across the system.

From the literature, especially Chandrasekhar [8], and from our analysis of the Silveston and Brown's data, we note the following behavior for these systems:

(1) Heat Dissipation Rate (transfer of heat between the plates) is a linearly increasing function of the gradient ΔT (See Figure 2A, recalling that Ra is proportional to ΔT).
(2) Entropy production rate (P) vs. ΔT increases in a nonlinear way (see Figure 2B).
(3) Exergy destruction rate (ø) vs. ΔT increases in a nonlinear way, the shape of the curve being the same as P vs. ΔT.
(4) Points (2) and (3) imply that as the gradient increases it is harder (requires more available work, that is, exergy) to incrementally increase the gradient. The further from equilibrium that the system is, the more it resists being moved further from equilibrium. In any real system there is an upper limit to the gradient which can be applied to the system.
(5) Once convection occurs, the temperature profile within the fluid is vertically isothermal outside the boundary layer, i.e., the

temperature in the convection cells is constant, thus effectively removing the gradient through most of the fluid (see Figure 1).

(6) As the gradient increases, further critical points are reached. At each critical point the boundary layer depth decreases.

(7) Point (1) is true because the rate of heat transfer is controlled (more or less) by the rate of heat flow across the boundary layer, i.e., by conduction which is a linear process. This process is also responsible for most of the entropy production, as there will be little production due to convection. The slope change is caused by the decrease in the boundary layer depth (l) at a mode change (critical point). (Recall $Q = k\Delta T/l$; thus as l decreases, slope (k/l) increases.)

(8) Nu is $Q/Qc = P/Pc = ø/øc$; that is, in Bénard Cells, the increase in dissipation at any point due to the emergent process is equal to the increase in degradation at any point due to the emergent process. This is true for any process which involves only heat transfer. Otherwise degradation $\neq$ dissipation. (This is why Prigogine at times mistakenly uses these terms interchangeably.)

(9) The principle governing these systems is not one of maximum entropy production but rather one of entropy production change being positive semi-definite as you increase the gradient. See Point (7) above. The interesting question is, how much structure emerges for a given gradient, and how much resistance exists to increasing the gradient further.

As the temperature difference increases, there are a number of further transitions at which the hexagonal cells re-organize themselves so that the cost of increasing the temperature gradient escalates even more quickly. Ultimately the system becomes chaotic and dissipation is maximum in this regime.

The point of this example is that in a simple physical system, new structures emerge which better resist the application of an external gradient. The Bénard cell phenomena is an excellent example of our nonequilibrium restated second law. As we will see later, other physical, chemical, and living systems observe similar rules.

The more a system is moved from equilibrium, the more sophisticated its mechanisms for resisting being moved from equilibrium. This behaviour is not sensible from a classical second law perspective, but is what is expected given the restated second law. No longer is the emergence of coherent self-organizing structures a surprise, but rather it is an expected response of a system as it attempts to resist and dissipate externally applied gradients which would move the system away from equilibrium. The term dissipative structure takes on a new meaning. No longer does it mean just increasing dissipation of matter and energy, but dissipation of gradients as well.

In this regard, it is important to distinguish between energy dissipation and energy degradation. Dissipation of energy means to move energy through a system, as in the Bénard cell. Dissipation may or may not destroy gradients. Degradation of energy means to destroy the ability of energy to set up gradients. The ability to set up gradients is measured by the availability or exergy of the energy. Thus, energy degradation means exergy destruction, that is, the degradation of the ability of the energy to produce gradients that can accomplish work. In simple systems involving only heat flow, energy degradation is via energy dissipation.

The gradient reducing nature of self-organizing systems is dramatically demonstrated by a simple experimental device sold as a toy in a nationwide scientific catalog (Edmunds). Their "Tornado in a Bottle" (not to be confused with "fusion in a bottle") consists of a simple plastic orifice that allows the connection of two 1.5 liter plastic soda bottles end to end. The bottles are connected and set on a level surface to drain with the upper bottle filled with water and the lower bottle empty. When set vertical, a thirty centimeter gradient of water exists.

Due to the orifice configuration, the bottle drains slowly, requiring approximately six minutes to empty the upper bottle (i.e., to reduce the gradient in the system). The experiment is then repeated with the bottles being given a slight rotational perturbation. A vortex forms, driven by the gravitational gradient within the system, and drains the upper bottle in approximately 11 seconds. The "tornado," a highly organized structure, has the ability to dissipate the gradient much faster, thus bringing the system to its local equilibrium more quickly! Here again is a manifestation of the restated second law, a macroscopic highly organized structure of 10^{23} molecules acting coherently to dissipate a gradient. The production of the highly organized system, the tornado, leads to more effective dissipation of the larger driving gradient, the gravitational differences in water levels between the bottles. As in the Bénard convection experiments, organized structures reduce gradients more quickly than random linear processes.

It should not be surprising that this simple experimental device mimics meteorological phenomena such as mesoscale weather patterns. The development of temperature gradients between a warm earth and a cooler overlying atmosphere results in highly organized convective cloud patterns which reduce the troposphere temperature gradient. The violent destructive power of tornadoes is a manifestation of the ability of these self-organizing structures to rapidly dissipate strong temperature and barometric gradients. Hurricanes are another example of mesoscale dissipative meteorological structures.

The global weather, wind and ocean circulation patterns are the result of the difference in heating at the equator relative to the poles. The general meteorological circulation of the earth, although affected by spatial, Coriolis and angular momentum effects, is driven by gradients and the global system's attempt to dissipate them and come to local equilibrium. Paltridge [21] has suggested that the earth-atmosphere, climate system configures itself into a state of maximum dissipation and that the global distribution of clouds, temperature and horizontal energy flows are governed by thermodynamic dissipative processes similar to those described above. We see that the earth-climate system, as well as other dissipative systems, do not reach a static equilibrium state because they are open thermodynamic systems constantly receiving a supply of external energy (i.e., from the sun), which drives them and maintains them in a non-equilibrium organized state.

So far we have focused our discussion on simple physical systems and how thermodynamic gradients drive self-organization. The literature is replete with similar phenomena in dynamic chemical systems. Prigogine and the Brussels School and others have documented the thermodynamics and behavior of these chemical reaction systems. Chemical gradients result in dissipative autocatalytic reactions, examples of which are found in simple inorganic chemical systems, in protein synthesis reactions, or phosphorylation, polymerization and hydrolytic autocatalytic reactions.

Autocatalytic reactions systems are a form of positive feedback where the activity of the system or reaction augments itself in the form of self-reinforcing reactions. Consider a reaction where A catalyzes the formation of B and B accelerates the formation of A; the overall set of reactions is an autocatalytic or positive feedback cycle. Ulanowicz [22] notes that in autocatalysis, the activity of any element in the cycle engenders greater activity in all the other elements, thus, stimulating the aggregate activity of the whole cycle. Such self-reinforcing catalytic activity is self-organizing and is an important way of increasing the dissipative capacity of the system. Cycling and autocatalysis is a fundamental process in nonequilibrium systems.

The notion of dissipative systems as gradient dissipators holds for nonequilibrium physical and chemical systems and describes the processes of emergence and development of complex systems. Not only are the processes of these dissipative systems consistent with the restated second law, it should be expected that they will exist wherever there are gradients.

Living Systems as Gradient Dissipators

We will now focus our attention on the role of thermodynamics in the evolution of living systems. The father of statistical thermodynamics, Boltzmann recognized the apparent contradiction between the thermodynamically predicted randomized cold death of the universe and the existence of a process (i.e., life) in nature by which systems grow, complexify, and evolve, all of which reduce their internal entropy. As early as 1886, Boltzmann observed that the gradient on earth, caused by the energy provided by the sun, drives the living process and suggested a Darwinian-like competition for entropy in living systems:

> Between the earth and sun, however, there is a colossal temperature difference; between these two bodies, energy is thus not at all distributed according to the laws of probability. The equalization of temperature, based on the tendency towards greater probability, takes millions of years, because the bodies are so large and are so far apart. The intermediate forms assumed by solar energy, until it falls to terrestrial temperatures, can be fairly improbable, so that we can easily use the transition of heat from sun to earth for the performance of work, like the transition of water from the boiler to the cooling instillation. The general struggle for existence of animate beings is therefore not a struggle for raw materials—these, for organisms, are air, water and soil, all abundantly available—nor for energy which exists in plenty in any body in the form of heat (albeit unfortunately not transformable), but a struggle for entropy, which becomes available through the transition of energy from the hot sun to the cold earth. In order to exploit this transition as much as possible, plants spread their immense surface of leaves and force the sun's energy, before it falls to the earth's temperature, to perform in ways yet unexplored certain chemical syntheses of which no one in our laboratories has so far the least idea. The products of this chemical kitchen constitute the object of struggle of the animal world [23].

As we noted in the introduction, Boltzmann's ideas were further explored by Schrödinger in *What is Life?* [1]. Schrödinger, like Boltzmann, was perplexed. He also noted that some systems, like life, seem to defy the classical second law of thermodynamics. However, he recognized that living systems are open and not the adiabatic closed boxes of classical thermodynamics. An organism stays alive in its highly organized state by taking energy from outside itself, that is, from a larger encompassing system, and processing it to produce a lower entropy state within itself. Life can be viewed as a far-from-equilibrium dissipative structure that maintains its local level of organization, at the expense of producing entropy in the larger system of which it is part.

If we view the earth as an open thermodynamic system with a large gradient impressed on by the sun, the thermodynamic imperative of the restated second law is that the system will strive to reduce this gradient by using all physical and chemical processes available to it. We have already shown that self-organizing processes are an effective means of reducing gradients. We have discussed that meteorological and oceanographic circulation degrades some of the energy gradients and disequilibrium from the 1580 watts/meter^2 of incoming solar energy. However there are still additional energy gradients requiring dissipation. We suggest that life exists on earth as another means of dissipating the solar induced gradient and, as such, is a manifestation of the restated second law. It is obvious that living systems are far-from-equilibrium dissipative systems and have great potential of reducing radiation gradients on earth [24]. Much of this dissipation is accomplished by the plant kingdom (less than 1% of it through photosynthesis, with most of the dissipation occurring through evaporation and transpiration).

An interesting exception to this are the deep sea vent ecosystems that derive their energy from temperature and chemical gradients emanating from the sea floor rather than from solar energy [25]. In these systems, life is most abundant where the chemical and temperature gradients are the greatest, thus the required gradient dissipation is the highest.

The Origin of Life

The origin of prebiotic life is the development of another route for the dissipation of induced energy gradients. Life with its requisite ability to reproduce, insures that these dissipative pathways continue, and it has evolved strategies to maintain these dissipative structures in the face of a fluctuating physical environment. We suggest that living systems are dynamic dissipative systems with encoded memories, the gene with its DNA, that allow the dissipative processes to continue without having to restart the dissipative process via stochastic events. Living systems are sophisticated mini-tornados, with a memory (its DNA), whose Aristotelian "final cause" may be the second law of thermodynamics. However, one should be clear not to overstate the role of thermodynamics in living processes. The restated second law is a necessary but not a sufficient condition for life.

This paradigm presents no contradiction with the neo-Darwinian hypothesis and the importance of the genetic process, and its role in biological evolution. We reject the selfish gene [26] as the only process in selection and would insert the gene or species as a component in competing autocatalytic ecosystems [27], competing to degrade available energy gradients.

The origin of life should not be seen as an isolated event. Rather it represents the emergence of yet another class of processes whose goal is the dissipation of thermodynamic gradients. Life should be viewed as the most sophisticated (until now) end in the continuum of development of natural dissipative structures from physical to chemical to autocatalytic to living systems. As we have discussed earlier, autocatalytic chemical reactions are the backbone of chemical dissipative systems. The work of Eigen [28], Eigen and Schuster [29] connect autocatalytic and self-reproductive macromolecular species with a thermodynamic vision of the origin of life. Ishida [30] describes Eigen's work:

> Eigen [28] has asserted that the first steps of biological self-organization occurred in a structureless soup, certainly involving functional macromolecular structures such as nucleic acids and proteins. In accordance with this assertion, he has considered a chemical system composed of a wide variety of self-reproductive macromolecules and many different energy-rich monomers required to synthesize the macromolecules. Each macromolecular and monomeric species are distributed uniformly throughout the available volume. The system is open and maintained in a nonequilibrium state by continuous flows of energy-rich and energy-deficient monomers. It is thus tremendously complex. In addition, metabolism, self-reproduction and mutability are necessary for each macromolecular species to yield the behavior of Darwinian selection at the molecular level. For each macromolecular species, Eigen has proposed a simple phenomenological rate equation materializing these three necessary prerequisites. This rate equation, which we shall now call Eigen's equation, is interpreted in terms of the terminology of chemical kinetics.

Eigen [28] himself says:

> Evolution appears to be an inevitable event, given the presence of certain matter with specified autocatalytic properties and under the maintenance of the finite (free) energy flow necessary to compensate for the steady production of entropy.

From the point of view of nonequilibrium thermodynamics of chemical reactions in a homogeneous system, Eigen has shown that the Darwinian selection of such macromolecular species can be linked to Glansdorff and Prigogine's stability criterion [6, 31]. By examining the entropy changes and entropy production of such systems Ishida has formulated a nonequilibrium thermodynamics of the hypercycles. This theory is used to discuss the stability of the "quasi-species" (a solution to Eigen's equations with nonvanishing mutation terms). In principle, it is possible, using nonequilibrium thermodynamics of open systems, to put forward scenarios for the development of biochemical machinery for the duplication and translation of nucleic acids and other macromolecules essential to life.

Biological Growth and Development

Our thesis is that growth, development, and evolution are the response to the thermodynamic imperative of dissipating gradients [32, 33] (which is not to say this is the only imperative governing life; survival is equally important). Biological growth occurs when the system adds more of the same types of pathways for degrading imposed gradients. Biological development occurs when new types of pathways for degrading imposed gradients emerge in the system. The larger the system, i.e., the larger the system flow activity [22], the more reactions and pathways (both in number and type) are available for gradient destruction. Clearly the above principle provides a criteria for evaluating growth and development in living systems. All else being equal, the better dissipative pathway is preferred. In what follows, we explore some examples of this principle.

Plant growth is an attempt, within the constraints of its genetic (and evolutionarily tested) make-up and its environmental boundary conditions, to capture solar energy and dissipate usable gradients. The gradient capturing aspects of plants can be seen in phototrophism and their symmetrical shapes, designed to capture and degrade sunlight. Although each tree species has its own genetically endowed form, the energy capturing aspect of an isolated tree leads to its magnificent symmetry. Canopies of plants of many species arrange themselves into leaf index assemblies to optimize energy capture and degradation. The gross energy budgets of terrestrial plants show that the vast majority of their energy use is for evapotranspiration, with 200–500 grams of water transpired per gram of fixed photosynthetic material. This mechanism is a very effective energy degrading process with 600 calories used per gram of water transpired [34].

Data synthesized by Currie and Paquin [35] show that the large scale biogeographical distribution of species richness of trees is strongly correlated with realized annual evapotranspiration and available energy. The more energy is partitioned among species, the more pathways there are available for total energy degradation. Trophic levels and food chains are based upon photosynthetic fixed material and further dissipate these gradients by making more higher ordered structures. Thus, we would expect more species diversity to occur where there is more available energy. Species diversity and trophic levels are vastly greater at the equator, where 5/6 of the earth's solar radiation occurs, and there is more of a gradient to reduce.

A Thermodynamic Analysis of Ecosystems

Ecosystems display the influence of thermodynamic principles in their patterns of growth and development. A thermodynamically based theory of ecology holds the promise of propelling ecology from a descriptive to a predictive science. Ecosystems are the result of the biotic, physical, and chemical components of nature acting together as a nonequilibrium dissipative process. As such ecosystem development increases, energy degradation thus follows the imperative of the second law. This hypothesis can be tested by observing the energetics of ecosystem development during the successional process or by determining their behavior as they are stressed or as their boundary conditions are changed.

As ecosystems develop or mature, they should increase their total dissipation, and should develop more complex structures with greater diversity and more hierarchical levels to abet energy degradation. Species which survive in ecosystems are those that funnel energy into their own production and reproduction and contribute to autocatalytic processes which increase the total dissipation of the ecosystem. In short, ecosystems develop in a way which systematically increases their ability to degrade the incoming solar energy.

In this sense, the development of ecosystem maturity via succession is the result of the system organizing itself to dissipate more incoming energy with each stage of succession. Thus, one would expect successional processes in ecosystems to result in systems with:

1. More energy capture:—because the more energy that flows into a system, the greater the potential for degradation.
2. More energy flow activity within the system:—again, the more energy that flows within and through a system, the greater the potential for degradation.
3. More cycling of energy and material:
   (A) Numbers of cycles:—more pathways for energy to be recycled in the system results in further degradation of the incoming energy.
   (B) The length of cycles:—more mature systems will have cycles of greater length, i.e., more nodes in the cycle. Each chemical reaction at or within a node results in entropy production; the more such reactions, the more complete the degradation.
   (C) The amount of material flowing in cycles (as versus straight through flow) increases. The ecosystem becomes less leaky, thus, maintaining a supply of raw material for energy degrading processes.
   (D) Turnover time of cycles or cycling rate decreases:—more nodes or cycles in a system will result in nutrients or energy being stored at nodes in the system resulting in longer residence time in the system. This phenomena is the same as the decrease in production/biomass (*P/B*) ratio decreases observed in maturing ecosystems. *P/B* is the residence time of material in the system. The residence time of a nitrogen atom in a simple bacterial cycle will be a matter of hours; however, the residence time of a similar nitrogen atom in a rainforest will be years.
4. Higher average trophic structure:
   (A) Longer trophic food chains:—energy is degraded at each step of the trophic food chain, therefore, longer chains will result in more degradation.
   (B) Species will occupy higher average trophic levels:—this will result in more energy degradation as energy at higher trophic levels has a higher exergy content.
   (C) Greater trophic efficiencies:—energy that is passed higher up the food chain will be degraded further than energy that is shunted immediately into the detrital food chain.
5. Higher respiration and transpiration:—transpiration and respiration result in energy degradation.
6. Larger ecosystem biomass:—more biomass means more pathways for energy degradation.
7. More types of organisms (higher diversity):—more types of organisms will provide diverse and different pathways for degrading energy.

This list of ecological attributes can be explained by ecosystems behaving in such a manner as to degrade as much of the incoming energy as possible [36, 37] and provides causality for most if not all the phenomenological attributes of maturing ecosystems developed by Odum in 1969 [38]. Lotka's [39] suggestion that living systems will maximize their energy flow, Odum's [40] maximum power principle for ecosystems, and Lieth's [41] maximum energy conductivity are all subsumed and explained by our theory.

If ecosystems develop into dynamic quasi-stable states, one would expect them to respond to changes in boundary conditions that perturb these states by retreating to configurations with lower energy degradation potential. Stressed ecosystems will retreat down their thermodynamic branch into more primitive systems with attributes opposite those presented above [42]. Stressed ecosystems often appear similar to earlier successional stage ecosystems and will reside at some distance closer to thermodynamic equilibrium.

We have recently analyzed a carefully collected data set for carbon-energy flows in two aquatic tidal marsh ecosystems adjacent to a large power generating facility on the Crystal River in Florida. The ecosystems in question were a "stressed" and a "control" marsh. The stressed stem is exposed to hot water effluent from the nuclear power station. The stress is an approximately 6°C water temperature increase. The "control" ecosystem is not exposed to the effluent but is otherwise exposed to the same environmental conditions. We wished to determine the effect of the change in environmental conditions on the structure of the stressed system.

Table 2 summarizes a set of ecosystem indicators. The I/O or Input/Output Measures indicate various aspects of the flows through the ecosystem. In absolute terms (the first three columns Table 2) all the flows dropped in the stressed ecosystem. Overall, the drop in flows was about 20%, in particular, the imported flows (that is, the resources available for consumption) dropped by 18% and the TST (the total system throughput, the total flow activity in the system) dropped by 21%. The biomass dropped by about 35%. The implication of these numbers is that the stress has resulted in the ecosystem shrinking in size, in terms of biomass, its consumption of resources, and its ability to degrade and dissipate incoming energy. .

If the flows are scaled by the import to the ecosystem from the outside (the last three columns in Table 2), the resulting numbers indicate how well the ecosystem is making use

Table 2. Ecosystem indicators for the Crystal River marsh gut ecosystems.

Crystal River	Mg/m^2/day			Scaled by import		
	Control	Stressed	Δ%	Control	Stressed	Δ%
Biomass	1,157,136	755,213	34.73%	157.50	125.49	20.32%
TOTAL I/O						
Imports	7,347	6,018	18.09%	1	1	0.00%
TST	22,768	18,055	20.70%	3.10	3.00	3.19%
Production	3,292	2573.9	21.82%	0.45	0.43	4.56%
Exports	952	872	8.37%	0.13	0.14	−11.86%
Respiration	6,400	5,148	19.57%	0.87	0.86.	1.80%
LIVING I/O						
Production	400	326	18.39%	0.05	0.05	0.37%
Exports	316	253	19.93%	0.04	0.04	2.24%
Respiration	3,566	3,078	13.69%	0.49	0.51	−5.37%
To Detritus	5,726	4,315	24.64%	0.78	0.72	7.99%
DETRITUS I/O				SCALED DETRITUS INPUT		
Input from Living	5,726		24.64%	1.00	1.00	0.00%
Production	2,893	2,248	22.29%	0.51	0.52	−3.11%
Exports	636	619	2.64%	0.11	0.14	−29.19%
Respiration	2,834	2,070	26.96%	0.49	0.48	3.09%
FOOD WEB						
Cycles	142	69	51.41%			
Trophic Levels	5	5	0.00%			

Table 3. The Radiation Equations

$$R_n = K^* - L^*,$$

where K^* = net flux of solar radiation (incoming), L^* = net flux of long wave radiation (outgoing)

R_n = net radiation flux absorbed at surface, (K^*, L^*, and R_n are all measured watts/m^2)

and

$$R_n = H + L_c + G,$$

where H = sensible heat flux, L_c = latent heat flux,
G = energy flux into the ground, R_n is the energy which is degraded from radiation into molecular motion.

Furthermore,

$$L_c = \varepsilon\,[\sigma(T)^4],$$

where ε = emissivity, σ = Stefan-Boltzmann constant,
T = surface temperature.

If K^* is constant, the smaller T, the smaller L^* and, hence, the larger R_n and, thus, the larger the amount of energy degraded.

Table 4. Radiative estimates from Thermal Infrared Multispectral Scanner for different ecosystem types in the H. J. Andrews Experimental Forest, Oregon. The data is presented from least (quarry) to most developed (400-year-old forest). Data from Luvall and Holbo [44].

	Quarry	Clearcut	Douglas Fir Plantation	Natural Forest	400-year-old Douglas Fir Forest
K^* (watts/m^2)	718	799	854	895	1005
L^* (watts/m^2)	273	281	124	124	95
R_n (watts/m^2)	445	517	730	771	830
T (°C)	50.7	51.8	29.9	29.4	24.7
R_n/K^* (%)	62	65	85	86	90

K^*= incoming net solar, L^* = net long wave outgoing, R_n= net radiation transformed into nonradiative process at the surface, R_n/K^* = percent of net incoming solar radiation degraded into nonradiative process. See Table 3 for the equations relating the variables.

of the resources it does capture. The percentage changes in these scaled flow rates reveal that, in total, the stressed ecosystem is, relatively speaking, exporting more. In other words, it is losing material more quickly than the control ecosystem. It is a leaky ecosystem. Looking at the living side, the big changes are an increase in respiration and a decrease in flow to the detritus. There was a small decrease in exports. These changes indicate that the consumed resources are being more effectively used by the living components. It also indicates the species are stressed. Looking at the detritus, a different picture emerges. There was a slight increase in production and a small decrease in respiration and a very large increase in exports. This analysis of the scaled flow rates indicates that the living components are somewhat stressed but that more importantly, there is a large breakdown in the ability of the ecosystem to recycle material through the detritus and thus retain and degrade its resources.

Examining the Food Web data further confirms this (the last three rows in Table 2). The number of cycles in the stressed ecosystem is 51% of the number in the control. Furthermore, overall these cycles are shorter in length. In the effective grazing chain, the number of trophic levels was not changed, but the trophic efficiencies were changed dramatically, as was the flow to the top trophic levels. These are all indicators of a stressed ecosystem [43].

Overall, the impact of the effluent from the power station has been to decrease the size of the ecosystem and its consumption of resources while impacting its ability to retain the resources it has captured. In short, the impacted ecosystem is smaller, has lower trophic levels, recycles less, and leaks nutrients and energy. All of these are signs of disorganization and a step backward in development. This analysis suggests that the function and structure of ecosystems follows the development path predicted by the behavior of nonequilibrium thermodynamic structures. A more complete analysis of these data and the implications of this hypothesis for ecosystem health or integrity are found in [36].

Ecosystems as Energy Degraders

The energetics of terrestrial ecosystems provides an excellent test of our thesis that ecosystems will develop so as to degrade energy gradients more effectively. More developed dissipative structures will degrade more energy. Thus, we would expect more mature ecosystems to degrade the exergy content of the energy they capture more completely than a less developed ecosystem. The exergy drop (i.e., gradient dissipation) across an ecosystem is a function of the difference in black body temperature between the captured solar energy and the energy reradiated by the ecosystem. If a group of ecosystems are bathed by the same amount of incoming energy, we would expect that the most mature ecosystem would reradiate its energy at the lowest exergy level, that is, the ecosystem would have the coldest black body temperature. (See Table 3.) The black body temperature is determined by the surface temperature of the ecosystem. Luvall and Holbo [44–46] and Luvall et al. [47] conducted experiments in which they overflew terrestrial ecosystems and measured surface temperatures using a Thermal Infrared Multispectral Scanner (TIMS). Their technique allows assessments of energy budgets of terrestrial landscapes, integrating attributes of the overflown ecosystems, including vegetation, leaf and canopy morphology, biomass, species composition and canopy water status. Luvall and his co-workers have documented ecosystem energy budgets, including tropical forests, mid-latitude varied ecosystems, and semiarid ecosystems. Their data shows one unmistakable trend, that when other variables are constant, the more developed the ecosystem, the colder its surface temperature and the more degraded its reradiated energy.

Table 4 portrays TIMS data from a coniferous forest in western Oregon, North America. Ecosystem surface temperature varies with ecosystem maturity and type. The warmest temperatures were found at a clearcut and over a rock quarry. The coldest site, 299K, some 26K colder than the clearcut, was a 400-year-old mature Douglas Fir forest with a three-tiered plant canopy. The 23-year-old naturally regrowing forest had the same temperature as the 25-year-old plantation of Douglas Fir. Even with the initial planting of the late successional species of fir, the natural system obtained a similar energy degrading capacity in a similar time.

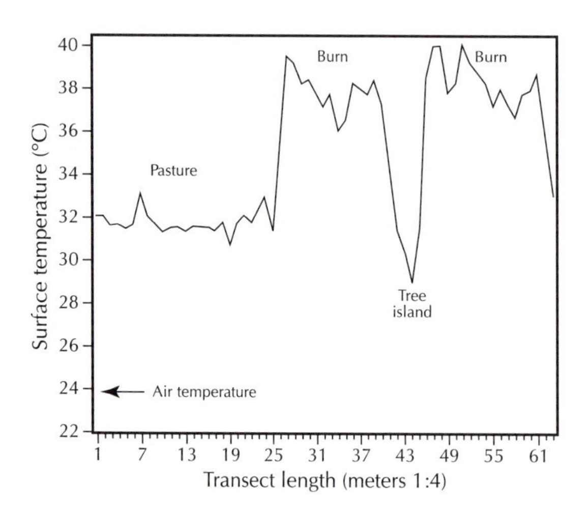

Figure 3. A Thermal Infrared Multichannel Scanner (TIMS) profile over a varied landscape in the Braulio Carrillo National Park in Costa Rica (done by Luvall et al. [47]). The TIMS is mounted on an aircraft measuring six thermal channels in the wave length channels of 8.2 to 12.2 pm and flown over a 400-meter transect. Temperatures are canopy temperature values calculated from net reradiated thermal energy. Temperatures showed that forest fire burn was the warmest system (it did not degrade as much incoming energy) and the tree island the coldest (the best degrader). The pasture degraded energy in an intermediate fashion. Airborne thermal imaging may hold great promise for rapid evaluation of terrestrial ecological integrity.

Luvall's data allows for an interesting calculation of the percent of the net solar radiation (K^*) that is degraded into energy in the form of molecular motion (R_n). Luvall points out that R_n is the net available energy dissipated through evaporation, sensible heat and storage. The ratio R_n/K^* is the percentage of the radiative fluxes converted to lower exergy thermal heat. The quarry degraded 62% of the net incoming radiation while the 400 year old forest degraded 90%. The remaining sites fell between these extremes, increasing degradation with more mature or less perturbed ecosystems.

These unique data sets show that ecosystems develop structure and function that degrades imposed energy gradients more effectively. Analysis of airborne collected reradiated energy fluxes appears to be a unique and valuable tool in measuring the energy budget and energy transformations in terrestrial ecosystems. According to proposals made in this paper, a more developed ecosystem degrades more energy. Thus, the ecosystem temperature or R_n/K^* may be excellent indicators of ecological integrity that can be formulated from first principles of thermodynamics and physics.

For these indices to be used to determine comparative or, hopefully, absolute measures of ecosystem integrity, much more research is needed on energy transformation in ecosystems, as well as method development to correct for system specific location dependent energy processes, i.e., rainfall and altitude. The potential for these methods to be used for remote sensed ecosystem classification and ecosystem health evaluation is apparent in Figures 3 and 4. Forest fire burn, pasture and trees have their own distinctive temperatures. More mature ecosystems are colder. Luvall has also tested algorithms to measure the thermal inertia of ecosystems, analyzed spectral variability, day night differences and evapotranspiration, all which show specific ecosystem characteristics.

These same phenomena are observed at a larger scale and exhibit the global nature of the biosphere in reducing the induced solar gradient on earth. Sato and his colleagues [481 have recently calculated the mean surface energy budget for four large regions of the earth for 50 days in the summer (see Table 5). The regions are:

(a) the Amazon basin which is uniformly covered by rainforest,
(b) central and eastern United States which consists mainly of cultivated land, grasslands and some mixed forest,
(c) Asia, a heterogeneous mix of tropical rain forest, forest, cultivated land and desert, and
(d) the Sahara desert.

Their data was obtained from the satellite derived Earth Radiation Budget Experiment and from the SiB climate biosphere model of Sellers et al. [49]. They measured insulation, albedo, net long and short wave energy absorbed at the earth's surface, net radiation or available energy and calculated important heat fluxes by modeling physiological and biological processes which influence radiation, momentum, mass, and heat transfer related to vegetation-surface atmosphere interactions. Their calculations are important because they integrate the atmosphere-biosphere as a net dissipative system and allow for the determination of the reduction of the solar induced gradient between a warm black body earth and the 2.8 K temperature of outer space.

The first four rows presented in Table 5 are arranged from the most developed ecosystem (Amazon rainforest) to the least. The re-radiated long wave radiation and the sensible heat flux represent energy which has not been degraded to the ambient and which will cause disequilibrium. Evapotranspiration represents energy which is dissipated and thus has been converted to lower quality exergy. On the basis

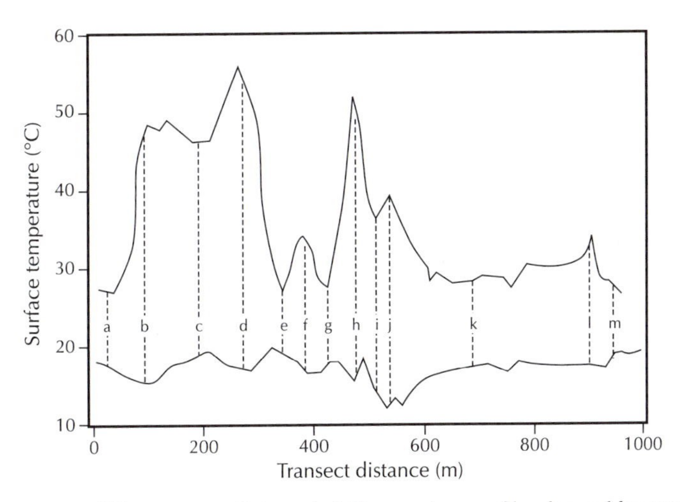

Figure 4. Surface temperature (°C) vs. transect distance (m). Temperature profiles along a 1 km transect in the H. J. Andrews Experimental Forest, Oregon [45]. The top profile is for noon and the bottom profile is for post-sunset. The data was collected using TIMS surface temperature acquisition system. The letters on the diagram correspond to surface features:

(a) edge of forest
(b) narrow road,
(c) somewhere in a clear cut,
(d) a wider road,
(e) one side of a small shelter wood of Douglas Fir
(f) a pond within the shelter wood,
(g) the other side of the shelter wood
(h) a wide road,
(i) trees along the road,
(j) in a flat part of a clear cut,
(k) somewhere in a 15-year-old Douglas Fir plantation,
(l) a trail,
(m) an old stand of Douglas Fir regrowth

Table 5. Energy absorbed (watts/m^2/day) at the surface of varied ecosystems, and the percentage of this energy which is re-emitted into space. Data was obtained from the satellite derived Earth Radiation Budget Experiment and from the SiB climate biosphere model of Sellers et al. [49]. The first four rows are arranged from the most developed ecosystem (Amazon rainforest) to the least. The reradiated long wave radiation and the sensible heat flux represent energy which has not been degraded to the ambient and which will cause disequilibrium. Evapotranspiration represents energy which is dissipated and will no longer cause disequilibrium. On the basis of the arguments presented in this paper, evapotranspiration should increase with ecosystem development, and long wave radiation and sensible heat flux should decrease. This is borne out by the results presented in the first four rows. The last two rows represent running the SiB model for the rainforest. Again it is seen that the more developed ecosystem is the better dissipative structure.

Area	Energy Absorbed at Surface	Energy Reradiated Long Wave	Sensible Heat Flux	Evapotranspiration
Amazon	184.7	17%	15%	70
U.S.	220.2	18%	19%	61
Asia	223.4	24%	26%	50
Sahara	202	41%	56%	2
Rain Forest	204	16%	22%	63
Grassland	186	22%	30%	48

of the arguments presented in this paper, evapotranspiration should increase with ecosystem development, and long wave radiation and sensible heat flux should decrease. This is borne out by the results presented in the first four rows of Table 5.

Moreover, the last two rows represent running the SiB model for the rainforest and then for the same location except assuming a grassland instead of a rainforest [50]. Again, this data shows that the more developed an ecosystem, the more of the energy impinging on it is degraded to lower quality energy.

Satellite derived earth radiation data developed for global climate analysis shows that this same phenomena may also be apparent at the global scale. The Climate Analysis Center and the Satellite Research Laboratory of the National Oceanic and Atmospheric Administration (NOAA) produce monthly maps of outgoing longwave radiation (OLR) collected from multiwave spectral scanners aboard polar orbiting satellites.

Long wave infrared emissions (*L*) from the earth are dependent on the surface temperature, greenhouse gases, i.e., H_2O and CO_2, and cloud cover. Like the greenhouse gases, cloud cover will tend to act as a blanket and trap long-wave infrared.

Figure 5 is a global OLR map for February, 1991. For the tropical rainforests of the Amazon, the Congo, and over Indonesia and Java, *L* is less than 200 watts/m^2. The deserts emit a net OLR of over 280 watt/m^2. Interestingly enough, the tropical rain forests with their coupled cloud system, with the sun directly overhead, have the same surface temperature as Canada in the winter. The low tropical rainforest OLR temperatures are due to the cold temperatures of the convective cloud tops which are generated by the underlying cooler forests.

As we have seen earlier, mature ecosystems can lower surface temperature by approximately 25°C. The low reradiation from the rainforest-cloud systems appears to be a

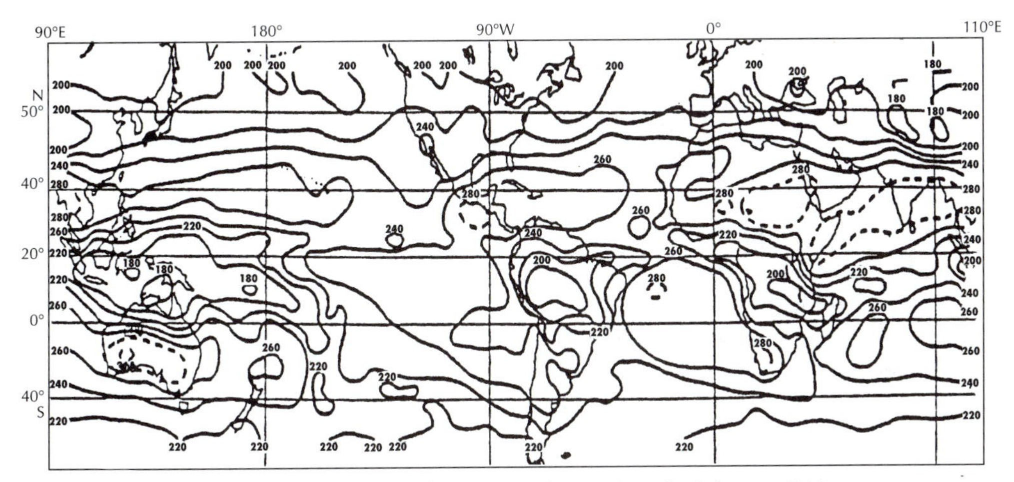

Figure 5. The mean monthly outgoing long wave radiation (OLR) is measured by the NOAA-9 polar orbiting satellite using the infrared channel. Contours are in watts/m^2. Note the low longwave emissions over the tropical rain forests and high values over the earth's deserts.

global scale signal of solar gradient degradation. Most of the energy degradation in terrestrial ecosystems is due to the latent heat production via evapotranspiration. Tropical rain forests produce a prodigious amount of water vapor via this process, and convective induced cooling produces high clouds which tends to reinforce the cooling of the rain forests. The coupled rain forest cloud system lowers the earth to space gradient even more than the forest alone.

This phenomena may appeal to supporters of Lovelock's [51] Gaia hypothesis, as it is an example of the coupled global biosphere-climate system. Without immersing ourselves into the Gaiaean controversy, the coupled system does decrease the apparent earth-space gradient. Furthermore, the rain forests with their attendant cloud formation act to cool the earth on a global basis and thus play an important role in global climate energy budgets. We have research underway to further quantify these propositions.

Discussion

In this paper, we have recast the second law of thermodynamics from the old statement of "entropy increase" into a statement that describes systems undergoing processes so that they will reach a unique state of equilibrium. This description draws on the work of Carathéodory, Hatsopoulos and Keenan, and Kestin [3–5]. It allows for the discussion of system behaviour in nonequilibrium situations. It overcomes the difficulty of describing nonequilibrium systems in terms of entropy, which can only be defined in the equilibrium state. We suggest that, in nonequilibrium situations, systems will take advantage of all available means to resist the gradients responsible for the nonequilibrium condition. Furthermore, the stronger the applied gradient, the greater the effect that the equilibrium attractor will have on the system. Emergence of coherent self-organizing structures are the expected response of systems as they attempt to resist and dissipate the external gradients that are moving them away from equilibrium.

We have shown these principles are manifested in the behaviour of physical dissipative systems (i.e., Bénard cells) and described how they apply to mesoscale meteorology. We apply these same principles to the origin of life, biological growth, the development of ecosystems, and biological evolution. Living systems are not only permissible under the restated second law of thermodynamics, but it is the restated second law which mandates living processes and is a necessary but not sufficient cause for life itself. We provide biology with a paradigm that not only describes the "why" of life but also describes the directions in which living systems will develop and evolve.

We have documented that ecological processes are driven and governed by thermodynamic imperatives; ecosystems develop and select energetic pathways that strive to degrade as much of the energy available to them as possible. We believe that these same principles extend to Darwinian selection. Margalef [52], Odum [38], Wicken [53] and Schneider [33] all observed that the strategy of succession, a short term process, is basically the same strategy as the long term evolutionary development of the biosphere. Margalef [52] noted that:

> Evolution cannot be understood except in the framework of succession. By the natural process of succession, which is inherent in every ecosystem, the evolution of species is pushed or sucked onto the direction taken by succession. Succession is in progress everywhere and evolution follows encased in succession's frame.

Zotin [54] studied the bioenergetic trends of the evolution of organisms and noted that evolution has progressed in a manner such that organisms evolved with increasing energetic dissipation rates, i.e., respiratory intensity. Evolution, like ecosystems, seems to select species and ecosystems that increase the global dissipation rate.

Our suggestions should not be confused with the work of Brooks and Wiley [55] Their "increase of entropy," observed in their cladograms of evolving species, refers to an information theoretic entropy and, as such, explains things in terms of microstate probabilities of information states. Our work examines the thermodynamic development of natural systems and, as such, explains things in terms of macroscopic behaviour of energy processing. The foundations of our work were certainly developed by Prigogine and Wiame [56] and Wicken [52, 57], but we circumvent their discussions of entropy and life, and instead rely on a more appropriate approach to nonequilibrium systems, that is, the development of self-organization as a means of dissipating gradients imposed on systems.

Lotka [39] and Odum and Pinkerton [40] have suggested that those biological systems that survive are those that develop the most power inflow and use it to best meet their needs for survival. Our work would propose that a better description of these "power laws" would be that biological systems develop in a manner as to increase their degradation rate, and that biological growth, ecosystem development and evolution represent the development of new dissipative pathways. This work is quite different than Jørgensen's Maximum Exergy Principle [58]. He has focused on exergy concentrations, while we are investigating exergy degradation. The two concepts may be interrelated but our work has not as yet shown the linkage.

Our proposals represent a new approach to the biological synthesis. We have not been able to discuss all the ramifications these principles have for our understanding of the growth and development of biological systems. We have intentionally not discussed the application of these principles to the development and operation of genetic processes. We note that our thermodynamic description of living systems fits comfortably into the present day molecular-genetic research program. The importance of the gene and its role in morphogenesis, speciation, and in carrying the message of evolution forward is vital.

Life represents a balance between the imperatives of survival and energy degradation. To quote Blum [59]:

> I like to compare evolution to the weaving of a great tapestry. The strong unyielding warp of this tapestry is formed by the essential nature of elementary nonliving matter, and the way in which this matter has been brought together in the evolution of our planet. In building this warp, the second law of thermodynamics has played a predominant role. The multi-colored woof which forms the detail of the tapestry I like to think of as having been woven onto the warp principally by mutation and natural selection. While the warp establishes the dimensions and supports the whole, it is the woof that most intrigues the aesthetic sense of the student of organic evolution, showing as it does the beauty and variety of fitness of organisms to their environment. But why should we pay so little attention to the warp, which is after all a basic part of the whole structure? Perhaps the analogy would be more complete if something were introduced that is occasionally seen in textiles—the active participation of the warp in the pattern itself. Only then, I think, does one grasp the full significance of the analogy.

What we have tried to do in this paper is to show the participation of the warp in producing the pattern of the tapestry of life. However this does not in any way diminish the importance of the woof, which records the struggle for survival.

To return to Schrödinger, life is comprised of two processes, "order from order and order from disorder." The work of Watson and Crick described the gene with its DNA, and solved the "order from order" mystery. Our hypothesis supports the "order from disorder" premise and connects biology with physics, thus providing a unifying macroscopic theory for living systems.

Footnotes

[1]Where Q is heat transfer in calories and T is in degrees Kelvin.

[2]Energy/Availability: A measure of available work content of energy [11–13]. It is a measure of the potential of energy to perform useful work. It reflects the quality of the energy. Irreversible processes destroy exergy.

References

1. Schrödinger, E. 1944. *What is Life?* Cambridge University Press.
2. Eddington, A. 1958. *The Nature of the Physical World.* University of Michigan Press, Ann Arbor, MI.
3. Carathéodory, C. 1976. Investigations into the foundations of thermodynamics. In: *The Second Law of Thermodynamics.* Kestin, J. ed., 5:229–256. (Benchmark Papers on Energy.) Dowden, Hutchinson, and Ross.
4. Hatsopoulos, G. and J. Keenan. 1965. *Principles of General Thermodynamics.* John Wiley.
5. Kestin, J. 1966. *A Course in Thermodynamics.* Blaisdell.
6. Nicolis, G. and I. Prigogine. 1977. *Self-Organization in Nonequilibrium Systems.* J. Wiley and Sons.
7. Nicolis, G. and I. Prigogine. 1989. *Exploring Complexity.* W.

H. Freeman.

8. Chandrasekhar, S. 1961. *Hydrodynamics and Hydromagnetic Stability.* Oxford University Press.

9. Fermi, E. 1956. *Thermodynamics.* Dover Publications.

10. Silveston, P. L. 1957. Warmedurchange in horizontalen Flassigkeitschichtem. Ph.D. Thesis. Techn. Hochsch., Muenchen, Germany.

11. Brzustowski, T. A. and P. J. Golem. 1978. Second law analysis of energy. In: Processes Part 1: Exergy—An Introduction. Transactions of the Canadian Society of Mechanical Engineers 4:209–218.

12. Ahern, J. E. 1980. *The Exergy Method of Energy Systems Analysis.* J. Wiley and Sons.

13. Moran, M. J. 1982. *Availability Analysis: A Guide to Efficient Energy Use.* Prentice-Hall.

14. Prigogine, I. 1955. *Thermodynamics of Irreversible Processes.* John Wiley.

15. Swenson, R. 1989. Emergent attractors and the law of maximum entropy production: Foundations to a theory of general evolution. Systems Research 6(3):187–197.

16. Katchalsky, A. and P. F. Curren. 1965. *Nonequilibrium Thermodynamics in Biophysics.* Harvard University Press, Cambridge.

17. Peusner, L. 1986. *Studies in Network Thermodynamics.* Elsevier, Amsterdam.

18. Mikulecky. D. C. 1984. Network thermodynamics: A simulation and modeling method based on the extension of thermodynamic thinking into the realm of highly organized systems, Math. Bioscience 72:157–179.

19. Glansdorff, P. and I. Prigogine. 1971. *Thermodynamic Theory of Structure, Stability, and Fluctuations.* Wiley-interscience.

20. Brown, W. 1973. Heat-flux transitions at low Rayleigh number, J. Fluid Mech. 69:539–559.

21. Paltridge, G. W. 1979. Climate and thermodynamic systems of maximum dissipation. Nature 279(5714):630– 631.

22. Ulanowicz, R. E. 1986. *Growth and Development: Ecosystem Phenomenology.* Springer-Verlag.

23. Boltzmann, L. 1974. The second law of thermodynamics. In: *Ludwig Boltzmann, Theoretical Physics and Philosophical Problems.* McGinness, B., ed. D. Reidel.

24. Ulanowicz, R E. and B.M. Hannon. 1987. Life and the production of entropy. Proc. R. Soc. Lond. B. 232:181–192.

25. Corliss, J. B. 1990. The dynamics of creation: The emergence of living systems in Archaean submarine hot springs. Under review.

26. Dawkins, R. 1976. *The Selfish Gene.* Oxford Press.

27. Weber, B. H., et al. 1989. Evolution in thermodynamic perspective: An ecological approach. Biology and Philosophy 4:373–405.

28. Eigen, M. 1971. Naturwissenschaften 58:465.

29. Eigen, M. and P. Schuster. 1979. *The Hypercycle: A Principle of Natural Self-Organization.* Springer-Verlag.

30. Ishida, K. 1981. Nonequilibrium thermodynamics of the selection of biological macromolecules. J. Theor. Biol. 88:257–273.

31 Peacocke, A. R. 1983. *The Physical Chemistry of Biological Processes.* Oxford University Press.

32 Kay, J. 1984. Self-organization in living systems. Ph.D. Thesis. Systems Design Engineering, University of Waterloo, Waterloo, Ontario, Canada.

33. Schneider, E. S. 1988. Thermodynamics, information, and evolution: New perspectives on physical and biological evolution. In: *Entropy, Information, and Evolution: New Perspectives on Physical and Biological Evolution.* Weber, B. H., D. J. Depew and J. D. Smith, eds. MIT Press. Pp. 108–138

34. Kimmins, J. P. 1987. *Forest Ecology.* Macmillan, New York.

35. Currie, D. J. and V. Paquin. 1987. Large scale biogeographical patterns of species richness of trees. Nature 329:326–327.

36. Kay, J. and E. Schneider. 1991. Thermodynamics and measures of ecosystem integrity. In: *Proceedings of a Symposium on Ecological Indicators.* Elsevier.

37. Kay, J. J., L. Graham and R. E. Ulanowicz., 1989. A detailed guide to network analysis. In: *Network Analysis in Marine Ecosystems.* Coastal and Estuarine Studies, Volume 32. Wulff, F., J. G. Field and K. H. Mann, eds. Springer-Verlag. Pp. 15–61.

38. Odum, E. P. 1969. The strategy of ecosystem development. Science 134:262–270.

39. Lotka, A. 1922. Contribution to the energetics of evolution. In: Proceedings of the National Academy of Science, U.S.A. Pp. 148–154.

40. Odum, H. T. and R.C. Pinkerton, 1955, Time's speed regulator. Ami. Sci. 43:321–343.

41. Lieth, H. 1976. Biophysical questions in ecology and environmental research, Rad. and Environm. Biophys. 13:337–351.

42. Kay, J. 1991. A nonequilibrium thermodynamic framework for discussing ecosystem integrity. Env. Mgmt. 15(4):483–495.

43. Ulanowicz, R. E. 1985. Community measures of marine food networks and their possible applications. In: *Flows of Energy and Materials in Marine Ecosystems.* Fasham, M. J. R., ed. Plenum, London. Pp. 23–47.

44. Luvall, J. C. and H. R. Holbo. 1989. Measurements of short term thermal responses of coniferous forest canopies using thermal scanner data: 1989. Remote Sens. Environ. 27:1–10.

45. Luvall, J. C. and H. R. Holbo. 1991. Thermal remote sensing methods in landscape ecology. In: *Quantitative Methods in Landscape Ecology,* Chapter 6. Turner, M. and R. H. Gardner, eds. Springer-Verlag.

46. H. R. Holbo and J. C. Luvall. 1989. Modeling surface temperature distributions in forest landscapes. Remote Sens. Environ. 21:11–24.

47. Luvall, J. C., D. Lieberman, M. Lieberman, G. S. Hartshorn and R. Peralta. 1990. Estimation of tropical for-

est canopy temperatures, thermal response numbers, and evapotranspiration using an aircraft-based thermal sensor. Photogrammetric Engineering and Remote Sensing 56(10):1393–1410.

48. Sato, N. et al. 1989. Effects of implementing the simple biosphere model in a general circulation model. Journal of the Atmospheric Sciences 46(18):2757–2782.

49. Sellers, P. J. and Y. Mintz. 1986. A simple biosphere model (SIB) for use within general circulation models. J. Atmos. Sci. 43:505–531.

50. Shulka, J., C. Nobre and P. Sellers. 1989. Amazon deforestation and climate change. Science 247:1322–1325.

51. Lovelock, J. 1990. *The Ages of Gaia.* Bantam Books, New York.

52. Margalef, R. 1968. *Perspectives in Ecological Theory.* University of Chicago Press, Chicago.

53. Wicken, J. S. 1987, *Evolution, Thermodynamics, and Information: Extending the Darwinian Program.* Oxford University Press.

54. Zotin, A. I. 1984. Bioenergetic trends of evolutionary progress of organisms. In: *Thermodynamics and Regulation of Biological Processes.* Lamprecht, I. and A. I. Zotin, eds. Walter de Gruyter and Co., Berlin. Pp. 451–458.

55. Brooks, D. R., and E. O. Wiley. 1986. *Evolution as Entropy.* University of Chicago Press, Chicago, IL.

56. Prigogine, I. and J. M. Wiame. 1946. Biologie et thermodynamique des phenomenes irreversible. Experientia 11:451–453.

57. Wicken, J. S. 1980. Thermodynamic theory of evolution. J. Theor. Biol. 87:9–23.

58. Jørgensen, S. E. and H. Mejer. 1979. A holistic approach to ecological modeling. Ecol. Modeling 7:169–189.

59. Blum, H. F. 1968. *Time's Arrow and Evolution.* Princeton University Press, Princeton, NJ.

60. Kay, J. J. 1989. A thermodynamic perspective of the self-organization of living systems. In: *Proceedings of the 33rd Annual Meeting of the International Society for the System Sciences, July, 1989.* Volume 3. Ledington, P. W. J., ed. Edinburgh. Pp. 24–30.

Kandis Elliot, biologist, author and scientific illustrator, selected and introduced two of the reprints for this book, which was prepared by her at the Institute of Implied Science, a studio of design, illustration and book compositing. The body text is New Baskerville and was set on Apple computers using QuarkXPress, Adobe Illustrator and Adobe Photoshop.